ADAPTATION

EDITED BY

Michael R. Rose

George V. Lauder

Department of Ecology and Evolutionary Biology
University of California
Irvine, California

ACADEMIC PRESS
San Diego   New York   Boston   London
Sydney   Tokyo   Toronto

Photo courtesy of the American Museum of Natural History Archives.

This book is printed on acid-free paper. ∞

Academic Press, Inc.
525 B Street, Suite 1900, San Diego, California 92101-4495, USA
http://www.apnet.com

Academic Press Limited
24-28 Oval Road, London NW1 7DX, UK
http://www.hbuk.co.uk/ap/

Library of Congress Cataloging-in-Publication Data

Rose, Michael R. (Michael Robertson), date.
    Adaptation / edited by Michael R. Rose, George V. Lauder.
        p.   cm.
    Includes bibliographical references and index.
    ISBN 0-12-596420-X (alk. paper : Hard cover)
    ISBN 0-12-596421-8 (alk. paper : Paperback)
    1. Adaptation (Biology) 2. Natural selection. 3. Evolution
(Biology)   I. Lauder, George V.
    QH546.R58   1996
    575--dc20   96-33932
                                                      CIP

PRINTED IN THE UNITED STATES OF AMERICA
96 97 98 99 00 01 QW 9 8 7 6 5 4 3 2 1

To Della and Susan

Contents

PART II: EMPIRICAL METHODS FOR STUDYING ADAPTATION

PART III: DIVERSITY OF ADAPTIVE PROCESSES

Contributors

Amundson, Ron Department of Philosophy, University of Hawaii at Hilo, Hilo, Hawaii 96720

Basolo, Alexandra L. School of Biological Sciences, University of Nebraska, Lincoln, Nebraska 68588

Chippindale, Adam K. Department of Biology, N.S. IV, University of California, Santa Cruz, California 95064

Frank, Steven A. Department of Ecology and Evolutionary Biology, University of California, Irvine, California 92717

Hudson, Richard R. Department of Ecology and Evolutionary Biology, University of California, Irvine, California 92717

Hurst, Laurence D. Department of Genetics, University of Cambridge CB2 3EH, United Kingdom

Kirkpatrick, Mark Department of Zoology, University of Texas, Austin, Texas 78712

Larson, Allan Department of Biology, Washington University, St. Louis, Missouri 63130-4899

Lauder, George V. Department of Ecology and Evolutionary Biology, University of California, Irvine, California 92717

Losos, Jonathan B. Department of Biology, Washington University, St. Louis, Missouri 63130-4899

Novacek, Michael J. Department of Vertebrate Paleontology, American Museum of Natural History, New York, New York 10024

Nusbaum, Theodore J. Department of Ecology and Evolutionary Biology, University of California, Irvine, California 92717

Reznick, David Department of Biology, University of California, Riverside, California 92521

Rose, Michael R. Department of Ecology and Evolutionary Biology, University of California, Irvine, California 92717

Seger, Jon Department of Biology, University of Utah, Salt Lake City, Utah 84112

Sinervo, Barry Department of Biology, Indiana University, Bloomington, Indiana 47405

Stubblefield, J. William Cambridge Energy Research Associates, Cambridge, Massachusetts 02138

Travis, Joseph Department of Biological Science, Florida State University, Tallahassee, Florida 32306

Vermeij, Geerat J. Department of Geology, and Center for Population Biology, University of California, Davis, California 95616

Wade, Michael J. Department of Ecology and Evolution, University of Chicago, Chicago, Illinois 60637

Preface

The first thing to be said is that this book has been a joint and equal effort on the part of the two editors. The order of names is a reflection only of the first editor's lack of eloquence, relative to the second editor, in persuading the latter to be first editor. This book has been an adventurous trek for us to edit, and we are grateful to many individuals for helping us along the way. First and foremost has been Chuck Crumly, our editor at Academic Press. We are also appreciative of Armand Leroi's initial efforts to bring the two of us together before this book was ever broached. Like a good matchmaker, he was persistent in his talking up the merits of each of us to the other, overcoming our coy reluctance to deal jointly with the problems of adaptation, the comparative method, and so on. If Chuck Crumly was the book's godfather, then Armand Leroi surely has been its godmother. Both editors also have benefited enormously from the financial support of the NIH and NSF which have fostered our empirical research bringing us face to face with the problem of adaptation.

We thank our authors, though they may now feel little inclined to accept our thanks. Let us at least proffer our apologies to them for our slow responses to their queries, our intemperate threats, and our many demands. We knew that a topic as large as adaptation could not be treated with justice by any small set of authors, requiring instead the extensive recruitment and herding that our editorial work has been. Allow us also to thank the production staff that has made this book possible, particularly Heidi Doan at the University of California, Irvine, and Ellen Caprio at Academic Press. Jay Phelan and Ted Nusbaum kindly assisted with proofreading.

Finally, we ask the readers for their indulgence as they use this book. It is long overdue, and we are aware that this subject matter is changing weekly, often in substantial ways. Nonetheless, we have essayed this presumption with the conviction that the topic has been neglected, dormant, and scattered for far too long. Adaptation deserves greater prominence within biology as a whole, even within evolutionary biology in particular. We hope that our efforts have advanced the field toward that goal.

Michael R. Rose
George V. Lauder

Post-Spandrel Adaptationism

MICHAEL R. ROSE

GEORGE V. LAUDER

A new adaptationism is creeping back into mainstream evolutionary biology. While the true believers of sociobiology and pan-selectionist molecular biology never lost faith in the adaptive significance of the characters they study, other varieties of evolutionary biologist have only recently resumed discussion of adaptation. This book is for them, as well as for their students who need to be introduced to contemporary adaptationism.

I. The Death of the Old Adaptationism

The old adaptationism started dying toward the end of the 1960s. A seminal event in this decline was the publication of "Adaptation and Natural Selection," the 1966 book by George C. Williams, which discredited the vague invocation of group selection and other infirmities of adaptationist reasoning. Williams (1966) emphasized that the concept of adaptation is "special and onerous" and should not be applied lightly. Many did not take his advice, leading to the final proclamation of death by Stephen Jay Gould speaking at a 1978 meeting of the Royal Society of London. This talk would later become widely known in the form of an article, "The Spandrels of San Marco and the Panglossian Paradigm: A Critique of the Adaptationist Programme" (Gould and Lewontin, 1979). In this article, Gould and Lewontin supplied a number of criticisms of the glib style of reasoning about adaptation that had become popular among many evolutionary biologists since World War II. In particular, they sought to discredit "adaptationism" as a style of research in evolutionary biology in which all features of organisms are viewed *a priori* as optimal features produced by natural selection specifically for current function. Instead, Gould and Lewontin advocated a more pluralistic view of evolutionary investigation, recognizing that traits may arise by other means than natural selection. They demanded that evolutionary biologists explicitly consider alternatives to a strictly selectionist

ADAPTATION

view of organismal design.

This paper had such a substantial impact on the fashions of evolutionary biology that the very term "adaptationism," and sometimes even "adaptation" itself, became pejorative. To a significant extent, the term adaptation was banished from the lexicon of evolutionary biology, for fear of being associated with the dread adaptationism. One of us attended a seminar in the early 1980s at which the speaker announced that he would not use the word adaptation in his talk. Rather, to avoid controversy and association with the negative implications of adaptationism, he would use the word "banana" whenever he meant adaptation. This approach was not without merit then, as merely using the word adaptation frequently brought forth strong feelings and arguments only marginally related to the seminar at hand. Other papers of the time reinforced concerns about adaptationism (Lewontin, 1977; Gould, 1982) and deepened the reluctance of evolutionary biologists to confront problems of adaptation head on.

Another factor in the decline of adaptationism in the late 1960s was Richard Lewontin's work on population genetics and electrophoresis (e.g., Lewontin and Hubby, 1966). Up until that time, it had been possible for figures like Theodosius Dobzhansky and Arthur Cain to plausibly, or at least conceivably, explain the differentiation of populations and genetic variation within populations almost entirely in terms of selection. Since then, successive developments in molecular biology have laid bare greater and greater amounts of both evolutionary differentiation between species and genetic variation within populations (e.g., Li and Graur, 1991). This astronomical amount of molecular genetic variation and differentiation made it implausible that each variant had been shaped by natural selection to have that particular configuration.

Recent molecular biology has gone further still. The old, Mendelian view of the genome as a string of genetic beads, each containing the information for making a discrete character, or enzyme, has been demolished. In most organisms, there are abundant quantities of noncoding DNA, DNA that also does not appear to have any regulatory function. Not only is this DNA present in the form of noncoding deserts between genes, it even interrupts genes as "introns" (Li and Graur, 1991). Furthermore, some of this DNA appears to replicate itself and insert copies randomly about the genome, and thus has been proposed to operate as "selfish DNA" (Doolittle and Sapienza, 1980; Orgel and Crick, 1980). The present conception of the genome has little to do with the old orderly library of adaptively constructed genetic information, the image of the genome conveyed to many of us as undergraduates.

But there have been other nails in the coffin of adaptationism, such as the growing understanding that natural selection is not a process that necessarily enhances adaptation. There are several facets to this point. The first is that even the classical overdominant, viability selection, random-mating, one-locus, deterministic model only leads to maximization of *mean* fitness, not the fitness of every genotype. A genetic load due to segregation remains, such that the fitness of any particular individual need not be near that of the maximum in the population. With nonlinear interactions in the determination of fitness, even simple fertility models do not necessarily maximize mean fitness; indeed, mean fitness can decrease in the course of selection (e.g., Pollak, 1978). The development of robust two-locus theory, especially by Karlin and Feldman (e.g., 1970), revealed that epistasis and linkage disequilibrium could undermine the "hill-climbing" effect of natural selection on mean fitness, assumed by both Wright and Fisher. Further developments in the realms of multilocus models and modifier theory have only produced further paradoxes of this kind. Even in the world of theory, the evolutionary attainment of adaptation may be problematic.

An integral component of Gould and Lewontin's spandrels of San Marco framework is the idea of structural or developmental constraints (see Gould, 1977; Alberch, 1982; but also Maynard Smith et al., 1985; Amundson, this volume). Sometimes this concept has been so overextended, constraints being discovered so promiscuously, that its conceptual content seems minimal. If everything is constrained in some way then the explanatory power of the idea of constraints is greatly diluted. However, there is an indubitable core to this criticism: the structure of the genotype-to-phenotype mapping (see Lewontin, 1974). A naive adaptationism might suppose that there are different genes for each aspect of the organism which can be separately molded by natural selection to the ends of optimal adaptation. The demonstrable existence of pleiotropy and epistasis, which connect up the expression of genetic variation among loci in nonlinear ways, makes this view untenable. Research in laboratory evolution, among other areas (cf. Loeschcke, 1987), strongly supports the notion that suites of characters evolve jointly, often in unpredictable ways (Rose et al., this volume). In addition, recent research in developmental genetics (e.g., Atchley and Hall, 1991) has amply demonstrated the complex ontogenetic linkages among characters and suites of characters. Adaptations are embedded in murky complexes of physiological constraints, constraints that may render the action of natural selection remarkably obscure.

Yet another factor in the demise of the old adaptationism was the rise of phylogenetic research. The immediate post-spandrels period of the 1980s was marked coincidentally by a decline in infighting over phylogenetic principles and techniques, battles that had been fought within the systematic community for nearly 20 years. Increased methodological harmony fostered a flowering of interest in applying phylogenetic methods to problems of form and function in organisms. The analysis of organismal design matured during this time from a simplistic search for how individual traits might be adaptive, to broader issues of phylogenetic trajectories, design constraints, and the analysis of intrinsic design elements and their historical consequences (Lauder, 1981; Emerson, 1988; Wake, 1982). Explicit mapping of characters on trees emphasized the history of traits or of character complexes and showed how previous hypotheses of adaptive significance could be refuted by demonstrating historically discordant patterns between structure and environmental change.

A final factor in the demise of the old adaptationism has been a developing understanding of the many levels of natural selection as a process. From the evolution of selfish DNA sequences (see Hurst, this volume) to interdemic selection (see Wade, this volume) to selection at the level of clades (Vermeij and Novacek, this volume), natural selection can operate in contexts that were only intermittently conceived before the 1970s, excluding perhaps the work of Sewall Wright (e.g., 1977, 1978). This is in some ways a spectacular testimony to the original intuitions of Darwin, but it poses some grave problems for the study of adaptation. It certainly does not need to be the case that selection at all levels operates with the same direction, much less magnitude. Formally, this must lead to "adaptive consequences" that cannot be predicted from the study of one level of selection on its own.

II. After the Funeral: The New Adaptationism

This litany of lamentation for adaptationism could go on, but funerals are to be endured for only so long. From a Celtic standpoint, what really matters is the wake, the party after the funeral. During the wake, the past is reviewed, and the loved one is praised and often criticized. In a sense, after the burial, the mourners find a way to go on with their lives. This book is a wake for adaptationism, in its old form. Here we celebrate some of its triumphs, but resignedly. We offer our own criticisms, but out of affection rather than hostility. But most importantly, we have to go on with our lives. We continue to study the problem of adaptation, a manifestly real biological phenomenon, after the Spandrels of San Marco.

So, it could be asked, what does post-spandrel adaptationism look like? What kind of future is there for evolutionary biologists who wish to study selection and its consequences? Is there anything left to the concept of adaptation that can still inform evolutionary research?

To a large extent, the chapters of this book are our answer, in themselves. However, it would be remiss of us not to attempt some sort of integrative reply to this question, beyond merely pointing to multiple lines of research in this area. Therefore, we venture a few opinions about the new adaptationism that is being born around us.

First, there are a variety of technical improvements on the old adaptationism that have not so much changed its direction as strengthened its force. For example, the more formal use of phylogenies in the comparative method has greatly improved the intellectual rigor of interspecific comparison (Brooks and McLennan, 1991; Felsenstein, 1985; Harvey and Pagel, 1991). No longer is it acceptable for species to be treated as independent data points and gross correlations with environment used as evidence of adaptation (see Losos and Larson, this volume). This does not remedy a number of profound problems for the comparative method, such as the legitimacy of inferences of selection mechanisms (Leroi et al., 1994), but it at least saves comparative biology from the kind of egregious errors that arise when one ignores history.

Likewise, Hudson, Kreitman, and colleagues are pushing at the limits of our ability to infer the action of selection from DNA sequences (e.g., Hudson et al., 1994; Kreitman and Hudson, 1991; Hudson, this volume). From comparisons of differentiation within and among species, they are developing some ability to demonstrate the action of balancing selection as opposed to neutral gene evolution. This work does not, however, readily lead to the inference of the particular nature of the selection mechanisms involved, as the continuing mystery concerning the action of selection at the *Adh* locus in *Drosophila melanogaster* illustrates (e.g., Laurie and Stam, 1994).

Second, the study of natural selection in the wild has been greatly improved, partly through the development of higher critical standards (e.g., Endler, 1986) and partly through the development of methodologies based on quantitative genetics theory (e.g., Arnold, 1983, 1986; Lande, 1978). Recent studies of natural selection in the wild (see Reznick and Travis, and Sinervo and Basolo in this volume) represent a striking advance over previous work. Such studies often incorporate field experimentation lasting many years to demonstrate that environmental and selective manipulation can replicate extant differences among populations. The combination of long-term field observations on multiple replicate populations, field manipulations, and laboratory-based genetic data is a powerful one for the study of selection.

Third, many disciplines which in the past made frequent and gratuitous use of the concept of adaptation have greatly restricted their inferences of adaptation and have focused instead on doing what they do best. For example, the disciplines of comparative morphology and biomechanics were shaped for years by the notion, implicit if not explicit, that virtually every character is molded in isolation by selection for its current role. Such views are no longer widely held, and these areas are now turning to focus on both the design principles of biological systems (Niklas, 1992; Thomason, 1995; Vogel, 1988) and the historical transformation of structure and function during evolution (e.g., Wake and Roth, 1989). Functional morphology and biomechanics, as disciplines whose goal is the analysis of biological design, may still have a critical role to play in arguments about adaptation (see Lauder, this volume), but their new focus both broadens the intellectual base of these disciplines and brings a new comparative and historical rigor to adaptationism. Furthermore, the study of organismal development, not long ago a mainly descriptive discipline, is now the subject of comparative evolutionary investigation using molecular techniques (e.g., Hall, 1992; Hanken and Thorogood, 1993; Raff et al., 1990), as well as important new quantitative genetic analyses (e.g., Atchley et al., 1990; Atchley and Hall, 1991). These approaches bring the promise of increased clarity to our understanding of how traits arise and of the intercorrelations among traits; both issues are critical to the study of adaptation.

In the area of laboratory evolution, work with *Escherichia coli* and *Drosophila* (see Rose et al., this volume) has greatly improved our opportunity to observe natural selection as close to the tips of our noses as possible. A notable trend in this area has been a significant increase in the number of replicate populations used per experiment (e.g., Lenski et al., 1991). Another trend has been the use of multiple, distinct, selection treatments.

In addition to these specific components of the adaptationist structure, the last decade has also seen repaired foundations for the structure. The books by Dawkins (1987), Brandon (1990), Sober (1984, 1989), Williams (1992), and Dennett (1995) all focus on conceptual issues critical to future progress in the analysis of adaptation. While this will remain more an area of discussion than simple progress, unlike for example recent work on the statistical use of phylogenies, such mortar-work will be essential as the edifice of adaptationism vaults farther within evolutionary biology and even beyond.

III. New tunes for the New Adaptationism

But there are other aspects of the study of adaptation that are more than variations on old themes. One of these is the growing role of selective adaptation in nonevolutionary realms, in artificial intelligence, automated design, and the like (see Frank, this volume). Ironically, while evolutionary biologists distanced themselves from the concept of adaptation during the 1980s, a number of scientists from other fields, even engineers, have embraced it enthusiastically as the most powerful way to solve key problems. Adaptation is a much bigger concept than evolutionary biologists, proponents and critics alike, may have realized.

A second novelty that is beginning to surface is the perversity or, if you will, the creativity of adaptation. This is apparent in the spectacular oddities of the fossil record, from the Burgess Shale (Gould, 1989) to the dinosaurs of the Gobi desert (see Novacek, this volume). But it is also apparent in the oddities and paradoxes of laboratory evolution, which inspired one of us to compare evolution to Alice's Wonderland (Rose et al., this volume).

A third challenge for the new adaptationism is the remarkable conservatism that has been found in the genetic mechanisms that underlie the design of even widely divergent organisms. It seems that nearly every week a new gene or gene family is isolated that has a common effect on developmental patterning in mammals and *Drosophila*. The extent of fundamental conservatism among structurally divergent clades sharing only distant common ancestry has been far greater than could have been conceived even fifteen years ago. The new adaptationism must deal with the fact that major components of extant organismal design may represent more a reshuffling of ancient parts than a combination of novel features arisen *de novo* in response to specific selection forces.

A final theme of recent note is the extent to which random events in the history of life may have radically changed the environmental and biological context for adaptation. It now seems likely, for example, that large meteor impacts may occur with reasonable frequency on the geological time scale, and that such events may be associated with significant changes in patterns of biological diversity. Such large-scale disruptions provide new opportunities for selection and reshuffle relative species diversity among major clades. Clades that survive may do so for reasons unrelated to traits that were deemed to have been "adaptations" prior to the disruption, and we have only begun to explore the effects of such changes on historical trajectories of organismal design.

Still other new themes are no doubt emerging in the minds of those who study adaptation. Now, exactly 30 years after the 1966 publication of "Adaptation and Natural Selection," post-Spandrel adaptationism is just beginning to develop. The certainties of the old adaptationism are gone. The new adaptationism is but an unruly toddler, exploring its environment with reckless curiosity, impatient to discover the secrets hidden about the evolutionary wonderland.

References

Alberch, P. (1982). Developmental constraints in evolutionary processes. *In* "Evolution and Development" (J.T. Bonner, ed.) pp. 313-332, Springer-Verlag, Berlin.

Arnold, S. J. (1983). Morphology, performance, and fitness. *Amer. Zool.* 23, 347-361.

Arnold, S. J. (1986). Laboratory and field approaches to the study of adaptation. *In* "Predator-Prey Relationships: Perspectives and Approaches from the Study of Lower Vertebrates" (M. E. Feder, and G. V. Lauder, eds.) pp. 157-179, Univ. of Chicago Press, Chicago.

Atchley, W. R., Cowley, D. E., Eisen, E. J., Prasetyo, H., and Hawkins-Brown, D. (1990). Correlated response in the developmental choreographies of the mouse mandible to selection for body composition. *Evolution* 44, 669-688.

Atchley, W. R., and Hall, B. K. (1991). A model for development and evolution of complex morphological structures. *Biol. Rev.* 66, 101-157.

Brandon, R. N. (1990). "Adaptation and Environment." Princeton University. Press, Princeton.

Brooks, D. R., and McLennan, D. A. (1991). "Phylogeny, Ecology, and Behavior: A Research Program in Comparative Biology." Univ. of Chicago Press, Chicago.

Dawkins, R. (1987). "The Blind Watchmaker." W. W. Norton, New York.

Dennett, D. C. (1995). "Darwin's Dangerous Idea: Evolution and the Meanings of Life." Simon and Schuster, New York.

Doolittle, W. F., and Sapienza, C. (1980). Selfish genes, the phenotype paradigm and genome evolution. *Nature* 284, 601-603.

Emerson, S. (1988). Testing for historical patterns of change: a case study with frog pectoral girdles. *Paleobiology* 14, 174-186.

Endler, J. A. (1986). "Natural Selection in the Wild." Princeton University Press, Princeton, NJ.

Felsenstein, J. (1985). Phylogenies and the comparative method. *Amer. Nat.* 125, 1-15.

Gould, S. J. (1977). "Ontogeny and Phylogeny." Harvard University Press, Cambridge, MA.

Gould, S. J. (1982). Darwinism and the expansion of evolutionary theory. *Science* 216, 380-387.

Gould, S. J. (1989). "Wonderful Life. The Burgess Shale and the Nature of History." W. W. Norton, New York.

Gould, S. J. and Lewontin, R. C. (1979). The spandrels of San Marco and the Panglossian paradigm. A critique of the adaptationist program. *Proc. Roy. Soc. Lond. B* 205, 581-98.

Hall, B. K. (1992). "Evolutionary Developmental Biology." Chapman and Hall, London.

Hanken, J. and Thorogood, P. (1993). Evolution and development of the vertebrate skull: the role of pattern formation. *Trends Ecol. Evol.* 8, 9-15.

Harvey, P. H., and Pagel, M. D. (1991). "The Comparative Method in Evolutionary Biology." Oxford University Press, Oxford.

Hudson, R. R., Bailey, K., Skarecky, D., Kwiatowski, J., and Ayala, F. J. (1994). Evidence for positive selection in the superoxide dismutase (*Sod*) region of *Drosophila melanogaster*. *Genetics* 136, 1329-1340.

Karlin, S., and Feldman, M. W. (1970). Linkage and selection: two-locus symmetric viability model. *Theor. Pop. Biol.* 1, 39-71.

Kreitman, M., and Hudson, R. R. (1991). Inferring the evolutionary histories of the *Adh* and *Adh-dup* loci in *Drosophila melanogaster* from patterns of polymorphism and divergence. *Genetics* 127, 565-82.

Lande, R. (1978). Evolutionary mechanisms of limb loss in tetrapods. *Evolution* 32, 73-92.

Lauder, G. V. (1981). Form and function: structural analysis in evolutionary morphology. *Paleobiology* 7, 430-442.

Laurie, C. C., and Stam, L. F. (1994). The effect of an intronic polymorphism on alcohol dehydrogenase expression in *Drosophila melanogaster*. *Genetics* 138, 379-85.

Lenski, R. E., Rose, M. R., Simpson, S. E., and Tadler, S. C. (1991). Long-term experimental evolution in *Escherichia coli*. I. Adaptation and divergence during 2000 generations. *Amer. Nat.* 138, 1315-1341.

Leroi, A. M., Rose, M. R., and Lauder, G. V. (1994). What does the comparative method reveal about adaptation? *Amer. Nat.* 143, 381-402.

Lewontin, R. C. (1974). "The Genetic Basis of Evolutionary Change." Columbia University Press, New York.

Lewontin, R. C. (1977). "Sociobiology - a caricature of Darwinism." PSA 2, Philosophy of Science Association.

Lewontin, R. C., and Hubby, J. L. (1966). A molecular approach to the study of genic heterozygosity in natural populations. II. Amount of variation and degree of heterozygosity in natural populations of *Drosophila pseudoobscura. Genetics* 54, 595-609.

Li, W-H., and Graur, D. (1991). "Fundamentals of Molecular Evolution." Sinauer, Sunderland, MA.

Loeschcke, V. (ed.) (1987). "Genetic Constraints on Adaptive Evolution." Springer-Verlag, Berlin.

Maynard Smith, J., Burian, R., Kauffman, S., Alberch, P., Campbell, J., Goodwin, B., Lande, R., Raup, D., and Wolpert, L. (1985). Developmental constraints and evolution. *Quart. Rev. Biol.* 60, 265-287.

Niklas, K. J. (1992). "Plant Biomechanics." Univ. of Chicago Press, Chicago.

Orgel, L. E., and Crick, F. H. C. (1980). Selfish DNA: the ultimate parasite. *Nature* 284, 604-607.

Pollak, E. (1978). With selection for fecundity the mean fitness does not necessarily increase. *Genetics* 90, 383-89.

Raff, R. A., Parr, B. A., Parks, A. L., and Wray, G. A. (1990). Heterochrony and other mechanisms of radical evolutionary change in early development. *In* "Evolutionary Innovations" (M. H. Nitecki, ed.). pp. 71-98, Univ. of Chicago Press, Chicago.

Sober, E. (1984). "The Nature of Selection: Evolutionary Theory in Philosophical Focus." MIT Press, Cambridge, MA.

Sober, E. (1989). "Reconstructing the Past: Parsimony, Evolution, and Inference." MIT Press, Cambridge, MA.

Thomason, J. J. (ed.) (1995). "Functional Morphology in Vertebrate Paleontology." Cambridge University Press, Cambridge, MA

Vogel, S. (1988). "Life's Devices: the Physical World of Animals and Plants." Princeton University Press, Princeton, NJ.

Wake, D. B. (1982). Functional and evolutionary morphology. *Persp. Biol. Med.* 25, 603-620.

Wake, D. B., and Roth, G. (eds.). (1989). "Complex Organismal Functions: Integration and Evolution in Vertebrates." John Wiley and Sons, Chichester.

Williams, G. C. (1966). "Adaptation and Natural Selection." Princeton University Press, Princeton.

Williams, G. C. (1992). "Natural Selection: Domains, Levels, and Challenges." Oxford University. Press, Oxford.

Wright, S. (1977). "Evolution and the Genetics of Populations." Vol. 3. Experimental Results and Evolutionary Deductions. Univ. of Chicago, Chicago.

Wright, S. (1978). "Evolution and the Genetics of Populations." Vol. 4. Variability within and among Natural Populations. Univ. of Chicago, Chicago.

PART I

Concepts and Theories of Adaptation

Adaptation is no longer something that can be safely assumed by evolutionary or other biologists. Indeed, the more one examines the concept, the more it comes to resemble a newly landed fish: slippery, slimy, obstreperous, but glittering with potential. There it is, flapping about, full of energy, but the significance of all the commotion is not clear. Perhaps the solution of some evolutionary biologists is best – just throw the damn thing back in the water. But of course our authors have not chosen that course of action, and we are left with the problem of what to do with the fish.

In the first chapter of this section, Amundson sorts out the basic features of the concept of adaptation over the course of its historical development. Reading this history serves to remind us that there is little new under the sun. The debates about structure, resemblance, and function *before* Darwin are strongly reminiscent of the spandrel controversy (chronicled in Chapter 2) that occurred during the 1980s.

The three subsequent chapters present contemporaneous views about three basic theoretical frameworks for understanding adaptation: design principles, optimization theory, and evolutionary genetic theory. Lauder analyzes the invocation of *a priori* design principles from engineering, physics, and the like. He cautions against making assumptions about the use of biological design and questions our ability to infer adaptation from design alone. Seger and Stubblefield, on the other hand, provide an eloquent defense of the use of optimization principles, with their strong design elements, in the study of adaptation. This contrast illustrates an important feature of this book: Our authors don't always agree. Finally, Kirkpatrick gives a wonderfully iconic summary of salient results from population genetics theory. We offer these different theories and perspectives to foster debate not to end it.

CHAPTER I

Historical Development of the Concept

of Adaptation

RON AMUNDSON

"The logic of [William Paley's *Evidences of Christianity*] and as I may add of his Natural Theology gave me as much delight as did Euclid. The careful study of these works, without attempting to learn any part by rote, was the only part of the Academical Course which, as I then felt and as I still believe, was of the least use to me in the education of my mind. I did not at the time trouble myself about Paley's premises; and taking these on trust I was charmed and convinced by the long line of argumentation" (Darwin, 1958, p. 59).

"The main task of any theory of evolution is to explain adaptive complexity, that is, to explain the same set of facts that Paley used as evidence of a Creator" (Maynard Smith, 1969, p. 82).

The phenomenon of adaptation is at the core of modern evolutionary biology. Natural selection, the mechanism universally regarded as the primary causal influence on phenotypic evolutionary change, is first and foremost an explanation of adaptation. Some of the most satisfying explanations of anomalous biological phenomena show them either to *be* adaptations or to have been produced as by-products of adaptive changes. In the 18th century, Newtonian physicists labored to demonstrate that oceanic tides were gravitational phenomena. In the 20th century, much labor among Darwinian evolutionists is directed toward revealing the underlying selective and adaptive nature of a variety of often–puzzling biological traits.

    As Maynard Smith reminds us above, interest in biological adaptation did not have its origin with Charles Darwin. Adaptation was the central focus of William Paley's natural theology, and the fit between organisms and their natural environment was already an ancient theme in Paley's day. The notion that *Nature does nothing in vain* has guided centuries of Western natural history, and biology is the natural home of that epigram. The modern evolutionary explanation of adaptation is so different from earlier ideas that modern biologists only occasionally acknowledge their affinities with pre–Darwinians (Cain, 1964; Maynard Smith, 1969; Dawkins, 1983; Mayr, 1983). Due perhaps to continued tension between religion and science, biologists usually stress the differences rather than the similarities between Darwinian and pre–Darwinian biology. They emphasize the *causes* of adaptation, beliefs about which separate modern from pre–Darwinian thinkers. Many prominent biologists are eager to isolate the vocabulary of biology from pre–Darwinian purposive terms such as teleology. In iconoclastic reaction, Lennox (1993) has declared that "Darwin *was* a teleologist". Lennox does not attribute to Darwin a belief in supernatural purposiveness, but rather a recognition of important similarities between the effects of his own mechanism of natural selection and of teleology as traditionally conceived. The present work will not adopt Lennox's terminology, but it will take seriously the biological thought of pre–Darwinians. Anathematizing the older vocabulary can lead to an underappreciation of many continuities of thought regarding the ubiquity of adaptation and its importance for the understanding of life itself.

Adaptation is a central topic of biological study, but it is not the only topic. Organic diversity and homology, for example, are not *prima facie* adaptive phenomena. It might of course be discovered (or argued) that such phenomena are adaptive or by-products of adaptation. After all, if the most important causal influence on evolutionary change is itself a cause of adaptations, we should expect adaptations to spring up in surprising places. The propriety of this train of thought has been questioned in recent days. A range of opinion exists today on the validity of emphasizing adaptation over other biological phenomena. It is argued that "adaptationist" research commitments may result in the neglect of important nonadaptive biological phenomena, and of nonadaptive causal influences on all biological phenomena. This, too, is not new. Throughout history, and especially during the early 19th century, advocates of the central importance of adaptation had their critics. The criticisms have sometimes proven groundless, but at other times have been important influences in the growth of biology. The term "adaptationism" itself was introduced during a recent episode of the controversy (Lewontin, 1978; Gould and Lewontin, 1979). While the term is sometimes taken as pejorative, many advocates of the study of adaptation have found it unobjectionable (Mayr, 1983). It will be used in the present essay as descriptive of a commitment to adaptation as the single topic of central importance to biology. In any case, to appreciate modern adaptationism in its historical context, we must also consider the history of anti–adaptationist thought.

The concept of adaptation has always been closely related to concepts like purpose and function, and in scholarly contexts to teleology and final causation. These concepts have clear applications to human actions and artifacts, but may also be considered intuitively applicable to natural entities such as the parts of organisms and (for that matter) the entire world system. In earlier times a reference to purpose in nature was presumed to invoke either a divine Mind or some special mind–like teleological force. The invocation of these special entities has been abandoned within biological theory, but the purposive terms still carry portions of their original significance and still properly refer to many of the same natural phenomena. In both ancient and modern times, an adaptation is not just one of the *products*, but one of the *fruits* of nature. To explain a trait as an adaptation is to provide it with a sort of *rationale* which connects the trait to some benefit which it produces (Dennett, 1987). This is true whether the explanation comes from a Darwinian or a natural theological source. Some of the same rationales serve equally well for either ancient or modern biologist. Most differences relate less to a trait's adaptive status than to the causal origins of that adaptation.

In their richest purposive sense, concepts like adaptation and function have dual implications. When applied to a trait of an organism or artifact, they first designate an end or benefit produced by that trait. (In intentional contexts they may designate an intended or expected benefit.) Secondly, it is implied that the trait exists because of the designated benefit. So both the historical origins and the current causal properties of a trait are addressed when labeling it as an adaptation, or ascribing to it a function. Such a trait appears to exist because it is good for some end. This is surely the source of human fascination with the apparently purposive aspects of the natural world. But it is also a target for skeptical criticism. Darwin's theory itself can be seen as partly a reaction to skepticism regarding adaptation in the organic world. While full-blown historical/Darwinian interpretations of the concepts of adaptation and function are in wide use today, some biologists have found grounds to weaken the implications of the concepts. Nonhistorical accounts both of adaptation and of function have been proposed. On these accounts, identifying a trait as an adaptation or stating its function is not intended to explain its existence. These accounts may be offered either as addenda or as replacements for more traditional and purposive interpretations; given these diverse alternatives, debate on the range and meaning of adaptation of course continues.

The present essay will first trace the historical development of the modern concept of biological adaptation and examine some of the methodological and philosophical debates which surrounded it. In the historical context, no embarrassment will be shown regarding theological involvement in the practice of science. Theological and scientific uses of purposive vocabulary will be distinguished only when there is a possibility of confusion. Surprisingly enough, it seldom matters.

The essay will trace conceptions of purposive order in the world from the ancient Greeks through the beginnings of modern science. We will see the role of Newtonian physics in clarifying the special nature of biological adaptation. We will consider how debates regarding adaptationism formed the theoretical context of biology in England in the period just before Darwin's (1859) *Origin of Species*. In the *Origin*, Darwin came down on the side of adaptation, strongly siding with Paley on the importance (but not on the causes) of biological adaptation. From that point on, but especially after the Modern Synthesis of the 1930s and 1940s, adaptation was seen from a new, naturalistic perspective (Mayr and Provine, 1980). Nevertheless, many of the old debates regarding the centrality of adaptation have survived Darwin, survived the Modern Synthesis, and continue in current biological discussion. The essay will conclude with a review of modern discussions of adaptationism and the proper definition of adaptation itself.

I. Early Concepts

Aristotle's concept of *final causation* was an early and extremely influential account of purposive, adaptive, or teleological styles of explanation. The final cause is the end or goal towards which a process is proceeding. The three other "causes" were material, formal, and efficient. The efficient cause comes closest to the modern concept of direct physical cause. Final and efficient causes concerned conditions of change or "becoming"; material and formal were conditions of "being" (Bhaksar, 1981). The four causes were best applied to biology and to human workmanship. The causes were not seen to be in conflict; all four would be found in a complete account of a given process. They could be seen as answers to four kinds of questions: What materials were involved? What forms were actualized? What end was being sought? What initiated the process?

The mentalist or intentional flavor of final causation has always been recognized. It is reflected in a logical peculiarity of final cause ascriptions. The state of affairs labeled as the "final cause" comes into existence only after the object which is the purported "effect" of that cause. So something in the future seems to be causing something in the past; the welfare of my eyes today is what caused my tear ducts to develop while I was an embryo. This violates intuitions on the necessity for a cause to precede its effects. Efficient causes always precede their effect. In the days before Darwin, believers in the reality of natural purpose and final causation had two ways to account for the temporal oddity of final causes. The first was to explain natural purpose by analogy with human intelligence, and regard phenomena which receive purposive explanations as having been contrived and constructed by an intelligent mind which can foresee the benefits which a contrivance will produce. This was an "external teleology" in that natural purpose exists because it was placed there by an external agent (e.g., a designing Intelligence). Plato and the British natural theologians accepted such a teleology. The alternative is an "imminent" or "internal teleology", in which the cosmos is itself goalward driven. Aristotle favored this view, and later versions can be recognized in Lamarck, Herbert Spencer, and the 19th century German teleomechanists (Lenoir, 1982).

It is helpful to think of the four Aristotelian causes as kinds of explanation rather than kinds of cause. Final and efficient causes do not differ in the way that gravitational causes differ from electromagnetic causes. The consistency among the four causes reflects a sort of relativity of explanation. It is easy to see that distinct explanations of the same phenomenon can be logically consistent; they are answers to distinct questions about the phenomenon. Ernst Mayr's important distinction between proximate and ultimate causation in biology resembles the four causes in this way (Mayr, 1961). Indeed the proximate/ultimate distinction may be seen as an evolutionary modernization of the distinction between efficient and final causation.

There is a point in describing purposes as *causes*. To assign a purpose to a trait is not merely to designate its beneficial effects. It is also to offer an explanation of why the trait exists at all — it exists *because of* the benefits it brings. But this account seems to confuse cause with effect. The benefits of a trait are its *effects*, and effects can only occur after the trait already exists. The *cause* of a trait must have existed before the trait came into being. This odd temporal circularity in purposive explanation derives from the need for the (future) benefit to serve as the (past) cause. Modern concepts of adaptation retain their similarity to earlier concepts because Darwin was able to describe a theoretical mechanism which would do both of these tasks — it would designate a trait's benefits, and assert that the trait exists today *because of* (ancestors' possession of) those benefits.

In the dialogue *Pheado*, Plato has Socrates express a preference for what amounted to final over efficient causes. On his deathbed Socrates recalls his excited anticipation of the writings of Anaxagoras, who was said to have explained all worldly phenomena by reference to Mind. When he read Anaxagoras's works, Socrates discovered no more than (purported) efficient or material causes. Worldly phenomena were consequences of the properties of elements like air, ether, and water. Socrates was greatly disappointed. He had expected the "causes" of phenomena to explain why the world was *better* as it is than if it had been any other way. "I thought that if he asserted that the earth was in the center, he would explain in detail that it was better for it to be there; and if he made this clear, I was prepared to give up hankering after any other kind of cause" (Plato, 1961, p. 79). Socrates contrasted this with an example of efficient causation, which he considered of little value. He offered two explanations of why he, Socrates, was then lying on that bed. One stated that "my body is composed of bones and sinews, and the bones are rigid and separated at the joints ... [these and other biological facts] enable me sometimes to bend my limbs, and that is the cause of my sitting here in a bent position." The other explanation was that the Republic had condemned him, and it was *better* that he die than that he violate the judgements of the Republic. Notably, Socrates did not deny the truth of the efficient causal explanation. He rather denied that it deserved the honorific label "cause." Just as Socrates' present situation should be explained by what is "better," so should the rest of the world be explained. Efficient causes in either case are not exactly false — they are rather less deserving than final causes. Socrates was not able to say what "good" all worldly phenomena served, but he valued final causal explanation far more than mere efficient causation.

Medicine provided some early details on the "good" of things. Galen, a physician of the 2nd century AD, considered all of the body's parts to be consequences of purposive causation. Discussing the anatomy of the ducts, cavities, and linings of internal organs, he remarks that the formative faculty of Nature does "everything for some purpose, so that there is nothing ineffective or superfluous, or capable of being better disposed" (Galen, 1952, p. 170). Nature does nothing in vain.

Just as Plato had inferred an intelligent creator from the "good" in things, the Greek Stoics inferred such an intelligence from other natural phenomena. A prominent example was the regularity of celestial motions. These were likened to the motions of a fleet of ships moving across the horizon (Hurlbutt, 1965, p. 108; Glacken, 1967, p. 56). Just as we would infer intelligent guidance for the ships from their regular motion, we should infer supernatural guidance for the heavens from the same phenomena. Here intelligent creation and guidance are inferred from regularity of pattern and do not exactly reveal the "good" in things. Some aspects of celestial motion do seem to benefit terrestrial life; celestial periodicities like seasons and days have organic significance. But in astronomy the Stoic inference of design was based on the pattern alone and did not require a separate proof of goodness.

From a modern perspective these examples conceal a dichotomy in the kinds of phenomena which called for a final cause. The "good" of which Plato speaks can be seen most easily in biological processes like nutrition, reproduction, and ecological fit. The regular patterns of the Stoic example are most obvious in astronomical phenomena, the ceaseless regularities of celestial movements. The two domains intersect in the periodic celestial motions of terrestrial life. Day/night cycles and yearly seasons provide a point of connection between the needs and purposes of life and the patterned movement of the heavens.

The ambiguity between goodness and pattern was to persist through medieval times. An illustration is the last of Thomas Aquinas's "Five Ways" to prove the existence of God. "The fifth way is taken from the governance of the world. We see that things which lack knowledge, such as natural bodies, act for an end, and this is evident from their acting always, or nearly always, in the same way, so as to obtain the best result. Hence it is plain that they achieve their end, not fortuitously, but designedly. Now whatever lacks knowledge cannot move towards an end, unless it be directed by some being endowed with knowledge and intelligence; as the arrow is directed by the archer. Therefore some intelligent being exists by whom all natural things are directed to their end; and this being we call God" (Aquinas, 1952, p. 13).

It is hard to know what Aquinas would consider a typical "natural body" here. If we consider an astronomical object like the moon, it certainly acts always in the same way. But how does that imply that it acts "for an end" and why are approximately 28–day phase cycles "the best result?" Socrates' old question: Why is it good? Biological phenomena provide better means/ends relationships in the suitability of body parts of animals for their ways of life. But organisms are not as regular in their action as an arrow from a skilled archer. Two ideas are entangled here — the means/ends relationships so apparent in biological phenomena, and the regularity of pattern so striking in astronomical phenomena.

This kind of argument is generically titled the Argument from Design. The ambiguity in the early versions of these ideas happens to match an ambiguity in the modern term "design." When we refer to a design, we sometimes merely mean a pattern. Gentle waves on beach sand or winter frost on a window can make interesting designs. In other cases design implies purpose and the contrivance of means to ends. The front end of an automobile may be *designed to* crumple on impact. Here the design involves a means (the crumpling and resulting absorption of the force of impact) and an end (reduction of impact on the passengers). The goodness in the second kind of design is a function of *how well* the means serve the end. A means/ends relation provides a clear interpretation of what "good" means. But what would make a mere pattern "good?"

The two versions of the Argument from Design remained conflated through the medieval period. One could look at the natural world and see purpose and pattern together, all one grand cosmos. There was no reason to tease apart the various kinds of natural evidence of intelligent creation. In the 17th century, physics and astronomy began making advances. Since the Argument(s?) from Design was based on the best knowledge of the natural world, new scientific developments were to be accounted for within that framework. The idea that stars in their motion were like ships on the horizon could hardly be maintained after Galileo, Kepler, and Newton. Newton unified the laws of motion on earth and in the heavens, giving astonishingly detailed explanations of astronomical phenomena. No similar revolutionary developments took place in biology. The contrast between Arguments from Pattern and Arguments from Purpose was becoming clearer. For the first time, careful distinctions were made between the astronomical and the organic bases for the inference to intelligent creation. This recognition was the dawn of a specifically biological adaptationism.

Before proceeding to modern science, we must note that the affection for final causes was not universal in early days. Many Greek and Roman thinkers rejected purposive interpretations of nature. Some were simply skeptical of the theological and metaphysical tone of final causation, but others gave detailed grounds for challenging attributions of purpose. Lucretius, a 2nd century AD follower of Epicurus, is an example. He cautioned against even common sense attributions of purpose to body parts. We should avoid the "mistake of supposing . . . that the forearms were slung to the stout upper arms and ministering hands given us on each side, that we might be able to discharge the needful duties of life. Other explanations of like sort which men give, one and all *put effect for cause* through wrongheaded reasoning; since nothing was born in the body that we might use it, but that which is born begets for itself a use: thus seeing did not exist before the eyes were born, nor the employment of speech ere the tongue was made" (Lucretius, 1952, p. 55). Lucretius accepted purposive attributions for intentionally designed artifacts because the needs to be served existed prior to the artifact's creation; drinking existed before cups were invented to assist in it. But final causes and purposes could never explain the existence of natural objects. Lucretius recognized the temporal circularity of purposive explanation, it put effects before causes. On that ground he rejected it. This kind of skepticism about natural purpose would be expressed by others in later days.

II. The Influence of Science: From Astronomical Pattern to Biological Purpose

Isaac Newton's achievements in the late 17th century in mathematics, physics, and astronomy are now viewed as major steps in the mechanization of the Western image of nature. Newton himself would have been horrified at this assessment. He opposed a mechanistic view of the universe, and was one of the strongest proponents of an astronomy-based natural theology. "When I wrote my treatise upon our Systeme [sic] I had an eye upon such Principles as might work with considering men for the beliefe [sic] of a Deity and nothing can rejoyce [sic] me more than to find it usefull [sic] for that purpose" (quoted in

Jacob, 1986, p. 244).

Newton believed that the facts of astrophysics gave evidence for divine contrivance of the universe. First, the principle of universal gravitation required action at a distance, a kind of remote causation considered by Newton and many others to be physically impossible. Gravity could not be an internal, essential property of matter. "Gravity must be caused by an agent acting constantly according to certain laws; but whether this agent be material or immaterial, I have left to the consideration of my readers" (quoted in Dijksterhuis, 1961, p. 488). Newton never suggested a material agency, and was pleased with the fact that an immaterial (divine) agency was thereby implicated. In contrast, Descartes' vortex theories were truly mechanical. Involving no action at a distance, they required no participation of a perhaps-divine agent. Newton also believed that particular features of the solar system required divine contrivance and participation. All of the planets and moons in the solar system traveled in nearly circular orbits, in the same direction, and in nearly the same plane. The stability of the solar system depends on the coincidence of these orbits. Newton considered that his physical principles explained the stability and orderliness which now exist in the solar system, but they could not explain how such a finely tuned system had come into existence. Note the similarity with organic adaptations, the benefits of which are apparent but the causes unobserved. The origin of the solar system, with its orbital paths adapted to long-term stability, was left in divine hands.

The stability of the solar system was not perfect, however. Jupiter was known to be slowly accelerating its orbital speed, while Saturn was decelerating. At this rate the system would eventually be disrupted. In the *Opticks*, Newton suggested that periodic divine maintenance work would be needed to "reset" the clockwork of the solar system lest the whole assembly fall apart (Hahn, 1986). Comets were a possible tool for these adjustments. But why should God have created a world in need of maintenance? The notion seemed absurd to many. Leibniz suggested that God's craftsmanship was demeaned by the need to remedy defects in His creation.

So Newton saw evidence for adaptive final causation, and even occasional divine intervention in the workings of the heavens. According to his theological voluntarism (implying God's eternal ability to act in the world), the divine maintenance of the solar system was no more objectionable than divine participation in gravitational attraction (Brooke, 1991). Newton believed that the remarkable simplicity of the laws of gravity and inertia would be a special boon to natural theology. He thought that the evidence of final causation could be seen more clearly in the simplicity of celestial physics than in the messy complexities of organic design.

Newton's achievement made necessary the first distinction between the theological reasoning based on astronomy and that based on biology. The distinction was between "astrotheology" and "physicotheology." ("Physiotheology" would be a better term for modern ears; physicotheology is based not on evidence from physics, but from biology.) The two approaches divided the old Argument from Design into an astronomical Argument from Pattern and a biological Argument from Purpose. But contrary to Newton's expectations, his achievements did not advance the standing of astronomy in natural theology. They did mark the beginning of a heyday of British natural theology — but of a natural theology self–consciously based on biology rather than astronomy. Once the scientific differences between Pattern and Purpose were clearly spelled out, the benefits of Purpose became apparent.

Two important scientists, the naturalist John Ray and the physicist/chemist Robert Boyle, set the tone for post–Newtonian natural theology. They and their followers began to marginalize astrotheology. Natural theology texts began to mention astronomy only in passing. Increased emphasis was placed on organic design, where purpose and the adaptation of means to ends were more apparent. Ray's *The Wisdom of God Manifested in the Works of Creation* appeared in 1691. Astronomy was used only to illustrate the grandeur and expanse of creation, not to prove design. Ray looked forward to the afterlife, during which in greater knowledge we might "clearly see to our great satisfaction and admiration, the Ends and Uses of these Things, which here were either too subtle for us to penetrate and discover, or too remote and inaccessible for us to come to any distinct view of, *viz.* the Planets, and fix'd Stars" (Ray, 1977, p. 171). In other words, astronomical phenomena must *have* a purpose, but that purpose is invisible to mortal eyes. Prior to Newton, the orderliness of astronomical motion seemed grounds enough to infer purposive design. After Newton, the already explained patterns of motion did not seem to demand further purposive explanation. More direct evidence of ends and uses was available, and it came not from astronomy but biology. Ray gave hundreds of examples. The biologizing of natural theology meshed nicely with the increasing British interest in natural history. No one had the impiety to fully abandon astrotheology. The heavens still "proclaimed the glory of God" but they ceased to prove His existence. The weight of natural theology was carried more and more by the broadening information base concerning the natural history of life on earth.

Commentators on this era of natural theology recognize the shift from astrotheology to physicotheology. It is usually seen as movement from a good argument to a better one. But for Boyle, at least, the shift was not a simple expansion of the basis of natural theology; it was a movement from shaky ground towards solid ground. Boyle was wary of that very physical simplicity which had inspired Newton. Charles Gillespie's assessment: "Boyle nonetheless feared that it might be just possible, even if not very probabl[e], that the inanimate objects of these comparatively simple systems might have 'after many essays ... cast one another in divers of those circumvolutions of matter' described by Epicurus or Descartes. ... In any event, 'the situations of the coelestial bodies do not afford by far so clear and cogent arguments of the wisdom and design of the Author of the world, as do the bodies of animals and plants.'" (Gillespie, 1987, p. 26).

So astrotheology was scarcely named in time for it to be neglected. In Aquinas's time it was easy to move from *always acts the same* to *acts for an end*, and thence to *achieves the best result*. The Arguments from Pattern and Purpose were not distinguished. When the astronomical patterns had been unexplained, prior to Newton, they were conflated with other more purposive aspects of the universe. But after Newton provided a deeper understanding of the celestial bodies' regular motions, it was far from obvious what end was being achieved and why the result was best. An elegant and simple nonpurposive explanation removed the patina of purpose from astronomy. After Pattern received a nonpurposive explanation, its usefulness to natural theology faded.

Detaching natural theology from astronomy turned out to be a wise choice. The action at a distance aspect of gravitation became less metaphysically worrying as the principle became more familiar. Newton's followers had more faith in the completeness of his system than Newton had; they did not despair of a stable solar system. The apparent instability of the orbits of Jupiter and Saturn was seen as a puzzle to be solved rather than evidence of divine intervention. And solved it was. In the late, 18th century Pierre Laplace demonstrated that the Jupiter/Saturn interaction would eventually reverse, and was a stable oscillation over the long run. Many natural theologians saw Laplace's stable universe as a great success. They agreed with Leibniz that a maintenance–free world testified to a God of greater power. Others were less impressed. The solar system's astrophysical self–sufficiency was consistent not only with natural theology, but also with deism, according to which God was *merely* the Creator and not an ongoing participant in the world.

Laplace's discussion of the historical origins of solar systems was more widely troubling to natural theologians. Beginning in 1796, Laplace developed and refined the Nebular Hypothesis, according to which solar systems coalesced from nebular clouds. This would account for the correspondences in plane and the direction of motion of the planets and moons. With long–term stability (a mixed blessing at best) and hypotheses of nonmiraculous historical origins of the celestial system, little remained of the specifics of Newtonian astrotheology. Naturalistic efficient–causal explanations removed the need for divine adaptive contrivance, just as they had for its periodic maintenance.

In contrast, new means–ends adaptations were continually being discovered in biology. William Paley's 1802 *Natural Theology* was the classic of its day. On astronomy Paley concurred with the century–old opinion of Boyle and Ray; astronomy "is *not* the best medium through which to prove the agency of an intelligent Creator" (Paley, 1831, p. 517).

The early 19th century brought other surprising scientific discoveries to public attention. The most distressing were the geological facts of organic extinctions and the great age of the earth. Interpretation of these facts was a theological challenge. After all, Paley had stated that life was so well adapted to the Earth's climate that not one species of plant had gone extinct since Creation (ibid., p. 512). To modernize the Paleyan arguments, the last Earl of Bridgewater in 1829 commissioned a set of Treatises intended to corroborate the truths of natural theology using material from contemporary science. Most Treatises dealt with matters related to biology, as might be expected. The challenge of the Bridgewater Treatise on Astronomy and Physics fell to William Whewell, a polymath with credentials in crystallography, mineralogy, and the study of tides who would eventually become Master of Trinity College, Cambridge. From the 1840s on he would write very influential works in history and philosophy of science. His 1833 Bridgewater Treatise can be read as the last gasp of astrotheology.

Of the astrophysical factors which Newton had cited as evidence of divine contrivance, very little remained for Whewell's use. The newfound stability of the solar system could be dealt with; it simply testified to a creation of greater workmanship than Newton had realized. But the adaptive contrivance Newton saw in the coincidence of planetary motions was of little use to Whewell. The

Nebular Hypothesis removed the need for contrivance, and Whewell accepted the hypothesis (at least provisionally). Whewell tried to save the Argument from Pattern by arguing that the Nebular Hypothesis does not disprove contrivance, but only pushes it back a step. Whence arose that cloud of stellar gas which became the collapsing Nebula, and whence arose the laws by which it coalesced? "What but design and intelligence prepared and tempered this previously existing [nebular] element, so that it should by its natural changes produce such an orderly system?" (Whewell, 1836, p. 101). If the solar system itself was not divinely contrived, then its nebular precursor must have been. And if a naturalistic account were given for the Nebula itself, the astrotheologian would simply take another step back and appeal to an even earlier divine contrivance. In the tide of new natural explanations of the physical world, Whewell the theologian is willing to relocate the act of divine adaptation to earlier and earlier events in the cosmological order.

This move kept divine contrivance in the picture, but at a cost. Under Whewell's treatment, astrotheology had lost contact with any specific empirical discoveries of astronomy. When phenomena receive naturalistic efficient–causal explanations, the astrotheologian simply accepts them and assigns divine contrivance to their predecessors. Whewell accepted the irrelevance of specific empirical discoveries by broadening the Argument from Pattern. He proposed a view of natural law which exploited the semantic ambiguity between natural and legislative law. Whewell claimed that a legislative law could only be produced by a Lawgiver, and he asserted the same of a natural law. God was described as a Legislator, and natural laws his pronouncements. Physical law no longer had to be shown to be *good*, since any law at all proved the Legislator's existence. The connection formerly asserted by natural theology between the *specific* discoveries of astronomy and the nature of the Creator had been cut. For Whewell's theological purposes, any astronomical discoveries would do as well as any other.

Whewell's *Treatise* did offer masses of astronomical information to his readers. From Whewell's astrotheological perspective such information could count only as illustration, not as data in support of divine contrivance. But he also discussed the ways in which astrophysical factors influenced organic life on earth. In this way, by poaching on the scientific territory of the biologists, he managed to give evidence of purposive contrivance.

Whewell returned to the age–old example of correlations between organic cycles and astronomical time periods. Seasons are produced by astronomical phenomena, plants' growing cycles by something internal to plants, and each corresponds to the other. "Why should the solar year be so long and no longer? or, this being of such length, why should the vegetable cycle be exactly of the same length? ... No chance could produce such a result. And if not by chance, how otherwise could such a coincidence occur than by a intentional adjustment of these two things to one another?" (ibid., p. 26). Whewell applied the same style of argument to specific organic adaptations. The arrangement by which eyes are able to perceive light could equally well be seen as a contrivance of the nature of light to be seen by eyes. Ears and the Earth's atmosphere were coadapted "by an Intelligence which was acquainted with the properties of both" (ibid., p. 72). Though he could find no clear cases of means/ends contrivance in astronomy per se, there are means/ends relations in the structure of an organism. Whewell simply emphasized the physical/environmental factors by use of which organic adaptations achieve their ends, thereby associating physical laws with organic adaptations.

Whewell's a prioristic "divine Legislator" argument removed astronomy from the empiricist program of natural theology. But his insinuation of astrophysical facts into purposive, biological, physicotheology was consistent with the British intellectual tradition. Surprising as it seems from a modern perspective, in Whewell's day there was no principled way of deciding the chicken–or–egg question of whether the physical world was adapted to the biological, or vice versa. Some implications of this problem will be addressed below. The important point is that purpose, contrivance, and adaptation are now fully the province of biology. The only observable purpose in the natural world is biological adaptation.

III. The 19th Century before the Origin

The drama of 19th century biology according to most popular narratives was the struggle between scientific evolution and special creation. Some recent historians of biology consider that version of events to be of limited historical value. Dov Ospovat was one of the early critics of the evolution versus creation account (Ospovat, 1978, 1981). He considers that dichotomy to ignore many important differences between nonevolutionists and to artificially separate people of very similar scientific orientations based merely on when they accepted evolution or (even worse) on their personal attitude towards Charles Darwin's theory.

Ospovat replaces the evolution/creation dichotomy with a distinction which more nearly reflects the dynamics of scientific debate during the period. He distinguishes between teleological and anti–teleological biologists. Teleologists were primarily advocates of mainstream natural theology, with its emphasis on organic adaptation. For them, biological explanation consisted in identifying the purposes of organic traits by identifying (or hypothesizing) the benefits they produced for the organisms; locate a trait and state why it is "good." Geological time and organic extinctions were scientifically accepted by this time, and the successions of similar fossil forms in geological strata were explained as successive adaptations to gradually changing environmental conditions. Evolutionary modification was not seen to be involved in these changes. The new forms were not adaptations *of* older forms, but were new creations adapted to the new environments. In contrast, anti–teleologists believed that the forms of animals (and their successions in geological time) depended on biological factors other than adaptation to conditions. This second opinion was supported by an emphasis on similarities of organic structure which appeared unrelated to adaptation or similarity of function.

Ospovat's teleologist/anti–teleologist distinction matches a division in approaches to organic form explored in depth by E. S. Russell. For the early 19th century Ospovat's and Russell's distinctions pick out the same scientists. Russell's (1916) contrast between *Form and Function* distinguishes schools of biology which emphasize the structural organization ("form") of organisms from schools which emphasize adaptation and purposive "function." This a dispute about causal or explanatory primacy — which came first, form or function? Ospovat's teleologists were the advocates of function — objects had their form in virtue of the function they were to serve. Their opponents, identified by Russell as morphologists, were nonfinalists in the explanation of form and often considered functions as little more than fortuitous by-products.

Russell traces the form and function approaches back to antiquity, but the real debates heated up at the end of 18th century with the theories of Johann Wilhelm von Goethe and George Cuvier. Cuvier was the advocate of function and adaptation. The anti–teleologists, students of biological form or structure, were anatomists, morphologists, and embryologists. This style of biology had developed in Germany and France. Its introduction into Britain after about 1830 set the stage for what Ospovat sees as the crucial debates of the period before 1859.

The nonteleological biology important in this period has variously been labeled "idealistic morphology," "transcendental anatomy," and sometimes "philosophical" or "higher anatomy" (Rehbock, 1983). When not ignored, it has often been misrepresented in historical discussions. Described as mystical, idealist (in the sense of anti–materialist), and essentialist (and therefore anti–evolutionary) with respect to species, it sounds like an altogether regressive program. The otherworldly tone of its labels encourages this view. These scientists actually held a broad range of metaphysical and religious commitments. Taking a sample of "transcendental morphologists" to include Goethe, Etienne Geoffroy St. Hilaire, and Richard Owen, we find ourselves with two evolutionists (Owen and Geoffroy), one apparent materialist (Geoffroy), and perhaps only one believer in a personal God (Owen). Louis Agassiz, not a major player in the debates, was probably the only prominent scientist to fit the stereotype of idealistic theistic anti–evolutionist. (It is thus unfortunate that Ernst Mayr has chosen Agassiz as typical of the movement; Mayr, 1964, 1976; Winsor, 1976). Charles Darwin's own work makes it clear that the morphological movement was not scientifically regressive, especially as compared with its adversaries.

Besides the multiple labels for these anti–teleologists, there are other ambiguities in the traditional contrast between *form* and *function*. For example, form should be understood as structure (not mere external shape), and function considered in its adaptive and not simply causal aspects (Amundson and Lauder, 1994). I will here refer generically to the anti–teleologists as "structuralists" in biology, and the teleologists as "adaptationists." As we have seen, organic adaptation was all that remained of the teleology of natural theology by this period.

A stimulus and precursor to the British version of these debates was the slightly earlier French exchange between Cuvier and Geoffroy (Appel, 1987). Cuvier emphasized the biological importance of the demands of "Conditions of Existence," the conditions necessary for the animal to survive and reproduce in an environment. His focus was more on internal functional integrity than on organism/environment interactions. British readers gave an environmentalist turn to Cuvier's adaptationism, but adaptationism it was. Cuvier asserted that biological similarities between species exist only because the species have similar adaptive needs; all anatomical similarities reflect adaptive similarities. So any "affinities" which seem to group organisms into genera or higher taxonomic groupings are by-products and consequences of adaptation.

In contrast, Geoffroy stressed nonpurposive biological patterns which ranged across forms of life, patterns which revealed a "Unity of Type." In the broader debate these involved serial repetition of parts, symmetry, shared structural patterns within types, and the common developmental patterns revealed by embryology. Unity of Type structuralism saw the central fact of biology as the commonality of structure which underlies and unifies the diversity of organic form. The functions of particular traits were made possible by their structures. The undeniable adaptedness of some traits is no more than a secondary modification of structurally determined body parts.

For the adaptationists, structure simply followed functional need. For the structuralists, function was merely the putting to use of the products of structural laws.

The theoretical basis of structuralism was the concept of *homology*. Expanding on earlier vague ideas of organic unity, Geoffroy and his colleagues developed a research program of locating and cataloguing points of identity in the anatomy of different species and higher groups. Homologous body parts were characterized as the *same* body part existing in different organisms under different conditions of modification. Geoffroy identified homologies primarily via the "principle of connection" whereby homologs were identified by their connectedness within the system of body parts (not by their "form" or shape alone). German morphologists developed similar ideas concerning serial repetition of parts within organisms, and embryology provided a third method of identifying organic "sameness" — embryological development of body parts of different species from similar embryonic precursors. Geoffroy and his followers were extravagant in some of their claims of unity, finding homologies among insect, vertebrate, and molluscan body plans. These and other proposed homologies proved fanciful, but the research program as a whole was robust.

Preevolutionary concepts of homology come so close to the modern concept that one must be careful to avoid assuming an evolutionary basis for the ideas. While many of the structuralists were (or became) evolutionists, the facts of homology were presented as objective observable facts of biology, independent of questions of organic origin. Scientific methodology was at the time strongly inductivist, and overly hypothetical explanations were discouraged. It was common and respectable for scientists to carefully describe phenomena for which they proposed no explanation (Hull, 1983). Geoffroy (1830) insisted that Unity of Type was a low–level law of nature, similar to Kepler's laws of planetary motion. It was not an explanatory or causal theory like Newton's theory of gravitation. This was the general public stance of most of the structuralists. Some hinted at a natural, materialist explanation of the Unity of Type; others claimed that the Unity must have a theistic basis. But even the most theistic (e.g., Agassiz) were adamant on adaptationism's inability to explain biological form. The label of "transcendentalist" calls attention to structuralists' common commitment to a unifying reality for organic existence which goes beyond the immediately observable functions and adjustments of organisms to their environment. The true nature of that reality was open to debate. The anti–materialist connotation of "transcendental" is in this context misleading.

Adaptationists included people like Whewell, the geologists Adam Sedgwick and John Buckland, and astronomer John Herschel, who themselves had earlier done battle against religious conservatism to establish modern nonbiblical geology in university curricula (Cannon, 1978, Ch. 2). They had won out over biblical literalism, and were very protective of the remaining scientific content of natural theology — the adaptation of organisms. The best biological adaptationist among them was Charles Bell, a renown physician and author of one of the most respected Bridgewater Treatises (Bell, 1933). Bell, like Cuvier, saw purpose and adaptedness in every part of the body, and had no patience for the "lovers of system" who organized the organic world on nonfunctional criteria. He lamented the "very extraordinary opinion in the present day that all animals consist of the same elements" (ibid., p. 134). He criticized the structuralist schools in ways that showed considerable familiarity with their approach. "Shall we follow a system which informs us that when a bone is wanting in the cavity of the ear we are to seek for it in the jaw, and which yet shall leave us in the contemplation of this class of animal [Batracian] which is deficient in 32 ribs without pointing out where they are to be found or how their elements are to be built up in other structures?" (ibid. p. 137). To Bell the ear/jaw homology looked especially foolish when "we find that the sense of hearing is enjoyed in an exquisite degree in birds, that the organ of the sense is not imperfect but is adapted to a new construction and a varied apparatus suited to the condition of the bird, and that there is no accidental dislocation or substitution of something less perfect than what we find in other classes of animals" (ibid., p. 139). If mammals and birds hear equally well, what can be gained by identifying ear bones in one with jaw bones in the other?

Signs of common type were not actually invisible to Cuvier, Bell, and other adaptationists. They were simply uninteresting. The "types" they picked out were adaptive types, and so were redundant to the adaptationist program. Once adaptations were identified, deciding on the "types" merely meant grouping together organisms with similar sets of adaptations. Any supposed structural identity was reducible to adaptive similarity. Cuvier would accept Geoffroy's identifications of homologous parts only when the parts were adapted to the same purpose. A body part's primary nature is fixed by its adaptive nature, and signs of "type" are simply generalizations over adaptations. As with mammals' ears and birds' jaws, purported homologies would be a threat to teleological biology only if they acted to the detriment of adaptation.

To be sure, adaptationists admitted that organs and body parts exist which have no known adaptive purpose. The universal stance on these items might be called the principle of *presumptive adaptation:* Never infer a lack of adaptation from a lack of knowledge of adaptation because it is always more probable that an unknown adaptive purpose exists than that no purpose exists. The presumption should be that the trait is adaptive, and that eventually its purpose would be discovered. Alexander Pope's words could be read as a biological doctrine: "And spite of pride, in erring reason's spite, One truth is clear, Whatever is, is right."

Social dimensions of the dispute also raised the level of acrimony. In the 1830s the continental ideas were championed by British radicals against the aristocratic medical power structure (Desmond, 1989). Adam Segwick complained about Geoffroy and his "dark school . . . his cold and irrational materialism" (Clark and Hughes, 1890, p. 86). Conservatives saw Unity of Type as materialist, raising the specter of a self–organizing Nature (and possibly a self–organizing society) in opposition to a world shaped by a benevolent Legislator–artisan and ruled from above. Structuralist radicals considered the Paleyan celebrations of the world's fitness as an unreasoned endorsement of the social status quo. By 1840 the threat from the radicals had diminished. In the late 1840s, Richard Owen, a conservative in politics and religion and a favorite of the adaptationist mainstream, suddenly announced a new and pious version of Unity of Type biology (Owen, 1848, 1849).

Owen presented elegant arguments against the sufficiency of adaptationist biology, especially in *On the Nature of Limbs* (1849). In one such argument he pointed out the structural diversity of transportation devices invented by humans (sailing ships, balloons, locomotives, etc.) and contrasted them with the structural identity of vertebrate limbs which served diverse transportive functions (bat wings, mole paws, manatee paddles, etc.). If vertebrate limbs were created on adaptive principles alone (as were the human devices), what accounts for the structural identity? On Owen's version, the structural identity is traced to the Vertebrate Archetype, which he likened to a Platonic Form, said to have existed in the mind of God prior to the incarnations of any of its modifications (real vertebrate species) on earth. The piety appealed to religious conservatives, and even Whewell finally acknowledged the reality of Unity of Type. But the small book concluded with a passage which would give conservatives pause. "To what natural laws or secondary causes the orderly succession and progression of such organic phaenomena [the introductions of new species] may have been committed we as yet are ignorant" (ibid., p. 86). Secondary causes are ordinary causal laws of nature, with God's creation of the world being seen as the "primary cause." Owen here alludes to a naturalistic explanation of the origin of species.

However much his self–described Platonism repels modern biologists, Owen managed to introduce Unity of Type into mainstream, pious, British natural history. Other structuralists expected materialist rather than idealist accounts of the "transcendent" reality behind the Unity of Type. Owen's interest in secondary causes shows some leaning in that direction. But whether materialist or idealist, Owen had brought structuralism to natural history's high table. One way or another, in the material world or in the ideal, homologous structures could be said to have preexisted their adaptive uses. Lucretius would have been gratified.

Before following this episode to its culmination with Darwin, let us consider just why these two approaches to biology were in such strife. We can pattern the dispute on Russell's dichotomy of form versus function. To put it crudely, which came first? Adaptationists were centrally interested in body parts as adaptations, and explained them teleologically. They naturally categorized body parts according to similarities in function. Structuralists claimed not to be studying adaptation (or for that matter maladaptation) and so one would think there would be no conflict between the two approaches. But structuralists claimed to have found *real identities* between body parts of different organisms even when those body parts had different functions. A fish air bladder and a tetrapod lung (first asserted by Owen to be homologs) or a mammalian ear ossicle and an avian jawbone shared an identity which was deeper than their "superficial" functional dissimilarity. If the structuralists' homologies were real, then structure *was* "prior to" function — the correct identification of a body part was the structural one. The part's function was a modification superimposed on its true identity. As in the days of astro– and physicotheology, we have one approach which emphasizes Pattern and one which emphasizes Purpose. But here the theories are in the same scientific domain and stand in considerable opposition to one another.

There is a tendency among commentators on transcendental anatomy to treat it as merely an idealist version of the traditional Argument from Design (Ruse, 1979; Bowler, 1977, 1988). This fits it nicely into the creationist side of the old evolution versus creation dichotomy. But as has been noted, many structuralists had nonreligious and noncreationist commitments to the science. Even in its religious interpretation, structuralism does not cite evidence of purpose and intention imminent in nature. At best it is, like astrotheology, an Argument from Pattern. Theological conclusions receive only the kind of support they had received from astronomy. If shoring up religion were really the intention of the religious structuralists, they did a poor job of it. If, on the other hand, their intention was to make the genuine advances of continental structuralism a religiously palatable part of British natural history, they succeeded. Advocates of form, followers of Owen or of the less pious continental morphologists, were winning the debate against function in the 1850s. The pattern–like theories of the structuralists were seen as more progressive than the purposive theories of the adaptationists. That is the context into which Darwin dropped the *Origin of Species*.

IV. Darwin's Achievement: One More Version

The 1859 publication of Charles Darwin's *Origin of Species* transformed the concept and the study of adaptation. The *Origin* can be read as having two parts, directed to two specific goals. Part I (Chapters I–XIII) introduced the principle of natural selection and presented evidence that the conditions necessary for its operation were present in both wild and domestic species. Natural selection could produce an indefinite amount of change through time, and the change which it produced typically *constituted adaptation.* Part II (Chapters IX–XIII) presented the evidence that evolutionary change had in fact occurred — that distinct species were in fact descended from common ancestors. Part II met its goal more decisively than Part I. The argument for the fact of evolution was strong enough to convince many who remained doubtful of natural selection as the chief mechanism of change.

Darwin's plan for his species work had included this pair of goals since his earliest speculations on evolution, back when Part I was merely "my theory" and natural selection had not even been conceived. The *Origin* was not divided into two parts as were the preliminary "Sketch of 1842" and the "Essay of 1844," but the format was followed. Material from Part I takes up about two-thirds of the *Origin*. Ospovat points out that in earlier documents Part II had always been the longer section, and Darwin had described it as the more important one (Ospovat, 1981, Ch. 4). The question of priority with Wallace on the concept of natural selection may have motivated Darwin to elaborate Part I in the 1859 work.

Part II relied on many of the discoveries of structuralist biology discussed in the previous section. Darwin showed how descent with modification would explain widely known facts of embryology, morphology, vestigial organs, and geographical distribution of related types. He reported the anti–adaptationism of Owen (1849), and continued "If we suppose that the ancient progenitor, the archetype as it may be called, of all mammals, had its limbs constructed on the existing general pattern, for whatever purpose they served, we can at once perceive the plain signification of the homologous construction of the limbs throughout the whole class" (Darwin, 1859, p. 435). The term "archetype" of course alluded to Owen's work, and the odd term "signification" was also carefully chosen. Owen had introduced it with some fanfare as a translation of the German morphologists' use of *Bedeutung*. In this way Darwin connected the well–known phenomena of structuralist biology with his hypothesis of descent with modification. With this skillful marshalling of evidence, Darwin's Part II won wide acceptance for the fact of evolution. Natural selection was a different story. Even though there would always be a core of Darwinian naturalists, the general biological acceptance of natural selection as the primary engine for evolutionary change did not occur until well into the 20th century (Bowler, 1983, 1988).

Though underappreciated even into this century, Part I reveals Darwin's unique genius. There had been other evolutionists, and sooner or later someone would have cataloged the evidence as persuasively as Darwin's Part II. Natural selection, on the other hand, was a major innovation in biology. In an important sense (to be discussed) it was the first and only fully naturalistic explanation of biological adaptation. But in a way it was not a culmination of the biology of the 1850s, but a reversion to the biology of the 1830s. The *Origin* once again treated biological adaptation as the most important feature of the organic world. This was a position of prominence which adaptationism had not enjoyed since the days of the Bridgewater Treatises. In 1874, Asa Gray praised Darwin for his "great service to natural science in bringing it back to Teleology; so that, instead of Morphology *versus* Teleology, we shall have Morphology wedded to Teleology" (Gray, 1963, p. 237). Ignoring the metaphysical resonance of the term teleology, Gray is correct about the return of adaptation. Even though Owen and other structuralists had acknowledged the reality of adaptation, they had treated it as secondary to a deeper Unity of Type. Darwin restored adaptation to its former glory (Ruse, 1979, p. 184).

Darwin's "wedding" of adaptation with Unity of Type was a complex affair. Recall Russell's structure versus function dichotomy. Darwin fully accepted the structuralists' patterns of homologies. This meant that structure actually did precede function in functionally distinct homologs, in the straightforward sense that structure *came first*. Ancestors possessed homologous body parts long before their descendants evolved functional specializations of that "same" body part. But in a different way adaptation (function) has priority in the *Origin*. Natural selection, an adaptive force, is the primary causal agent for change. Which priority is deeper, the temporal priority of structure or the causal priority of function?

Darwin sided with function. Unlike many of his evolution–leaning contemporaries (including his champion T. H. Huxley), Darwin had always seen adaptation as primary. His autobiography, quoted above, fondly recalls his college readings of Paley. He realized early in his evolutionary thought that evidence for Part II must come from nonadaptational sources, but seemed almost to regret it. From 1837, Notebook B: "The condition of every animal is partly due to direct adaptation and partly to hereditary taint" (Darwin, 1987, p. 182). A structuralist might have described the animal as partly an expression of its type and partly *adaptive* taint. Darwin establishes the priority of function over structure with a short argument at the end of Chapter VI. "Hence, in fact, the law of the Conditions of Existence is the higher law; as it includes, through the inheritance of former adaptations, that of Unity of Type" (Darwin, 1859, p. 206).

Darwin was not an extremist on adaptation and never insisted on its ubiquity. The *Origin* names only six or seven specific traits as putative "adaptations," and four of them are dismissed as illusory (ibid., p. 197). Nevertheless, the "higher law" argument served an important methodological function. It maintained a focus on adaptation in the presence of phenomena which are *prima facie* nonadaptive. Darwin could not pooh–pooh Unity of Type as Bell had, even if he had wanted to. It contained his best evidence for community of descent. Strictly speaking the "higher law" argument does not claim that *all* inheritance is of former adaptations, only that inheritance *includes* former adaptations. Nonadaptive (that is, neveradaptive) traits could also be inherited. Darwin accepted that some traits had nonadaptive causes, for example, those which arose from the "laws of growth." But these are not mentioned in the discussion of "higher law." There Darwin directed our attention instead to the cases in which functionally diverse (or even nonfunctional) homologs had arisen from a common adaptive origin and been passed on through inheritance. In structuralist hands, these homologies had supported the Unity of Type. Darwin reduced the structuralist implications of the homologies by calling attention to hypothesized adaptive origins. The "higher law" argument was the first of many interpretive devices which would be involved in the development of a naturalistic adaptationist evolutionary research program.

One achievement of Darwin's explanation of adaptation is so obvious that it is likely to be overlooked. Recall Whewell's Bridgewater claim that either the seasons are adapted to plant cycles or plant cycles are adapted to the seasons. Natural theology had no way to determine the polarity, as it were, of adaptation. God could have just as easily adapted either phenomenon to the other. Darwin's theory removed that dilemma. Darwin had at one time read a reference to a similar Whewell comment concerning the adaptation between the human sleep cycle and the 24-hour day, and was astonished: "whole universe so adapted!!! & not man to Planets. Instance of arrogance!!" (Darwin, 1987, p. 347). To be fair, Whewell had been noncommittal about the direction of adaptation. But Darwin had a mechanism which explained how organisms adapt to their environments, and it did not allow reversals. Coadaptation could occur between different species (e.g., insects and flowers), but not between a species and its physical environment.

A more generally remarkable feature of natural selection is its metaphysical naturalism. This feature was poorly perceived by most early readers. Chauncey Wright, an American mathematician, philosopher, and

Darwin correspondent, fully appreciated the naturalistic implications of selection. In 1870 he pointed out that natural selection explains adaptation without "making the cause to be engendered by the effect." In other words, selection avoids the implication of reversed causation which marked Aristotelian final cause explanations — that implication which had seemed to require a designing intelligence. Wright also saw the implications for the status of biology within natural theology. "The organical sciences lose their traditional and peculiar value to the arguments of Natural Theology, and become only a part of the universal order of nature, like the physical sciences generally, in the principles of which philosophers have professed to find no sign of a divinity" (Wright, 1877, p. 101). Wright misstated what philosophers profess; Whewell had professed to see divinity in *any* natural law. But Wright puts his finger on exactly the threat which natural selection posed for natural theology. Astrotheology had whithered in the wake of Newton's success. Physicotheology now faced the same fate.

It has even been argued that natural selection is the only possible naturalistic explanation of adaptative complexity; that the choice is either final causation or natural selection. But what about the "Lamarckian" mechanism of use–inheritance? Let us compare two commentaries on this mechanism. Richard Dawkins (1983) has argued that the inheritance of acquired characteristics is conceptually inadequate to account for adaptation. Inheritance of acquired characteristics explains phylogenetic adaptive change by reference to ontogenetic adaptive change. Exercised muscles increase in strength; abraded skin thickens. But how is ontogenetic adaptation itself explained? And what determines that only adaptive ontogenetic changes are inherited, not every little scar and broken tooth? These phenomena remain unexplained by "Lamarckian" inheritance alone. Dawkins concludes that "there has to be a deep Darwinian underpinning even if there is a Lamarckian surface structure . . . Lamarckian mechanisms cannot be fundamentally responsible for adaptive evolution" (ibid., p. 409).

Dawkins' insight on the insufficiency of "Lamarckian" inheritance is not new. Paley had a similar understanding of Lamarck, and for that very reason declined to label his theory "atheistical." He recognized that Lamarck's inheritance theory did away with final causes in the structures and uses of the body parts, but it did not explain inheritance and reproduction itself, nor did it explain ontogeny's "original propensities, and the numberless varieties of them" (Paley, 1831, p. 528). Paley understood Lamarck to have invoked (not Darwin but) a divine Creator to engineer ontogeny's adaptive propensities, such as the propensity for exercised muscles to increase in strength. Whewell too recognized this gap. He more accurately reported Lamarck's full theory, which included continuous spontaneous generation (to fill all levels of the Scale of Being) and a universal "tendency towards progressive improvement," a temporalized version of an Aristotelian imminent teleology (Whewell, 1863, p. 566, a passage unchanged from the 1837 1st edition).

Lamarck's need for these additional assumptions was not lost on Darwin. A passage in Notebook E (1838–1839) refers to Whewell's discussion of Lamarck, claiming "the non–necessity of the <<so–called>> progressive tendency law" for Darwin's own theory. Marginalia on the above cited page in Darwin's copy of Whewell reads, "These are not assumptions, but consequences of my theory, & not all are necessary" (Darwin, 1987, p. 415 n. 70–1). Darwin saw adaptation, progression, and diversity as following from natural selection without the need for a separate law of progress; spontaneous generation was simply unnecessary.

It is difficult to judge from a modern perspective what would have counted as a metaphysically naturalistic hypothesis in the early 19th century. The law of progressive tendency may have seemed to Lamarck no more supernatural than the law of gravitation; Herbert Spencer certainly regarded it as such. (Remember that Newton had considered even the law of gravitation to implicate some possibly supernatural agency.) But Lamarck's evolutionary theory required the progressive law together with an unexplained propensity for ontogenetic adaptation. In contrast, natural selection derives phylogenetic adaptive change from the interplay of phenomena *which are not individually adaptive*. Indeed, natural selection can explain the origin of even ontogenetic adaptability. If Lamarck's was a naturalistic theory, then it was extremely limited in scope. Dawkins was right about the insufficiency of "Lamarckian" inheritance to explain adaptation. So was Paley. Dawkins and Paley take adaptation very seriously. Such thinkers can spot the shortcomings of Lamarck's treatment.

Let us consider Darwin's report of illusory adaptations (1859, p. 197). Four traits are listed in the paragraph. Each is judged by comparative data not to be an adaptation. Related species either have the trait but cannot make use of it or they lack the trait even though it would be of similar usefulness. 1) Darwin says that the green color of a woodpecker species would be universally identified as camouflage if all woodpeckers were green. Knowing that there are many black and pied woodpeckers, he has "no doubt that the color is due to some quite distinct cause, probably to sexual selection." 2) Caution is expressed regarding whether the naked heads of vultures are adaptations for "wallowing in putridity," since "clean–feeding" male turkeys are also bald. 3) Hooks on the branches of a climbing species of bamboo are rejected as adaptations because, even though "of the highest service," they are also found in nonclimbing relatives. They probably arise from "unknown laws of growth, and have been subsequently taken advantage of by the plant undergoing further modification and becoming a climber." 4) Sutures in the skulls of young mammals had been suggested as an adaptation to aid parturition. Darwin rejects this because birds and reptiles share the sutures even though they travel no birth canal. Again the laws of growth are cited, even though the sutures "no doubt facilitate, and may even be indispensable for" mammalian parturition. (This example too was taken from Owen 1849, a very helpful book for Darwin.)

Two details emerge from this list. First, Darwin was only willing to label traits as adaptations when they had been shaped by selection to do the task being attributed to them. Even traits which were indispensable for the organism, and traits (like the bamboo hooks) which permitted other adaptive specialization were nonadaptations, since they were not formed by natural selection for the purpose hypothesized. Second, sexual selection did not produce adaptations. It was an alternative to natural selection, not a special case of it. Both of these points are matters of contention today. The proper criteria for a trait to be called an adaptation have exercised philosophers and biologists, and will be discussed presently. Sexual selection, after a long period of disuse, is back in the toolbox of evolutionary biology, though its nature and status are under examination (Cronin, 1992; Spencer and Masters, 1992).

Darwin's achievement is usually assessed in terms of the modern dichotomy of evolution versus creation. He is a singular historical figure on that scale; he gave the first publicly convincing argument that evolution had occurred. Assessed from Ospovat's dichotomy of adaptive versus structural biology, Darwin's position is also unique, but a bit more complex. The *Origin of Species* accomplished two noteworthy goals. First, it gave a unified account of the phenomena of *both* adaptationist and structuralist biology; Darwin wedded Morphology to Teleology. (For the semantically squeamish, he wedded structuralist to adaptationist biological concerns.) Richard Owen and others had earlier acknowledged both sets of phenomena, but offered no explanation of how the two sets related to one another or arose from the same source, at least none that was free from question-begging. Darwin gave that explanation. Second, Darwin gave naturalistic accounts of both adaptive and structural biological facts. (This was more surprising for adaptivity than structure; Unity of Type had always seemed materialistic to natural theologians.) Darwin considered adaptivity to be a "higher law" than structure. But he had full command of structuralist concepts, and used them in Part II to demonstrate the fact of evolution. Seen from the perspective of adaptive versus structural biology, Darwin's work provided a real wedding of the two traditions. As we will see, the wedding was not permanent.

The differences between natural theological and natural selective studies of adaptation are profound. Natural theology in the 1830s and 1840s showed some embarrassment with old–fashioned optimism and human–centered "purposes." But no principles were suggested which could rule out a hypothesized adaptive purpose. The problem of polarity of adaptation, which allowed Whewell to postulate that air was adapted to ears, seemed to infect all adaptationist biology. Natural selection changed all of that. The requirement for a selective explanation places strong conditions on who or what is the beneficiary of the hypothesized "good." It does not solve all problems, as the recent debates on "units of selection" indicate (Brandon and Burian, 1984). But it does require that plants' seasons are adapted to the Earth's orbital period, and not vice versa. By giving a naturalistic theory of the cause of adaptation, Darwin allowed research into *factors* of that cause which would otherwise seem irrelevant to adaptation. Phenotypic variation within populations, the sizes of breeding populations, and modes of migration and dispersal all become implicated in the study of adaptation. Natural theologians had always studied adaptation, but only (so to speak) after the fact. On theological understanding the cause of adaptation (divine contrivance) had occurred at inaccessible times and places. Darwinians could study adaptation as a *process* and inspect the bits and pieces of which that process was made. Darwin did not invent the study of adaptation, but he invented a way to study adaptation *from within*.

V. The Challenge of Chance

Between the publication of the *Origin of Species* and the Modern Synthesis of the 1930s and 1940s a wide range of non–Darwinian theories of evolution shared the biological spotlight. Neo–Lamarckian and orthogenetic theorists and (after the turn of the century) Mendelian mutationists were among the adversaries of Darwinism. One challenge to the centrality of Darwinian adaptation arose during that period which has persisted to the present day. Its roots are in the diffuse movement sometimes called the Probablistic Revolution (Kruger et al., 1987), a movement in which the concept of natural selection itself played a role. Probablistic thinking in 19th century biology was allied to the new concept of organisms as *sources* of variation, not merely as reactors to outside stimuli. One example is the concept of trial and error learning, with trials spontaneously or randomly emitted. This idea seems obvious and commonplace today. Surprisingly, trial and error learning was first elucidated by Alexander Bain at about the time Darwin proposed natural selection. It was opposed by traditional empiricist epistemologists like John Stuart Mill. Mill favored a more passive and associationist account of learning, and Bain's suggestions of spontaneous exploratory trials may have offended Mill's causal determinism. The discoveries in Germany of spontaneous (unstimulated) activity in the nervous system occurred at about the same time (Boakes, 1984). Prior to these findings, organisms had been conceived not as spontaneous actors, but only as reactors to environmental stimuli. Darwin's account of natural selection involved a similarly spontaneous source of heritable variability. Like other early proposals of spontaneity, Darwin's were not happily received. John Herschel reacted to the presumed spontaneity by referring to natural selection as "the law of higgeldy–piggeldy" (Hull, 1983, p. 67).

Darwin himself did not consider spontaneity to be an important theoretical *discovery*. It was merely a necessary condition for the operation of natural selection. In Darwin's view the chancy spontaneity of variation was closely supervised by the deterministic process of environmental selection. But suppose the environment were "inattentive," so to speak, and the variability went unsupervised? Could genuinely spontaneous, nonadaptive traits result? Unlike Unity of Type phenomena, the idea of random variations of form had not been proposed as a challenge to adaptationism in preevolutionary natural history. But soon after Darwin the notion of spontaneous organic change became an alternative to adaptation, and random genetic drift contrasts with adaptation to the present day.

The first theory in which randomness or spontaneity played an important directional role was that of John T. Gulick. Gulick was the son of a missionary to Hawaii. He had become an evolutionist in the early 1850s after reading Darwin's expedition report commonly called the *Voyage of the Beagle*. Gulick had been impressed by Darwin's description of his 1835 visit to the Galapagos. Darwin had not argued for evolution in that book, and little of the Galapagos material was incorporated into the *Origin*. The Galapagos, like Hawaii, were an isolated archipelago of islands under an consistent climate. This similarity, and a research topic which Darwin had hinted at, stimulated Gulick's interest. Darwin had found the diversity of the Galapagos surprising. "I never dreamed that islands, about 50 or 60 miles apart, and most of them in sight of each other, formed of precisely the same rocks, placed under a quite similar climate, rising to a nearly equal height, would have been differently tenanted; but we shall soon see that this is the case" (Darwin, 1937, p. 398). This diversity included extreme variation among closely related species of tortoises and finches under identical environmental conditions. Darwin expressed regret that he could not have studied it further.

Given an evolutionary reading, the Galapagos report from the *Voyage* hints at geographical causes for speciation, with populations spontaneously varying even under identical environments. This was indeed Darwin's view during the early 1840s. But between 1844 and 1857 he developed his principle of divergence, which placed more emphasis on organic than physical environment in forming a selective regime. In 1859 the *Origin* was published with no mention of geographical speciation. Darwin proposed that the organic environments had varied on the different Galapagos islands even though their physical climates were identical. This gave a toehold to natural selection as a cause of speciation, and the diversity of Galapagos species was adaptationally explained. In the meantime, however, Gulick had gathered a huge collection of Hawaiian tree snails, mostly *Achatinella*, and logged their locations in the valleys of Oahu. Before the *Origin* was published Gulick had convinced himself that neither physical nor organic environmental differences could explain the variation patterns of the Hawaiian *Achatinella*. His scientific correspondence and his snail collection were the basis for an influential isolationist challenge to adaptationist theories of speciation (Amundson, 1994b; Provine, 1986).

That debate began during the 1880s. It was primarily an in–house affair among Darwinians and did not involve the many other now–abandoned evolutionary views. Alfred Russel Wallace took the adaptationist side. George J. Romanes, Darwin's protege in his late years, allied himself with Gulick on the side of nonadaptive variation and speciation. Geographical isolation, or at least breeding segregation, was seen by Romanes and Gulick as crucial for speciation and the generation of much of the observed diversity. They made use of the fact that small isolates are statistically more likely to diverge from a population mean than large isolates. The debate was intense, often quite subtle, and of course inconclusive. It was the source of English and American biological recognition of the importance of isolation (Lesch, 1975) and of the possibility of what we now call random drift.

Darwinism was in a somewhat precarious position in the early 20th century. The rediscovery of Mendel's laws had strengthened a mutationist alternative to natural selection. An excellent expression of the state of the theory was Vernon Kellogg's 1908 *Darwinism To–Day*. Kellogg discussed and evaluated the many current attacks on Darwinism. In his concluding chapter he observed that natural selection was uniquely successful as an explanation of adaptation. "However mightily the scientific imagination must exert itself to deliver certain difficult cases into the hands of selection, and however sophisticated and lawyer–like the argument from the selection side may be for any single refractory example, the fact remains that the selectionist seems to be able to stretch his explanation to fit all adaptations with less danger of finding it brought up against positive adverse facts than is possible to the champion of any other proposed explanation" (ibid., p. 381). After Karl Popper's critiques of unfalsifiable theories, scientists are now less eager to boast of the "stretchability" of their theories, but selection retains this distinction *vis a vis* adaptation.

On the other hand, Kellogg acknowledged that the Darwinian explanation of speciation seemed inadequate. He discussed geographical theories, including Romanes and Gulick's nonselective and Wagner's selective theories, in a chapter on friendly "auxiliaries" to Darwinism. Provine has traced Gulick's influence through Kellogg to Sewall Wright. "Seventy five years after he read this account of Gulick's data and conclusions Wright had a keen recollection of how deeply impressed he had been with it" (Provine, 1986, p. 228). While Wright has expressed various views from time to time on the importance of chance and random drift, he is regarded as the primary proponent of drift among the founders of the Modern Synthesis.

The tension between drift and adaptation has been remarkably persistent from the 1880s to the present day. Following Gulick, land snails have often been central to the debates. Provine gives a history, including a narrative of the "Great Snail Debate" from 1950 to about 1980 (1986, p. 437 ff.). The empirical details were complex, as was the question of just what counted as evidence for random drift versus selective control. The rhetorical issue often concerned what biologists' default position ought to be on traits of no known adaptive value. A. J. Cain asserted that traits should only be described as adaptive or uninvestigated (never as nonadaptive) since "every supposed example of random variation *that has been properly studied* has been shown to be non-random" (Cain, 1951, emphasis added). This expresses the evolutionary version of the natural theologians' principle of presumptive adaptation. In reply G. S. Carter seemingly distrusted Cain's assessment of propriety and chose to "regard [drift] and selection as equally possible explanations when neither is proved" (Carter, 1951). Adaptationism was in a strong position at this time in the debate. Cain was able to cite impressive demonstrations of geographic clinal stability of variants. Those demonstrations were especially forceful because they had been directed at traits previously cited as examples of drift. Clinal stability showed that the traits were influenced by selection, though not that they were themselves the targets of selection. That is, Sober's (1984) *selection of*, but not *selection for*, the traits had been shown. The difficulty of demonstrating nonadaptivity required that Carter and other advocates of drift base their positions more on theoretical grounds than on field studies.

The drift debates have long outlasted most challenges to Darwinian adaptationism. NeoLarmarckian theories, orthogenesis, and mutationism were put to rest early in the century. One factor in the persistence of drift is its consistency with the Modern Synthesis. The statistical foundation of population genetics is fully agnostic with respect to drift. Whether natural selection or drift controlled the evolution of a trait is a matter of the real–world values of selection coefficients, effective population size, and so on. Other proposed evolutionary mechanisms lack a natural interpretation within population biological theory. Neo–Lamarckism, for example, was not a simple victim of experimental refutation. Its demise was due to the rise of Mendelian genetics, and especially to the incorporation of genetics into the Modern Synthesis. To be sure, recalcitrant empirical data have contributed to the persistence of the drift/adaptation debates, as has the massive complexity of population dynamics in nature. After all, if selective explanations had been easily confirmable for every observed trait, we would have long since abandoned the possibility of drift.

VI. The Challenge of Constraint

Debates regarding the relative importance of adaptation and genetic drift have simmered ever since the Modern Synthesis. A different kind of challenge to Synthesis adaptationism has arisen in the past two decades. The first influential version of the critique was a paper by S. J. Gould and R. C. Lewontin entitled "The Spandrels of San Marco and the Panglossian Paradigm: A Critique of the Adaptationist Programme" (1979). The paper is remarkable in its influence, especially because its core assertions probably continue to be minority opinions among evolutionary biologists. Challenges raised by "the spandrels paper" and associated literature are not ignored today, even by the adaptationists who most strongly reject them. Although it was published in a technical journal, the paper has become well known in nonscientific circles. An indication of this broad influence is the publication of a volume of essays of rhetorical criticism dedicated to examining the spandrels paper alone (Selzer, 1993).

The spandrels paper popularized the term "adaptationism" (following Lewontin, 1978) and the term has been accepted even by those it was targeting for criticism. The paper appeared in the midst of the late 1970s debates on human sociobiology. Sociobiology receives special criticism, and its authors elsewhere argued against the social implications which had been drawn from sociobiology. This political context tainted the paper in the view of some readers. But from a historical perspective it is not surprising. Adaptationist biology does assign evolutionary rationales to current traits, and thus has the potential use of rationalizing the social status quo. Political progressives have naturally opposed some of these implications. Similar cases during the 19th century include the British Unity of Type radicals against the Paleyan natural theologians (Desmond, 1989), Gulick against Social Darwinism (Amundson, 1994b), and to some extent Geoffroy against Cuvier (Appel, 1987).

The spandrels paper describes adaptationists as single mindedly ignoring nonadaptational explanations of biological phenomena, and as assuming on insufficient grounds the adaptive status of virtually all traits. What Kellogg had termed the "stretchability" of natural selection was now under attack. Adaptationists were said to be overeager to accept plausible (adaptationist) scenarios and unwilling to consider nonadaptive explanations of the same phenomena. Belief in the ubiquity of adaptation was described as Panglossian, after Voltaire's Doctor Pangloss who believed that all that happened was for the best. Random drift and complex consequences of selection were among the possible alternatives which adaptationists ignored. But the most prominent alternative, described as "Another, unfairly maligned, approach to evolution," was the study of phyletic, developmental, and architectural constraints on organic design (Gould and Lewontin, 1979, p. 159).

A great deal of debate swirled around the spandrels paper, and the dust has not completely settled. Many authors challenged the accuracy with which adaptationist method was described; did any real–life adaptationist actually reason like that? Others defended the principle of presumptive adaptation, arguing for the propriety of searching only for adaptive explanations on grounds similar to those of Cain. Only after *all* adaptive possibilities have been ruled out should nonadaptive explanations be considered (Cain, 1951; Mayr, 1983). The issue of falsifiability was discussed, and the distinction between its application to individual adaptationist hypotheses and to adaptationism as a research program. The paper continues to be heavily cited, although a sizable portion of the citations are what Gould calls "'honorary' mentions from adaptationists who want to acknowledge an alternative for fairness sake, but have no intention of considering the arguments explicitly" (Gould, 1993, p. 331). D. A. Winsor (1993) examined a random sample of 28 recent articles citing the spandrels paper and found only one explicitly hostile to Gould and Lewontin. This is not an unqualified victory for the critics of adaptationism, however. Seventeen of the articles arrived at adaptationist conclusions and cited Gould and Lewontin (1979) only to assure the reader that the present paper avoided the problems there discussed. While there is surely no consensus on the *correctness* of the paper's allegations, Gould now feels that the days of hardline adaptationism are past, and is pleased that the spandrels paper contributed to its critique (Gould, 1993). It is not unusual for even committed adaptationists to acknowledge that such criticisms have forced adaptationists to "sharpen our logic and definitions" (Reeve and Sherman, 1993, p. 27).

The implications of the "Pangloss" label warrant some discussion. Dr. Pangloss was Voltaire's (rather unfair) parody of the philosopher Leibniz, who had argued on metaphysical grounds that this was the best of all possible worlds. Pangloss was able to dream up fatuous benefits (always benefits to humans) of horrific phenomena and crudely violated even the most obvious polarities of adaptation — he considered legs to be adapted to fit trousers rather than the reverse. Gould and Lewontin's paper alleged that adaptationists are overeager to accept adaptive explanations, so there is at least that parallel to Pangloss. But taken seriously, the Pangloss label does belittle Darwin's and later evolutionists' contributions to the understanding of adaptation. Modern adaptationists work with a theory which places a great many more burdens on adaptive explanations than Pangloss or the natural theologians accepted. Modern adaptationists accept all of the causal conditions Darwin placed on adaptive explanation and on its polarity, they adopt his skepticism about the global and objective "good" of the products of adaptive evolution, and they are typically committed in addition to individual (rather than group) selectionist mechanisms. They may *still* be overeager to identify adaptations. But it is not surprising that the rhetorical flourish of comparing them to an unconstrained, optimistic, human–centered fantasist like Pangloss has drawn heated objections.

Adaptationists may have become more circumspect after the criticisms of Gould, Lewontin, and others. But the integration of phyletic and developmental constraints into mainstream (adaptationist) evolutionary theory has been limited. The advocacy of these constraints in the spandrels paper had been inspired by Gould's historical research for his book *Ontogeny and Phylogeny* (1977), which brought him into contact with continental European traditions of morphology. Coincident with the spandrels paper, critiques of mainstream evolutionary theory were coming from within modern embryology and developmental biology. A main topic was the constraints which the processes of ontogenetic development placed on evolutionary processes (Maynard Smith et al., 1985).

While both random drift and developmental constraint were invoked in the spandrels paper, the latter was the new issue. Morphology and embryology had been an important part of evolutionary studies during the 19th century and up until the Modern Synthesis. But those fields of study did not participate in the great integration the Synthesis achieved. V. Hamburger (1980) reported that Synthesis biologists treat embryology as a black box, the contents of which were irrelevant to evolutionary studies. E. Mayr (1991, p. 8) claims that it was embryologists' own disinterest which kept them out of the Synthesis. Both may be correct. Mendelian transmission genetics was crucial to the Synthesis, while neither embryology nor developmental genetics was a pressing need. The Weismannian sequestration of the germ line, and later the "Central Dogma" of modern genetics, can be used by Synthesis biologists to argue the irrelevance of ontogenetic development to evolutionary change (Wallace, 1986). In response, developmentalists argue that embryology (but not transmission genetics) is able to explain the generation of form and thereby explain the range of phenotypes which might conceivably occur to be selected among.

This is a significant clash of biological traditions, with much deeper methodological contrasts than the debates pitting drift against adaptation. A new conflict of Purpose versus Pattern, the debate shows an eerie similarity to the situation in Britain in the 1850s. Natural theologians, like the modern adaptationists, found adaptations to be the point of central interest in biology and the central target of explanation. Transcendental anatomists, like modern structuralists or developmentalists, took more interest in homology and the patterns of development. Neither side could summon much interest in the focal topic of the adversary. The modern debate is an evolutionary version of that same debate, so the conceptual landscape changes somewhat. History, phylogeny, and the mechanisms of change play new and important roles in the 20th century. But the marks of similarity (one is tempted to say the "logical homologies") between the debates remain. Adaptation is the key to understanding biology according to one side of the debate; structure and homology is the key for the other side. The processes of ontogenetic development are central to the structuralist critics of adaptation, both in the 19th and 20th century versions of the debate. E. S. Russell's 1916 *Form and Function* could be easily updated to include the current debate on adaptation versus developmental constraint.

Gould has buttressed the conclusions of the spandrels paper with historical commentary which can be used to illustrate the contrast between the approaches to evolutionary explanation (Gould, 1983). Gould claims that the Modern Synthesis was pluralistic in its views of evolutionary mechanisms during its early days, but by 1950 the adaptationism of major Synthesis biologists was "hardening." He supports this thesis by comparing successive editions of evolutionary texts by Dobzhansky, Simpson, and Wright. This analysis is not universally accepted (Mayr, 1988, p. 528), and anyhow adaptationism was in an objectively stronger evidentiary position in 1950 than earlier during the Synthesis (Provine, 1986, p. 411). Nevertheless, Gould shows some striking examples of 1950s adaptationism. One is from the 1951 3rd edition of Dobzhansky's *Genetics and the Origin of Species*. Dobzhansky applied Sewall Wright's adaptive landscape to the question of the discontinuous distribution of organic forms. "Each living species may be thought of as occupying one of the available peaks in the field of gene combinations. The valleys are deserted and empty" (Dobzhansky, 1951, p. 10). Dobzhansky went on to explain discontinuities between species and even taxonomic relations among higher groups on the basis of the ecologically determined landscape. "Thus, the ecological niche occupied by the species 'lion' is relatively much closer to those occupied by tiger, puma, and leopard than to those occupied by wolf, coyote, and jackal. The feline adaptive peaks form a group different from the group of the canine 'peaks.' But the feline, canine, ursine, musteline, and certain other groups of peaks form together the adaptive 'range' of carnivores, which is separated by deep adaptive valleys from the 'ranges' of rodents, bats, ungulates .... The hierarchical nature of the biological classification reflects the objectively ascertainable discontinuity of adaptive niches ..." (ibid.)

Gould rejects this interpretation: "But this cannot be, for surely the cluster of cats exists primarily as a result of homology and historical constraint. All felines are alike because they arose from a common ancestor shared with no other clade. That ancestor was well adapted, and all its descendants may be. But the cluster and the gap reflect history, not the current organization of ecological topography. All feline species have inherited the unique cat *Bauplan*, and cannot deviate far from it as they adapt, each in its own particular (yet superficial) way. Genealogy, not current adaptation, is the primary source of clumped distribution in morphological space." (Gould, 1983, p. 80)

Dobzhansky's passage does express a "hardened" form of adaptationism, showing no interest in explaining phenotypes by ancestry. One wonders why we should consider felines a monophyletic group. If the adjacency of their adaptive niches is sufficient to account for their similarities, perhaps they all converged! Gould considers the passage an overenthusiastic mistake on Dobzhansky's part (ibid., p. 79). An alternative interpretation of Dobzhansky will be suggested in the following section when we consider divergent definitions of adaptation.

The contrast between Gould and Dobzhansky here mirrors the classic contrast between form and function, morphology and adaptation. A single phenomenon, the hierarchy of discontinuous groupings, is given distinct and apparently exclusive structuralist (Gould) and adaptationist (Dobzhansky) explanations. But surely neither of these stories can tell the *whole* truth. Gould referred to the adaptedness of felines (ancestral and descendent) as an obvious condition in need of no further discussion and declared constraint to be the "primary" reason for clumping. Dobzhansky most likely considered the monophyly and retained homologies of the group as similarly obvious but uninteresting, and came to the mirror–image conclusion that adaptation was primary. The divergent explanatory interests are revealed when Gould referred to the "clumped distribution in *morphological* space." Dobzhansky had been thinking of ecological space. Morphological and ecological space are quite different realms (Amundson, 1994b). The clumpiness of morphological space is a main theoretical interest of structuralists, but not of adaptationists. For an adaptationist, morphology (form) is presumed to follow adaptive function, with function the real topic of interest. With a deeper interest in morphology, Gould naturally sees historical homology as its proper explanation. Note again the parallel to the natural theologians and transcendental anatomists. Dobzhansky defines the feline type adaptively, by the species' clustered adaptive peaks, just as Cuvier and Bell insisted on similarity of adaptation as the only basis for defining homologs and types. Gould defines the feline group in terms of history and homology, and considers adaptation to be a secondary phenomenon. Similarly, the transcendental anatomists regarded homology as the true basis of the categories of the organic world, even without connecting homology with evolutionary history. The debate about the primacy of form versus function continues to thrive. The Darwinian revolution changed its terms, but did not put it to rest.

A significant number of developmental biologists currently share the opinion that their field has more to contribute to evolutionary theory than mainstream adaptationists recognize. Some call for a new developmental or embryological synthesis with implications as dramatic as those of the Modern Synthesis (Gilbert, 1991; Horder, 1989). This synthesis will be difficult to achieve given the differences between the explanatory strategies of the two approaches.

One example of conceptual divergence may illustrate the problem. Much of the debate on the importance of history turns on the question of *constraints*. Constraints appear to have become an important topic not because they are at the core of nonadaptationist evolutionary theorizing, but because they are the developmental topic mostly closely related to adaptation. Nevertheless, developmentalists and adaptationists have subtle but quite distinct concepts of constraint (Amundson, 1994a). Each party considers constraint to be acting on the phenomena of central interest to their own tradition. Adaptationists consider constraints to act on adaptation; to them a constraint is a reduction of the degree of adaptation a species might otherwise achieve. Developmentalists consider constraints to act on biological form; to them a constraint places bounds on the range of variability which can be generated by the particular mode of embryological development within a lineage. These two concepts are obviously related, but they are not identical. Adaptationists do recognize the existence of constraints on adaptation and (often in response to the spandrels paper) they can list many such constraints. Since such a constraint is defined in terms of adaptation (i.e., as limits on adaptation), its recognition does not necessarily involve the understanding of any embryological processes. For example, foraging theorists can identify the perceptual limitations of a species as a "constraint" on its optimal foraging (Stephens and Krebs, 1986). Adaptationists are understandably perplexed when the acknowlegement of these constraints (on adaptation) fails to satisfy the critics. Developmentalist critics of adaptationism are not satisfied by the acknowledged constraints on adaptation because they are primarily concerned with the generation of biological form, not of adaptations. They attempt to discover constraints on form by investigating mechanisms of development, and do not bother to check on the maladaptedness of the developmentally prohibited forms. The real interest of the new developmentalist critics is to apply embryological knowledge to evolutionary problems, and not just to get adaptationists to admit that adaptation is imperfect. When adaptationists define constraints as imperfect adaptedness and explain it as a by-product of adaptive "tradeoffs," they are measuring constraint using an adaptationist's yardstick. Since these techniques avoid opening the "black box" of embryological development, they fail to impress developmentalist critics.

In historical perspective the symmetry of the opposing sides is striking. The adaptationist understanding of constraint calls to mind Bell's rejection of the ear/jaw homology on the grounds that birds' hearing is not reduced by their shortage of ear ossicles; what does not matter to adaptation does not matter to biology. Gould's and the developmental critics' contrary insistence that history, homology, and the processes of development are "primary" echoes the transcendentalists' commitment to homology.

In recent days, constraints have become a common topic of study in biology, although the variety of forms is dazzling (Antonovics and van Tienderen, 1991). There are attempts to combine adaptationist and developmentalist approaches in nonantagonistic ways. But the prospect for a smooth developmentalist reform of adaptationist evolutionary theory does not look promising. As can partly be seen in the Gould/Dobzhansky example above, the explanatory interests of the two traditions are quite distinct. An approach which integrates both sets of interests is hard to picture. Just as in the 19th century, the modern adaptationist and structuralist camps each have many ways to argue for the primary importance of their own concerns, and the secondary importance of their adversaries'. Darwin's argument that Conditions of Existence is a "higher law" than Unity of Type was the first such move by an evolutionist. Similar dismissive arguments are available to both sides in the modern debate. My readings in the history of such disputes makes me skeptical about the prospects for a smooth synthesis. The 19th century conflict was put to an end by no less an event than the publication of Darwin's *Origin of Species*.

It may or may not turn out that developmental facts become central to evolutionary understanding. Perhaps detailed accounts of pleiotropy will make embryological knowledge redundant within evolutionary studies. Arguments for the importance of development, like arguments for the importance of drift, pleiotropy, or for adaptation itself, amount to predictions about the probable success of distinct research programs. If presumptive adaptation remains biologically respectable, it will at least have survived significant challenges.

As a final note on the history of disputes about adaptation, the critics do have some claim to the title of pluralist. Lucretius denied the purposiveness of animal body parts, but no later critic has been so bold. Francis Bacon, Voltaire, Hume, Owen, and many others have argued against the *overuse* of teleological and purposive explanations on the grounds that they were too easy to devise and they distracted attention from other scientific studies. They did not deny that body parts had adaptive purposes. Modern critics of adaptation make similar arguments. There are said to be other important biological phenomena to discover, and nonadaptive explanations ought not be ignored in favor of adaptive ones. Adaptation is no longer considered a rare biological phenomenon. What is at issue is the relative importance of other phenomena.

VII. Definitions

Etymology has been kind to the term adaptation in its evolutionary context. The Oxford English Dictionary (OED) traces the root "apt" to Latin "apt–us" meaning "fitted, suited, appropriate," derived from "ap–ere, to fasten, attach" for example as applied to a garment. Even before it entered biology, adaptation was related to fitness. Both the verb adapt and the noun adaptation are from French and late Latin, with first cited English uses in 1610 and 1611.

As evolutionists are well aware, adaptation refers both to a process and to its product. This semantic fact is not surprising; many process terms also apply to the results of the processes. The process of description results in a description, and an object which undergoes modification acquires new traits themselves called modifications. The same is true of adaptation. In the OED at least, the process is primary. Adaptation is "1) The action or process of adapting, fitting or suiting one thing to another, . . . 2) The process of modifying a thing so as to suit new conditions." As the term refers to the results of the process it is "3) The condition or state of being adapted. ... 4) A special instance of adapting."

Nothing said so far discriminates between phylogenetic and ontogenetic adaptations, or between biological adaptations and those produced by human beings through intentional actions. Increased chest capacity is one ontogenetic adaptation of humans to high altitudes; those who beat their swords into plowshares presumably adapt one edge of the sword to a plow's moldboards. Our present interests are in phylogenetic adaptations, but the meanings we give to the concept in the domain of phylogenetic change are not free from associations with the word in other contexts.

While the distinction between the process and the product of adaptation is commonly recognized, the distinction between senses 3) and 4) above is not often noted. The process of adaptation gives rise to a *generic state* of overall adaptedness, and also to *specific traits* which were adaptive modifications and are for that reason called adaptations. So the process yields both adaptation and adaptations. Most uses of the term in Darwin's *Origin* were the generic sense, as in "the degree of adaptation of species to the climates." As we have seen, Darwin used the trait sense of the term just as frequently to deny as to affirm that a specific trait was an adaptation.

Whether conceived as a process, a generic state, or an individual trait, adaptation is a relational concept. The process of adaptation is the "fitting" of one thing to another. The generic state is a relation between an organism and (roughly) an environment to which it is adapted. *An* adaptation (adaptation *qua* individual trait) is a modified part of an organism which performs a biological function for the organism and thus contributes to the organism's state of adaptation. Apart from complexities regarding what counts as the environment (for example, in the evolution of complex functional integration), these points are uncontroversial.

Evolutionary biologists take Darwin's view on the two roles of natural selection in the evolutionary process. It is simultaneously the primary cause of change and the cause of (most or all) adaptation. Depending on how broadly natural selection is conceived, it might just *be* the process of adaptation. (This is the case if natural selection is seen to include the generation of variation.) The process is causally responsible for the high degree of generic adaptation to be found in current populations, and also for the numbers of individual adaptations which accumulatively account for the generic adaptation. Again, I take this to be uncontroversial with respect to the definition of adaptation.

Gould and Vrba (1982) discussed the relation between the process of adaptation and a current trait being designated *an* adaptation. They pointed out the presence of two distinct adaptation concepts in the literature, one historical and one nonhistorical, associated respectively with the writings of Williams (1966) and Bock (1980; Bock and von Wahlert, 1965). According to the historical concept of adaptation, a current trait is an adaptation for a current use just in case it arose by natural selection for that use. On the nonhistorical definition, a trait is an adaptation just in case it contributes to current fitness. Gould and Vrba endorsed the historical definition, calling support from Darwin's discussion of the nonfused skull sutures of infant mammals. They then nominated the term "aptation" to replace the nonhistorical concept of adaptation, designating a feature which contributes to fitness irrespective of its origin. "Exaptations" were introduced as aptations which were nonadaptations, traits which currently provide benefits for which they were not selected. The similarity between exaptation and Simpson's older term "preadaptation" was acknowledged; a preadaptation was an exaptation considered before the trait was exapted. Since adaptations were (by definition) selected for their beneficial effects, preadaptations should be labeled "preaptations" on the new scheme; they would become aptations before they became adaptations.

Philosophers of biology have meanwhile been engaged in similar analytic projects. "A is an adaptation for task T in population P if and only if A became prevalent in P because there was selection for A, where the selective advantage of A was due to the fact that A help perform task T" (Sober, 1984, p. 208). At present the historical definition of adaptation may be more widely held among philosophers of biology than among biologists (Brandon, 1981; Burian, 1983; Griffiths, 1992; Sober, 1993). Brandon (1990, p. 186) calls it the "received view". The specification of task T in Sober's definition matches the need for some specific current use in Gould's and Vrba's. The simpler expression "A is an adaptation" would simply mean that there exists a task T such that the definition applies. One convention which does not match Gould and Vrba is the use of the term "adaptedness" nonhistorically, so that an organism's adaptedness is a matter of the current fitness it confers, whereas its status as adaptation is a matter of selective history (Brandon, 1978). Gould and Vrba would presumably speak of "aptness" rather than adaptedness in this context. The asymmetry created by the historical definition between adaptation and fitness is recognized. Adaptation looks to the past and a trait's selective history, while fitness points to future reproductive successes (Sober, 1984, p. 210). It is occasionally argued that the historical definition should include traits that were maintained in a population by selection even if they did not originate by selection (Griffiths, 1992; Sober, 1993). But while the nonhistorical concept of adaptation is occasionally acknowledged by philosophers, no one to my knowledge has argued in its favor.

Among biologists, on the other hand, specific arguments can be found for the superiority of the nonhistorical concept of adaptation. Some of these arguments are also intended to defend adaptationism against its critics. The historical definition does place higher standards than the nonhistorical definition on claims that a trait is an adaptation. Advocates of adaptationism should be expected to prefer a less demanding conception just as critics might prefer a more demanding one. But other challenges to the historical definition seem neutral with respect to the adaptationism debates. One such concern is that the historical definition of adaptation confuses the product with the process of adaptation (Endler, 1986; Endler and McLellan, 1988; Fisher, 1985; Gans, 1988; Reeve and Sherman, 1993). All agree that the process of adaptation (natural selection) is a cause of states of adaptation. But codifying that fact by embodying it in the definition of states of adaptation is, it is argued, going too far. Fisher gives the strongest version of this argument. He claims that if we *define* adaptations as products of natural selection it becomes impossible to *explain* them by citing their selective causes. "Defining the state of adaptation in terms of its contribution to current fitness, rather than origin by natural selection, is essential if natural selection is to be considered an *explanation* of adaptation" ... "Darwin's (1859) intent ... was clearly to offer the process of natural selection as an *explanation* for features and relationships that can be observed in the world today. ... It is quite a different matter to assert by *definition* that adaptations are created by natural selection for current function" (Fisher, 1985, p. 120, 123).

Fisher appears to see a sort of tautology in the historical definition of adaptation that is similar to the notorious tautology regarding the expression "survival of the fittest." If we define the "fittest" as "those that survive" then "the fittest will survive" carries no empirical content. This is because "fittest" by definition has no other measure than survival. Does the historical definition of adaptation leave natural selection similarly empty and explanatorily useless? It does not. Not all definitions create causal circularities. The historical definition proposes that "an adaptation" be defined as "a fitness enhancing trait which was produced by natural selection for its current fitness benefits." Biologists are presumably able to distinguish the fitness values of traits independently of discovering their selective history. So the statement "this fitness enhancing trait was produced by natural selection for its current benefit" is not an empty tautology, since some fitness enhancing traits are fortuitous. To be sure, "this adaptation was produced by natural selection" *is* a tautology on the historical definition, just like "this bachelor is unmarried." Adherents of the historical definition would presumably never bother uttering the words. Nevertheless, it is still meaningful to study the causes of fitness–enhancing traits, referring to them as "aptations" after Gould and Vrba or "contributors to adaptedness" after Brandon. When the causes are discovered to be natural selection, the traits would be pronounced adaptations.

The fitness tautology would exist only if there were no other measure than survival to determine fitness. Analogously, Fisher's suggested adaptation tautology would exist only if there were no other measure than selective history to determine the current fitness enhancement conferred by a trait. But current fitness enhancement (rather than selective history) is exactly how the advocates of the nonhistorical definition wish to define "adaptation." Indeed, the practical applicability of their definition is one of its attractions. So the historical definition is tautologous only if current fitness enhancement is unmeasurable. But if this is the case, both definitions are impossible to apply, since each requires the assessment of current benefits to fitness. So the historical definition of adaptation is tautologous only if the nonhistorical definition is impossible to apply.

Its critics notwithstanding, nothing in the historical definition reduces the power of natural selection to causally explain states or traits of adaptation. But other considerations may still weigh against it. More compelling support for a current–fitness concept of adaptation comes from observations on the interests of distinct research domains. Reeve and Sherman distinguish questions of *evolutionary history* from questions of *phenotype existence.* Practitioners of the latter, behavioral and evolutionary ecologists, "ask why certain traits predominate over conceivable others in nature, irrespective of the precise historical pathways leading to their predominance, and then infer evolutionary causation based on current utility" (Reeve and Sherman, 1993, p. 2). Endler and McLellan (1988) use the term adaptation to indicate current fitness, including exaptations, on the grounds that "as soon as a new function for a trait occurs, natural selection will affect that trait in a new way and change the allele frequencies that generate that trait" (p. 409). The history of a heritable variant does not matter in this kind of study. What matters is its current selective status. A trait with a long adaptive pedigree gets no special treatment by the contingencies of selection. The dynamics of a population depend on the actual fitness of the traits present, which depends on the functions they are currently serving. How long they have been serving those functions is irrelevant.

This contrast in approaches to evolution reveals the distinction between *equilibrium* and *transformational* biological studies (Lewontin, 1969; Lauder, 1981). Adaptationist ecology takes the equilibrium approach, presupposing at least a momentary equilibrium between traits and selective forces. Reeve and Sherman describe adaptation as a state of equilibrium. They liken ecologists to astronomers, who study spatial arrangements of stars and galaxies with no particular interest in the historical details of each system's arrangement because "for a given set of conditions, a large number of distinct individual histories tend to converge on a small number of stable states" (p. 15). As with other equilibrium studies, this style of ecology implies very little about the causal histories of those systems which are in equilibrium. Equilibrium biology does have what we might call a shallow concept of history, in that it is considered legitimate to infer that a trait of especially high fitness relative to alternatives must at least have been *maintained* in the population by selection. But remote history is irrelevant. "Whatever is important about a trait's history is already recorded in the environmental context and the biological attributes of the organism" (Reeve and Sherman, 1993, pp. 9–10.)

Transformational studies, for example in paleontology, comparative anatomy, and perhaps developmental biology, address themselves first and foremost to historical sequences. The two concepts of adaptation track the interests of the two kinds of study. Interested only in current population dynamics, equilibrium biology designates "adaptations" in terms of current causal forces. Looking to the historical processes which produced modern organisms, transformational biology labels "adaptations" according to their distinct selective origins. The distinct adaptation concepts mark distinct interests in the practice of biology.

Another semantic feature of the dispute meshes well with the notion that different disciplinary interests are involved. Exaptation was intended to replace Simpson's term "preadaptation," which many consider to have an unfortunate forward–looking teleological tone (Ghiselin, 1966; Gould and Vrba, 1982). But the future–directed semantics of preadaptation may have had a point. "'[E]xaptation' is a static description, while 'protoadaptation' and 'preadaptation' look to the possible" (Endler, 1986, p. 47). This forward–looking aspect of the term preadaptation accounts for its teleological aftertaste, but perhaps also marks its suitability within a forward–looking program of population genetics. With a current–fitness concept of adaptation and no interest in historical trajectories, "preadaptation" conjures up only coincidence, not prescience.

It is possible to interpret Dobzhansky's apparent abandonment of feline phylogeny on these grounds. The section on adaptive peaks had been inserted in the 1951 edition of Dobzhansky's text just before a section entitled "Evolution" which remained little changed from the 1937 edition. There Dobzhansky reported "two distinct approaches to evolutionary problems." One is historical and studies the "actual course which the evolutionary process took in the history of the earth." The other approach "emphasizes studies on the mechanisms that bring about evolution, causal rather than historical problems, phenomena that can be studied experimentally rather than events which happened in the past" (Dobzhansky, 1951, p. 11). Roughly speaking, these are the transformational and equilibrium approaches. Dobzhansky's preference for the "causal" (population genetic, equilibrium) problems over issues of history was surely involved in his accounting for the feline type by the (supposed) adjacency of the adaptive niches of feline species. Gould is correct that Dobzhansky's adaptationism comes out more dramatically in that 1951 allusion. But the seeds of the approach were present in 1937. A few pages after the 1937 "Evolution" section, Dobzhansky gave what seems to have been the first population genetic definition of evolution itself. "Since evolution is a change in the genetic composition of populations, the mechanisms of evolution constitute problems of population genetics" (Dobzhansky, 1982, p. 11). By 1951, perhaps, the conception of the changing gene pool as an exhaustive characterization of evolution had distracted Dobzhansky from the nested sets of homologies which traditionally characterize the feline type. Gene pools can easily be conceived along equilibrium lines. The present dynamics of the pool are the product of current distributions and forces only. But the concept of homology is transformationally defined, as is historical adaptation. Neither homologies nor historical adaptations (*qua* historical adaptations) play causal roles in the dynamics of the gene pool.

If we conclude that the historical and nonhistorical concepts of adaptation simply reflect the difference in theoretical approach between equilibrium and transformational biology, the question remains of why philosophers (and others) have so preferred the historical concept. A possible answer is that the historical concept of adaptation is nearer to the common sense concept of purpose, and so allows a better interpretation of biological or evolutionary purpose. Recall that purposive concepts play the dual explanatory role of specifying the benefits of an item and of explaining the item's existence as a consequence of those benefits. The nonhistorical concept of adaptation simply fails to carry the second implication. "How did that adaptation arise" might well be answered "We have no idea" (or "It makes no difference to our study"). The question of trait origin is loaded against the ecologist, of course. Trait maintenance is more easily dealt with than trait origin on an equilibrium account. Ecologists consider it justified to infer that natural selection has at least *maintained* a trait in a population if the trait has the highest fitness among reasonable alternatives. This conclusion has far less purposive punch than the historical definition. Consider trait A which was selectively shaped to perform task T and which continues to thereby enhance fitness to a degree appropriate to infer that it is maintained by selection for that function. Calling A a current–fitness–adaptation can imply no more than that it has been maintained by selection for T. Calling it a historical–adaptation explains its origins. Ecologists will point out that its origins are causally irrelevant to the dynamics of its current population. But even if this is true, the additional explanatory content of the historical–adaptive assessment remains. It explains how the trait came into existence, whether or not ecologists have a scientific use for that explanation. After all, states of equilibrium are not the only things we want explained. Even in astronomy we often desire historical rather than equilibrium explanations of certain phenomena. For example, where did the Earth's moon *come from?* The Earth and its moon are an unusual pair in the solar system, where planets are generally much larger than their satellites. The observation that the Earth and moon are in an equilibrated orbit does not begin to satisfy our curiosity about the origin of the system.

The historical origins of traits may not be of interest to the practice of ecology, but they interest other biologists. Explaining a item's origin by its consequences roughly coincides with common sense views of purpose and purposive explanation. If the concept of Darwinian biological purpose is coherent, the historical but not the current–fitness concept of adaptation carries its meaning.

Many of the defenses of the current–fitness concept of adaptation are also defenses of adaptationism against its critics. The nonhistorical concept of adaptation appears to be incidental to those defenses. Fisher's (1985) argument against the historical definition is among the strongest, but he favorably discusses explanatory contributions from studies of development and theoretical morphology. Reeve and Sherman, on the other hand, offer a series of arguments to demonstrate the irrelevance of constraint–style explanations to the understanding of evolution. These are modern analogs of Darwin's argument that Conditions of Existence is a "higher law" than Unity of Type. One example is their commentary on Wake (1991). Wake had argued that certain patterns of homoplasy involving digit reduction in salamanders could be explained by reference to features of plethodontid developmental processes. Reeve and Sherman did not judge Wake's explanation to be false. Rather they claimed that he did not even offer an explanation. "Since Wake offers no evidence of the relative fitnesses of small four–toed and five–toed individuals within plethodontid taxa, design constraints offer at best a description, not an explanation, of the occurrence of four–toedness" (Reeve and Sherman, 1993, p. 22). To committed adaptationists, an explanation with no reference to adaptation is no explanation at all. No doubt Wake would disagree with this principle. Like the two concepts of constraint, the divergence between adaptationist and structuralist standards of explanation echoes as far back as Bell's rejection of the ear/jaw homology.

While the current–fitness concept of adaptation might require a rephrasing of the adaptationism debates, it does not end them. Reeve and Sherman did reject structuralist alternatives to adaptationism, but their rejection depended on a series of arguments separate from the current–fitness definition itself.

Given the divergence in theoretical interests between the two sides of the dispute, a resolution will not be simple.

There is one further troubling issue for the historical definition of adaptation. It is the epistemic question of its applicability. Just how often will we be justified in declaring a trait to be an adaptation, historically defined, for a specifiable function? It has recently been argued that the answer is very seldom (Leroi et al., 1994). These authors argue that actual information about phylogenetic distribution of a trait, its functional performance, and the selective regime in which it evolved radically underdetermines its status as a (historical) adaptation. Additional information, such as a trait's fitness relative to the trait it replaces, is seldom available for interesting traits. In the absence of such information, "phylogenetic patterns will often suggest that a trait is an adaptation when in fact it is not and suggest that it is not, when in fact it is" (ibid., p. 383). If these conclusions are valid, it would seem that practically no individual identification of a historical adaptation will receive reasonable empirical support. The arguments have some similarities to those of Gould and Lewontin (1979). But unlike the authors of the spandrels paper, Leroi et al. (1994) have no more hope for structuralist than adaptationist historical explanations. The problem is not that historical adaptationists fail to consider constraint alternatives. The problem is that the evidence available will never decide between constraint and adaptation, let alone settle upon one of the possible adaptive or structuralist alternatives. The authors argue not for pluralism but agnosticism with respect to evolutionary mechanisms.

Leroi et al. (1994) do not conclude that transformational evolutionary studies must cease. Analyses of the patterns of character evolution might continue which are not founded on estimates of evolutionary forces, as for example in functional morphology. It is significant that functional morphology itself uses a concept of function defined without reference to fitness or selective history, in contrast to the adaptive and/or historical notions of function popular in philosophy and in other fields of biology (Amundson and Lauder, 1994). Transformational studies of functional morphology will not identify the selective forces responsible for evolutionary change. So if we require a strong epistemic warrant for ascriptions of historical adaptation there may be very few cases to discuss. We may be left with the shallow histories supported by equilibrium studies of current fitness, together with transformational but nonadaptive studies of character change.

On the other hand, the feeling is irresistible that comparative and functional studies give us *some* insight into the adaptive origins of existing traits. Can one remain fully agnostic with respect to selective origins and still be impressed by the work of Kingsolver and Koehl (1985)? Surely their work did not *demonstrate* that the homologous precursors of modern insect wings were (historical) adaptations for thermoregulation. But it was relevant to that hypothesis. Perhaps the best we can hope for is circumstantial evidence of the likelihood of one or another hypothesis about the remote causal origins of biological traits.

VIII. Conclusions

The usual approach to the history of ideas regarding adaptation has been to compare the *causes* which were believed to produce adaptation in the world. This divides the history of ideas into two main epochs, the finalistic (usually seen as the theological) and the Darwinian. Coincidentally the topic of organic origins is usually considered to classify history into the same two epochs. The history sketched in this essay focuses more on *examples* of adaptation in the world than on its causes — it asks not only what was seen as *causing* adaptations, but what kinds of phenomena were seen to *be* adaptations. That history is more complex, though again Darwin plays a central role in the drama.

From ancient days until the end of the 17th century the celestial and the organic realms together supplied the most striking cases of design or adaptation in the world. Skeptics about adaptation, always a minority, attributed the appearances of adaptation to blind necessity. The achievements of Newton and Laplace, somewhat ironically, gave comfort to the skeptics. The marvelously simple foundations of Newtonian physics eroded the adaptive appearance of celestial phenomena. This forced a distinction between the relatively oblique astrotheological Argument from Pattern and the robust biological Argument from Purpose. Adaptation now belonged fully to biology, and most biologists considered it the central and most important feature of life.

During the early 19th century, a new challenge arose to adaptationist biology. The Unity of Type biologists argued that adaptation was an insufficient basis for understanding life. The Pattern–like influences of Type were claimed to be deeper and more basic than organic adaptation. These new skeptics on adaptation varied widely in their theological commitments, but were unanimous on the insufficiency of adaptation to explain the organic realm. This was the context for Darwin's achievement.

With respect to concepts of adaptation, Darwin's *Origin* was both conservative and radical. It subsumed and defused the Unity of Type objections to adaptation by accepting them as evidence for the fact of evolution. Adaptation regained its former glory at the center of biological interest. But this conservative result was achieved in the most radical possible way — by naturalizing the concept of adaptation, thus removing from it all implications of finalism, reversed causation, and intelligent contrivance. Adaptation was restored, but in a way which made it useless to its old theological champions.

Many traits of organisms could be seen as adaptations under either finalistic or Darwinian concepts. For these traits, Darwin simply added a historical explanation. But earlier adaptationists had no clear criteria by which to rule out many additional putative adaptations. It seemed possible to identify almost any beneficial effect as a divinely intended adaptation. These included now–absurd reversals of adaptive polarities, supposed "adaptations" possessed by one species but which benefitted a different species, and rampant assertions of natural phenomena which were said to exist "for" human benefit. None of these categories could be countenanced under natural selection. So in naturalizing the cause of adaptation, Darwin allowed adaptation to be delimited in ways never previously possible.

Newtonian physics had removed the adaptive status of celestial phenomena, and Darwin pruned back the scope of organic adaptations. Adaptationist biology survived Darwin's naturalistic pruning and survived the collapse of the natural theological program which had previously been its home. Darwinism was only one of several evolutionary theories for many decades, but even among competitors natural selection was the favored explanation of most cases of adaptation. The Modern Synthesis returned Darwinian adaptationist biology to prominence in a new statistical and Mendelian form.

Modern versions of adaptationism retain some similarities to their historical counterparts. These are especially visible in the recent "adaptation versus constraint" debates. The similarity of these to 19th century disputes raises the possibility of as yet unresolved conflicts between functional and structural modes of explanation in biology.

Many other complexities in the study of adaptation arise not from unresolved, but from newly discovered perspectives and technologies, and from the sheer complexity of the domain. The organic realm is hierarchically structured in several ways (geneologically from species through higher taxa, physiologically from enzyme through cell, tissue, and organ — and even ecosystems are often conceived hierarchically). Individual studies of adaptation which are directed at one level (of one hierarchy) will relate in complex ways to phenomena at other levels. It is virtually impossible to speculate how the explosion of knowledge of molecular biology will influence our understandings of adaptation in "higher" levels of the hierarchies.

As shown in the preceding section, distinct research interests can lead to the development of quite distinct concepts of adaptation itself. Darwin had pruned back the excesses of speculative theological adaptationism. Under modern science, recognized examples of adaptation have again increased far beyond the bounds of theological imagination. As study proceeds, not only will we increase our understanding of *cases* of adaptation, we will also learn more about the nature — or natures — of adaptation itself.

References

Amundson, R. (1994a). Two concepts of constraint: adaptationism and the challenge from developmental biology. *Philosophy of Science* 61, 556–578.

Amundson, R. (1994b). John T. Gulick and the active organism: adaptation, isolation, and the politics of evolution, *In* "Darwin in the Pacific" (P. Rehbock and R. MacLeod, eds.), University of Hawaii Press, Honolulu.

Amundson, R., and Lauder, G. V. (1994). Function without purpose: the uses of causal role function in evolutionary biology. *Biology and Philosophy* 9, 443–469.

Antonvics, J., and van Tienderen, P. H. (1991). Ontoecogenophyloconstraints? The chaos of constraint terminology. *Trends in Ecology and Evolution* 6, 166–169.

Appel, T. A. (1987). "The Cuvier–Geoffroy Debate: French Biology in the Decades before Darwin." Oxford University Press, New York.

Aquinas, T. (1952). Summa Theologica. *In* "Great Books of the Western World" (R. M. Hutchins, ed.), Vol. 19, Encyclopedia Brittanica, Inc., Chicago.

Bell, C. (1933). "The Hand Its Mechanism and Vital Endowments as Evincing Design." Bridgewater Treatise, Treatise IV. William Pickering, London.

Bhaksar, R. (1981). Aristotle's theory of cause. *In* "Dictionary of the History of Science" (W. R. Bynum, E. J. Browne, and R. Porter, eds.), pp. 26–28, Princeton University Press, Princeton.

Boakes, R. (1984). "From Darwin to Behaviorism." Cambridge University Press, Cambridge.

Bock, W. J. (1980). The definition and recognition of biological adaptation. *American Zoologist* 20, 217–227.

Bock, W. J., and von Wahlert, G. (1965). Adaptation and the form–function complex. *Evolution* 19, 269–299.

Bowler, P. J. (1977). Darwinism and the argument from design: suggestions for a reevaluation. *Journal of the History of Biology* 10, 29–43.

Bowler, P. J. (1983). "The Eclipse of Darwinism." Johns Hopkins Press, Baltimore.

Bowler, P. J. (1988). "The Non–Darwinian Revolution." Johns Hopkins Press, Baltimore.

Brandon, R. N. (1978). Adaptation and evolutionary theory. *Studies in History and Philosophy of Science* 9, 181–206.

Brandon, R. N. (1981). Biological teleology questions and explanations. *Studies in History and Philosophy of Science* 12, 91–105.

Brandon, R. N. (1990). "Adaptation and Environment." Princeton University Press, Princeton, NJ.

Brandon, R. N., and Burian, R. M. (eds.) (1984). "Genes, Organisms, Populations." MIT Press, Cambridge, MA.

Brooke, J. H. (1991). "Science and Religion." Cambridge University Press, Cambridge.

Burian, R. M. (1983). Adaptation. *In*"Dimensions of Darwinism." (M. Grene, ed.), Cambridge University Press, Cambridge.

Cain, A. J. (1951). So–called non–adaptive or neutral characters in evolution. *Nature* 168, 424.

Cain, A. J. (1964). The perfection of animals. *In* "Viewpoints in Biology." (J. D. McCarthy and C. L. Duddington, eds.), Butterworths, London.

Carter, G. S. (1951). Non–adaptive characters in evolution. *Nature* 168, 700–701.

Cannon, S. F. (1978). "Science in Culture: The Early Victorian Period." Dawson and Science History Publications, New York.

Clark, J. W., and Hughes, T. M. (1890). "The Life and Letters of Adam Sedgewick." Cambridge University Press, Cambridge.

Cronin, H. (1992). Sexual selection: historical perspectives. *In* "Keywords in Evolutionary Biology" (E. F. Keller and E. A. Lloyd, eds.), Harvard University Press, Cambridge, MA.

Darwin, C. (1859). "On the Origin of Species." John Murray, London.

Darwin, C. (1937). "Voyage of the Beagle." Reprint of 1845 2nd edition, P. F. Collier and Son, New York.

Darwin, C. (1958). "The Autobiography of Charles Darwin." W.W. Norton and Company, New York.

Darwin, C. (1987). "Darwin's Early Notebooks, 1836–1844." P. H. Barrett, P. J. Gautrey, S. Herbert, D. Kohn, and S. Smith (eds.), Cornell University Press, Ithaca, NY.

Dawkins, R. (1983). Universal Darwinism. *In* "Evolution from Molecules to Men" (D.S. Bendall, ed.), Cambridge University Press, Cambridge.

Dennett, D. C. (1987). "The Intentional Stance." MIT Press, Cambridge, MA.

Desmond, A. (1989). "The Politics of Evolution." University of Chicago Press, Chicago.

Dijksterhuis, E. J. (1961). "The Mechanization of the World Picture." Oxford University Press, Oxford.

Dobzhansky, T. (1951). "Genetics and the Origin of Species." Third Edition, Revised, Columbia University Press, New York.

Dobzhansky, T. (1982). "Genetics and the Origin of Species." facsimile reprint of 1937 First Edition, Columbia University Press, New York.

Endler, J. A. (1986). "Natural Selection in the Wild." Princeton University Press, Princeton, NJ.

Endler, J. A., and McLellan, T. (1988). The processes of evolution: towards a newer synthesis. *Annual Review of Ecology and Systematics* 19, 395–421.

Fisher, D. C. (1985). Evolutionary morphology: beyond the analogous, the anecdotal, and the ad hoc. *Paleobiology* 11, 120–138.

Galen (1952). On the natural faculties. "Great Books of the Western World" (R. M. Hutchins, ed.; A. J. Brock, M. D., trans.), Vol. 10, Encyclopedia Brittanica, Inc., Chicago.

Gans, C. (1988). Adaptation and the form–function relation. *American Zoologist* 28, 681– 697.

Geoffroy St. Hillaire, E. (1830). On the philosophy of nature. (report from *Le Globe*), *Edinburgh New Philosophical Journal* 8, 152–155.

Ghiselin, M. T. (1966). On semantic pitfalls of biological adaptation. *Philosophy of Science* 33, 147–153.

Gilbert, S. F. (1991). "Developmental Biology, Third Edition." Sinauer Associates, Inc., Sunderland, MA.

Gillespie, N. C. (1987). Natural Order, Natural Theology, and Social Order: John Ray and the 'Newtonian Ideology'. *Journal of the History of Biology* 20, 1–47.

Glacken, C. J. (1967). "Traces on the Rhodian Shore," University of California Press, Berkeley, CA.

Gould, S. J., and Lewontin, R. C. (1979). The spandrels of san marco and the panglossian paradigm: a critique of the adaptationist programme. *Proceedings of the Royal Society of London* B205, 581–598.

Gould, S. J. (1977). "Ontogeny and Phylogeny." Harvard University Press, Cambridge, MA.

Gould, S. J. (1983). The hardening of the modern synthesis, *In* "Dimensions of Darwinism" (M. Grene, ed.), Cambridge University Press, Cambridge.

Gould, S. J. (1993). Fulfilling the spandrels of world and mind. In "Understanding Scientific Prose" (J. Selzer, ed.), University of Wisconsin Press, Madison, WI.

Gould, S. J., and Vrba, E. S. (1982). Exaptation — a missing term in the science of form. *Paleobiology* 8, 4–15.

Gray, A. (1963). "Darwiniana; Essays and Reviews Pertaining to Darwinism" (A. H. Dupree, ed.), Harvard University Press, Cambridge, MA.

Griffiths, P. (1992). Adaptive explanation and the concept of a vestige. *In* "Trees of Life" (P. Griffiths, ed.), Kluwer Academic Publishers, Dordrecht.

Hahn, R. (1986). Laplace and the mechanistic universe, In "God and Nature" (D. C. Lindberg and R. L. Numbers, eds.), University of California Press, Berkeley, CA.

Hamburger, V. (1980). Embryology and the modern synthesis in evolutionary theory. *In* "The Modern Synthesis" (E. Mayr and W. Provine, eds.), Harvard University Press, Cambridge, MA.

Horder, T. J. (1989). Syllabus for an embryological synthesis. *In* "Complex Organismal Functions Integration and Evolution in Vertebrates" (D. B. Wake and G. Roth, eds.), John Wiley and Sons, Chichester.

Hull, D. L. (1983). Darwin and the nature of science. *In* "Evolution from Molecules to Men" (D. S. Bendall, ed.), Cambridge University Press, Cambridge.

Hurlbutt, R. H. (1965), "Hume, Newton, and the Design Argument." University of Nebraska Press, Lincoln, NE.

Jacob, M. C. (1986). Christianity and the newtonian worldview, *In* "God and Nature" (D. C. Lindberg and R. L. Numbers, eds.), University of California Press, Berkeley, CA.

Kellogg, V. L. (1908). "Darwinism To–Day." Henry Holt and Company, New York.

Kingsolver, J. G., and Koehl, M. A. R. (1985). Aerodynamics, thermoregulation, and the evolution of insect wings differential scaling and evolutionary change. *Evolution* 39, 488-504.

Kruger, L., Daston, L. J., and Heidelberger, M. (1987). "The Probablistic Revolution." 2 vols. MIT Press, Cambridge, MA.

Lauder, G. V. (1981). Form and function: structural analysis in evolutionary morphology. *Paleobiology* 7, 430–442.

Lennox, J. G. (1993). Darwin *was* a teleologist. *Biology and Philosophy* 8, 409–421.

Lenoir, T. (1982). "The Strategy of Life." University of Chicago Press, Chicago.

Leroi, A. M., Rose, M. R., and Lauder, G. V. (1994). What does the comparative method reveal about adaptation? *American Naturalist* 143, 381-402

Lesch, J. E. (1975). The role of isolation in evolution: george j. Romanes and john t. Gulick. *ISIS* 66, 483–503.

Lewontin, R. C. (1969). The bases of conflict in biological explanation. *Journal of the History of Biology* 2, 35-45.

Lewontin, R. C. (1978). Adaptation. *Scientific American* 249, November 212–222.

Lucretius (1952). On the nature of things. *In* "Great Books of the Western World" (R. M. H., ed.; H. A. J. Munro, trans.), Vol. 12, Encyclopedia Brittanica, Inc., Chicago.

Maynard Smith, J., Burian, R., Kauffman, S., Alberch, P., Campbell, J., Goodwin, B., Lande, R., Raup, D., and Wolpert, L. (1985) developmental constraints and evolution. *The Quarterly Review of Biology* 60, 265–287.

Maynard Smith, J. (1969). The status of neo–darwinism. *In* "Towards a Theoretical Biology" (C. H. Waddington, ed.), University Press, Edinburgh.

Mayr, E. (1961). Cause and effect in biology. *Science* 134, 1501–1506.

Mayr, E. (1964). Introduction. *In* "On the Origin of Species" (Darwin, Charles, Reprint of First Edition), Harvard University Press, Cambridge, MA.

Mayr, E. (1976). Agassiz, Darwin, and evolution. *In* "Evolution and the Diversity of Life." Harvard University Press, Cambridge, MA.

Mayr, E. (1983). How to carry out the adaptationist program? *American Naturalist* 121, 324–334.

Mayr, E. (1988). "Toward a New Philosophy of Biology." Harvard University Press, Cambridge, MA.

Mayr, E. (1991). An overview of current evolutionary biology. *In* "New Perspectives on Evolution" (L. Warren and H. Koprowski, eds), John Wiley and Sons, New York.

Mayr, E., and Provine, W. (1980). "The Evolutionary Synthesis." Harvard University Press, Cambridge, MA.

Ospovat, D. (1978). Perfect adaptation and teleological explanation. *Studies in History of Biology* 2, 33–56.

Ospovat, D. (1981),"The Development of Darwin's Theory." Cambridge University Press, Cambridge.

Owen, R. (1848). "On the Archetype and Homologies of the Vertebrate Skeleton." John Van Voorst, London.

Owen, R. (1849). "On the Nature of Limbs." John Van Voorst, London.

Paley, W. (1831). "The Works of William Paley; Complete in One Volume." Thomas Nelson and Peter Brown, Edinburgh.

Plato (1961). Phaedo. *In* "The Collected Dialogues of Plato, Including the Letters" (E. Hamilton and H. Cairns, eds.), Bollingen Series 71, Pantheon Books, New York.

Provine, W. B. (1986). "Sewall Wright and Evolutionary Biology" University of Chicago Press, Chicago.

Ray, J. (1977). "The Wisdom of God as Manifested in the Works of the Creation." facsimile of 1717 Edition, Arno Press, New York.

Reeve, H. K., and Sherman, P. W. (1993). Adaptation and the goals of evolutionary research. *The Quarterly Review of Biology* 68, 1–32.

Rehbock, P. F. (1983). "The Philosophical Naturalists." University of Wisconsin Press, Madison, WI.

Ruse, M. (1979). "The Darwinian Revolution." University of Chicago Press, Chicago.

Russell, E. S. (1916). "Form and Function." John Murray, London.

Selzer, J. (1993). "Understanding Scientific Prose." University of Wisconsin Press, Madison, WI.

Sober, E. (1984). "The Nature of Selection." MIT Press, Cambridge, MA.

Sober, E. (1993). "Philosophy of Biology." Westview Press, Boulder, CO.

Spencer, H. G., and Masters, J. C. (1992). Sexual selection: current debates. *In* "Keywords in Evolutionary Biology" (E. F. Keller and E. A. Lloyd, eds.), Harvard University Press, Cambridge, MA.

Stephens, D. W., and Krebs, J. R. (1986). "Foraging Theory." Princeton University Press, Princeton.

Wake, D. B. (1991). Homoplasy: the result of natural selection, or evidence of design limitations? *American Naturalist* 138, 543–561.

Wallace, B. (1986). Can embryologists contribute to an understanding of evolutionary mechanisms? *In* "Integrating Scientific Disciplines" (W. Bechtel, ed.), Martinus Nijhoff Publishers, Dordrecht.

Whewell, W. (1836). "Astronomy and General Physics, Considered with Reference to Natural Theology, Bridgewater Treatises, Treatise III." A New Edition, Carey, Lea & Blanchard, Philadelphia, PA.

Whewell, W. (1863). "History of the Inductive Sciences: From the Earliest to the Present Times." Third Edition, with additions, in two volumes, v. 2, D. Appleton and Company, New York.

Williams, G. C. (1966). "Adaptation and Natural Selection", Princeton University Press, Princeton.

Winsor, D. A. (1993). Constructing scientific knowledge in Gould and Lewontin's 'The spandrels of San Marco'. *In* "Understanding Scientific Prose" (J. Selzer, ed.), University of Wisconsin Press, Madison, WI.

Winsor, M. P. (1976). Louis Agassiz and the species question. *Studies in History of Biology* 3, 89–117.

Wright, C. (1877). "Philosophical Discussions." Henry Holt and Company, New York.

CHAPTER 2

The Argument From Design

GEORGE V. LAUDER

I. Introduction

Let us suppose that one day, while out for a walk, I have the good fortune to come upon a Rolex watch lying on the ground. At first I am pleased at the prospect of possessing an improvement to the inexpensive plastic watch that I am currently wearing. But as I continue my walk and examine the Rolex more closely, I begin to suspect that this watch may not be all that it seems. The gold on the casing has started to flake off, and the band has a suspiciously flimsy feel to it. I come to suspect that the watch I have found might be a counterfeit, discarded by an embarrassed purchaser who uncovered the deception. Just how good is the watch that I have found? How does this "Rolex" compare to my inexpensive digital watch in timekeeping and resistance to shock? How does a Rolex keep time in comparison to the U.S. National Bureau of Standards atomic clock? The watch I have found may run slow due to lack of a precise fit among the parts. The watch may not run at all due to a broken spring. Or the watch may actually be a counterfeit Rolex, made by a clever watchmaker, which looks well made but only runs a few days before breaking. We can only discover the quality of a watch we have found by taking it apart, examining its mechanics, and testing its performance against alternative designs.

The central theme of this chapter is that we have been too assumptive in the study of organismal design. Too often we have been willing to assume, like the purchaser of a counterfeit Rolex, that the watchmaker has done a good job as a result of looking at the external features of the watch instead of focusing on mechanical and performance evaluation. We also have been assumptive in presuming that some features of organismal design are accidental by-products of the method of construction, without a design analysis. This is not so much the fault of the watchmaker as it is a case of "buyer beware." Evolutionary biologists interested in organismal design must experimentally assess the mechanical quality of a watch they have discovered in comparison to other designs, and not restrict themselves to an analysis of the face alone.

This chapter addresses two specific issues. First, I suggest that in our desire to draw conclusions about biological design and to support theoretical views of how organisms are built, we have been too willing to make assumptions about the relationship between structure and mechanical function, about the patterns of selection that have acted on components of design, and about the process of design construction itself. Second, I argue that we have not often conducted the mechanical and performance tests needed to assess the average quality of organismal design. Perhaps we have been blinded by a few exemplary cases, but if the variance in quality is high, these few best cases will give a misleading picture of the range and variation in design. Additionally, conclusions that specific phenotypic features may be accidental by-products of the design process may also suffer from the lack of a rigorous design analysis.

I begin by reviewing the argument from design in its classical and modern formulations. I then consider four reasons why we might be cautious about accepting an argument from design as *prima facie* evidence of the past action of selection in generating adaptive traits. My analysis of the argument from design will use primarily concepts and examples from the study of animal morphology and physiology. The investigation of animal design is the focus of research in the disciplines of functional morphology, biomechanics, and comparative physiology, and many issues relevant to the argument from design can be seen with special clarity via the study of specific features of animal structure. Two major means by which we recognize design in organisms are by applying engineering and mechanical principles, and by analogies to man-made devices. In addition, both the original formulations of the argument from design and their modern counterparts often rely heavily on anatomical examples and analogies. While the argument from design is certainly much broader than this and has been applied to numerous other phenotypic features, such as behaviors and host-parasite relationships (e.g., Cronin, 1991; Dawkins, 1987; Williams, 1966, 1992), this chapter will focus on the application of the argument from design as it relates to the specific areas of anatomical structure and physiological function.

One advantage of restricting the analysis to this area is that the mechanistic emphasis of biomechanics and physiology allows explicit specification of performance metrics such as efficiency, force output, and velocity of motion (Lauder, 1991b, 1994). Also, structural configuration can be quantified with considerable precision, and the physiological function of these structures can be measured directly to permit a direct evaluation of design performance. It is often more difficult to conduct comparative analyses of behavioral or ecological designs and their relative efficiencies. In addition, the causal implications of differences in design among species are often easier to determine in a mechanical system (where lever arms, masses, etc. can be measured precisely) than in ecological or behavioral comparisons, and we can thus minimize the problem of correlative results not possessing causal information.

II. The Argument From Design

A. The Classical Argument

The charm of the argument from design (or AFD) is considerable, and in its classical formulation, the logic of the AFD is as follows. The very existence of any object of complex structure or function implies the existence of a purposeful maker of that object. Hence, Paley (1836), in an oft-cited example, explained that finding a watch on the ground implies directly the existence of a watchmaker. It would be inconceivable to attribute the presence of a watch on the ground to the spontaneous assembly of materials into the exact configuration needed to make a working watch. However, should we happen to find a stone on the ground, we need not invoke any maker of the stone as it is common for stones to arise as a result of numerous geological processes. In the case of the stone, we can clearly identify a mechanical causal agent that could place the stone in the position that

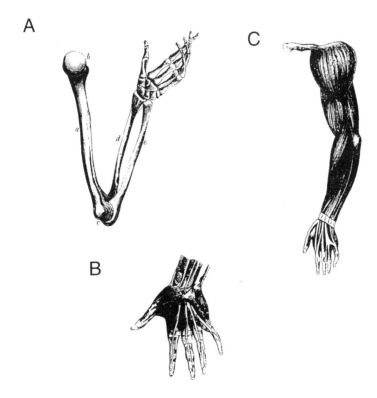

Figure 1   Representative plates from Paley (1836) who used the mammalian forelimb to illustrate biological design. (A): Human arm bones, including the humerus (a), radius (d), ulna (e), and hand. (B): Human hand showing the superficial musculature and flexor tendons to the fingers. (C): Human arm with skin removed to show the musculature.

we happen to find it, without the need to resort to a "designer" to construct the stone. A watch is considerably more complex and intricate than a stone, and Paley argues that it is this complexity that requires the invocation of a maker or "designer" of the watch.

Paley then compares the complexity of the watch to the complexity of organisms (Fig. 1; Table 1). He argues that just as the intricate watch required a designer, so does a complex organism. Organisms, then, provide evidence of a Creator and are one manifestation of His work.

There are differing interpretations of the logic of Paley's argument from design. If one accepts that Paley was constructing an argument by analogy, suggesting an analogy between how we conceive of the intricacy a watch and how we perceive the complexity of organisms, then Hume (1779) is often held to have demolished this argument more than 50 years prior to Paley. Thus, Hurlbutt, (1985, p. 169) was able to remark that "... I shall assume what I feel to be quite evident: the proposition that Hume's analysis destroyed the rational foundations of the design argument as it was commonly presented in the seventeenth and eighteenth centuries." In fact, Hume's "Dialogues Concerning Natural Religion" (1779) are valuable reading for modern biologists interested in the logical foundation of the argument from design. In his Dialogues, Hume describes a conversation among three individuals: Philo, Demea, and Cleanthes. Cleanthes argues in favor of the proposition that God is manifest in the world and that such manifestation is evident in the design of organisms (Table 1). That the Designer of the world is intelligent and magnanimous is shown by the intricacy and harmonious nature of organic design. Cleanthes directly uses an argument from analogy: human designers built a mechanical world, organisms are mechanical in design, so a Creator must have constructed organisms. A key feature of Cleanthes' argument is that the parts of organisms appear to be adapted to their functions as "means to ends"; how else could this occur without a Designer? It is the role of Philo in the Dialogues to destroy Cleanthes' argument, and he does so by showing that the argument from analogy is logically flawed.

Sober (1993, pp. 30-36) argues that Paley was in fact not arguing by analogy (hence Hume's critique has no force) and that Paley actually used two arguments as follows. First, a watch is a complex design for keeping time; either the watch was produced by a designer or by a random physical process. Second, organisms are of complex design which aids in survival and reproduction; either organisms have been constructed by a Designer or organisms are the result of random physical processes. Sober suggests that the Paley AFD amounts to a general statement that if you accept the first argument then it is plausible for you to accept the second. Since no one believes that the watch was the product of a random physical process, then you should be inclined to accept that organisms are the product of a Creator.

B. The Darwinian Argument

This dispute over the logical nature of Paley's AFD notwithstanding, it is clear that publication of the *Origin of Species* (Darwin, 1859) ushered in a new way of thinking about the design of organisms. Richard Dawkins has most clearly

TABLE I

Selected Statements from the Literature Illustrating Both the Teleological and Modern Argument from Design

Reference	Comment on the argument from design
1. Hume (1779, p. 109), writing as Cleanthes	"Look around the world: contemplate the whole and every part of it: you will find it to be nothing but one great machine, subdivided into an infinite number of lesser machines..." "All these various machines, and even their most minute parts, are adjusted to each other with an accuracy, which ravishes into admiration all men, who have ever contemplated them." "The curious adapting of means to ends, throughout all nature, resembles exactly ... the productions of human contrivance..."
2. Paley (1836, p. 18)	"...for, in the watch which we are examining, are seen contrivance, design, and end, a purpose, means for the end, adaptation to the purpose."
3. Rudwick (1964, pp. 34-35)	"...the detection of any adaptation in a fossil organism must be based on a perception of the machine-like character of its parts, and on an appreciation of their mechanical fitness to perform some function in the presumed interest of the organism."
4. Williams (1966, p. 10)	"A frequently helpful but not infallible rule is to recognize adaptation in organic systems that show a clear analogy with human implements. " "At other times the purpose of a mechanism may not be apparent initially, and the search for the goal becomes a motivation for further study. Adaptation is assumed in such cases...on the indirect evidence of complexity and constancy."
5. Pinker and Bloom (1990, p. 707)	"Evolutionary theory offers clear criteria for when a trait should be attributed to natural selection: complex design for some function, and the absence of alternative processes capable of explaining such complexity."
6. Thornhill (1990, p. 31)	"...adaptations are the long-term consequences of evolution by selection and thus understanding the functional design of an adaptation is synonymous with understanding how evolution by directional selection worked ..."
7. Williams (1992, p. 40)	"Adaptation is demonstrated by observed conformity to *a priori* design specifications." "The hand is an adaptation for manipulation because it conforms in many ways to what an engineer would expect, *a priori*, of manipulative machinery..."

expressed the Darwinian formulation of the argument from design in his book entitled "The Blind Watchmaker" (Dawkins, 1987). Dawkins argues cogently in favor of the view that the process of evolution by natural selection may act like a blind watchmaker. Basic biological principles of variation, heredity, and selection operate to generate biological design and complexity of structure without requiring a purposive guiding hand or a Designer with a view to the end product. Biological design arises as a result of selection and is modified in succeeding generations as selection acts on those designs that result from previous episodes of selection. Although mutation, which generates new variation, may be random with respect to currently available variation within populations, natural selection is not random. Selection acts on existing variation within the context of the current environment to build up biological design, modifying what has come before and not acting to generate design *de novo* in each generation. The potential transformation of any biological design is thus constrained by its past history of variation and selection. The role of Paley's purposive Creator/Designer/ watchmaker has been replaced by selection acting on heritable variation. Darwin's mechanistic explanation of organic design was surely a great intellectual achievement.

One might suppose that the presence of a mechanistic explanation for biological design such as that proposed by Darwin would have marked the end of the argument from design in modern biology. After all, given the sufficiency of a purely statistical/mechanical explanation for design, it would seem unnecessary to even present arguments derived from the way organisms are constructed.

C. The Modern Application

But in fact the AFD is alive and well, although certainly different in its particulars, if not in character, from what Paley invoked. In its modern formulation, the AFD centers around the recognition of adaptive characters and the inference that these traits have been produced by the process of natural selection. Instead of attempting to infer the action of a Creator from the manifest complexity of organisms, many evolutionary biologists seek instead to infer the action of selection from organismal design. Often it is the complexity of design itself that is held to provide evidence of the past action of selection (Table 1; Pinker and Bloom, 1990). A second criterion for the inference of adaptation is the resemblance of organic design to man-made design. The selection of comments from Rudwick (1964) and Williams (1966, 1992) presented in Table 1 illustrates this widespread view. If we can recognize in organisms a configuration reminiscent of how a human engineer might have designed a structure (enabling us to call the structure a "design"), then according to the modern AFD we can reasonably infer that selection acted to produce the structure and that the structure serves an adaptive function, aiding survival and reproduction.

There are many conceptual similarities between the modern AFD and the classical version presented by Paley. Indeed, some modern advocates of the AFD use Paley explicitly as a suggested guide for the study of adaptation. Thus, Williams (1992, p. 190) suggests that the works of Paley be studied as a means of understanding how adaptation may be recognized, and he approvingly reproduces portions of Paley's Natural Theology as a methodological guide for comparing biological design to man-made structures. A significant overall similarity in all arguments from design is the inference of a process (either natural selection or the action of a Designer) from a pattern (organismal design, however recognized). Indeed, the validity and the utility of the modern AFD depend critically on the extent to which one can reliably make such an inference.

Not all evolutionary biologists have accepted the modern version of the AFD or even use the term "design" to reflect the result of a mechanistic process that generates complex structure (whether by a Designer or by natural selection). Dawkins (1987) subtitled his book "why the evidence of evolution reveals a universe without design," implying that the very concept of "design" has been made obsolete for biological systems. Continued use of the argument from design in ethology prompted Ollason (1987) to proclaim in frustration that "...the idea that animals are designed is dead, killed by Hume, buried, perhaps unwittingly, by Darwin, but however comprehensively it is disposed of, like the walking dead it haunts us still."

Far more than simply haunting us, the modern version of the AFD is so pervasive that in many ways it may be said to be the most commonly used means of inferring the past action of selection (Table 1; Thornhill, 1990, 1996; Williams, 1966, 1992). And far from abandoning the use of the term "design," evolutionary biologists have instead relied heavily on this concept as a means of inferring selective forces.

In order to understand the logical structure of the modern AFD, it is necessary to consider what is meant by three key terms: adaptation, function, and design. Because each of these is used in many different ways in the literature, clarifying these concepts will be critical to understanding both the promise and the difficulties inherent in the modern formulation of the AFD.

1. Adaptation

While it has been widely noted that the term "adaptation" may refer to both a trait (an adaptation) and the process of becoming adapted (Brandon, 1990; Futuyma, 1986), this distinction is necessarily conflated in most discussions of adaptation (see Lauder *et al.*, 1993; Leroi *et al.*, 1994; Sober, 1984). Many workers accept some variation of the following historical definition of adaptation, and this is the definition I will use here: an adaptation is a trait that enhances fitness and that arose historically as a result of natural selection for its current biological role (Arnold, 1994b; Baum and Larson, 1991; Brooks and McLennan, 1991; Coddington, 1988; Gould and Vrba, 1982; Greene, 1986; Harvey and Pagel, 1991; Lauder *et al.*, 1993; Mishler, 1988; Sober, 1984). Thus, adaptation as a state is defined in part by the action of a mechanistic process, natural selection, acting directly on that trait currently deemed to be an adaptation. One implication of this definition is that neither a trait that resulted from correlated responses to selection on a *different* character nor a trait that results from random genetic drift would be considered as adaptations. Such traits might have fitness effects and they might even form an integral part of the phenotype, but they would not be adaptations. [A number of other terms have been proposed to describe traits of this type (Baum and Larson, 1991; Gould and Vrba, 1982; Griffiths, 1992), some of which will be considered later in this chapter]. To be an adaptation a trait must have resulted from direct selection for that trait. Not all workers accept this approach, however: other definitions of adaptation emphasize that an adaptive trait is one that enhances current fitness without regard to the mechanism that gave rise to the trait (Fisher, 1985; Reeve and Sherman, 1993) and consider the comparative fitness of extant variation in a trait as the means to identifying adaptation.

2. Function

The term "function" has an equally confused history in the literature, with biologists in different research areas using this term with very different meanings. Recently, a number of authors have attempted to sort out some of this conceptual confusion and clarify the multiple uses of this term (Allen and Bekoff, 1995; Amundson and Lauder, 1994; Bekoff and Allen, 1995; Griffiths, 1992; Lauder, 1994). Most evolutionary biologists use the term function as synonymous with selective advantage, suggesting that the function of a trait is that property of the trait that resulted in the trait having been selected for. Thus, the function of long limbs in ungulates might be given as "escape from predators." By this definition, a function is recognized as an effect of the process of selection [and hence may be referred to as the selected-effect (SE) definition of function (Amundson and Lauder, 1994)] and by the performance/fitness advantage incurred by the trait. Implicit in this definition is that the history of selection needs to be known in order to make an informed statement about function, and hence the SE definition of function is inherently historical.

On the other hand, functional morphologists and physiologists often use the term function to mean the use, action, or mechanical role of phenotypic features (Amundson and Lauder, 1994; Bock and von Wahlert, 1965; Lauder, 1994). Thus, a bone might have the mechanical function of stiffening the limb against gravity even though this may not be the effect or performance advantage on which selection acted during the origin of the bone. (For example, the bony components of the limb could have originated in an aquatic ancestor of the terrestrial taxon now under consideration.) This is a nonhistorical use of the term "function" that does not depend on identifying selection forces. The concept of anatomical function is essential to our discussion of the argument from design because it is the measurement (or estimate) of how structures are used by animals that most often is the basis by which inferences about adaptation and selection are made. The biomechanical analysis of anatomical structure and function provides the basis for comparing biological phenotypes with man-made devices, and thus is a tool by which one could possibly detect adaptation in biological design. It is analysis of the "machine-like character" of traits (Rudwick, 1964), "complex design for some function" (Pinker and Bloom, 1990), and "conformity to *a priori* design specifications" (Williams, 1992) that encourages the inference of adaptation and selection, and hence constitutes a key component of the modern argument from design (Table 1).

This chapter uses the phrase "mechanical function," "anatomical function," or "physiological function" to distinguish this nonhistorical meaning from the selected-effect use more common in evolutionary biology ("evolutionary function" or "adaptive function"). This will be a convenient way to describe the action of complex morphology in our analysis of the AFD below, without making any assumptions about selected functions of those structures. The biological context of structure and function is referred to as the "biological role" of a structure or mechanical function (Bock and von Wahlert, 1965; Lauder, 1994).

3. Design and the Argument from Design

The concept of design as used in the fields of biomechanics and functional morphology refers to an organized biological system which performs one or more mechanical functions (Lauder, 1982, 1995; Amundson and Lauder, 1994). Design in this sense is not necessarily equivalent to structure, but is a more general concept of structure in relation to function. However, the term design is often also used as shorthand for complex structure that appears to have some, possibly yet to be determined, function.

In evolutionary biology, the notion of function often has a teleological connotation that is closely related to the concept of "design" (Allen and Bekoff, 1995; Ayala, 1970; Griffiths, 1992). The "design" of an organism in evolutionary parlance often includes the adaptive (selected-effect) function of those traits that we wish to call a "design" (Dennett, 1995). Although the purposive connotations of design have engendered some controversy (e.g., Ollason, 1987), in general there is widespread recognition of the fact that natural selection acts to alter biological organization based on currently available variation and not with anticipation of an endpoint in the future.

The modern evolutionary incarnation of the argument from design is based either directly or indirectly on several propositions that are required to perform the research program suggested by the AFD. At its core, the AFD is an inference of process from pattern: the process and the action of natural selection are inferred from the complexity and configuration of structure. Most of the discussion about the AFD concerns allied research programs, as there are several means of analyzing biological design. How, empirically, do we recognize biological adaptation? How can we analyze complexity of organization and the relationship between biological and human design? What can we infer about the action of selection from an analysis of the phenotype alone? If the argument from design is to be more than a theoretical construct, we must be able to successfully conduct an empirical research program.

There are four critical components to the analysis of biological design. Not all of these will be present in any one analysis, and any given investigation may use aspects of more than one of these components.

First, we must be able to make accurate inferences of mechanical function from structure. Many authors writing about the AFD have stressed the relevance of understanding how organisms function as machines (Table 1). It is through such biomechanical analyses that we come to understand how organisms are constructed and how various components of the phenotype interact to generate behavior. And it is via an understanding of the relationships between mechanical function and structure that we propose hypotheses about evolutionary function, adaptation, and selection, key elements of the modern AFD.

Second, if organic design is to be compared to man-made implements as a means of inferring evolutionary function, then we must be able to specify design criteria. That is, given a mechanical function such as "detect prey in the water at a distance of 10 cm" as a goal, we must be able to abstract from the desired mechanical function basic elements of design that will meet that goal. Then we need to evaluate the extent to which one or more biological designs match those design criteria and thus are likely to have performed the hypothesized function. In addition, it would be enormously helpful in formulating AFD hypotheses to construct alternative designs and to assess the relative performance of those configurations. Are there several designs with equivalent performance that could perform the needed function?

Third, we must be able to infer evolutionary function from analysis of the phenotype alone. Many AFDs are based exclusively on a phenotypic analysis of a few individuals or a few groups of related taxa. Central to the modern AFD is the ability to make inferences of patterns of selection on traits so that we can say that a given trait is an adaptation for some specific evolutionary function. Indeed, we must be able to go further than this and distinguish between *direct* selection for a trait and alternative explanations for that trait: correlated response to selection on another character, drift, unmodified inheritance from the ancestral condition, and the like.

Fourth, a key to the successful implementation of the AFD is the ability to atomize complex collections of phenotypic features into relevant component parts. This is necessary in order to analyze individual components of design. If all parts of an organism function in such an interconnected and integrated manner that we cannot consider components in isolation, then we cannot apply the AFD to less than the entirety of the organism, the latter not normally being a feasible research program.

III. Problems With Applying The Argument From Design

This section treats the four issues discussed above in detail and evaluates the empirical sufficiency of each issue as it might relate to our ability to apply the AFD to a specific case. The general theme of this section is that there are considerable difficulties associated with all four empirical components of the AFD, and that these difficulties place considerable constraints on applying this argument in practice.

A. Structure And Mechanical Function Are Not Tightly Matched

The notion that structure and function are closely linked has a long history in biology dating from at least the time of Aristotle (Russell, 1916). Indeed, some of the leading figures in the history of biology, such as Cuvier, have been strong believers in the matching of structure to function, and this view is held by many to this day [(Arber, 1950; Dullemeijer, 1974; Rudwick, 1964); also see papers in Thomason (1995)]. Certainly the ability to predict the mechanical function of a structure from an analysis of the structure alone has a strong appeal, if only because we could then avoid having to measure the physiological function of structures directly.

Unfortunately, as functional morphologists have begun experimental investigation of the relationship between structure and function, a large number of complexities have been discovered that render the structure-function relationship more obscure than previously supposed (Gans, 1983; Lauder, 1995).

Much of the discussion of structure-function relationships tends to be highly qualitative and assumptive in character, drawing on expected patterns of function or on hypothetical relationships that appear likely to be true based on knowledge from human engineers. In order to evaluate the precision of the relationship between structure and function it is vital to conduct experimental analyses and directly test predicted relationships.

I have previously argued that the predictability of mechanical function from structure may be scale dependent (Lauder, 1995). At a histological level of analysis, the relationship between cellular structure and function appears to be reasonably predictable. For example, myofibrillar cross-sectional area can be used with a high degree of accuracy to predict maximal muscle contractile tension; changes in membrane surface area have predictable effects on diffusion rates. At the general level of behavior and ecology, numerous examples exist of well-founded general inferences of traits such as habitat (arboreal or aquatic) or diet (hard or soft prey) based on organismal structure. Although we may in fact be ignorant of many ecological complexities when we make such broad comparisons, good success has been obtained in inferring general dietary characteristics. It is at an intermediate level, of which musculoskeletal function acting to generate behavior is one good example, where the relationship between structure and function becomes much more obscure. Yet it is at this level where anatomical structures may be studied with relative ease in both living and fossil taxa and predictability is most often desired. It is also at this gross anatomical level where the argument from design, from Paley (1836) to Williams (1992), has often been made.

Here I briefly present two case studies from the experimental functional morphology of vertebrates that exemplify the complex character of structure-function relationships.

1. Intraspecific Differentiation in Structure and Function

Many vertebrates that eat hard prey possess molariform teeth which act as crushing surfaces for the application of forces from the (usually) hypertrophied jaw muscles. Sunfishes (Centrarchidae) provide a nice example of durophagy (consumption of hard prey) as several species are known to eat snails and to possess both molariform teeth and hypertrophied pharyngeal muscles. In addition, we know from experimental studies of muscle function in which direct electromyographic recordings were made of the relevant jaw muscles, as both hard and soft prey were being eaten, just when the muscles are activated by the nervous system during the crushing behavior (Lauder, 1983a, b). In addition to measuring the structures associated with durophagy, we also quantified the pattern of muscle activity used by fishes in different populations to discover if there was differentiation in the mechanical function of the muscles that perform the crushing behavior. One might expect to find a good correlation between the presence of hard prey in the diet (in this case, snails), hypertrophy of jaw muscles, and changes in function of the hypertrophied muscles across populations that vary in the proportion of hard prey in the diet. Is there, at this interpopulational level, a positive relationship between structure and function?

There is indeed a clear relationship between the morphology of the teeth and muscles and the ecological character of the presence of snails in the diet in

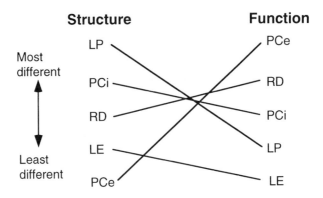

Figure 2   Comparison of structure and function in the pharyngeal jaw bones and muscles of the pumpkinseed sunfish (*Lepomis gibbosus*). The left column lists several different muscles that control the crushing behavior of the pharyngeal jaws. Two populations were compared that differ in the extent to which snails are an important part of the diet, and the muscles are ranked within each column by the magnitude of the difference among populations. Thus, the LP muscle differed most in structure between the snail-eating and the nonsnail-eating populations, while the PCe differed least between these two populations. A similar ranking was performed for muscle function. Note that little correlation exists between structural and functional differentiation in the muscles of these two populations. Muscle abbreviations: LE, levator externus; LP, levator posterior; PCe, pharyngocleithralis externus; PCi, pharyngocleithralis internus; RD, retractor dorsalis.

different populations of these fishes (Wainwright *et al.*, 1991a, b). We found that the two populations studied were significantly different in their muscle activity patterns (Wainwright *et al.*, 1991a). However, the surprise was that the pattern of differentiation in structure and function among muscles was not consistent between populations: muscles that were most different in structure were least differentiated in function. This result may be visualized (Fig. 2) by ranking the level of differentiation between populations in muscle structure in one column (from most different at the top to least at the bottom) and comparing it to the level of differentiation in mechanical function ranked similarly in a second column [see Wainwright *et al.* (1991a) for details]. Connecting homologous muscles across columns shows clearly the discordance between structure and function. This result was surprising because it suggests that even in an intraspecific study there is little predictability of function from structure at the level of gross anatomical muscle and bone function.

2. Musculoskeletal Design and the Causes of Behavior

Behavior results from patterned output from the central nervous system to musculature. This output, in conjunction with physiological properties of the musculature and the mechanics of the arrangement of muscles and bones, determines the observed pattern of movement that we call behavior. The analysis of musculoskeletal function provides one way of addressing the causal

A

Primitive behavior: flexion then extension

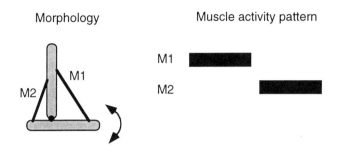

B

Derived behavior: extension then flexion

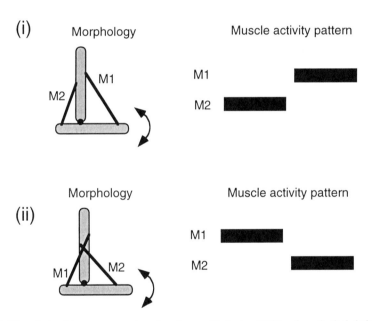

Figure 3 The relationship among structure, function, and behavior. Within a hypothetical clade, both primitive (A) and derived (B) behaviors are observed. The behavior primitive for this clade is "flexion then extension" which is achieved by the activation of muscles (M1 and M2). Muscle M1 is activated first (indicated by the black bar) to effect the flexion movement of the arm, and then muscle M2 becomes active to extend the arm. The upper arrow indicates flexion movement and the lower arrow extension. The derived behavior shown in B is "extension then flexion" which may be achieved by novelty in either (i) the function (timing) of muscle activity or (ii) the topology of muscles. In B(ii), activation of muscle M1 now causes extension while activation of muscle M2 results in flexion. Modified from Lauder and Reilly (1996).

(mechanistic) bases of evolutionary changes in behavior and of illustrating the complexity of the relationship among structure, function, and behavior.

If two species differ in some aspect of their behavior, then these differences might have resulted from changes in either musculoskeletal structure (topology) or in neural output to the muscles that produce the behavior. Figure 3 illustrates this concept using a simplified vertebrate forelimb (humerus and forelimb bones and associated muscles). Given a monophyletic clade, we might investigate the pattern of movement (behavior) in each of the species and determine that the primitive condition in the clade is the behavior "flexion then extension": that is, the forearm flexes by moving dorsally (up) and then extends by moving ventrally (down). Thus, the forelimb in most species in this clade moves dorsally around its articulation with the humerus in a flexion motion, followed by ventral rotation in an extension movement (Fig. 3A). This behavior is caused by the activation of two muscles (M1 and M2) in sequence. Now imagine that within this clade we discover a new species that exhibits the novel behavior of "extension then flexion." This novel behavior could be produced by either of two possible mechanisms. Changes in muscle *function* (the timing of activation) certainly could generate the new behavior by altering the sequence in which each muscle is activated and hence generate the novel behavior [Fig. 3B(i)]. But changes in musculoskeletal *structure* could also produce the same new behavior. The primitive pattern of muscle activity in conjunction with a new arrangement of muscles (in which muscle M1 now attaches to the posterior of the forearm, while the muscle M2 insertion has moved anteriorly) also could generate the new behavior [Fig. 3B(ii)]. Muscle M1 now acts to move the forearm ventrally in a flexion motion and is activated first, while muscle M2 elevates the forearm and is activated next.

This theoretical example illustrates the point that changes in either musculoskeletal structure or function may occur and that changes in structure and function may be dissociated from each other during the evolution of novel behaviors. *There is no obligatory historical coupling between novelty in structure and function; there is no reason why structure and function at any one level must evolve in concert.*

Here I provide a case study of musculoskeletal function and its relationship to behavior by summarizing the results of a study of the jaw muscles and bones of salamanders and their physiological function during aquatic prey capture. Some background is needed before the major conclusions can be presented. Aquatic feeding occurs in several salamander families and involves the coordinated action of many cranial muscles to cause a rapid expansion of the head and the subsequent creation of a current of water (carrying the prey) into the mouth (Lauder, 1985; Lauder and Shaffer, 1985; Reilly and Lauder, 1992). There is considerable diversity in the structure of the head of salamanders in different clades that exhibit aquatic feeding, and the goal of this analysis was to study four taxa to determine if interspecific patterns of variation in mechanical function and feeding behavior matched structural differentiation [details are provided in (Lauder, 1995) and (Reilly and Lauder, 1992)]. We quantified the structure of the musculoskeletal system of the head, the function of the head muscles (by recording muscle activity electromyographically), and behavior (by obtaining high-speed video recordings of prey capture); each class of traits was quantified by measurement of a number of variables. We were thus able to make a multivariate assessment of behavior, function, and structure and to assess the extent to which

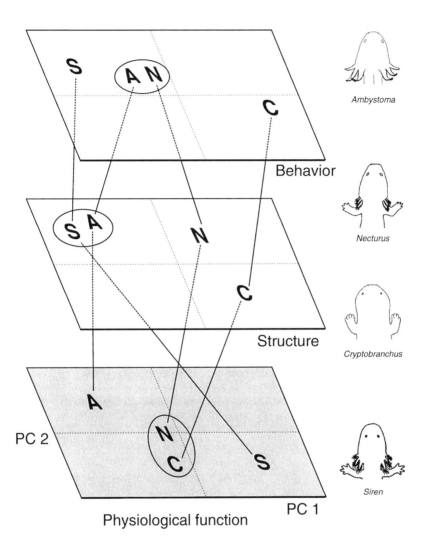

Figure 4 The results of three principal components analyses of physiological, structural, and behavioral data on four taxa of salamanders: A, *Ambystoma*, C, *Cryptobranchus*, N, *Necturus*, and S, *Siren*. The principal component plots of PC 2 versus PC 1 summarize the pattern of variation for each data set. Dorsal views of the head of each taxon are shown on the right. The symbol plotted on each plane for each taxon represents the mean for all the individuals studied in that taxon. Lines connect taxa on adjacent planes to show the mapping across levels of analysis. Note that there is no clear relationship among the positions of taxa in a single plane to their positions on other planes, making the prediction of function and behavior from structure alone difficult. Modified from Lauder and Reilly (1996).

structure and function coevolve in these salamander taxa. To enable visualization of the relationships among characters and taxa, the relationships among characters at any one level (function, structure, or behavior) have been reduced to two dimensions by conducting a principal components analysis on each class of data separately.

Figure 4 illustrates the pattern obtained from these data. Each plane represents the results of the analysis of one class of characters, and the position of the four taxa in each plane is determined by the mean value for all the individuals studied for that taxon. Taxa that are circled occupy positions in principal component space that are not significantly different from each other (based on a MANOVA of PC scores), and lines connect taxa in adjacent planes to illustrate the mapping across levels. Thus, in the muscle activity (motor pattern) plane at the bottom of Fig. 4, *Necturus* (N) and *Cryptobranchus* (C) possess patterns of muscle activity during prey capture that are not significantly different from each other. *Siren* (S) and *Ambystoma* (A) both possess novel muscle activity patterns that are significantly different from each other as well as from N and C together. The principal component analysis summarizing the morphological data shows a different pattern of interspecific differentiation in structure from that seen in the function plane. Here, *Siren* and *Ambystoma* possess similar structures of the cranial musculoskeletal system, while *Necturus* and *Cryptobranchus* are morphologically different from both each other and the *Siren–Ambystoma* group. Finally, at the behavioral level, a third pattern of interspecific differentiation is observed, where *Ambystoma* and *Necturus* show similar patterns of feeding movements.

The data summarized in Fig. 4 illustrate three results of special interest. First, *Ambystoma* and *Siren* possess different feeding behaviors but similar morphology. Due to the mechanistic nature of musculoskeletal function, this could only be true if these two taxa differed in the pattern of muscle activity: given similar structure, differences at the level of physiological function must exist to generate different behaviors. This is indeed the case, as the motor patterns of *Siren* and *Ambystoma* are significantly different. Second, two taxa, *Ambystoma* and *Necturus*, show different patterns of grouping in each of the three levels. These two taxa differ in both morphology and muscle function, but these differences interact mechanically to produce similar behaviors. Third, these data illustrate that there is not a simple mapping across levels as might be anticipated if the structural and functional components of musculoskeletal design evolved in a predictable, regular manner. It is clear that an analysis of any one level alone is an insufficient description of the design of the feeding system in salamanders, and that prediction of behavior or physiological function from structure alone in this case study is effectively impossible.

3. Conclusion

Numerous other examples could have been chosen to illustrate the point that structure and function exhibit a complex relationship that makes prediction of function from structure a very hazardous proposition (e.g., Gans, 1983, 1988; Lauder, 1991a, 1994, 1995; Vermeij and Zipser, 1986). Without an assessment of mechanical function independent of structure, we cannot hope to avoid circularity in our efforts to understand organismal design. And yet, what structures actually do, how they are used, and how they perform is an integral component of arguments about the history of organismal design and about the mechanistic causes of design.

B. Relevant a Priori Design Criteria are Rarely Identified

1. Engineering Models and Mechanical Function

The claim that relevant design criteria can be specified *a priori* to allow the analysis of biological design amounts to a claim that we can specify in advance the problem or problems that the design is supposed to solve. Although it is almost always possible to specify *some* design criterion, the more complex the design, the less likely is it that we will be able to determine what the relevant performance and mechanical functions are that any given structure needs to solve. And furthermore the less likely is it that we will be able to meaningfully weigh alternative performance goals. Models of optimal foraging, sex ratio proportions, and life history evolution have used a few carefully chosen design criteria such as minimal foraging time as the performance goal to evaluate ecological design (Beatty, 1980; Orzack and Sober, 1994a, b; Parker and Maynard Smith, 1990), and analyses of plant biomechanics have been successful in using engineering criteria to evaluate alternative designs (Niklas, 1986, 1992). While even some of these efforts have been criticized (e.g., Rose *et al.*, 1987), such endeavors have been still less successful in other arenas (Lauder, 1995).

The major problem facing an anatomist in specifying design criteria is that the anatomical system of interest is often so complex and the "degrees of design freedom" so high that a virtually infinite number of possible performance goals may be specified to explain the very large number of anatomical traits. In cases where biological structures operate in primarily two dimensions and the number of elements is low (reducing the number of degrees of freedom), it will be easier to use engineering theory to determine design criteria and hence to evaluate performance. In practice, the necessary design specifications are most often simply assumed, and based on this assumption the argument from design is used to suggest an evolutionary function. A particularly good example of this is the human hand. Williams (1992, p. 40), for example, with reference to human anatomy, asserts that "the hand is an adaptation for manipulation" because it conforms to an engineer's *a priori* specification for a manipulative machine (Table 1). Let us examine this claim.

First, what are the structural components of human hand design that call for explanation? The human hand possesses 26 separate bones (Fig. 5; for our purposes I define hand structure as those elements distal to the radius and ulna, although there are of course structures such as nerves, blood vessels, and tendons that pass from the forearm into the hand). Some of these bones articulate with each other to form five separate fingers, elongate combinations of phalangeal skeletal elements that articulate with each other in ball and socket joints and which are capable of some independent movement. Five other bones, the metacarpals, articulate both with the phalanges and a collection of seven carpal bones grouped together at the base of the wrist (Fig. 5). Numerous ligaments restrict motion of the hand bones. For example, the transverse metacarpal

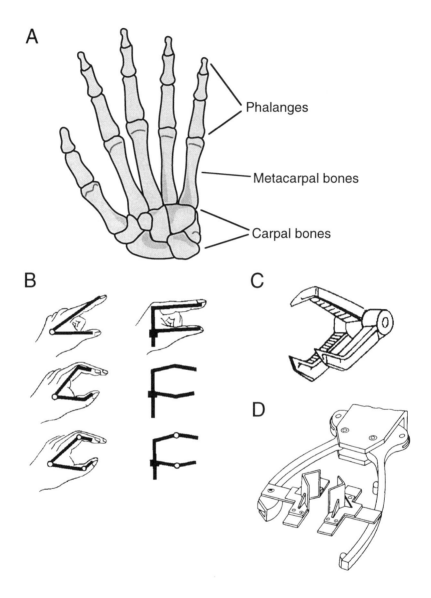

Figure 5 (A): The major osteological components of hand design. This diagram shows the human left hand in ventral view. (B): Diagram of a human hand on the left and a mechanical schematization on the right to show one possible mechanical abstraction of the human hand. (C): Design of a mechanical hand with three fingers. (D): Design of a mechanical hand with two fingers which move to grip cylindrical items in the center. Panels B, C, and D are modified from Kato (1982). Reproduced by permission. All rights reserved.

ligaments tie the distal tips of the metacarpals to each other, while the collateral capsular ligaments of the phalanges restrict motion at the knuckle joints to primarily one plane.

The hand also contains about 18 intrinsic muscles, limited to the hand itself, which interconnect skeletal elements. The various interosseus muscles, for example, act to abduct, adduct, flex, and extend the fingers. The lumbrical muscles of the hand take their origin from tendons of the flexor digitorum profundus (whose belly lies in the forearm) and these muscles both flex and extend the fingers at separate joints. Additional muscles abduct and adduct the thumb and little finger. Finally, the hand contains an extensive network of sinovial sheathes and bursae (fluid-filled sacs) through which tendons run.

Second, exactly which components of hand structure described above are designed specifically for manipulation? Has the hand in its entirety resulted from selection for manipulative function, or, have only some components of this design been selected for manipulation? If only some components were so designed, which ones? Is it perhaps only the muscles, bones, and sinovial sheaths of the thumb that have been designed by selection for manipulation, or is it the combination of the palmar interosseus muscles, the dorsal intercarpal ligament, and the triquetral carpal bone that we should focus on? We cannot gloss over the problem of identifying those elements that were subjected to selection for manipulative function because we must have a biological design to compare to our engineering model: the closer the fit, the more likely our argument from design is correct.

Third, which engineering model are we to choose to represent the function of manipulation? We might choose to develop a specific mechanical model based on a series of rigid elements, articulating with each other, controlled by a series of cables to permit motion of the various "joints," and capable of movement in opposition (such as our thumb and fingers) so that individual objects may be picked up. Or, we might simply specify general engineering principles: that manipulation can be accomplished by any design with independently movable parts connected in a way that permits objects to be grasped and controlled. But surely we must go farther than this and specify the model precisely enough to generate a predicted design that one would expect *a priori* to exist when manipulation of objects is a problem to be solved.

Fourth, we must ask if there are other ways of designing a manipulative structure. What is striking about the mechanical engineering literature on manipulative function is the extent to which the engineering solutions themselves have been dictated by existing *human* anatomy and used multiple jointed elements as key features of the design for manipulation (Fig. 5: B, C, D). Everything from remote manipulation arms in deep-sea research submarines to artificial forelimbs appear derived from a human model (Kato, 1982), and it is thus circular to argue that mechanical hands provide evidence of "good" biological design. It could be argued that this similarity is in fact evidence for the engineering design of the human hand itself, as there may only be one "good" design. But if alternative designs for manipulation exist, then this argument fails.

Octopus arms are an excellent manipulation device that embody well-understood engineering design concepts using the principle of a muscular hydrostat and layers of circular and longitudinal muscle (Kier and Smith, 1985). There is little about this manipulative design that can be claimed *a priori* as being

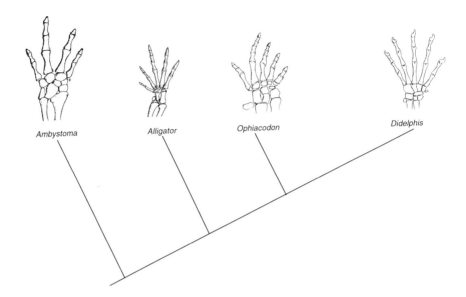

Figure 6 Phylogenetic relationships of several selected hand designs to show that many design elements of the human hand are in fact ancient features of the vertebrate forelimb. Such ancient features (such as multiple jointed digits that are capable of largely independent motion) cannot be used to argue for adaptive design of those features in the human hand. Note that even taxa as phylogenetically disparate as the salamander *Ambystoma*, the pelecosaur *Ophiacodon*, and the opossum *Didelphis* show many similarities in hand design. The diagrams show the left hand in dorsal view [diagrams modified from THE VERTEBRATE BODY, Fifth Edition by Thomas S. Parsons and Alfred S. Romer, copyright © 1977 by Saunders College Publishers, reproduced by permission of the publisher].

irrelevant to the development of a general engineering model for manipulation: octopus arms contain no specialized anatomical elements that could not in theory develop in any multicellular organism. In addition, octopus arms do not contain jointed bony elements and yet are capable of fine manipulative function, especially when several arms are used in concert. If we choose the octopus arm model, then it will indeed be difficult to argue that the human hand is the only possible engineering design for manipulation. If there are alternative designs for manipulative function, then which of the available models will we choose? And how will we ensure that we have not biased our choice of model by the knowledge we already have about the biological design that we are trying to explain?

Fifth, on what basis has it been decided that manipulation is the problem to be solved? Perhaps various components of the human hand have resulted from selection for grasping function (perhaps the grasping of branches during arboreal locomotion). Some of the anatomical features that characterize the design of the human hand today may have been incorporated into hand design to "solve" problems that were experienced by ancestral populations. For example, many of the ligaments that bind hand bones together might be designed to solve problems relating to ancestral locomotor styles. Finally, perhaps certain aspects of current hand structure evolved as a result of selection for more precise communication using hand signals. Generating *a priori* engineering designs that distinguish between grasping and manipulation will be a challenge, as will ruling out other possible *a priori* functions that might be proposed for the hand.

Despite the manifold difficulties that occur when attempting to specify in advance the engineering problem to which any anatomical design is a solution, there is at least one method that can be used to rule out particular structural features as possible design elements for current function. A phylogenetic analysis of tetrapod hand structure, for example, can at least provide some guidelines as to which structural components of the human hand are *not* likely to be adaptations to manipulative function. A highly simplified phylogeny with representative schematic illustrations of hand osteology is shown in Fig. 6. From this phylogenetic pattern is it clear that possession of independently mobile jointed elements ("fingers") is not a design component that could be linked to any specific function that is unique to the human hand: fingers are an ancient design feature of the vertebrate forelimb (Coates and Clack, 1990) and occur in many animals that do not have the manipulative abilities of the human hand. Neither can the presence of carpal bones be adduced as a design feature of the human hand that arose by selection for manipulative function as these too are ancient elements. Indeed, the basic structure of the tetrapod forelimb predates fully terrestrial vertebrate life (Eaton, 1960; Edwards, 1989) and cannot even be used to support a claim that forelimb design arose as a result of selection for support against gravity and locomotion on land.

With a phylogenetic analysis of hand design in humans and closely related outgroup taxa, we could at least narrow down the possible set of design features for which explanation is required: traits unique to those taxa which share the purported mechanical function. That is, the relevant anatomical structures should possess a concordant phylogenetic distribution with the proposed function. But this would not alleviate the requirement of (1) demonstrating experimentally that the function was in fact present in the relevant taxa, (2) producing an engineering model that demonstrates that the design features actually do solve the functional problem, and (3) rejecting other plausible models of function.

None of these arguments change the fact that the human hand might actually be *used* for manipulation. But our observation that the human hand is currently used for manipulation in no way supports an *a priori* engineering argument that the hand was *designed* for manipulation (or any other function for that matter) as a result of natural selection for that function. In fact, no engineering argument has been presented at all, alternative functions have not been ruled out, and manipulation as the engineering "problem" that must be "solved" by natural selection is simply assumed.

2. Spandrels and the Assumption of Design Constraints

The problems in specifying design criteria for biological structure outlined above are not restricted to cases in which we might like to argue that some design resulted from selection for one specific physiological function. A similar difficulty exists in attempting to claim that a particular trait is an accidental by-product of construction and in fact has not arisen as a result of selection for a specific function.

Perhaps the most famous analysis of a design that is held to have arisen as a by-product of construction, one that spawned numerous rebuttals, much soul

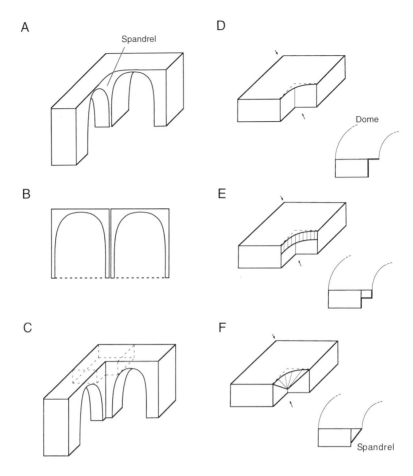

Figure 7 The construction of a spandrel [more properly called a pendentive, Mark (1996)]. In (A) and (C) the domed roof has been omitted for clarity. (A): A spandrel is formed by the space between two perpendicular adjacent arches which is filled in and frequently decorated. These arches arise in walls perpendicular to the floor with a curved dome above forming the roof. (B): Two arches juxtaposed within rectangles in the same plane. (C): Bending the two arches from the previous panel into a right-angle intersection leaves an upper corner space (dotted lines) that does not contain a spandrel. Note that the walls containing the arches meet at a right angle from floor to ceiling. (D), (E), and (F) show three alternative designs for putting a domed roof onto a corner intersection of two walls (whether they contain arches or not). In the cross-sectional views shown in panels D, E, and F thick lines accentuate the surface facing into the room. The portion of the wall illustrated is that outlined by the dashed lines in C; the dome is again not shown for clarity. The inset for each of these panels shows a crosssection through the region supporting the dome (in the plane defined by the two arrows) and one possible configuration of a domed roof. (D): By curving the supporting surface above the corner, a curved dome can be attached above a right-angle junction of two walls. (E): The curved addition onto the corner can be extended toward the floor, increasing the amount of supporting material under the edge of the dome and providing a surface for decoration (vertical lines) that can be seen from the room below. (F): A spandrel may be constructed by adding material in the corner in a cone shape that tapers from the curved interface with the dome above to a point below. This design also allows for decoration. A spandrel is thus not an obligatory element arising necessarily from building two arches at right angles to each other: spandrels are added design elements and do not arise from mechanical necessity.

searching, new analyses (Bock, 1980; Brandon, 1990; Coddington, 1988; Dennett, 1983; Gans, 1988; Garland and Adolph, 1994; Harvey and Purvis, 1991; Lauder *et al.*, 1993; Mayr, 1983; Mitchell and Valone, 1990; Reeve and Sherman, 1993; Ridley, 1983; Wenzel and Carpenter, 1994), and even a book on literary critical analysis of the paper itself (Selzer, 1993), is the engineering case presented by Gould and Lewontin (1979) of the spandrels of the San Marco church in Venice. In brief, Gould and Lewontin use spandrels to argue that the traditional optimization research program in evolutionary biology does not allow for design features that may arise as purely constructional artifacts. These design features may indeed be coopted by organisms (and architects) and subsequently put to some (adaptive) use. But Gould and Lewontin suggest that we need to develop a research program that enables us to recognize that the primary origin of some components of design may have nothing to do with selection for a specific function.

The architectural example used by Gould and Lewontin (1979) to illustrate their point is the spandrel. When two arches are brought together with their bases at right angles, a triangular area is formed between the top of each arch and a domed roof (Fig. 7A). Note that the arches arise perpendicular to the floor, and that the dome sits on top of these vertical walls which contain arches. This triangular space, which is decorated to great effect in the church of San Marco, is the spandrel (or pendentive; Mark, 1996). In addition, note that the surface area of the spandrel is not parallel to either of the walls containing the arches and is at roughly a 45 degree angle to the plane of each arch. The surface of the spandrel faces out into the room, and the spandrel actually occupies a triangular volume extending from the wall behind on each side out to the decorated surface. Gould and Lewontin argue that spandrels arise from architectural or constructional constraints. Spandrels must be formed when two arches are brought together, and it would be incorrect to view them as primary (adaptive) design features of the church: they are a "secondary epiphenomenon" (Gould and Lewontin, 1979, p. 584) that should not receive an adaptive explanation. Any attempt to argue adaptation from spandrels would be erroneous. It is important to realize that Gould and Lewontin's analysis of spandrels is primarily an engineering one: they suggest that there is simply no other way to design a dome mounted on two arches without forming a spandrel. The term spandrel has even passed into the biological literature to mean a trait that arose nonadaptively as a by-product of constructional necessity [see Pinker and Bloom (1990) and the discussion articles that follow].

But the argument that spandrels are secondary by-products of constructional principles is similar to the argument discussed above that the function of the human hand is manipulation. Both rely not on actual engineering models or comparative mechanical analyses which the reader can evaluate, but on the *assumption* of a mechanical function (or the lack thereof). Let us examine the claim that spandrels arise from constructional necessity.

Even a simple design analysis reveals that spandrels are not an obligatory architectural construct associated with two arches [as first pointed out by Dennett (1995) who discusses other architectural implications of the Gould and Lewontin paper]. Figure 7B shows (in two dimensions) two arches aligned in the same plane but with some construction material such as concrete filling in the spaces between each arch and the edge of a rectangle enclosing the arch. Pushing the two arches together so that one edge of each rectangular area meets results in an expanse of concrete between the two arches [a proper spandrel as defined by Mark (1996)]. If we now bend the two arches so that they form a right angle to each other and view the result in three dimensions (Fig. 7C), we can see that the two walls containing arches can meet at a right angle with no necessary triangular space being formed. The narrower the arches, the greater the cross-sectional area of the column supporting the roof and thus the less stress (force per unit area) on that column. Widening each arch to reduce the surface area of concrete between the arches (and hence the cross-sectional area of the column) does not produce a necessary triangular space, but rather a smaller corner.

Now consider the ways in which a domed roof might be added above the walls by focusing on the region of the corner marked in dotted outline in Fig. 7C. We might simply add a dome-shaped roof by curving the region where the dome meets the walls, leaving a small ledge in the corner (Fig. 7D). This would create a dark area of shadow (if light emanated from lights or skylights in the dome) under the curved region above the corner. Alternatively, we could design the dome to curve around the corner with a supporting shelf (Fig. 7E). This too leaves a ledge in the corner, but now at least there is a curved surface facing out that could be used for decoration. There is still an area under the ledge that might be dark and unsuitable for decoration. Finally, we could fashion a decorative spandrel-shaped region in the corner (by adding material below the dome) with the base of the dome forming the curved top of the spandrel (Fig. 7F). To orient the surface of the spandrel so that the decorated surface might be seen from below, we might taper the lateral edges of the spandrel so that they meet at the junction of the two walls (Fig. 7F).

Each of these designs will have engineering strengths and weaknesses, but they do represent valid alternative designs to the "problem" of designing the space formed by a roof and two walls. It is hardly true that a spandrel *must* be an incidental artifact of construction (Dennett, 1995). Indeed, as Fig. 7F shows, construction of a spandrel requires deliberately *adding* material to the corner to build up a surface that faces down and into the room. In fact, Mark (1996) argues that this additional material behind the curving face of the pendentive is functionally significant in supporting the dome itself. Hence, a spandrel could be a structure designed to allow decoration of what would otherwise be a dark corner, and might be a functional necessity for dome support. The spandrel would in this case be an "adaptation". Dennett (1995, p. 273) even suggests that the spandrel might represent the minimal surface area that could be stretched across the corner space, and hence represent an optimal solution (using minimal added material) to a decorative problem.

I have discussed both the spandrel and hand examples at some length to illustrate a general feature of the design literature from two perspectives: *a priori* design specifications are usually assumed and rarely defined and analyzed. This is not surprising, as deriving an appropriate engineering model, even if one can divine the functional problems to be solved, is no easy task given the complexity of biological design. Even in the relatively simple case of the spandrels, where there are few structural elements, we have not exhausted the range of possible configurations, nor have we conducted any mechanical analyses of force transmission from the dome to the supporting columns. It may well be that the configurations illustrated in Fig. 7 (panels D, E, and F) have differing performance under load and that the performance ranking of the designs might change as we change testing parameters (such as altering the configuration or weight of the roof).

In practice, using the criterion of *a priori* design specification is likely to be applicable under only certain restricted circumstances. Even Williams (1992, p. 41), who strongly advocates the construction of *a priori* designs as a key method for analyzing adaptation, admits that "...those who wish to ascertain whether some attribute of an organism does or does not conform to design specifications are left largely to their own intuitions...." This is hardly a desirable situation for a key methodological underpinning of the argument from design.

C. Inferring the Retrospective Action of Selection from the Phenotype Alone is Difficult

Successful application of the argument from design requires us to be able to make accurate inferences about the process of selection from an analysis of the phenotype. After all, a component of design is an adaptation only if it enhances fitness and arose historically as a result of natural selection on that trait for its current function. Hence, if we are to recognize design by selection we must be able to determine if the process of selection acted on the trait of interest.

Under the best of circumstances, evolutionary biologists can indeed make strong inferences about the action of selection. The best of circumstances include the availability of longitudinal data on extant populations (Arnold, 1983, 1986; Endler, 1986), study of selection in laboratory evolution coupled with genetic analysis of intertrait correlations (Lande, 1978, 1979; Rose, 1984a, b), and the execution of replicated field evolution experiments (Reznick, 1989; Reznick and Bryga, 1987). These conditions do not often hold when the argument from design is made. In fact, the AFD is often applied retrospectively and to taxa in which genetic data and manipulative laboratory and field experiments cannot be done. Imagine, for example, that we have discovered a new species of fish in a geological formation of Mesozoic age and wish to ascertain if a particular structure (such as a conspicuously enlarged caudal fin) is an adaptation or not. How are we ever to know if selection acted directly on the caudal fin? Specifically, there are four major reasons why such retrospective analyses of selection pressures present difficulties for the inference of selection.

First, if the trait of interest is novel, with few available analogs in extant taxa, then we will have considerable difficulty in designing mechanical/engineering models to guide us in estimating the mechanical or evolutionary function of that trait. In cases where structural analogies can be made with extant taxa, then alternative hypotheses of mechanical function can at least be developed by using engineering approaches. But even such analyses are fraught with difficulty because of the large number of assumptions that must be made about the nature of the functional "problem" to be solved. Many structures may have been designed for such problems as signaling during courtship which do not permit easy mechanical modeling. Also, argument by analogy to extant taxa has conceptual problems: we are forced to rely on structures that are analogous to infer analogous functions, which assumes that structure and function are closely linked. I have argued above that this is a hazardous proposition [also see Lauder (1995)]. We cannot reason from the stronger base of homology of structure.

Second, many structures have more than one mechanical function and may have been subjected to multiple selection pressures in the past. If we can demonstrate using an engineering model that a given structure might have had several mechanical functions, then how are we to choose which function (or functions) should be the one(s) to which we ascribe evolutionary significance by labeling the structure an adaptation? For example, let us say that we wish to investigate the potential significance of the enlarged caudal fin of our fossil fish and propose two likely mechanical explanations: large fin area might increase maneuverability during slow speed swimming (by allowing longer wavelength undulations of the fin rays) or it might enhance escape performance by increasing surface area and hence thrust. If a design analysis shows that both are possible mechanical functions, how can we choose which problem or problems the structure was designed by selection to solve and which was the incidental function?

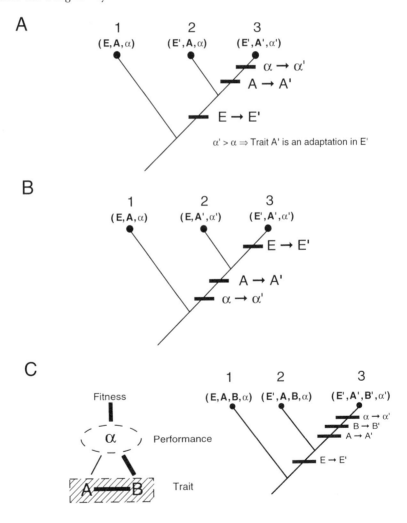

Third, selection may not have been responsible for the structures of interest in the first place, and if so we cannot assume that these structures are adaptations. Lande (1976, p. 333), for example, examined the extent of phenotypic change in tooth traits of Tertiary mammals and concluded that, "Using statistical tests, it is found that the observed evolution of these mammalian tooth characters could have occurred by random genetic drift in rather large populations, with effective sizes in the tens or hundreds of thousands." If a statistical model of drift cannot be ruled out as an explanation of a trait, then claiming that the trait is an adaptation will require a substantial body of ancillary evidence to justify adducing selection as the causal agent of design.

Fourth, even if we can with some confidence estimate a set of possible mechanical functions of a trait, how are we to know that selection acted on *that trait* and not on some other character with which the trait of interest is correlated. Thus, the enlarged caudal fin might have been purely a correlated response to selection for increased body size, selection for increased pectoral fin size, selection for reduced vertebral number, or a response to selection on any one of a myriad of other characters correlated either positively or negatively with caudal fin area. If so, the large caudal fin cannot be called an adaptation for either locomotor escape performance or for maneuverability, *even if* a biomechanical analysis shows the large caudal fin to have performance benefits. The issue of correlated characters is not merely of academic interest. There is increasing evidence that even traits with major fitness effects can evolve as a correlated response to selection on other characters (Price and Langen, 1992).

One possible means of avoiding some of these difficulties is to use comparative phylogenetic analysis to evaluate the adaptive significance of characters, and in recent years a great many comparative methods have been

Figure 8   The comparative historical approach to studying adaptation and one difficulty with that approach. This clade contains three taxa, 1, 2, and 3, and character states for each taxon are listed beneath in parentheses. (A): The comparative method for uncovering adaptation suggests that by studying the phylogenetic distributions of traits (A and A'), environments or selective regimes (E and E'), and the performance of traits (α and α') one can determine if a trait is an adaptation. If a derived trait (A') is phylogenetically correlated with derived performance (α') and both occur within derived selective regime (E'), then A' is held to be an adaptation that resulted from natural selection for improved performance within the new environment. (B): Phylogenetic methods may also be used to reject hypotheses of adaptation. A study of taxon 3 may suggest that trait A' is an adaptation that resulted from selection for increased performance (α') in environment E'. But a broader comparative perspective would show that trait A' arose prior to the derived selective regime (E') and hence could not be an adaptation in E'. (C): Character correlations pose problems for comparative methods. The diagram on the left shows that traits A and B are both positively correlated with each other (indicated by the thick line connecting the two traits) and that only trait B has a positive correlation with performance and fitness. Thin lines connect genetically uncorrelated traits such as trait A and performance. Trait A' might be deemed an adaptation in the absence of knowledge about trait B because trait A' occurs phylogenetically in correlation with α' and within the derived selective regime E'. But trait A' only occurs at this position phylogenetically due to its correlation with B and B'. Selection acted on B and not on A; trait A' cannot then be viewed as an adaptation. Panels A and C modified from Lauder *et al.* (1993).

advocated as providing data on the adaptive significance of traits (Arnold, 1994b, 1995; Baum and Larson, 1991; Brooks and McLennan, 1991; Coddington, 1988; Garland and Adolph, 1994; Harvey and Pagel, 1991; Harvey and Purvis, 1991; Losos and Miles, 1994; Miles and Dunham, 1993; Pagel and Harvey, 1989; Wenzel and Carpenter, 1994). These methods all require comparative data for at least three taxa on the environment (or problem to be solved), the trait of interest, and performance of that trait in each taxon. These methods do not then alleviate all of the problems described above, but they promise improved inferential ability in some cases where selection cannot be measured directly.

In brief, phylogenetic methods operate by developing a historical view of the pattern of acquisition of traits, performances, and environments (Fig. 8A); arguments about the adaptive character of traits are then made based on the order of appearance of these characters. For example, if trait A' is to be deemed an adaptation for increased performance in climbing trees (α') as a result of selection for escape from terrestrial predators (E'), then to be consistent with this hypothesis trait A' must have evolved in a clade in which both the increased performance is found and in which the environmental shift to a new selective regime has occurred. Also fundamental to a phylogenetic approach to adaptation is the notion that comparisons must be made between the derived states A', α', and E' and the primitive states of these characters (A, α, and E; Fig. 8A). As described by Baum and Larson (1991, p. 6), "To constitute an adaptation, a trait must have enhanced utility relative to its antecedent state, and the evolutionary transition must be found to have occurred within the selective regime of the focal taxon...." In the example presented in Fig. 8A, the transition from trait A to trait A' is correlated with a performance increase from α to α', and this transition occurred within the derived selective regime of E' which characterizes taxon 3 (because the transition from E to E' characterizes a larger clade of taxa 2 and 3).

A schema such as that in Fig. 8A can also be used to explain a variety of other terms that have been applied to characters, performance, and environments [see discussions in Arnold (1994a, 1995), Brooks and McLennan (1991), Coddington (1988), Funk and Brooks (1990), Gould and Vrba (1982), Greene (1986), Harvey and Pagel (1991)]. For example, Baum and Larson (1991) term a trait in which performance decreases in the face of a new selective regime ($\alpha'<\alpha$) a "disaptation," while no change in performance ($\alpha'=\alpha$) results in a trait being a "nonaptation." An "exaptation" is a trait which retains its performance advantage despite a change in environment after its origin.

One area in which the phylogenetic approach can be of use is to reject hypotheses of character adaptation by evaluating the expected phylogenetic pattern that must minimally hold if the proposed pattern of selection actually produced the postulated performance advantage (Basolo, 1990; Björklund, 1991; Brooks and McLennan, 1991, 1993; Meyer *et al.*, 1994). Figure 8B shows a situation in which we might wish to reject the hypothesis that trait A' resulted from selection in environment E' for performance advantage α' (making trait A' an adaptation). This hypothesis initially might have resulted from a population-level study of taxon 3 where all three derived traits are present. A broader comparative phylogenetic analysis reveals, however, that trait A' arose prior to the change in environment in correlation with the performance increase (α to α'). This phylogenetic pattern suggests that trait A' might actually have resulted from other patterns of selection in the lineage leading to taxa 2 and 3, but cannot be an adaptation to E'.

Enthusiasm for the phylogenetic approach to adaptation has been considerable but not universal, and several authors have pointed out hidden assumptions and difficulties that may limit the ability of phylogenetic analysis to confirm that a character is indeed an adaptation (Frumhoff and Reeve, 1994; Lauder *et al.*, 1993; Leroi *et al.*, 1994; Wenzel and Carpenter, 1994). While endorsing the general value of a phylogenetic approach to the problem of analyzing character evolution, Lauder *et al.* (1993) and Leroi *et al.* (1994) pointed out several theoretical and practical difficulties of such methods.

Correlations among characters, already noted above as a potential liability for inferences of past selection, also may confound phylogenetic analysis (Fig. 8C). If, for example, trait A is positively genetically correlated with trait B, and selection acted on trait B to effect a transformation to B' (which confers performance advantage α'), then trait A' will be interpreted as an adaptation when it was not in fact the target of selection. As long as correlations among characters are not recognized, errors of this sort are likely to be common. In addition, strong *directional* selection may occur to prevent a trait D (which confers performance advantage and which is genetically correlated with another character E) from changing as a result of selection on character E. Character D will appear to be phylogenetically static, but is so only as a result of directional selection on E.

One practical difficulty that often arises with applications of phylogenetic methodology to the study of adaptation is that data are not often available on patterns of selection and performance for all taxa of interest. Morphological or structural components of design are relatively accessible, and these data are most often the source for hypotheses of adaptive design. But obtaining relative performance data for many taxa is a task not often accomplished (but see Arnold, 1995; Losos, 1990; Losos and Miles, 1994), and obtaining data on the selective regime in relation to the specific trait of interest across a range of taxa is almost never performed. To mitigate this situation, relative performance is often simply assumed, and gross environmental differences (such as arboreal or terrestrial) are substituted for quantified patterns of selection. Baum and Larson (1991) present an analysis of salamander hand morphology that uses environmental surrogates of this type (e.g., scansorial, terrestrial) for a demonstrated selective regime. But as our coding of the environmental character (E in Fig. 8) becomes broad, the number of possible traits and performances that might be associated with such an environment becomes huge, and the possible existence of characters correlated with the trait of interest nearly certain. A final problem with analyses of this sort, discussed in more detail by Lauder *et al.* (1993, p. 296), is the conflation of performance and fitness. Performance is often used as a surrogate of fitness as it is rarely possible to measure fitness directly in comparative phylogenetic studies. But performance and fitness are not the same thing, and measurement of performance, especially the relative performance of two taxa that differ in many traits (not just the one of interest), cannot be equated with the fitness of those taxa.

One implication of the difficulties noted above in inferring retrospectively the action of selection is that much of the terminology so useful for *theoretical* explication of possible patterns of evolution is likely to have limited *practical* utility. For example, if we are to demonstrate empirically that a trait is an exaptation following Gould and Vrba (1982), then we must be able to provide evidence for, and not merely assume, (1) current utility of the trait (a performance advantage),

(2) selection for that specific trait in its current environment, (3) previous utility of the trait in an ancestral taxon or lineage with a *different* performance advantage than the current one, and (4) natural selection for that trait (and not for a genetically correlated trait) in the ancestral environment. To successfully identify an exaptation we must be successful in our retrospective inferences of natural selection and of the trait(s) that it acts on. If an exaptation is claimed to have arisen as a by-product of selection on other aspects of organismal design (perhaps as a correlated character), then we must be able to demonstrate this pattern of selection and not simply assume that selection acted in accordance with our wishes.

D. Complex Features Cannot Easily Be Atomized

The final problem with implementing the argument from design arises when we must decide on those traits and characters that are the subject of selection. Innumerable such traits exist, from the whole arm (Fig. 1) to various parts of the hand (Figs. 5 and 6). Is the finger a trait? Is the terminal phalanx a trait? Or, are the opponens digiti minimi muscle and the deep transverse metacarpal ligament together one character? Clearly if we cannot meaningfully atomize design in some way then it will be difficult to characterize those components of the organism that are the target of the hypothesized selection pattern, and difficult to conduct performance tests on alternative designs.

   The problem of how we dissect the organism into component parts has been widely recognized (e.g., Beatty, 1980; Cronin, 1991; Lewontin, 1978; Sober, 1990), and it is useful to view the problem of atomism as arising from couplings among characters at two (related) levels: phylogenetic, and genetic/developmental.

   Phylogenetically, characters arise and are integrated into organisms so that our *a posteriori* phylogenetic reconstructions show nested sets of characters, some of which we interpret as homologies and others of which may arise multiple times convergently across a clade. But the pattern of character evolution is not random, in the sense that on average characters persist after arising, allowing us to use them as synapomorphies to recognize monophyletic clades. Some traits persist for a remarkably long time, such as the jaw-opening mechanical system in fishes which, after its origin, persists to characterize a clade of over 20,000 species (Lauder and Liem, 1989). The jaw-opening system in a fish species that is a member of the derived clade Percomorpha consists of elements that have arisen at various phylogenetic levels, and each element forms one component of the final hierarchical design. Biological designs are built piecemeal (Lauder, 1990, 1991a) and yet we must typically disassemble them to apply the argument from design in ways that do not mirror the pattern of construction.

   I view this phylogenetic coupling among traits as a fundamental (and mostly unrecognized) aspect of the problem of atomizing organic design and one that does not correspond well to our mechanical intuitions based on human design. If we remove the spark plugs from a car engine, we do so without caring if in the process of engine assembly at the factory the spark plugs were the last component installed. Although we recognize that different parts of a car engine must work together in order for the engine to perform mechanical work, we do not expect that the order of assembly *per se* will influence our analysis of why any particular engine does not function. On the other hand, biological design is the result of a historical process of assembly, in which each component was integrated into an existing design prior to the addition of subsequent design elements (Lauder, 1990). Removal of components of such a design in ignorance of the process of assembly is likely to lead to considerable difficulty: experimental manipulation of different components will involve altering nonequivalent historical couplings with other characters. While experimental manipulation will continue to be a valuable tool for the study of function, an increased awareness of the historical pattern of construction will limit the confounding effects of history on experimental interpretation.

Pleiotropy and epigenetic effects may also result in correlations among characters that make it difficult to atomize a complex design. Pleiotropic effects are well known and give rise to complex patterns of character intercorrelation (Price and Langen, 1992) which greatly complicate any attempt to isolate any one feature of design for analysis. As noted in the previous section, the widespread effects of pleiotropy make it difficult to assert that any trait could be the single target of selection and to dissociate features into meaningful independent units. Some progress has been made in attempting to use quantitative genetic methods as a basis for defining independent evolutionary associations of characters (Shaffer, 1986), but the vast majority of cases in which functional design has been studied lack such information.

Epigenetic effects arise in ontogeny as a functional connection among traits due to the action of developmental processes such as induction or the effects of hormones on disparate tissues (Atchley and Hall, 1991). For example, if the release of thyroid hormone into the bloodstream affects growth in both bone and muscle cells of the hindlimb, then to what extent are we justified in considering the ontogeny of bone and muscle to be independent of each other (or in analyzing just muscle ontogeny as though muscle were an isolated biological tissue)? As summarized by Atchley and Hall (1991, p. 143), "If an epigenetic factor (hormone, maternal effect, etc.) has a significant impact on development of a morphological structure, then a non-zero covariance results between the epigenetic factor and the morphological structure." Note that a significant evolutionary association between two traits may also arise via epigenetic mechanisms even if the associated structures share no genes in common.

Despite the many difficulties involved in discerning the linkages among characters and the natural biological units into which we might divide morphology, several excellent analyses have shown that progress can be made by combining morphological, developmental, and genetic analyses. For example, Atchley and Hall (1991) summarize current understanding of mouse mandible ontogeny and have managed to identify five developmental units of the lower jaw skeleton. Comparison of the mouse mandible to that of other mouse strains and other species of rodents holds the promise of providing a relatively complete understanding of mandible structure and how intraspecific and interspecific variation in design has been achieved by the transformation of fundamental developmental units.

While the practical consequences of the atomism problem have often been ignored or dismissed (e.g., Cronin, 1991), Sober (1990) suggests that instead of worrying about such difficulties we focus on the properties of traits and ask which of those properties might have resulted from selection. Under this view, there may be many ways of analyzing any given design, and any one of these ways will be adequate for a specific analysis. Perhaps what really matters are the properties of a specific atomization, regardless of whether or not the dissection into component parts falls along genetic or phylogenetically meaningful lines. However, only the analysis of designs in which phylogenetic and developmental genetic correlations are known will permit us to evaluate empirically the consequences of alternative views of organismal components. At present we are largely operating blind in most comparative studies, with little idea of the historical or genetic patterns that underlie chosen sets of characters.

IV. The Future of the Argument from Design

We have been far too assumptive about the design of organisms. All too often we have been willing to make assumptions about the process of design from the fact of complexity alone. The literature is filled with statements about the mechanical function of this or that trait, or about the "superiority" of this or that design without a single experiment to back up the claim. In addition, facile claims that particular traits cannot have had some specific function (or any function at all) often rest on equally thin ice, a lack of function simply having been assumed. Further, numerous papers elaborate on the evolutionary function of some design without even the pretense of performance analysis, a methodology for breaking down the design into biologically relevant component parts, data on genetically correlated characters, or a quantitative estimate of patterns of selection. It is one thing to simply assume that any complex aspect of the phenotype *must* have resulted from natural selection in its entirely and in its present configuration, but it is quite another (more interesting) thing to demonstrate the complexity of current design, the stepwise historical acquisition of components of that design, and the diverse evolutionary and developmental processes that likely have contributed to any complex phenotypic trait.

As we have seen, the empirical application of the argument from design does not rest on very solid ground. However, the AFD is not without theoretical support: most evolutionary biologists believe that complex organic designs are either partially or largely built by patterns of selection acting on various components of existing designs because there is extensive evidence demonstrating the power of selection to alter biological design in response to environmental demands. Rather, the point of this chapter is that the experiments and analyses necessary to demonstrate specific patterns of selection, especially on gross aspects of the phenotype, are rarely conducted: complexity of design alone is simply assumed to be sufficient evidence. To my mind, assumptive approaches are undesirable largely because they do not promote further research: if we are willing to assume that the hand is designed for manipulation, then we are unlikely to embark on the interesting research program of discovering the historical origin and phylogenetic pattern to hand structure, testing possible models of hand design and their performance on various tasks such as grasping or manipulation, or examining genetic or phylogenetic correlations among components of hand design.

Perhaps empirical standards could be raised if, when we encounter an "argument from design" in research papers or seminar presentations, we ask one or more of the following questions.

- Where are the physiological or biomechanical experiments to support the interpretation that a structure has the proposed mechanical function? For a comparative phylogenetic analysis, such experiments should be conducted for all members of the clade and outgroups relevant to the interpretation of traits of interest.

- Have alternative designs been compared for performance on a common criterion? It is all to easy to conduct experiments on one design and then simply assert that this design is better than alternatives.

- Do analyses of traits as adaptations or exaptations provide supporting data on (1) patterns of selection (either in the present or past, as appropriate), (2) character correlations to demonstrate that selection either did or did not act directly on the trait of interest, and (3) comparative performance tests to demonstrate the utility or lack thereof of the trait? In short, what evidence supports the inferred evolutionary function claimed for a trait?

- Has the author of an analysis of design explained the criteria by which a design was atomized into component parts and justified the analysis of only a few components of that design?

Let us refashion the argument from design. Instead of aiming to infer processes retrospectively from a design endpoint, we could choose to focus on the process of building the design: an *argument of construction*. How are complex biological designs constructed from the combination of direct selection for certain components, the correlated transformation of others, drift and historical contingency, with each stage in the process of construction influenced by the history of past design? This argument of construction points to a multifaceted research program in which biomechanics, population biology, quantitative genetics, and phylogenetic analysis all have roles to play in research to reveal the process of Darwinian fabrication. The extent to which such research will contribute to our understanding of biological design will be a function of how willing we are to abandon assumptive practices about evolutionary mechanisms to focus instead on patterns of biological design and the many possible mechanisms by which such patterns may have arisen.

Acknowledgments

Preparation of this chapter was supported by NSF IBN 9507181. I thank Gary Gillis, Heidi Doan, Miriam Ashley-Ross, Bruce Jayne, Alice Gibb, Amy Cook, Armand Leroi, Michael Rose, and Steve Frank for helpful comments and discussion.

References

Allen, C., and Bekoff, M. (1995). Function, natural design, and animal behavior: philosophical and ethological considerations. *In:* "Perspectives in Ethology, Vol. 11. Behavioral Design" (N. S. Thompson, ed.), pp. 1-46. Plenum Press, New York.

Amundson, R., and Lauder, G. V. (1994). Function without purpose: the uses of causal role function in evolutionary biology. *Biology and Philosophy* 9, 443-469.

Arber, A. (1950). "The Natural Philosophy of Plant Form." Cambridge Univ. Press, Cambridge.

Arnold, E. N. (1994a). Investigating the evolutionary effects of one feature on another: does muscle spread suppress caudal autotomy in lizards? *J. Zool., London* 232, 505-523.

Arnold, E. N. (1994b). Investigating the origins of performance advantage: adaptation, exaptation and lineage effects. *In:* "Phylogenetics and Ecology" (P. Eggleton and R. I. Vane-Wright, eds.), pp. 123-168. Academic Press, London.

Arnold, E. N. (1995). Identifying the effects of history on adaptation: origins of different sand-diving techniques in lizards. *J. Zool., London* 235, 351-388.

Arnold, S. J. (1983). Morphology, performance, and fitness. *Am. Zool.* 23, 347-361.

Arnold, S. J. (1986). Laboratory and field approaches to the study of adaptation. *In:* "Predator-Prey Relationships: Perspectives and Approaches from the Study of Lower Vertebrates" (M. E. Feder and G. V. Lauder, eds.), pp. 157-179. Univ. of Chicago Press, Chicago.

Atchley, W. R., and Hall, B. K. (1991). A model for development and evolution of complex morphological structures. *Biological Reviews* 66, 101-157.

Ayala, F. J. (1970). Teleological explanations in evolutionary biology. *Philosophy of Science* 37, 1-15.

Basolo, A. (1990). Female preference predates the evolution of the sword in swordtail fish. *Science* 250, 808-810.

Baum, D. A., and Larson, A. (1991). Adaptation reviewed: a phylogenetic methodology for studying character macroevolution. *Systematic Zoology* 40, 1-18.

Beatty, J. (1980). Optimal-design models and the strategy of model building in evolutionary biology. *Philosophy of Science* 47, 532-561.

Bekoff, M., and Allen, C. (1995). Teleology, function, design, and the evolution of animal behavior. *Trends in Ecology and Evolution* 10, 253-255.

Björklund, M. (1991). Evolution, phylogeny, sexual dimorphism and mating system in the grackles (*Quiscalus* spp.: Icterinae). *Evolution* 45, 608-621.

Bock, W., and von Wahlert, G. (1965). Adaptation and the form-function complex. *Evolution* 19, 269-299.

Bock, W. J. (1980). The definition and recognition of biological adaptation. *Am. Zool.* 20, 217-227.

Brandon, R. N. (1990). "Adaptation and Environment." Princeton Univ. Press, Princeton.

Brooks, D. R., and McLennan, D. A. (1991). "Phylogeny, Ecology, and Behavior: A Research Program in Comparative Biology." Univ. of Chicago Press, Chicago.

Brooks, D. R., and McLennan, D. A. (1993). "Parascript: Parasites and the Language of Evolution." Smithsonian Institution Press, Washington.

Coates, M. I., and Clack, J. A. (1990). Polydactyly in the earliest known tetrapod limbs. *Nature* 347, 66-69.

Coddington, J. A. (1988). Cladistic tests of adaptational hypotheses. *Cladistics* 4, 3-22.

Cronin, H. (1991). "The Ant and the Peacock." Cambridge Univ. Press, Cambridge.

Darwin, C. (1859). "On the Origin of Species by Means of Natural Selection, or, the Preservation of Favored Races in the Struggle for Life." John Murray, London.

Dawkins, R. (1987). "The Blind Watchmaker." W. W. Norton, New York.

Dennett, D. C. (1983). Intentional systems in cognitive ethology: the "Panglossian paradigm" defended. *Behavioral and Brain Sciences* 6, 343-390.

Dennett, D. C. (1995). "Darwin's Dangerous Idea: Evolution and the Meanings of Life." Simon and Schuster, New York.

Dullemeijer, P. (1974). "Concepts and Approaches in Animal Morphology." Van Gorcum, The Netherlands.

Eaton, T. H. (1960). The aquatic origin of tetrapods. *Trans. Kans. Acad. Sci.* 63, 115-120.

Edwards, J. L. (1989). Two perspectives on the evolution of the tetrapod limb. *Am. Zool.* 29, 235-254.

Endler, J. (1986). "Natural Selection in the Wild." Princeton Univ. Press, Princeton.

Fisher, D. C. (1985). Evolutionary morphology: beyond the analogous, the anecdotal, and the ad hoc. *Paleobiology* 11, 120-138.

Frumhoff, P. C., and Reeve, H. K. (1994). Using phylogenies to test hypotheses of adaptation: a critique of some current proposals. *Evolution* 48, 172-180.

Funk, V. A., and Brooks, D. R. (1990). "Phylogenetic Systematics as the Basis of Comparative Biology." Smithsonian Institution Press, Washington, D. C.

Futuyma, D. J. (1986). "Evolutionary Biology." (2nd ed.) Sinauer, Sunderland, MA.

Gans, C. (1983). On the fallacy of perfection. *In:* "Perspectives on Modern Auditory Research: Papers in Honor of E. G. Wever" (R. R. Fay and G. Gourevitch, eds.), pp. 101-112. Amphora Press, Groton, CT.

Gans, C. (1988). Adaptation and the form-function relation. *American Zoologist* 28, 681-697.

Garland, T., and Adolph, S. C. (1994). Why not to do 2-species comparisons: limitations on inferring adaptation. *Physiological Zoology* 67, 797-828.

Gould, S. J., and Lewontin, R. C. (1979). The spandrels of San Marco and the Panglossian paradigm: a critique of the adaptationist programme. *Proceedings of the Royal Society of London. Series B.* 205, 581 - 598.

Gould, S. J., and Vrba, E. S. (1982). Exaptation – a missing term in the science of form. *Paleobiology* 8, 4 -15.

Greene, H. W. (1986). Diet and arboreality in the Emerald Monitor, *Varanus prasinus,* with comments on the study of adaptation. *Fieldiana: Zoology n.s.* 31, 1-12.

Griffiths, P. (1992). Adaptive explanation and the concept of a vestige. *In:* "Trees of Life" (P. Griffiths, ed.), pp. 111-131. Kluwer Academic, Dordrecht.

Harvey, P. H., and Pagel, M. D. (1991). "The Comparative Method in Evolutionary Biology." Oxford Univ. Press, Oxford, U.K.

Harvey, P. H., and Purvis, A. (1991). Comparative methods for explaining adaptations. *Nature* 351, 619-624.

Hume, D. (1779). "Dialogues Concerning Natural Religion." 1991. Routledge, edited by S. Tweyman, London.

Hurlbutt, R. H. (1985). "Hume, Newton, and the Design Argument." Univ. of Nebraska Press, Lincoln.

Kato, I. (Ed.). (1982). "Mechanical Hands Illustrated." Hemisphere Publishing Corp., New York.

Kier, W. M., and Smith, K. K. (1985). Tongues, tentacles, and trunks: the biomechanics of movement in muscular hydrostats. *Zool. J. Linn. Soc. Lond.* 83, 307-324.

Lande, R. (1976). Natural selection and random genetic drift in phenotypic evolution. *Evolution* 30, 314 - 334.

Lande, R. (1978). Evolutionary mechanisms of limb loss in tetrapods. *Evolution* 32, 73-92.

Lande, R. (1979). Quantitative genetic analysis of multivariate evolution, applied to brain: body size allometry. *Evolution* 33, 402-416.

Lauder, G. V. (1982). Historical biology and the problem of design. *Journal of Theoretical Biology* 97, 57-67.

Lauder, G. V. (1983a). Functional and morphological bases of trophic specialization in sunfishes. *J. Morph.* 178, 1-21.

Lauder, G. V. (1983b). Neuromuscular patterns and the origin of trophic specialization in fishes. *Science* 219, 1235-1237.

Lauder, G. V. (1985). Aquatic feeding in lower vertebrates. *In:* "Functional Vertebrate Morphology" (M. Hildebrand, D. M. Bramble, K. F. Liem, and D. Wake, eds.), pp. 210-229. Harvard Univ. Press, Cambridge.

Lauder, G. V. (1990). Functional morphology and systematics: studying functional patterns in an historical context. *Annual Review of Ecology and Systematics* 21, 317-340.

Lauder, G. V. (1991a). Biomechanics and evolution: integrating physical and historical biology in the study of complex systems. *In:* "Biomechanics in Evolution" (J. M. V. Rayner and R. J. Wootton, eds.), pp. 1-19. Cambridge Univ. Press, Cambridge.

Lauder, G. V. (1991b). An evolutionary perspective on the concept of efficiency: how does function evolve? *In:* "Efficiency and Economy in Animal Physiology" (R. W. Blake, ed.), pp. 169-184. Cambridge Univ. Press, Cambridge.

Lauder, G. V. (1994). Homology, form, and function. *In:* "Homology: The Hierarchical Basis of Comparative Biology" (B. K. Hall, ed.), pp. 151-196. Academic Press, San Diego.

Lauder, G. V. (1995). On the inference of function from structure. *In:* "Functional Morphology in Vertebrate Paleontology" (J. J. Thomason, ed.), pp. 1-18. Cambridge Univ. Press, Cambridge.

Lauder, G. V., Leroi, A., and Rose, M. (1993). Adaptations and history. *Trends in Ecology and Evolution* 8, 294-297.

Lauder, G. V., and Liem, K. F. (1989). The role of historical factors in the evolution of complex organismal functions. *In:* "Complex Organismal Functions: Integration and Evolution in Vertebrates" (D. B. Wake and G. Roth, eds.), pp. 63-78. John Wiley and Sons, Chichester.

Lauder, G. V., and Reilly, S. M. (1996). The mechanistic bases of behavioral evolution: comparative analysis of musculoskeletal function. *In:* "Phylogenies and the Comparative Method in Animal Behavior" (E. Martins, ed.). Oxford Univ. Press, Oxford.

Lauder, G. V., and Shaffer, H. B. (1985). Functional morphology of the feeding mechanism in aquatic ambystomatid salamanders. *J. Morph.* 185, 297-326.

Leroi, A. M., Rose, M. R., and Lauder, G. V. (1994). What does the comparative method reveal about adaptation? *Am. Nat.* 143, 381-402.

Lewontin, R. C. (1978). Adaptation. *Sci. Am.* 239, 156-169.

Losos, J. B. (1990). The evolution of form and function: morphology and locomotor performance in West Indian *Anolis* lizards. *Evolution* 44, 1189-1203.

Losos, J. B., and Miles, D. B. (1994). Adaptation, constraint, and the comparative method: phylogenetic issues and methods. *In:* "Ecological Morphology: Integrative Organismal Biology" (P. C. Wainwright and S. M. Reilly, eds.), pp. 60-98. Univ. of Chicago Press, Chicago.

Mark, R. (1996). Architecture and evolution. *American Scientist* 84, 383-389.

Mayr, E. (1983). How to carry out the adaptationist program? *Am. Nat.* 121, 324-334.

Meyer, A., Morrissey, J. M., and Schartl, M. (1994). Recurrent origin of a sexually selected trait in *Xiphophorus* fishes inferred from a molecular phylogeny. *Nature* 368, 539-542.

Miles, D. B., and Dunham, A. E. (1993). Historical perspectives in ecology and evolutionary biology: the use of phylogenetic comparative analyses. *Annual Review of Ecology and Systematics* 24, 587-619.

Mishler, B. D. (1988). Reproductive ecology of bryophytes. *In:* "Plant Reproductive Ecology" (J. L. Doust and L. L. Doust, eds.), pp. 285-306. Oxford Univ. Press, Oxford, U.K.

Mitchell, W. A., and Valone, T. J. (1990). The optimization research program: studying adaptations by their function. *Quarterly Review of Biology* 65, 43-52.

Niklas, K. J. (1986). Computer simulations of branching patterns and their implication on the evolution of plants. *Lect. Math. Life. Sci.* 18, 1-50.

Niklas, K. J. (1992). "Plant Biomechanics." Univ. of Chicago Press, Chicago.

Ollason, J. G. (1987). Artificial design in natural history: why it's so easy to understand animal behavior. *In:* "Alternatives: Perspectives in Ethology" (P. P. G. Bateson and P. H. Klopfer, eds.), Vol. 7, pp. 233-257. Plenum, New York.

Orzack, S., and Sober, E. (1994a). How (not) to test an optimality model. *Trends in Ecology and Evolution* 9, 265-267.

Orzack, S. H., and Sober, E. (1994b). Optimality models and the test of adaptationism. *Am. Nat.* 143, 361-380.

Pagel, M. D., and Harvey, P. H. (1989). Comparative methods for examining adaptation depend on evolutionary models. *Folia Primatologica* 53, 203-220.

Paley, W. (1836). "Natural Theology." American Tract Society, New York.

Parker, J. A., and Maynard Smith, J. (1990). Optimality theory in evolutionary biology. *Nature* 348, 27-33.

Pinker, S., and Bloom, P. (1990). Natural language and natural selection. *Behavioral and Brain Sciences* 13, 707-784.

Price, T., and Langen, T. (1992). Evolution of correlated characters. *Trends in Ecology and Evolution* 7, 307-310.

Reeve, H. K., and Sherman, P. W. (1993). Adaptation and the goals of evolutionary research. *Quarterly Review of Biology* 68, 1-32.

Reilly, S. M., and Lauder, G. V. (1992). Morphology, behavior, and evolution: comparative kinematics of aquatic feeding in salamanders. *Brain, Behavior, and Evolution* 40, 182-196.

Reznick, D. N. (1989). Life-history evolution in guppies. 2. Repeatability of field observations and the effects of season on life histories. *Evolution* 43, 1285-1297.

Reznick, D. N., and Bryga, H. (1987). Life-history evolution in guppies (*Poecilia reticulata*). 1. Phenotypic and genetic changes in an introduction experiment. *Evolution* 41, 1370-1385.

Ridley, M. (1983). "The Explanation of Organic Diversity: The Comparative Method and Adaptations for Mating." Clarendon Press, Oxford.

Romer, A. S., and Parsons, T. S. (1986). "The Vertebrate Body." Saunders, New York.

Rose, M. R. (1984a). Artificial selection on a fitness component in *Drosophila melanogaster. Evolution* 38, 515-526.

Rose, M. R. (1984b). Laboratory evolution of postponed senescence in *Drosophila melanogaster. Evolution* 38, 1004-1010.

Rose, M. R., Service, P. M., and Hutchinson, E. W. (1987). Three approaches to trade-offs in life history evolution. *In* "Genetic Constraints on Adaptive Evolution" (V. Loeschcke, ed.), pp. 91-105. Springer-Verlag, Berlin.

Rudwick, M. J. S. (1964). The inference of function from structure in fossils. *Brit. J. Phil. Sci.* 15, 27-40.

Russell, E. S. (1916). "Form and Function: A Contribution to the History of Animal Morphology." Reprinted in 1982 with a new Introduction by G. V. Lauder. John Murray, London.

Selzer, J. (ed.) (1993). "Understanding Scientific Prose." Univ. of Wisconsin Press, Madison.

Shaffer, H. B. (1986). Utility of quantitative genetic parameters in character weighting. *Syst. Zool.* 35, 124-134.

Sober, E. (1984). "The Nature of Selection: Evolutionary Theory in Philosophical Focus." MIT Press, Cambridge, MA.

Sober, E. (1990). Atomizing the rhinoceros. *Behavioral and Brain Sciences* 13, 764-765.

Sober, E. (1993). "Philosophy of Biology." Westview Press, Boulder.

Thomason, J. J. (ed.) (1995). "Functional Morphology in Vertebrate Paleontology." Cambridge Univ. Press, Cambridge.

Thornhill, R. (1990). The study of adaptation. *In:* "Interpretation and Explanation in the Study of Animal Behavior." (M. Bekoff and D. Jamieson, eds.), Vol. II, pp. 31-62. Westview Press, Boulder.

Thornhill, R., and Gangestad, S. W. (1996). The evolution of human sexuality. *Trends in Ecology and Evolution* 11, 98-102.

Vermeij, G. J., and Zipser, E. (1986). Burrowing performance of some tropical pacific gastropods. *Veliger* 29, 200-206.

Wainwright, P. C., Lauder, G. V., Osenberg, C. W., and Mittelbach, G. G. (1991a). The functional basis of intraspecific trophic diversification in sunfishes. *In:* "The Unity of Evolutionary Biology" (E. C. Dudley, ed.), pp. 515-529. Dioscorides Press, Portland.

Wainwright, P. C., Osenberg, C. W., and Mittelbach, G. G. (1991b). Trophic polymorphism in the pumpkinseed sunfish (*Lepomis gibbosus*): effects of environment on ontogeny. *Funct. Ecol.* 5, 40-55.

Wenzel, J. W., and Carpenter, J. M. (1994). Comparing methods: adaptive traits and tests of adaptation. *In:* "Phylogenetics and Ecology" (P. Eggleton and R. Vane-Wright, eds.), pp. 79-101. Linnean Society of London, London.

Williams, G. C. (1966). "Adaptation and Natural Selection." Princeton Univ. Press, Princeton.

Williams, G. C. (1992). "Natural Selection: Domains, Levels, and Challenges." Oxford Univ. Press, Oxford.

CHAPTER 3

Optimization and Adaptation

JON SEGER

J. WILLIAM STUBBLEFIELD

"A biologist who posited adaptation would be like a physicist who posited that bodies fall. Competent biologists treat the occurrence of adaptation or maladaptation as contingent in the same way that competent physicists treat the rising and falling of bodies as contingent. Adaptation has to be hypothesized and tested like everything else in science... . The new adaptational biology is neither Panglossian nor pluralistic, but tests broad, general hypotheses against hard data and is not satisfied until all contradictions have been purged from the system. This paradigm ... is Darwin's paradigm, revived and modernized." (Ghiselin, 1983)

I. Summary

The optimization approach to the study of adaptation uses phenotypic models to ask which character states from a specified range of alternatives (the strategy set) are expected to be most fit or "optimal" under a suitable measure of fitness (the objective function), given other relevant assumptions about the biological situation being considered. To ask what is the best available character state, an investigator must confront the often critical issues of costs, benefits, constraints, and trade-offs that limit the possible or feasible phenotypic alternatives. Optimization models have been applied most often and most successfully to quantitative traits such as clutch sizes and sex ratios, where resource allocations subject to well-defined trade-offs have large effects on fitness. The productivity of the optimization approach derives from its requirement that the strategy set, the objective function, and other assumptions be clearly defined, and from its tendency to raise questions about the connections between different aspects of a species' biology. Optimization models assume that sufficient heritable variation will arise to permit the phenotypes under study to evolve toward their equilibria under the specified constraints, but this does not mean that they necessarily ignore or deny historical contingency. Often there are constraints on the evolution of a trait that can be understood only as outcomes of the unique history of a lineage. The optimization approach to adaptation is not based on the assumption that organisms are "optimal" in any global or metaphysical sense, and in fact it often reveals that they are far less than "perfectly" adapted.

II. Introduction

Living things give the appearance of having been designed for a purpose. Darwin's great discovery was that designs more intricate than any produced by human artifice can emerge from the mindless process of natural selection which simply compares heritable variants statistically with respect to their effects on reproduction. The selectively guided exploration of alternatives in the "neighborhood" of an existing design allows new variants with higher than average fitness to supplant older variants that were themselves fitter than those that they supplanted. A succession of such incremental substitutions can quickly improve a design, and eventually transform it radically (Dawkins, 1986; Dennett, 1995). Human engineers have come to appreciate the power of this process, which they use in the form of "genetic algorithms," to solve complex and otherwise difficult maximization problems (Holland, 1992, 1995; see also Chapter 14). In engineering, as in evolution, the fittest attainable solution is often a compromise, owing to *constraints* on the feasible design options and *trade-offs* among different benefits to be achieved by the design. The selective accumulation of variants that confer increased fitness therefore tends to *optimize* the parameters of the design with respect to the organism's (or engineered system's) overall state of local adaptation.

Optimization is about constraints and trade-offs, not "perfection." The fittest conceivable organism would live forever, reproducing continually at an infinite rate. The fittest possible organisms do far worse than this because they are constrained either absolutely or effectively by laws of physics, by deeply rooted evolutionary legacies, and by complex trade-offs among individually attainable but mutually incompatible character states. For example, two beneficial activities such as eating and watching for predators may be incompatible simply because they cannot be pursued at the same time. In a famous passage from the *Origin*, Darwin (1859) asks us to appreciate that "natural selection is daily and hourly scrutinizing, throughout the world, every variation, even the slightest; rejecting that which is bad, preserving and adding up all that is good; silently and insensibly working, whenever and wherever opportunity offers, at the improvement of each organic being in relation to its organic and inorganic conditions of life." This passage vividly evokes the relentlessness of selection but only hints that the difference between "bad" and "good" designs may often hinge on trade-offs.

Some trade-offs take similar forms wherever they arise, even though the fittest compromise may vary with aspects of the species' biology and environment. For example, parents usually face a trade-off between the number of offspring they can produce and the sizes of individual offspring, because their total reproductive effort is limited (Smith and Fretwell, 1974; Charnov and Downhower, 1995; Charnov et al., 1995). Other trade-offs derive as much from a species' history as from its current ecology. Selection chooses only among current alternatives, with no foresight, so adaptive pathways taken at one time may cast very long shadows into the future of a lineage, affecting what "opportunity offers" (Darwin's subtle acknowledgment of contingency in the quote above) at any given time. For example, the basic skeletal organization of amphibians, reptiles, birds, and mammals poses countless optimization problems today that owe their forms in large part to ancient adaptations for swimming like a fish. But whether relatively universal and recurring or relatively particular and historically conditioned, constraints that establish trade-offs set the stage for a diversity of solutions that may evolve through locally optimizing exploration of the set of accessible character-state combinations. Ecological specialization and evolutionary diversification may therefore be driven in large part by trade-offs that force species to choose, in effect, among restricted combinations of character states that will be simultaneously both compatible and competitive (e.g., Joshi and Thompson, 1995).

Biologists seeking to understand how the designed features of organisms were assembled, how they reflect underlying constraints, and how well they serve their functions are engaged in what amounts to a grand reverse-engineering project (see Dennett, 1995). Since the engineering was accomplished by cumulative selection rather than by conscious planning and experiment, biologists may have little choice but to begin by asking how the features under study might have been optimized for one or more functions, under one or more constraints. In a general sense, then, optimization is a fundamental principle of evolutionary biology, especially of the study of adaptation. However, to say that optimization permeates almost every problem in biology is not to say that it answers every question, nor is it to say that characters (even ones with strong effects on fitness) are necessarily expected to be found in optimal states.

This distinction between the *process* of optimization and the *state* of optimality is important. The process is ubiquitous and powerful, but the state can be elusive. In an unchanging environment, with an adequate supply of variation, most characters that affected fitness would be at least locally optimal in the sense that for all practical purposes selection acting in the here-and-now could not improve them given the prevailing constraints on the joint ranges of attainable character states (Charnov, 1989; Charlesworth, 1990). A species' phenotype would then tell us a great deal about its environment because there would be a direct mapping (mediated by genotypes) between its selective regime and its phenotype. But of course the world is nowhere near this simple. Environments may change, deleterious mutations may accumulate, and the ranges of accessible character states may be limited in complex, seemingly arbitrary ways that have more to do with a species' history, genetics, and developmental biology than with the demands and opportunities created by its way of life. Thus in the real world, the state of near-perfect adaptation, optimality, could be relatively rare, while the process of optimization is improving phenotypes everywhere and at all times (subject to prevailing constraints), by changing gene frequencies in ways that increase fitness.

The optimization approach to the study of adaptation can be very productive even where the attainment of optimality cannot be assumed. Indeed, optimization models often yield valuable insights when they fail by raising previously unasked questions about connections among particular aspects of the biologies of the species under study. This happens because optimization models focus attention on processes that give rise to selection pressures and on constraints that modulate a population's evolutionary responses to selection. The specific questions and hypotheses that emerge often motivate empirical tests that advance our understanding of the ways in which adaptive designs both evolve and fail to evolve.

In what follows we first outline the structure of an idealized optimization study, emphasizing the differences between the model itself, its analysis, and its tests. Next we review the origins of the optimization approach, comment on its aims and limits, and critically examine the elements of the model: the actors, currencies, objective functions, strategy sets, and constraints of many kinds. We conclude with some thoughts about the future evolution of the approach, including its possible application to emerging problems in genetics and development.

Various aspects of this subject have been reviewed, from different perspectives, by many authors including Williams (1966a, 1992), Levins (1966, 1968), Leigh (1971), Schoener (1971), Cody (1974), Pyke et al. (1977), Rapport and Turner (1977), Lewontin (1978, 1979, 1984), Maynard Smith (1978), Oster and Wilson (1978), Gould and Lewontin (1979), Horn (1979), Beatty (1980), Alexander (1982), Charnov (1982, 1993), Mayr (1983), Emlen (1984), Maynard Smith et al. (1985), Stephens and Krebs (1986), various authors in Dupré (1987), Mitchell and Valone (1990), Parker and Maynard Smith (1990), Thornhill (1990), Grafen (1991), Lessells (1991), Sibly and Antonovics (1992), Stearns (1992), Moore and Boake (1994), Orzack and Sober (1994a,b), Roff (1994), Bulmer (1994), Dennett (1995), and others. We have absorbed many ideas from these works, both consciously and unconsciously. The papers and books we cite in this chapter provide useful starting points for readers who want to explore on their own, but they inevitably reflect our personal interests, biases, and limited mastery of the huge literature on this complex subject. We apologize for the slights we will inevitably commit, unintentionally, to authors both cited and uncited.

III. The Structure of an Optimization Study

An optimization study is motivated by questions derived from some perceived problem concerning the form or taxonomic distribution of a possible adaptation. The study itself consists of three main parts: (1) a *model* that embodies explicit assumptions about the biological situation under study, casting it in a simplified, abstract form and defining the quantity that selection is expected to optimize; (2) an *analysis* that deduces the optimum, which may vary as a function of the model's parameters; and (3) a *test* that asks how well the model's assumptions and the predicted optimum agree with relevant empirical data. As in other applications of the hypothetico-deductive method in science, this process may cycle repeatedly back to the beginning, with modified forms of the model being reanalyzed and retested in a search for sets of assumptions that "work" in the sense that they "explain" the biological patterns that motivated the study (and others uncovered during the study) in a consistent and otherwise plausible way.

The model contains a number of distinct elements. First, it identifies the *actor*, typically an individual such as a foraging bird or a growing plant that expresses the phenotype(s) whose adaptive evolution we seek to understand. Sometimes the actor is a less inclusive entity such as a gene, and in principle it could be a more inclusive entity such as a local population. In any case, defining the actor establishes the scope of the model and the assumed locus of control (metaphorically, the point of view or interest) with respect to which the optimization is to occur.

Second, the model identifies a *currency* in which the fitnesses of alternative phenotypes can be evaluated, the *control variable(s)* that quantify the states of those phenotypes, and an *objective function* that describes how fitness (in units of the currency) depends on the control variable(s) and other aspects of the biological situation. For example, if the currency were "number of offspring reared to weaning" and the control variable were "number of zygotes conceived," then the objective function would give the number reared to weaning for each possible number of zygotes conceived.

Third, the model specifies a *strategy set* that describes the actor's options for manipulating the control variables. There is often much more to the strategy set than just a range of accessible values of the control variables. For example, the strategy set may allow the control variables to depend on states of the physical or social environment, or on the actor's individual circumstances such as age or size. Trade-offs and constraints of various kinds may appear either in the objective function or in the strategy set, depending on their origin. For example, a trade-off between the number of zygotes conceived and their probabilities of surviving to the age of weaning might naturally be built into the objective function, and the strategy set might then consist of all zygote numbers from zero to infinity. Or in a sex-ratio problem, the parent's fecundity might be fixed by random factors that vary among individuals, and the strategy set might include all sex ratios (proportions of male offspring, r) between 0 and 1, to be determined after the parent knows its own fecundity and possibly the fecundities of other individuals in its immediate neighborhood (e.g., Stubblefield and Seger, 1990).

These three kinds of assumptions define the model. When they have been fully specified the model can be analyzed to reveal which of the accessible strategies for manipulating the control variables gives the largest value of the objective function. Many different mathematical techniques are used to perform such analyses, depending on the structure of the model. In the simplest cases the objective function can be maximized by setting its derivatives with respect to the control variables equal to zero and solving the resulting system of equations. If the phenotype involves a temporal sequence of decisions, with early decisions affecting the consequences of later decisions (as in models of growth and reproduction), then dynamic programming techniques may be required. Where the fitness of a given phenotype depends on the distribution of other phenotypes in the population (as in sex-ratio problems), various kinds of stability analyses are used to find the "unbeatable" phenotype or distribution of phenotypes. Some models are so complex that general closed-form solutions giving the optimal phenotype as a function of the model's parameters cannot be found. In such cases numerical methods are typically used to find particular solutions for given sets of parameters.

The results of an analysis may be of interest in their own right, since they describe inevitable consequences of the assumptions and these may have been unknown or unappreciated. In this sense all correctly analyzed models are "true" (as theory), whether they explain anything about the real world or not (Parker and Maynard Smith, 1990). But as always, a model becomes more interesting scientifically to the extent that its assumptions are plausible and its predictions unexpected. Interesting models therefore invite tests that evaluate their relevance to our understanding of the evolution of the (real) phenotypes in question.

These points are illustrated concretely by the following highly simplified account of work on the problem of clutch-size variation in birds. David Lack (1947, 1948, 1954) asked why there should be so much variation both among and within bird species in the numbers of eggs that parents lay and attempt to rear in a single bout of reproduction. Species produce average clutches ranging from one egg to more than a dozen, and even within a single species many parents may produce clutches far above and below the mean. Lack reasoned that selection should adapt the egg-laying behavior of parent birds to produce, at least on average, clutch sizes close to those that yield the largest number of surviving, reproductively successful offspring. Clutch sizes too small would yield fewer surviving offspring than the parents might have produced, and clutch sizes too large might also yield fewer than the maximum number of surviving offspring, perhaps by overtaxing the parents' ability to feed the nestlings prior to fledging. This might cause early mortality or low fledging weights that would impair the young birds' expectations of subsequent survival or successful reproduction as adults.

Lack's hypothesis defines a model in which the actors are parent birds, the currency is offspring number, the control variable is number of eggs laid, and the objective function gives the number of offspring surviving to reproduce as a function of the number of eggs laid. The most obvious constraint is the limited amount of food that parents can bring back to the nest during the period of nestling development. This constraint gives rise to a trade-off between the number of offspring reared and the average amount of food that each one can receive, and thereby to a well-defined relationship, in units of the currency, between the number of offspring reared and the probability that any one will survive to breed successfully as an adult. The model can therefore be formalized as $W = n \cdot P(n)$, where W is parental fitness, n is clutch size, and $P(n)$ is the probability that an egg will give rise to a reproductive adult offspring, as a function of n.

The analysis reveals that if $P(n)$ declines rapidly enough at larger values of n, then W will have a "humped" or "domed" shape and there will be an optimal value of n (denoted n^*) that maximizes parental fitness under the trade-offs embodied in $P(n)$. For example, if we suppose for purposes of illustration that $P(n) = 1 - an$, then it is easy to show by elementary calculus that $n^* = 1/2a$ (Fig. 1). This almost absurdly simple model makes an interesting and potentially testable qualitative prediction: in species or populations where the increase in offspring mortality with increasing clutch size is relatively fast, average clutches

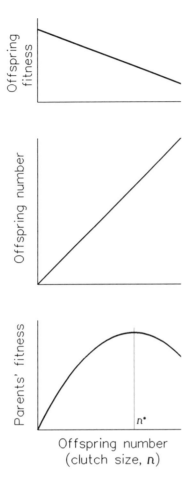

Offspring fitness

Offspring number

Parents' fitness

n^*

Offspring number
(clutch size, n)

Figure I Lack's model of clutch-size regulation. The fitness of each individual offspring is assumed to decline as clutch size increases, due mainly to limits on the parents' ability to feed nestlings. The parents' fitness (bottom) can therefore be expressed as a product of the average offspring fitness (top) and the number of offspring produced (middle). This function has a well-defined intermediate maximum (giving the optimal clutch size, n^*) for many plausible forms of the offspring fitness function (for example, where individual fitness declines monotonically and relatively more quickly at larger clutch sizes, as in the simple linear case shown here).

should be relatively small; where it is slow, they should be large. And of course within a population most parents should produce the "Lack clutch," which is the clutch that maximizes the number of offspring recruited (subsequently) into the adult breeding population.

This and many more elaborate versions of Lack's hypothesis have been tested over the last 40 years in many species of birds, perhaps most extensively in a population of great tits (*Parus major*, a close relative of North American chickadees) that nests in Wytham Wood near Oxford, England. A frequency distribution of clutch sizes for nearly 4500 clutches (distributed over 23 years) is

shown in Fig. 2, with the mean number of recruits produced by each clutch
size from 4 to 14 (Boyce and Perrins, 1987). The most productive clutch is 12,
but the commonest clutches are 8 and 9. Parents therefore appear to lay
smaller clutches, on average, than those that would maximize their fitness.
Similar patterns have also been found in other species. Thus Lack's hypothesis
does not seem to be supported, at least not in this simple form and with this
simple analysis of the data.

As the existence of systematic discrepancies between the Lack clutch and
the most common clutch became evident, it was realized that parents should be
selected to maximize their lifetime production of recruits, and that doing so
might not involve maximizing their production in any one year (e.g., Williams,
1966b; Charnov and Krebs, 1974). If the effort involved in rearing a large brood
this year so exhausts a parent that it is less likely to survive to breed again next
year, or less able to feed a large brood next year, then there will be a between-
years trade-off, and the clutch size that maximizes lifetime fitness may be
smaller than the one that maximizes fitness in any one year. Such trade-offs
have been found in several species (e.g., Gustafsson and Pärt, 1990; see
Lessells, 1991; Roff, 1992; Sibly and Antonovics, 1992; Stearns, 1992), but not in
the Wytham Wood population of great tits. A number of other modifications of
the hypothesis have been suggested (reviewed by Boyce and Perrins, 1987 and
the preceeding references) and several may provide part of the explanation.

The most important additional factor for great tits in Wytham Wood seems
to be that parents respond adaptively to their own highly variable individual
circumstances. Those that find themselves able to rear larger than average
broods (for reasons of health, territory quality, early pair formation, and the
like) lay larger than average clutches, while those that find themselves in

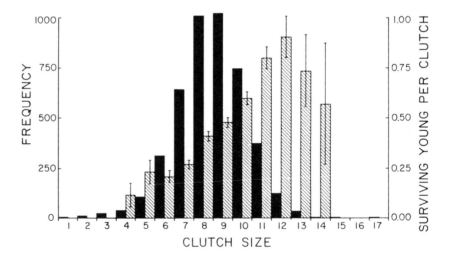

Figure 2 Clutch-size distribution and clutch-size-specific recruitment rates for great tits (*Parus major*) in Wytham Wood, 1960-1982. Solid bars show numbers of clutches of sizes from 1 to 17 eggs (*N*=4489). Hatched bars show mean numbers of young per clutch that survived to at least 1 year of age (±1 s.e.). From Boyce and Perrins (1987) courtesy of the authors and *Ecology*.

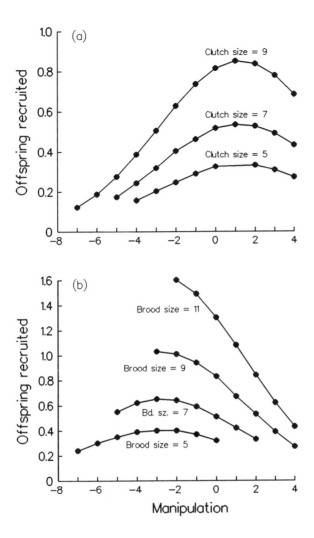

Figure 3 Offspring recruitment as a function of clutch-size manipulations for great tits in Wytham Wood. In both panels, the average number of offspring surviving to 1 year of age is estimated from a log linear model. (a) Numbers recruited are shown for *initial* clutch sizes of 5, 7, and 9 eggs. Thus the upper right-hand point represents clutches of 9 that were augmented by 4 newly hatched chicks, to a total (barring natural mortality) of 13. (b) Here the number recruited is shown for *final brood sizes* of 5, 7, 9, and 11 chicks. Thus the upper point represents a clutch of 13 eggs that was reduced by 2 to a final brood size of 11. Redrawn from Pettifor, Perrins and McCleery (1988) courtesy of the authors and *Nature.*

relatively poor condition lay smaller clutches. Thus the aggregate distribution shown in Fig. 2 is composed of many individual distributions for birds of different overall condition who have different Lack clutch sizes. How was this shown? Various hypotheses (including this one) had suggested that it would be interesting to know the success rates of broods from clutches that had been experimentally enlarged or reduced in size. Newly hatched chicks were therefore transferred between nests in Wytham Wood for many years. The resulting data most strongly support the individual-optimization hypothesis (Pettifor et al., 1988) (see Fig. 3). It appears, then, that parents can assess their condition and adjust their attempted brood size to their actual brood-rearing ability in a given year.

This conclusion does not rest simply on the fact that the clutch-size manipulations support a prediction of the individual-optimization hypothesis. Many other kinds of data contribute to the overall support for this interpretation of great tit clutch sizes in Wytham Wood and to related hypotheses (often different in detail) for other populations and other species. Such data support the models by confirming their key assumptions. For example, data on fledgling weight as a function of clutch size and clutch manipulation show that parents are at or near the limits of their foraging abilities during nestling development and that the growth of nestlings is limited by food. Data on survival and subsequent reproductive success as functions of fledgling weight show that an offspring's prospects are indeed related to the food it received from its parents (see Lindén and Møller, 1989; Newton, 1989; Godfray et al., 1991; Lessells, 1991). In general, then, our confidence in models of clutch-size evolution derives as much or more from tests of their assumptions as it does from tests of their predictions about the clutch-size distributions themselves.

The study of clutch size in birds was thoroughly quantitative from the beginning and it has become very sophisticated (e.g., Daan et al., 1990). Its style of analysis has been applied to an ever wider range of taxa, including plants and invertebrates (see Godfray et al., 1991; Lessells, 1991), and it has served as a model for studies of other allocation problems. With this classic example as background, we will briefly consider the origins of the optimization approach, and its aims and domain of application, before taking a closer look at the key elements of a typical optimization model.

IV. The Roots and Branches of Optimization

Evolution is a dynamical process without a goal or even a definite stopping point. Whether viewed at the level of phenotypes or of underlying genotypes, evolutionary trajectories just keep going and going as long as the population survives. This elementary but very important fact is appreciated better today than ever, due to many kinds of recent advances, including long-term studies of selection in nature (see Endler, 1986; Grant, 1986; Grant and Grant, 1989, 1993), a heightened concern with biotic (hence evolutionarily reactive) components of the environment (see Crawley, 1992; Thompson, 1994), and a rapidly accumulating wealth of molecular genetic data (see Gillespie, 1991).

In contrast, optima are fixed points that make no necessary reference to time or history; if you are there, you are there, no matter how you arrived. Looked at this way, optima seem counter-evolutionary, but they are connected to evolutionary trajectories in a way first made clear by Fisher's (1930) "Fundamental Theorem of Natural Selection", which shows that under suitable simplifying assumptions, gene frequencies will change so as to carry a population along the path that tends to maximize the current rate at which mean fitness increases. The Theorem embodies a seductive and influential view of evolution in which selection appears as a tireless hill-climbing algorithm that inexorably moves a population along the line of locally steepest ascent on the "fitness surface" that defines the current "adaptive landscape" (Wright, 1931). In this population-genetic formulation, Darwin's relentless scrutinizer of variations becomes a blind but efficient optimizing daemon. If there is a peak in the fitness surface (a point from which fitness declines in all directions), then a nearby population will be attracted to it. Once on top, the population will remain there until the landscape deforms (i.e., the environment changes) in a way that creates an ascending route of egress.

This beautiful if oversimple vision of evolution implicitly identifies optima as the phenotypes associated with fitness maxima, under a model of the relationship between genotypes and phenotypes in which the genetic variation needed to produce the fitness-maximizing phenotypes is readily available to the population. Under the assumptions of this model, a population will move remarkably quickly to the vicinity of a locally attracting fitness maximum if the path of ascent is even moderately steep (which is to say, if selection on the character states in question is at all strong). A very important implication is that, on average, populations should be close to fitness peaks with respect to traits that strongly affect fitness, to the extent that the assumptions of the model are met. The two most important of these assumptions are, again, that the fitness landscape has not recently changed too dramatically and that genetic variation for the traits in question is adequate to allow them to evolve toward the optima. We will examine these critical assumptions below; here we note that for certain kinds of situations they can be defended both theoretically and empirically.

Darwin and Fisher understood clearly that selection tends to increase individual fitness, not group or species fitness. But the Fundamental Theorem does not embody this distinction because it assumes that the fitness associated with a given phenotype is independent of its frequency in the local population. Thus under the assumptions of the Theorem, an optimum phenotype can be found by maximizing the population's mean fitness. This procedure is harmless under the assumptions of the Theorem, but not where fitnesses are frequency dependent. During the first half of the 20th century many biologists began speaking of selection as a force that "acts for the good of the species." This largely unconscious retreat from individual selection culminated in 1962 with the publication of V.C. Wynne-Edwards' *Animal Dispersion in Relation to Social Behaviour*, which explicitly interprets mating aggregations as population-assessment mechanisms that enable individuals to adjust their reproduction downward at times of high population density so as to prevent their species from outstripping its resources. Such behavior might be "optimal" from the point of view of the species, but it would not maximize individual reproductive success (the criterion identified by Darwin with natural selection), and in fact it could evolve only under extreme assumptions.

Wynne-Edwards' book was criticized, most notably by Lack (1966) and by G.C. Williams, whose *Adaptation and Natural Selection* (1966a) demolished naive group-selectionist explanation in biology and restored natural selection as the principal agent of adaptive evolution. Williams argued that although selection is an awesomely powerful process, adaptation remains an "onerous concept" to be invoked only for good reason. It is impossible to overstate the importance of this intellectual housecleaning. By bringing rigor to the analysis of adaptation and by establishing imperfection and conflict as ever present possibilities, *Adaptation and Natural Selection* set the stage for a flowering of optimization studies that could not possibly have occurred within the earlier non-Darwinian species-advantage framework.

Although the optimization approach as currently understood and practiced dates only from the middle 1960s, it has roots going back at least to Darwin, who appreciated not only the power of selection but also (if only at an impressionistic level) the possible importance of the kinds of constraints that now figure prominently in studies of allocation problems. For example, in his discussion of the evolution of plant sexual systems, Darwin (1877, pp. 280-281) invokes "the law of compensation" in an astonishingly modern hypothesis about how dioecy could evolve in an originally hermaphroditic species: A few individuals begin making seeds that are larger than average, and this is "highly beneficial" in the current environment; "but in accordance with the law of compensation we might expect that the individuals which produced such seeds would, if living under severe conditions, tend to produce less and less pollen, so that their anthers would be reduced in size and might ultimately become rudimentary"; then as the large-seeded proto-females became common, other individuals would "produce a larger supply of pollen, and such increased development would tend to reduce the female organs through the law of compensation, so as ultimately to leave them in a rudimentary condition; and the species would then become dioecious."

Many problems now studied within optimization frameworks have similarly deep roots. For example, Fisher saw how life history evolution could involve resource-based trade-offs between growth and reproduction: "There is something like a relic of creationist philosophy in arguing from the observation, let us say, that a cod spawns a million eggs, that therefore its offspring are subject to Natural Selection; and it has the disadvantage of excluding fecundity from the class of characteristics of which we may attempt to appreciate the aptitude. It would be instructive to know not only by what physiological mechanism a just apportionment is made between the nutriment devoted to the gonads and that devoted to the rest of the parental organism, but also what circumstances in the life-history and environment would render profitable the diversion of a greater or lesser share of the available resources toward reproduction" (Fisher, 1930, 1958, p. 47).

Fisher also explained why 1:1 sex ratios should be common in a brilliant but cryptic first application of the kind of analysis that would later become evolutionary game theory (1930, 1958, pp. 158-160). Darwin (1871, p. 399) was famously baffled by sex-ratio evolution, even though he seems to have come very close to grasping how sex allocation gives rise to frequency-dependent fitness payoffs, as suggested by the dynamics in his scenario for the evolution of dioecy (quoted above).

Darwin (1871) both framed the problem of sex differences and outlined the theory of sexual selection. Fisher (1915, 1930) greatly admired this work and extended it in ways that would later prove important. Otherwise, there were few significant contributions to sexual selection until the 1960s, aside from some pioneering empirical work on mate choice in *Drosophila* by Bateman (1948) and others (reviewed by Andersson, 1993, pp. 17-19). Discussions of sexual selection seldom employ the language of optimization, even though the subject abounds with trade-offs involving aggression, advertisement, and mate choice, on the one hand, *versus* survival and offspring production, on the other. There are many possible reasons for this, including that secondary sex characters are obviously not "optimal" from a species-benefit point of view and that the dynamics of many kinds of sexual selection have long been understood to be inherently frequency dependent and therefore logically treated in explicitly game-theoretic kinds of frameworks.

Darwin, Fisher, and Haldane also anticipated, in different ways, the outlines of what is now known as inclusive-fitness theory or kin selection (Hamilton, 1964), another major branch of the optimization approach that emerged in the 1960s and grew explosively thereafter. Here the trade-off to be optimized, as a function of ecological (including social) circumstances, is one between direct reproduction and indirect or vicarious reproduction *via* relatives.

Why did these problems and ideas that exemplify the optimization approach remain nearly dormant for most of a century, and then burst into luxuriant growth at very nearly the same time? This fascinating question clearly needs more attention from historians of science, but some elements of the answer seem fairly obvious. We have mentioned the implict group-selectionist outlook that would have forestalled the analysis of situations involving conflicts of interest. More generally, this outlook may have deflected attention away from factors associated with variation in lifetime individual reproductive success (see Clutton-Brock, 1988; Newton, 1989). It may be relevant that the pioneers of evolutionary theory in the first half of the 20th century (geneticists, naturalists, paleontologists, and systematists alike) were preoccupied with questions about elementary mechanical aspects of evolution, in an environment where the field's status was nowhere near as secure as it is today. Until this work was largely finished, many of the kinds of questions addressed today through optimization approaches might have seemed of secondary importance to those who happened to think about them.

Writing in the very early 1960s, Orians (1962) argued that ecology had been for many years an almost defiantly nonevolutionary discipline concerned mainly with mechanisms of population regulation, but that the separation between ecology and evolutionary biology was at last breaking down. Orians' paper is called "Natural Selection and Ecological Theory," and its main purpose is to celebrate the marriage that would give birth to evolutionary ecology, including the movement in community ecology associated with G.E. Hutchinson and R.H. MacArthur. Orians is explicit about the need to explain diversity rather than merely taking it as a given parameter of some other problem. The concepts of the niche and of limiting similarity were applied, in the first instance, to closely related, ecologically similar species that differed mainly in quantitative aspects of their phenotypes, such as sizes, or times and places of foraging. Thus the evolutionary changes at issue posed no mechanistic difficulties because they consisted almost entirely of simple adjustments to the means and variances of some ordinary quantitative characters that varied in similar ways within each of the species in question. We would guess that this fusion of ecology and evolution was as important to the development of the optimization approach as was the exorcism of naive group selection. It encouraged both ecologists and evolutionists to think seriously about the causes and consequences of variation in the structural, behavioral, and reproductive parameters of sets of related species in their real ecological settings and it forced them to attempt to explain the evolution of those parameters primarily in terms of ecological opportunities and constraints.

In his approach to the problem of clutch-size variation in birds, Lack (1947, 1948, 1954, 1966) was far out in front of the trends discerned by Orians. The clutch-size problem was one of the first applications of the optimization approach to give rise to well integrated bodies of theory and data. It now defines a field of considerable size and sophistication, with well developed links to a range of general problems in foraging, life-history evolution, and even developmental biology (e.g., Newton, 1989; Daan et al., 1990; Gustaffson and Pärt, 1990; Meijer et al., 1990; Dhondt et al., 1991).

The modern evolutionary theory of senescence has roots going back to another early and remarkably advanced application of optimization thinking (Medawar, 1946, 1952; Williams, 1957; Hamilton, 1966; Charlesworth, 1980, 1994; reviewed by Rose, 1991; Partridge and Barton, 1993; see Chapter 7). This example is important for many reasons, including that the phenomenon to be explained (aging) appears at first to reflect a failure of adaptation. On closer examination this turns out to be only half true since there may be phenotypes that increase fitness early in life at the expense of fitness later in life, and *vice versa*: "A gene or combination of genes that [favors early reproduction at the expense of later reproduction] will under certain numerically definable conditions spread through a population simply because the younger animals it favours have, as a group, a relatively large contribution to make to the ancestry of the future population" (Medawar, 1946). Thus a shortened lifespan may evolve, in principle, by optimizing selection in the face of age-related trade-offs (see Bell and Koufopanou, 1986; Kirkwood and Rose, 1991).

Clutch sizes and patterns of senescence are specific elements of *life histories*, which can be defined as age-specific schedules of growth, reproduction, and mortality. Attempts to derive general theories of life-history evolution that could make sense of the great diversity of life histories seen within many groups of plants and animals have proceeded from the idea that schedules of growth, reproduction, and mortality should reflect evolutionary compromises among competing demands, under constraints that limit the possible combinations of vital rates (Gadgil and Bossert, 1970; Schaffer, 1974; Stearns, 1976). The models have become increasingly sophisticated and there has been much effort devoted to identifying, in comparative data, general patterns that can be used to test and refine the models (Lessells, 1991; Roff, 1992; Sibly and Antonovics, 1992; Stearns, 1992; Charnov, 1993).

The study of animal behavior has been transformed in recent decades by "strategic" analyses that extend and generalize the optimization approach. We have already mentioned sexual selection and kin selection, which gave rise to vigorous fields of theoretical and empirical research on mate choice and altruism. Many kinds of social behavior have adaptive landscapes that cannot be visualized and analyzed in the usual way because they are inherently frequency dependent. This is true, for example, of many kinds of conflict (Maynard Smith and Price, 1973) and reciprocity (Trivers, 1971). Evolutionary game theory (Maynard Smith, 1982) was developed as a framework in which to analyze such situations where there is technically no "optimum" at all. Instead, there is usually an "unbeatable" phenotype (or mixture of phenotypes) that resists invasion by others. As was mentioned earlier, sex ratios also fall into this category, and they were explicitly recognized as having a game-like character by Hamilton in 1967. Sex-ratio evolution has also developed into a huge, multifaceted field of research that exemplifies many features of the optimization approach.

Foraging was one of the first kinds of behavior to be treated explicitly as an optimization problem (Emlen, 1966; MacArthur and Pianka, 1966; Schoener, 1971; Charnov, 1976; reviewed by Stephens and Krebs, 1986), and in many respects it remains the "classic" application of the approach. It involves a wealth of straightforward, realistic trade-offs that are usually not frequency dependent (e.g., size of prey items *versus* the time needed to find or retrieve or process them), and it has been extremely successful in helping to elucidate the constraints under which animals operate in attempting to secure nutrition for themselves and their offspring. Indeed, its success has been so great that the words "optimal" and "foraging" have fused in many peoples' minds, producing a new term, "optimalforaging," that denotes what animals do when they get hungry.

What do all these applications of optimization have in common? One answer is that they are about phenotypes of a kind that did not concern the Reverend Paley. Habitat-patch selection rules, numbers of eggs in a clutch, and age-specific schedules of growth, fertility, and mortality are not like watches. They do not shout "Look here! I'm an intricate, improbable, and therefore onerous *adaptation*! Explain me if you can!" Darwin's puzzlement about the sex ratio derives from his realization that it must be subject to selection, even though he could not see how to assess the adaptedness of a given ratio of males to females among the progeny of a reproducing individual.

Little theory is needed to recognize that the vertebrate eye embodies a great deal of adaptation. Here the problem (and the main purpose of theory) is to explain the manifest adaptation of the eye. A sex ratio, on the other hand, is not obviously a problem at all. We need theory to reveal the problem, by explaining in what ways sex ratios might be adapted (Thornhill, 1990). In other words, an important difference between eyes and sex ratios is that the presence of design is obvious for eyes but not for sex ratios, where the initial problem is to find a way to recognize the presence of design and to assess what would constitute a more or less well-designed sex ratio. Much the same can be said for most of the examples discussed above, and others like them, where the role of theory is subtly but importantly different than in examples like that of the eye. Of course optimization approaches can be used (and have been used) to study the designs of eyes (Goldsmith, 1990) and other structures from molecules through organ systems of many kinds, to the shapes of entire organisms. But optimization models serve significantly different functions in the two kinds of situations, and they tend to play much more central and conspicuous roles in the analysis of sex-ratio-like phenomena than they do in the analysis of eye-like phenomena.

V. The Aims and Limits of Optimization

For convenience and for the sake of brevity we have tended to speak of "the optimization approach" as though it were a well-defined method that could be applied in much the same way to almost any problem in the analysis of adaptation, but of course this is not so. In some applications, the approach amounts to little more than asking pointed questions about the selective regime and the potentially important constraints: What's the function of this structure? How does the observed variation in its form affect the fitnesses of individuals? Why does it take slightly different forms in different species, and in the sexes? Are there costs involved in producing different forms of the structure? Do the costs vary with environmental circumstances? What historical legacies might be limiting the range of accessible variation? And so on. At this level, the approach is mainly heuristic and could be described simply as "thinking like a biologist." Answers to the questions are plainly speculative and would never be viewed as more than guides to further empirical description or experimentation. Of themselves, they would never be mistaken for results or conclusions.

But when we move beyond this rather casual level of thinking about the evolution of an interestingly constrained phenotype that varies and that might have significant effects on fitness, and begin to construct more formal, mathematical models that identify specific optima as functions of given parameters, the relationship between an optimization model and a scientific belief becomes much more confusing. Of course a model is just a model, and the world is the world. But we cannot simply leave it at that because the whole point of the exercise is to use models to illuminate the world. How is this connection to be made, and how do we know what we have learned as a result?

A standard answer (suitable for scientists if not philosophers) is that models are intentional caricatures or cartoons, whose purpose is to help us strip away irrelevant complications so as to gain some insight about how a small number of key variables might interact. On this view a model is literally a toy world that we can manipulate and dissect in ways that we cannot directly manipulate or dissect the real world. Having understood the behavior of the toy world, we can ask whether it seems to mimic the real world in ways that suggest it embodies (in its highly abstracted and simplified way) interactions *like* those that occur in reality. To the degree that we persuade ourselves that it does, and that the modeled interactions are general, we have learned something potentially important about how the world might actually work (see Levins, 1966; Beatty, 1980; Grafen, 1991).

But can't we do better than this? Can't we validate models by testing their predictions? This is where the trouble begins because the answers are not straightforward. An optical model of an eye might be tested decisively because it embodies a fairly complete description of those aspects of the system at issue and because all of its parameters can be estimated directly. Such models have in fact been constructed and used to show that the vertebrate eye achieves a remarkably large fraction of its theoretical maximum optical efficiency on various measures of performance (Goldsmith, 1990). This is fascinating, but it does not answer the kinds of questions we are likely to be most interested in as students of the evolution of adaptation. After all, vertebrate eyes would still seem almost miraculous (and Paley's arguments would be just as important, qualitatively) if they were only 10% or even 1% efficient. Models that address the questions we are most concerned about, as evolutionists, are generally not subject to such decisive tests because they do not incorporate all the relevant parameters, or because not all the parameters can be estimated accurately, or both.

What, then, are the questions we hope to answer with the help of optimization models? It is sometimes stated that such models are made on the assumption, or for the purpose of showing, that the phenotypes under study are "optimal," or that natural selection produces optimal adaptations, or that all of the characteristics of organisms are adapted (e.g., Lewontin, 1978, 1979, 1984; Gould and Lewontin, 1979; Dupré, 1987; Orzack and Sober, 1994a,b). This "program" is often referred to as "adaptationism." In fact, the aims of real optimization studies are rarely to see how close a phenotype is to its optimal state. Herre's (1987) study of sex-ratio adjustment in 13 species of fig wasps is one of the few examples we know of (Fig. 4). Here the model can be assumed to capture the relevant biology with enough quantitative accuracy to allow the fitness consequences of a given sex ratio to be evaluated. And for reasons

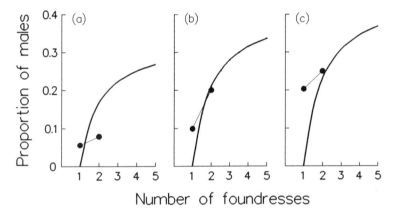

Figure 4 Predicted and observed sex ratios for three species of fig wasps in Panama. The species pollinate (a) *Ficus paraensis*, (b) *F. popenoei*, and (c) *F. trigonata*. In (a), 97% of all figs contain the offspring of just a single female. In (b), 36% of broods are from single females. In (c), only 7% of broods are from single females. The solid curves show predicted ESS sex ratios for each species as a function of the number of foundresses contributing offspring to the mating group inside a fig. The curves differ slightly due to differences in the average intensity of inbreeding in the three species. The filled circles show *observed* sex ratios in single-foundress and two-foundress figs. Note that in (a) and (c) mother wasps produce sex ratios near the prediction in the kind of fig they more often encounter, but not in the kind of fig they seldom encounter. However, in (b), a species that frequently encounters both single-foundress and two-foundress figs, sex ratios are fairly near the predicted ESS in both conditions. This suggests that selection in (a) and (c), has been too weak to maintain an appropriate sex-ratio response to a situation that females seldom encounter. Ten other species of fig wasps were studied in the same way. On average, the magnitudes of the adjustments made by females of a given species were positively related to the variance of foundress number in the field, consistent with the hypothesis that the precision of an adaptation such as this sex-ratio adjustment (which is absolutely advantageous in all species) will be higher in species that gain more from making precise adjustments than in species that gain less because they less often encounter figs with foundress numbers different from the mean for their species. Redrawn from Herre (1987) courtesy of the author and *Nature*.

explained in the figure legend, "how close" is a genuinely interesting question in this case. However, this is not a typical use of optimization models. The hypothesis under test is usually not that a phenotype is optimal, but instead that the specific assumptions embodied in the model (e.g., the sources of selection, and the constraints) could account, at least in principle, for the evolution of the phenotypes under study (see Parker and Maynard Smith, 1990). "The role of optimization theories in biology is not to demonstrate that organisms optimize. Rather, they are an attempt to understand the diversity of life" (Maynard Smith, 1978).

Formal optimization models figure prominently in the study of allocation of nutrients, time, and other resources to growth, reproduction, and other activities, but for several reasons they figure less prominently in the study of structures such as eyes. As we mentioned earlier, such models provide a means of detecting and evaluating the designed features of allocations, which are processes rather than architectures in the usual sense. In addition, such allocations often involve well-defined trade-offs that lend themselves to mathematical analysis. Allocations are also usually continuous or nearly continuous quantitative traits that can reasonably be expected to be under polygenic control and sufficiently heritable to permit their evolution in any relevant direction, if subject to selection. And most importantly, the kinds of allocations studied in this way are usually closely connected to fitness, so they can be assumed to be subject to fairly strong selection. Thus attention can be focused on attempting to understand the nature of that selection and of the constraints that modulate its effects. If a model is constructed and found to give a poor fit to data, suspicion reasonably falls on the model's specific biological assumptions, not on the assumption that the allocation affects fitness. Under these circumstances, the purely phenotypic and relatively ahistorical optimization approach makes sense, and it demonstrably works.

There is more to the diversity of life than resource allocation, of course. Optimization models will remain relatively unimportant in many areas where the diversification we seek to understand involves evolutionary transitions of a relatively qualitative kind, or where selection was not the dominant evolutionary force, or where there is no natural way to parameterize the trade-offs of interest in a common currency. In such situations optimization analyses may be impossibly difficult or simply beside the point. Even within a single complex structure or behavior, certain features may show clear evidence of optimization that can usefully be studied with the aid of formal models, while other features show little evidence of optimization and resist such analysis. For example, the vertebrate eye is stunningly optimized in some respects, yet it also has features that are conspicuously nonadaptive or maladaptive, as illustrated in the following passages from Williams (1992):

> As [Paley] claimed, the eye is surely a superbly fashioned optical instrument. It is also something else, a superb example of maladaptive historical legacy. The retina consists of a series of special layers in the functionally appropriate sequence. A layer of light-sensitive cells (rods and cones) stimulate nerve endings from one or more layers of ganglion cells that carry out initial stages of information processing. From these ganglia, nerve fibers converge to form the main trunk of the optic nerve, which conveys the

information to the brain. All layers are served by blood capillaries that provide their metabolic requirements. Unfortunately for Paley's argument, the retina is upside down. The rods and cones are the bottom layer, and light reaches them only after passing through the nerves and blood vessels. ... That we do not ordinarily perceive these shadows [cast by the larger blood vessels] is the result of minute involuntary eye movements, which keep the blood-vessel shadows moving, and of our brains recording the flux of images as continuous pictures. The reality of the shadow of the *vascular tree*, and the seriousness of the problem it presents, can be demonstrated with a flashlight and instructions from a visual physiologist.

Williams describes some other "stupidly designed" features of the vertebrate eye, and then offers a historical scenario that shows how our presently maladaptive retinal orientation seems to have been assembled, in a sequence that could have been driven by optimizing selection at every step.

The vertebrate eye originated in a tiny transparent ancestor that had no blood corpuscles and formed no retinal images. The retinas arose as light sensitive regions on the dorsal side of the anterior end of the nervous system. Evolutionary conversion of a flat to a tubular nervous system put the future retinas inside. In subsequent evolution, the photosensitive layer pushed outward from the brain to become part of the complex optical instrument known as the eye. All through history this layer has retained its position beneath the other layers of the retina. ... Unlike that of a vertebrate, the retina of a squid is right side up. Molluscan eyes evolved independently of vertebrate eyes, and show an entirely different suite of historical legacies. (Williams, 1992, pp. 72-74)

In summary, the outlook and techniques of optimization modeling are applied most often and most profitably to the analysis of phenotypes that involve the "allocation" of a continuous or nearly continuous "resource" of some kind to several "competing" activities with direct effects on fitness. Such phenotypes tend to show clear-cut trade-offs and continuous variation in parameter values. They can be modeled in fairly natural ways, are typically subject to strong selection, and are expected to respond in appropriate ways to such selection.

VI. How Do We Know What Matters and What Is Feasible?

When we considered the evolution of clutch size in birds we did not even ask whether we had found appropriate definitions of the actor (parent birds), the strategy set (any number of eggs from 1 to n), and the objective function (the number of offspring surviving to reproductive age). The analysis was equally straightforward: find the number of eggs n^* that maximizes the number of offspring that survive to one year of age. We did not worry about these matters because they seemed obvious, and because after we took account of a few important complications such as parental variation in condition, the result was a coherent and seemingly correct (or approximately correct) explanation of how one group of birds regulates clutch size. But there are cases in which these central elements of the model are not so easily identified. Indeed, a major purpose of optimization modeling is often to help answer the questions posed by the title of this section.

When two or more imperfectly related actors are involved in a common endeavor, their optima may differ. For example, if the eggs already laid in a clutch were allowed to decide when their parents should stop laying, they would usually have the parents settle on a smaller clutch than was optimal from the parents' point of view because each egg values its own future reproduction more than that of its siblings. At some point adding an additional sibling would reduce each egg's expected fitness (by reducing its fledging weight) more than it would contribute indirectly through the sibling's reproduction. Of course it is hard to imagine an egg imposing its will on a parent, but hatchlings can interfere with each other in ways that advance their interests at the expense of the interests of their siblings and their parents (e.g., Godfray and Parker, 1991; Mock and Parker, 1996).

There are many other situations where *parent-offspring conflict* (Trivers, 1974) implies that an allocation or other decision will not be optimal for some members of a family. For example, in many social Hymenoptera the equilibrium sex allocation is 1:1 from the queen's point of view, but female biased (up to 3:1) from the workers' point of view, due to the workers' closer genetic relationship to the colony's female reproductives than to its males (Trivers and Hare, 1976). A pattern of biased allocation ratios supports the view that the interests of workers largely prevail in many species. This model has been extended in various ways. For example, one extension predicts that bimodal sex-ratio variation should sometimes occur among colonies within a species as a response by workers to inter-colony variation in the magnitude of the relatedness asymmetry (Boomsma and Grafen, 1990). Both the prediction (bimodal sex-ratio variation) and the assumed causal factor (relatedness variation) have been confirmed in several species of ants (e.g., Sundström, 1994; Evans, 1995) (see Fig. 5).

A mammalian infant is imperfectly related to its mother and its present and future siblings, and it was equally imperfectly related to them as a fetus *in utero*. There is, as a consequence, no reason to assume that the intimate physiological relationship between mother and fetus will be purely cooperative and "optimally" designed for the efficient production of a healthy neonate, although this has long been an accepted view. Once it is appreciated that what is optimal for the fetus may differ from what is optimal for the mother (with respect to at least some aspects of the relationship), then evidence of pervasive design for conflict, on both sides, begins to emerge from a critical reevaluation of fetal and placental development and physiology (Haig, 1993 and personal communication).

There may even be conflicts within a fetus, between the maternally and paternally inherited halves of its genome, stemming from the unequal genetic relationships of maternally and paternally inherited genes to their homologs in siblings of the fetus (Haig, 1992). Thus the optimal pattern of fetal gene expression may frequently differ for genes inherited maternally and paternally. Differential gene expression of the expected kind appears to occur in diverse taxa, mediated by the mechanism of genomic imprinting (Haig and Graham, 1991).

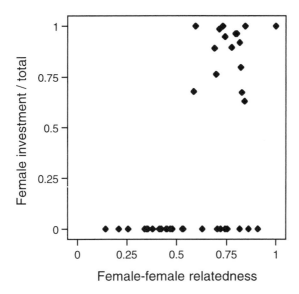

Figure 5 Proportional investment in female reproductives as a function of the average coefficient of relatedness among females within nests of the facultatively polygynous ant *Myrmica tahoensis*. Only nests with high relatedness (on average, near the maximum possible value of 0.75 for outbred haplodiploid full sisters) produced female reproductives, and if they did so, they produced *mainly* female reproductives. All nests with average relatedness coefficients below 0.5 and some with higher values produced only male reproductives. In this species, the considerable variation in average relatedness within nests is caused almost entirely by variable numbers of variably related queens; most queens mate with just one male, so all their daughters are full sisters. Redrawn from Evans (1995).

It is ironic that the "optimization" approach to the study of adaptation has revealed a wealth of irreconcilable conflicts of interest that often give rise to wasteful squabbles. Conflicts of interest tend to be noticed by investigators pursuing the optimization approach because the approach demands that actors and objective functions be explicitly identified.

In situations where the relative fitnesses of alternative phenotypes are frequency dependent, the appropriate objective function may not be subject to optimization in the usual sense. As was mentioned earlier, this often occurs in situations involving conflict, mate choice, mimicry, sex allocation, and other interactions where rarity tends to be advantageous. In these situations, optima are replaced by *evolutionarily stable strategies* (ESS) (Maynard Smith and Price, 1973), which are phenotypes or mixtures of phenotypes that cannot be invaded by other members of the strategy set. The ESS is often absolutely worse, for all actors, than another more "cooperative" equilibrium would be, but the cooperative equilibrium is evolutionarily unstable because it is subject to invasion by "defectors." The formal game most often used to illustrate this point is the prisoners' dilemma, in which both parties stand to gain the highest fitness by cooperating, and the lowest by defecting, but defection turns out to be (tragically) the only ESS (Maynard Smith, 1982). Balanced sex ratios are often tragic in a similar way. For most species in which males do not help to rear young, the species would do best if it had a strongly female-biased sex ratio because its potential rate of population increase is a function of the number of females, not the number of males. But such globally "optimal" sex ratios do not persist because they can be invaded by genotypes that incline their bearers to make only male offspring. What, then, is the sex ratio a design for doing? This question repays careful thought.

Sex-ratio evolution stumped Darwin because he could not identify an appropriate currency and objective function. Fisher's breakthrough was to realize that for purposes of this problem, there is a qualitative difference between "number of children" (which does not lead to a correct analysis) and "number of grandchildren" (which does, at least in most cases). With this key insight about how to assess the fitnesses of alternative phenotypes, and with an appreciation of frequency dependence, sex allocation suddenly becomes understandable. It has provided rich opportunities for successful application of the optimization approach because the trade-offs involved tend to have large effects on fitness and because the key variables can often be estimated fairly directly (see Charnov, 1982).

The most appropriate measure of fitness is seldom as easy to define as it often is for problems like clutch size and sex allocation. For example, theorists concerned with the evolution of senescence and related life-history phenomena have spent much effort trying to understand how measures such as age-specific intrinsic rates of increase can be used to evaluate the relative long-term fitnesses arising from life-history parameters in age-structured populations (see Rose, 1991; Charlesworth, 1980, 1994). Fitness can also be remarkably difficult to measure in temporally varying environments, especially for age-structured populations. Cohen showed in 1966 that partial germination of seeds could be favored by selection in annual plants that experience drastic year-to-year variation of environmental quality, when doing so maximizes their geometric mean fitness (reviewed by Seger and Brockmann, 1987; Philippi and Seger, 1989). But no similarly general fitness measure for age-structured populations (e.g., perennial plants) has yet been discovered. Some recent theoretical explorations suggest that age structure in the context of temporal variation may sometimes support unexpectedly complex polymorphisms involving "quantized" alternatives (Sasaki and Ellner, 1995; Ellner and Sasaki, 1996). Such findings would undoubtedly be easier to explain if we understood at a fairly general level how to construct natural objective functions for these kinds of populations.

The objective-function problem becomes even more difficult where the phenotypes under study are "performances" (such as that of the eye) that cannot easily be tied to fitness in a quantitative way. In such cases it may be almost impossible to parameterize the costs and benefits realistically. What is the cost of making a certain bone slightly thicker and thereby stronger? What exactly is the benefit? It may be possible to frame the issue in qualitative terms (e.g., a thicker bone will decrease maximum running speed and impose increased energetic costs of transport), but even where these functional costs can be estimated empirically, there is no obvious way to map them onto fitness differences. Without such a mapping, it is impossible to estimate the trade-off structure under which optimization is expected to take place. [See Chapter 2.]

As a consequence of this problem, the optimization approach tends to be applied in a different way to phenotypes relatively far removed from reproduction (for example, many aspects of morphology) than it is to clutch sizes, sex ratios, and the like. Where realistic, fully parameterized models cannot be constructed, the approach tends to emphasize induction from comparative data rather than hypothetico-deductive reasoning from first principles. For example, many comparative morphologists and physiologists are now concerned with fascinating questions about margins of safety in the design of organ systems that may fail catastrophically under unusually heavy loads that occur only rarely, if at all, in the life of a typical individual. "Excess capacities" should certainly be subject to optimization if they cost anything, which they undoubtedly do. But to estimate those costs directly would be extremely difficult if not impossible. However, where the excess capacities can be measured their allometries can be studied comparatively (e.g., Niklas, 1994a,b), and the resulting patterns can be used to refine hypotheses about the evolution of safety factors. In cases such as this, optimization modeling is more heuristic than analytical, but it may nonetheless be valuable as an agent of biological intuition and imagination, and it may thereby contribute to the refinement of the questions addressed by empirical research.

One of our favorite examples of a good question derived from heuristic optimization reasoning concerns the relative testis sizes of primates. Like many kinds of structures, testes show regular allometric relationships with body size, but not all species lie close to the line describing the overall trend of the data. Harcourt et al. (1981) decided to ask whether the deviations might be explained, in part, by variation in the mating systems of the species. What made them ask such a question? They realized that in species where estrous females commonly mate with more than one male, the fitnesses of males would be determined in part by sperm competition. Thus, other things being equal, males that produced large quantities of sperm should realize larger benefits from doing so in such species than in those where females typically mate with only one male. If sperm production were costly, then the optimal level of sperm production would be higher in species with multi-male social groups than in monogamous species and harem-polygynous species such as the gorilla. On the assumption that testis size and sperm production are correlated, Harcourt et al. reasoned that multimale species should tend to fall above the average allometric relationship (with testes large for their body sizes), while single-male species should tend to fall below the line. This is exactly what they found. The result is of interest for many reasons, including that it shows how different aspects of sexual dimorphism may exhibit very different patterns of association with each other and with ecological and social variables (see Harvey and Harcourt, 1984). For example, male gorillas are large and "masculine," both absolutely and relative to female gorillas, but they have testes that are absolutely and relatively tiny. Similar patterns occur in other taxa such as whales (Brownell and Ralls, 1986). For example, right whales have a spectacularly promiscuous mating system, and they have the largest testes recorded for any animal (both absolutely and relatively, at more than 1% of body mass). The mating systems of most large whales are unknown, so whale biologists interpret the observed allometric deviations as hints about the patterns of behavior to be expected in these species.

The difficulties involved in defining the actors and objective functions may sometimes be substantial, but they are seldom as great as those involved in defining the constraints and strategy sets. One problem is that the constraints are of many different kinds that are not always clearly distinguished. For example, many physical and ecological constraints correspond fairly closely to those that arise in engineering, but there are other, more "internal" developmental and genetic constraints that reflect historical legacies of a kind that loom much larger in biology than they do in engineering, where it is often possible to "leap" from one peak to another in the adaptive landscape. Evolution can only "tinker" with what already exists (Jacob, 1977). If a lineage is to survive then there must be a continuously viable and selectively favorable path between each successive state of any phenotype that is subject to even moderately strong selection. This pervasive dynamical constraint is at the root of many others in biology, but it has only weak analogs in engineering. The jet engine did not evolve gradually from the reciprocating engine, nor did the 747 evolve gradually from the Wright brothers' flyer, least of all in midair! Yet this is equivalent to what organisms do routinely. Among the consequences are "stupid" designs like those of the vertebrate retina, as described above.

Historically derived developmental and genetic constraints are of special concern because they may give rise to hidden components of the trade-off structure that prevent a phenotype from evolving in a direction that we might otherwise reasonably expect to see. The existence of such a constraint may be suggested by a striking lack of intra- or interspecific variation. For example, the number of cervical vertebrae is highly variable across tetrapods as a whole, but in some groups the number is remarkably constant. All turtles appear to have exactly eight cervical vertebrae, even though the necks of turtles are very diverse in other respects. Similarly, all mammals have seven cervical vertebrae, except for manatees (six) and three-toed sloths (eight or nine). Even "neckless" mammals such as whales and dugongs, and long-necked mammals such as giraffes, have seven. Why such constancy? It seems very hard to believe that eight is currently optimal for all turtles, and seven for almost all mammals from shrews to whales. Indeed, there are reasons to think that other numbers might be of great value to some turtles or mammals. Sloths exhibit an owl-like ability to rotate their heads through almost a complete circle, and a similar range of motion might well be of advantage to many other mammals. But there is evidently something about the developmental programs of turtles and mammals that makes it very difficult for the number of cervical vertebrae to change. Not all tetrapod groups show such conservatism. For example, the plesiosaurs included species with short necks and few vertebrae (e.g., 13 in *Brachauenius*), and others with elongated, flexible necks and many vertebrae (e.g., 76 in *Elasmosaurus*) (Romer, 1966).

Similar instances of extreme conservatism in some taxa and great variability in others can be found throughout the living world. For example, the parasitic Hymenoptera show a wide range of numbers of segments in the antennae, but all aculeate Hymenoptera (the familiar ants, bees, and true wasps) have 13 antennal segments in males and 12 in females. The only exceptions are a few species in which males, like females, have 12 segments. Again, it seems inconceivable that this pattern represents a uniform optimum, with respect to antennal function, for tens of thousands of aculeate species. Instead, it seems more likely that such patterns reflect other functions that impinge, in other ways, on the number of antennal segments. This can happen, for example, when particular structures or the developmental processes that give rise to them become necessary antecedents to the development of other structures that make an important contribution to fitness (see Maynard Smith et al., 1985). Thus the number of antennal segments may indeed be under stabilizing selection, but for reasons that have nothing to do with antennal performance as such. As a consequence, it may be more feasible to radically modify the sizes and morphologies of individual antennal segments (or cervical vertebrae, etc.) than to change their number.

In constructing a strategy set, it is important to ask whether the feasible phenotypic options may be limited (effectively if not absolutely) by deep-seated constraints whose origins and influence on the phenotypes of interest are less than obvious. Sometimes the existence of such unappreciated constraints is revealed by apparent "failures" of optimization. For example, annual plants seem to do some things exactly as optimization models say they should, such as adjusting their flower lifetimes in relation to construction and maintenance costs and the behavior of pollinators (Ashman and Schoen, 1994). But they chronically fail to show the kinds of sudden switches from vegetative growth to reproduction that simple life-history models often predict. This lack of fit between models and reality has led in recent years to the view that fundamental aspects of plant architecture and development may often make it most profitable for a plant to grow and reproduce simultaneously for much of the season (see Fox, 1992). Analogous temporal constraints on offspring production in bees may force a revision of standard models for the evolution of hymenopteran eusociality. These models treat reproduction as an instantaneous process, and for that reason proved unable to account for patterns of genetic relatedness within colonies of a very well studied sweat bee (Richards et al., 1995).

The principal goal of an optimization study is often, at least implicitly, to describe and understand the constraints and strategy sets. We want to know why organisms are limited in the ways they are, for example, to making their livelihoods only in certain ways, at certain places, and with finite lifespans. It is therefore no accident that the word "constraint" appears in the titles of increasing numbers of papers in which optimization approaches are used to study adaptation (e.g., Goldsmith, 1990; Partridge and Sibly, 1991; Arnold, 1992; Jones et al., 1992; Moran, 1994). Some constraints are highly particular and historically contingent, but others may be more or less universal. When the latter give rise to trade-offs that affect many species in similar ways, they may reveal themselves as allometric patterns in large-scale comparative data sets (e.g., Harvey and Pagel, 1991). The study of such patterns may lead to the discovery of general rules about important classes of trade-offs (for example, ones affecting particular aspects of life-history evolution) and hence, eventually, to a deeper understanding of the causes of the underlying constraints (Charnov, 1993).

VII. Conclusions

Natural selection can be viewed as a process that evaluates alternative designs by increasing the frequencies of those that are better adapted for reproduction in the current environment. Students of adaptation mimic this process by constructing and analyzing optimization models. Such models are among the most drastic simplifications tolerated anywhere in biology, and they are sometimes criticized for this reason. For example, they notoriously ignore the issues of evolutionary dynamics entailed by the genotypic control of phenotypes. But this simplification serves an important function. By focusing attention on a few key aspects of the situation under study, it elicits questions that might otherwise go unasked. Genetic models are most productive when they connect genotypes to phenotypes in ways that make biological sense, for example, by embodying appropriate constraints and by not preventing unbeatable genotypes to emerge. Thus optimization approaches can be viewed as important components of genetic approaches to the study of adaptation (see Chapter 4).

Optimization approaches work best where the phenotypes under study can change without forcing many other aspects of the organism to change at the same time ("quasi-independence"; Lewontin, 1984). The "mysterious laws of the correlation of growth" troubled Darwin (1859) for the same reasons that they trouble modern students of adaptation, but in fact such correlations do not interfere, in the end, with many kinds of evolutionary change. It seems extremely interesting and important that real organisms are only modestly integrated in this sense; they show "natural planes of cleavage" among organ systems, biochemical pathways, life stages, behaviors, and the like, which are at least sufficiently smooth to allow some character states to respond to selection without too seriously degrading the adaptation of other character states. What determines where these planes of cleavage are located and how deep and smooth they are? For understandable reasons, biologists have long treated them as unfathomable "givens." We suggest that it may now be possible to begin working toward a "theory of organic articulations" that would give insight into the "laws of correlation." Genome sequencing projects are already allowing us to glimpse, for the first time, not only the outlines of complete genetic "parts lists" for entire organisms, but just as importantly, the histories of those parts. A few theorists have begun to ask questions about the spontaneous emergence of "subsystems" in abstract functional networks subject to adaptive evolution (see Kauffman, 1993), but it should soon be possible to add some realism to such models, and it should certainly be possible to ask experimentally whether certain kinds of maladaptive pleiotropies and genetic correlations tend to be selected downward and, if so, then to find out how the response was achieved. In short, we suspect that the early 21st century will see the emergence of a field of study focused on questions about the evolution of organ systems, networks of genetic regulation, and other aspects of the "partly compartmented" design of organisms.

While biologists were borrowing classical optimization techniques from economics and engineering, workers in those fields were turning to biology for the inspiration behind genetic algorithms, which solve optimization problems by simulating evolution. This exchange should lead eventually to an enriched understanding of adaptation. Insights about the properties of successful genetic algorithms could illuminate the biological processes that inspired them. Genetic algorithms became interesting and practical after high-speed computing became widely available and virtually free. For biology, effectively unlimited computing power opens up the possibility of a previously impractical style of "experimental" theoretical study, in which model genetic systems underlying (initially) specified developmental systems are set up and allowed to evolve in large populations for very large numbers of generations. With increasing knowledge of development, we might come full circle and study adaptation by modeling the evolutionary process directly. For a long time we evaded the dynamical complexities of genetic evolution by reducing it to a black box with just a few parameters (as in quantitative genetics) or by simply assuming it away and directly optimizing the phenotypes themselves. Perhaps we will soon begin to understand at a deeper level why it has so often been possible to get away with this.

This chapter has pointed out repeatedly that few character states may ever be in truly "optimal" states, although most may be (metaphorically) "pursuing" optimality. If maladaptation is a chronic condition, shouldn't organisms be adapted to it? Could the phenomena long studied under the rubric of "developmental canalization" be manifestations, in part, of mechanisms that evolved to ease the burdens of imperfection?

Acknowledgments

We thank Ric Charnov and the editors for encouragement and for many helpful suggestions.

References

Alexander, R. McN. (1982) "Optima for Animals." Edward Arnold, London.

Andersson, M. (1993) "Sexual Selection." Princeton University Press, Princeton.

Arnold, S. J. (1992) Constraints on phenotypic evolution. *Am. Nat.* 140(supplement), S85-S107.

Ashman, T.-L., and Schoen, D. J. (1994) How long should flowers live? *Nature* 371, 788-791.

Bateman, A. J. (1948) Intra-sexual selection in *Drosophila. Heredity* 2, 349-368.

Beatty, J. (1980) Optimal-design models and the strategy of model building in evolutionary biology. *Phil. Sci.* 47, 532-561.

Bell, G., and Koufopanou, V. (1986) The cost of reproduction. *Oxf. Surv. Evol. Biol.* 3, 83-131.

Boomsma, J. J., and Grafen, A. (1990) Intraspecific variation in ant sex ratios and the Trivers-Hare hypothesis. *Evolution* 44, 1026-1034.

Boyce, M. S., and Perrins, C. M. (1987) Optimizing Great Tit clutch size in a fluctuating environment. *Ecology* 68, 142-153.

Brownell, R. L., and Ralls, K. (1986) Potential for sperm competition in baleen whales. *Rep. Int. Whal. Commn. Special Issue* 8, 97-112.

Bulmer, M. (1994) "Theoretical Evolutionary Ecology." Sinauer, Sunderland, MA.

Charlesworth, B. (1980) "Evolution in Age-Structured Populations." Cambridge University Press, Cambridge.

Charlesworth, B. (1990) Optimization models, quantitative genetics and mutation. *Evolution* 44, 520-538.

Charlesworth, B. (1994) "Evolution in Age-Structured Populations, 2nd ed." Cambridge University Press, Cambridge.

Charnov, E. L. (1976) Optimal foraging: attack strategy of a mantid. *Am. Nat.* 110, 141-151.

Charnov, E. L. (1982) "The Theory of Sex Allocation." Princeton University Press, Princeton NJ.

Charnov, E. L. (1989) Phenotypic evolution under Fisher's fundamental theorem of natural selection. *Heredity* 62, 113-116.

Charnov, E. L. (1993) "Life History Invariants." Oxford University Press, Oxford.

Charnov, E. L., and Downhower, J. F. (1995) A trade-off-invariant life-history rule for optimal offspring size. *Nature* 376, 418-419.

Charnov, E. L., Downhower, J. F., and Brown, L. P. (1995) Optimal offspring sizes in small litters. *Evol. Ecol.* 9, 57-63.

Charnov, E. L., and Krebs, J. R. (1974) On clutch-size and fitness. *Ibis* 116, 217-219.

Clutton-Brock, T. H. (ed) (1988) "Reproductive Success: Studies of Individual Variation in Contrasting Breeding Systems." University of Chicago Press, Chicago.

Cody, M. L. (1974) Optimization in ecology. *Science* 183, 1156-1164.

Cohen, D. (1966) Optimizing reproduction in a randomly varying environment. *J. Theor. Biol.* 12, 110-129.

Crawley, M. J. (ed) (1992) "Natural Enemies." Blackwell, Oxford.

Daan, S., Dijkstra, C., and Tinbergen, J. M. (1990) Family planning in the kestrel (*Falco tinnunculus*): the ultimate control of covariation of laying date and clutch size. *Behaviour* 114, 83-116.

Darwin, C. (1859) "On the Origin of Species by Means of Natural Selection." John Murray, London.

Darwin, C. (1871) "The Descent of Man, and Selection in Relation to Sex." John Murray, London.

Darwin, C. (1877) "The Different Forms of Flowers on Plants of the Same Species." John Murray, London.

Dawkins, R. (1986) "The Blind Watchmaker." Norton, New York.

Dennett, D. C. (1995) "Darwin's Dangerous Idea: Evolution and the Meanings of Life." Simon & Schuster, New York.

Dhondt, A. A., Adriaensen, F., Matthysen, E., and Kempenaers, B. (1991) Non-adaptive clutch sizes in tits: evidence for the gene flow hypothesis. *Nature* 348, 723-725.

Dupré, J. (ed) (1987) "The Latest on the Best: Essays on Evolution and Optimality." MIT Press, Cambridge MA.

Ellner, S., and Sasaki, A. (1996) Patterns of genetic polymorphism maintained by fluctuating selection with overlapping generations. *Theor. Pop. Biol.* (in press).

Emlen, J. M. (1966) The role of time and energy in food preference. *Am. Nat.* 100, 611-617.

Emlen, J. M. (1984) "Population Biology: The Coevolution of Population Dynamics and Behavior." MacMillan, New York.

Endler, J. (1986) "Natural Selection in the Wild." Princeton University Press, Princeton, NJ.

Evans, J. E. (1995) Relatedness threshold for the production of female sexuals in colonies of a polygynous ant, *Myrmica tahoensis*, as revealed by microsatellite DNA analysis. *Proc. Natl. Acad. Sci. USA* 92, 6514-6517.

Fisher, R. A. (1915) The evolution of sexual preference. *Eugenics Review* 7, 184-192.

Fisher, R. A. (1930) "The Genetical Theory of Natural Selection." Clarendon Press, Oxford.

Fisher, R. A. (1958) "The Genetical Theory of Natural Selection." second revised edition, Dover, New York.

Fox, G. A. (1992) Annual plant life histories and the paradigm of resource allocation. *Evol. Ecol.* 6, 482-499.

Gadgil, M., and Bossert, W. H. (1970) Life historical consequences of natural selection. *Am. Nat.* 104, 1-24.

Ghiselin, M. (1983) Lloyd Morgan's canon in evolutionary context. *Behav. Brain Sci.* 6, 362-363.

Gillespie, J. H. (1991) "The Causes of Molecular Evolution." Oxford University Press, New York.

Godfray, H. C. J., and Parker, G. A. (1991) Clutch size, fecundity and parent-offspring conflict. *Phil. Trans R. Soc. Lond. B* 332, 67-79.

Godfray, H. C. J., Partridge, L., and Harvey, P. H. (1991) Clutch size. *Ann. Rev. Ecol. Syst.* 22, 409-429.

Goldsmith, T. H. (1990) Optimization, constraint, and history in the evolution of eyes. *Q. Rev. Biol.* 65, 281-322.

Gould, S. J., Lewontin, R. C. (1979) The spandrels of San Marco and the Panglossian paradigm – a critique of the adaptationist program. *Proc. R. Soc. Lond. B* 205, 581-598.

Grafen, A. (1991) Modeling in behavioural ecology. *In* "Behavioural Ecology: An Evolutionary Approach" (J. R. Krebs, and N. B. Davies, eds.), 3rd ed., pp. 5-31. Blackwell Scientific Publications, Oxford.

Grant, B. R., and Grant, P. R. (1989) "Evolutionary Dynamics of a Natural Population: The Large Cactus Finch of the Galapagos." University of Chicago Press, Chicago.

Grant, B. R., and Grant, P. R. (1993) Evolution of Darwin's finches caused by a rare climatic event. *Proc. R. Soc. Lond.* B 251, 111-117.

Grant, P. R. (1986) "Ecology and Evolution of Darwin's Finches." Princeton University Press, Princeton.

Gustaffson, L., and Pärt, T. (1990) Acceleration of senescence in the collared flycatcher *Ficedula albicollis* by reproductive costs. *Nature* 347, 279-281.

Haig, D. (1992) Genomic imprinting and the theory of parent-offspring conflict. *Sem. Dev. Biol.* 3, 153-160.

Haig, D. (1993) Genetic conflicts in human pregnancy. *Q. Rev. Biol.* 68, 495-532.

Haig, D., and Graham, C. (1991) Genomic imprinting and the strange case of the insulin-like growth factor II receptor. *Cell* 64, 1045-1046.

Hamilton, W. D. (1964) The genetical evolution of social behaviour I, II. *J. Theor. Biol.* 7, 1-52.

Hamilton, W. D. (1966) The moulding of senescence by natural selection. *J. Theor. Biol.* 12, 12-45.

Hamilton, W. D. (1967) Extraordinary sex ratios. *Science* 156, 477-488.

Harcourt, A. H., Harvey, P. H., Larson, S. G., and Short, R. V. (1981) Testis weight, body weight and breeding system in primates. *Nature* 293, 55-57.

Harvey, P. H., and Harcourt, A. H. (1984) Sperm competition, testes size, and breeding systems in primates. *In* "Sperm Competition and the Evolution of Animal Mating Systems" (R. L. Smith, ed.), pp. 589-600. Academic Press, Orlando, FL.

Harvey, P. H., and Pagel, M. D. (1991) "The Comparative Method in Evolutionary Biology." Oxford University Press, Oxford.

Herre, E. A. (1987) Optimality, plasticity and selective regime in fig wasp sex ratios. *Nature* 329, 627-629.

Holland, J. H. (1992) "Adaptation in Natural and Artificial Systems." MIT Press, Cambridge, MA.

Holland, J. H. (1995) "Hidden Order, How Adaptation Builds Complexity." Addison-Wesley, Reading, MA.

Horn, H. S. (1979) Adaptation from the perspective of optimality. *In* "Topics in Plant Population Biology" (O. T. Solbrig, S. Jain, G. B. Johnson, and P. H. Raven, eds.), pp. 48-61. Columbia University Press, New York.

Jacob, F. (1977) Evolution and tinkering. *Science* 196, 1161-1166.

Jones, J. S., Ebert, D., and Stearns, S. C. (1992) Life history and mechanical constraints on reproduction in genes, cells and waterfleas. *In* "Genes in Ecology" (R. J. Berry, T. J. Crawford, and G. M. Hewitt, eds.), pp. 393-404. Blackwell Scientific Publications, Oxford.

Joshi, A., and Thompson, J. N. (1995) Trade-offs and the evolution of host specialization. *Evol. Ecol.* 9, 82-92.

Kauffman, S. A. (1993) "The Origins of Order: Self-Organization and Selection in Evolution." Oxford University Press, New York.

Kirkwood, T. B. L., and Rose, M. R. (1991) Evolution of senescence: late survival sacrificed for reproduction. *Phil. Trans. R. Soc. Lond.* B 332, 15-24.

Lack, D. (1947) The significance of clutch-size. Part I.–Intraspecific variations; Part II.–Factors involved. *Ibis* 89, 302-352.

Lack, D. (1948) The significance of clutch-size. Part III.–Some interspecific comparisons. *Ibis* 90, 25-45.

Lack, D. (1954) "The Natural Regulation of Animal Numbers." Oxford University Press, Oxford.

Lack, D. (1966) "Population Studies of Birds." Oxford University Press, Oxford.

Leigh, E. G. Jr. (1971) "Adaptation and Diversity." Freeman, Cooper and Co., San Francisco.

Lessells, C. M. (1991) The evolution of life histories. In "Behavioural Ecology: An Evolutionary Approach" (J. R. Krebs, and N. B. Davies, eds.), 3rd ed., pp. 32-68. Blackwell Scientific Publications, Oxford.

Levins, R. (1966) The strategy of model building in population biology. *Am. Sci.* 54, 421-431.

Levins, R. (1968) "Evolution in Changing Environments." Princeton University Press, Princeton, NJ.

Lewontin, R. C. (1978) Adaptation. *Scient. Am.* 239, 156-169.

Lewontin, R. C. (1979) Fitness, survival, and optimality. *In* "The Analysis of Ecological Systems"

(D. J. Horn, G. R. Stairs, and R. D. Mitchell, eds.), pp. 3-21. Ohio State University Press, Columbus, OH.

Lewontin, R. C. (1984) Adaptation. *In* "Conceptual Issues in Evolutionary Biology" (E. Sober, ed.), pp. 237-251. MIT Press, Cambridge, MA.

Lindén, M., and Møller, A. P. (1989) Cost of reproduction and covariation of life history traits in birds. *TREE* 4, 367-371.

MacArthur, R. H., and Pianka, E. R. (1966) On optimal use of a patchy environment. *Am. Nat.* 100, 603-609.

Maynard Smith, J. (1978) Optimization theory in evolution. *Ann. Rev. Ecol. Syst.* 9, 31-56.

Maynard Smith, J. (1982) "Evolution and the Theory of Games." Cambridge University Press, Cambridge.

Maynard Smith, J., Burian, R., Kauffman, S., Alberch, P., Campbell, J., Goodwin, B., Lande, R., Raup, D., and Wolpert, L. (1985) Developmental constraints and evolution. *Q. Rev. Biol.* 60, 265-287.

Maynard Smith, J., and Price, G. R. (1973) The logic of animal conflict. *Nature* 246, 15-18.

Mayr, E. (1983) How to carry out the adaptationist program? *Am. Nat.* 121, 324-334.

Medawar, P. B. (1946) Old age and natural death. *Mod. Quart.* 1, 30-56.

Medawar, P. B. (1952) "An Unsolved Problem of Biology." Lewis, London.

Meijer, T., Daan, S., and Hall, M. (1990) Family planning in the kestrel (*Falco tinnunculus*): the proximate control of covariation of laying date and clutch size. *Behaviour* 114, 117-136.

Mitchell, W. A., and Valone, T. J. (1990) The optimization research program: studying adaptations by their function. *Q. Rev. Biol.* 65, 43-52.

Mock, D. W., and Parker, G. A. (1996) "The Evolution of Sibling Rivalry." Oxford University Press, Oxford.

Moore, A. J., and Boake, C. R. B. (1994) Optimality and evolutionary genetics: complementary procedures for evolutionary analysis in behavioural ecology. *TREE* 9, 69-72.

Moran, N. A. (1994). Adaptation and constraint in the complex life cycles of animals. *Annu. Rev. Ecol. Syst.* 25, 573-600.

Newton, I. (ed) (1989) "Lifetime Reproductive Success in Birds." Academic Press, London.

Niklas, K. J. (1994a) "Plant Allometry: The Scaling of Form and Process." University of Chicago Press, Chicago.

Niklas, K. J. (1994b) The allometry of safety-factors for plant height. *Am. J. Bot.* 81, 345-351.

Orians, G. H. (1962) Natural selection and ecological theory. *Am. Nat.* 96, 257-263.

Orzack, S. H., and Sober, E. (1994a) How (not) to test an optimality model. *Trends Ecol. Evol.* 9, 265-267.

Orzack, S. H., and Sober, E. (1994b) Optimality models and the test of adaptationism. *Am. Nat.* 143, 361-380.

Oster, G. F., and Wilson, E. O. (1978) "Caste and Ecology in Social Insects". Princeton University Press, Princeton, NJ.

Parker, G. A., and Maynard Smith, J. (1990) Optimality theory in evolutionary biology. *Nature* 348, 27-33.

Partridge, L., and Barton, N. H. (1993) Optimality, mutation and the evolution of ageing. *Nature* 362, 305-311.

Partridge, L., and Sibly, R. (1991) Constraints in the evolution of life histories. *Phil. Trans. R. Soc. Lond.* B 332, 3-13.

Pettifor, R. A., and Perrins, C. M., McCleery, R. H. (1988) Individual optimization of clutch size in Great Tits. *Nature* 336, 160-162.

Philippi, T. E., and Seger, J. (1989) Hedging one's evolutionary bets, revisited. *TREE* 4, 41-44.

Pyke, G. H., Pulliam, H. R., and Charnov, E. L. (1977) Optimal foraging: a selective review of theory and tests. *Q. Rev. Biol.* 52, 137-154.

Rapport, D. J., and Turner, J. E. (1977) Economic models in ecology. *Science* 195, 367-373.

Richards, M. H., Packer, L., and Seger, J. (1995) Unexpected patterns of parentage and relatedness in a primitively eusocial bee. *Nature* 373, 239-241.

Roff, D. A. (1992) "The Evolution of Life Histories." Chapman & Hall, New York.

Roff, D. A. (1994) Optimality modeling and quantitative genetics: a comparison of the two approaches. *In* "Quantitative Genetic Studies of Behavioral Evolution" (C. R. B. Boake, ed.), pp. 49-66, University of Chicago Press, Chicago.

Romer, A. S. (1966) "Vertebrate Paleontology" 3rd ed. University of Chicago Press, Chicago.

Rose, M. R. (1991) "Evolutionary Biology of Aging." Oxford University Press, New York.

Sasaki, A., and Ellner, S. (1995) The evolutionarily stable phenotype distribution in a random environment. *Evolution* 49, 337-350.

Schaffer, W. M. (1974) Selection for optimal life histories: effects of age structure. *Ecology* 55, 291-303.

Schoener, T. W. (1971) The theory of feeding strategies. *Ann. Rev. Ecol. Syst.* 2, 369-404.

Seger, J., Brockmann, H. J. (1987) What is bet-hedging? *Oxf. Surv. Evol. Biol.* 4, 182-211.

Sibly, R., Antonovics, J. (1992) Life-history evolution. *In* "Genes in Ecology" (R. J. Berry, T. J. Crawford, and G. M. Hewitt, eds.), pp. 87-122. Blackwell Scientific Publications, Oxford.

Smith, C. C., and Fretwell, S. D. (1974) The optimal balance between size and number of offspring. *Am. Nat.* 108, 499-506.

Stearns, S. C. (1976) Life-history tactics: a review of the ideas. *Q. Rev. Biol.* 51, 3-47.

Stearns, S. C. (1992) "The Evolution of Life Histories." Oxford University Press, New York.

Stephens, D. W., and Krebs, J. R. (1986) "Foraging Theory." Princeton University Press, Princeton.

Stubblefield, J. W., and Seger, J. (1990) Local mate competition with variable fecundity: dependence of offspring sex ratios on information utilization and mode of male production. *Behav. Ecol.* 1, 68-80.

Sundström, L. (1994) Sex ratio bias, relatedness asymmetry and queen mating frequency in ants. *Nature* 367, 266-268.

Thompson, J. N. (1994) "The Coevolutionary Process." University of Chicago Press, Chicago.

Thornhill, R. (1990) The study of adaptation. *In* "Interpretation and Explanation in the Study of Animal Behavior" (M. Bekoff, and D. Jamieson, eds.), Volume II: Explanation, Evolution, and Adaptation, pp. 31-62. Westview Press, Boulder, CO.

Trivers, R. L. (1971) The evolution of reciprocal altruism. *Q. Rev. Biol.* 46, 35-57.

Trivers, R. L. (1974) Parent-offspring conflict. *Am. Zool.* 14, 249-264.

Trivers, R. L., and Hare H. (1976) Haplodiploidy and the evolution of the social insects. *Science* 191, 249-263.

Williams, G. C. (1957) Pleiotropy, natural selection, and the evolution of senescence. *Evolution* 11, 398-411.

Williams, G. C. (1966a) "Adaptation and Natural Selection." Princeton University Press, Princeton, NJ.

Williams, G. C. (1966b) Natural selection, the costs of reproduction, and a refinement of Lack's principle. *Am. Nat.* 100, 687-690.

Williams, G. C. (1992) "Natural Selection: Domains, Levels, and Challenges." Oxford University Press, New York.

Wright, S. (1931) Evolution in Mendelian populations. *Genetics* 16, 97-159.

Wynne-Edward, V. C. (1962) "Animal Dispersion in Relation to Social Behaviour." Oliver and Boyd, Edinburgh.

CHAPTER 4

Genes And Adaptation: A Pocket Guide to the Theory

MARK KIRKPATRICK

On a good day, evolution's blind watchmaker has a swift and steady hand. A famous example of the power of natural selection combined with Mendelian inheritance comes from Haldane's (1956) study of selection in the peppered moth. In the late 19th century, a melanic allele swept through some populations of *Biston betularia* in industrialized Britain. Haldane estimated a selective advantage of about 20%, sufficient to spread the gene through most of a local population in only a dozen generations. The moral from that analysis was clear: selection can cause large evolutionary change on geologically trivial time scales.

Modern evolutionary theory, however, allows us to go far beyond this kind of generalization to answer important quantitative questions. What is the smallest selective advantage to which evolution can respond? When will populations jump across adaptive valleys to reach higher adaptive peaks? How localized can adaptation be?

This chapter looks at seven questions about adaptation from the perspective of theoretical population genetics. The aim is to outline for nontheoreticians some of the major quantitative conclusions about adaptation. The emphasis on simple rules, where "simple" here means things that can be evaluated in a few seconds with a basic calculator. Some mathematical precision has been sacrificed in this presentation, but this may be forgivable in light of the shakiness of our estimates for many of the numbers that go into the equations. Readers who are interested in more complex (and perhaps realistic) assumptions, who need more precise answers, or who are curious about how the results were obtained are encouraged to consult the original papers cited in this chapter.

Fitness is a quantitative concept, but adaptation is a subjective quality. Whether adaptation is the dominant feature of evolution or not depends on how adaptation is defined. Since there is no quantitative definition, the title of this chapter is something of an oxymoron. I will concentrate here on how selection and other evolutionary forces affect rates of evolution and the fitness of populations. The final word on what theory says about adaptation, however, is a problem to be resolved by philosophers, professional and otherwise.

I. How Fast Is Adaptation?

Selection can be an overwhelmingly powerful force. The speed of evolution observed under strong natural and artificial selection is many orders of magnitude larger than the average rates found over macroevolutionary time scales. To quantify this impression, it is useful to use a measure that allows different traits to be compared and that lets us link our knowledge of microevolution with rates of macroevolution. Both considerations lead to the measurement of the evolutionary rate of change as a proportion of a trait's mean value.

Population genetic theory tell us that the proportional change in the mean of a quantitative trait across one generation is

$$R = \frac{\Delta \overline{z}}{\overline{z}} = CV_P \, h^2 \, \tilde{\beta} \,. \tag{1}$$

This equation is adapted from classical quantitative genetics (Falconer, 1981), with terms rearranged to be more useful to evolutionists. R is defined here in terms that are dimensionless. In Equation (1), $\overline{z}$ is the trait's mean and $\Delta \overline{z}$ its evolutionary change in one generation; CV_P is the phenotypic coefficient of variation; and h^2 is the heritability. The final term, $\tilde{\beta}$, is the standardized selection gradient (Lande and Arnold, 1983), which measures the force of directional selection. [The standardized selection gradient, $\tilde{\beta}$, is equivalent to the intensity of selection, i, that is often used in applied breeding.] While Equation (1) comes from quantitative genetics, in fact it also applies to allele frequencies at a single locus.

This equation says that the response to selection is determined by three factors: the amount of phenotypic variation on which selection can act, the fraction of that variation that is heritable, and the intensity of directional selection. What are typical values for these quantities? The phenotypic coefficients of variation for linear dimensions of mammals are often on the order of 10% (Yablokov, 1974). Heritabilities vary widely, but often fall in the range 0.3 to 0.7 for morphological traits (Mousseau and Roff, 1987). Selection gradients, of course, vary tremendously. At the upper end of the spectrum is a selection gradient of $\tilde{\beta} = 0.43$ that acted on the bill depth of medium ground finches in the Galápagos during the great drought of 1977 (Price et al., 1984). This intensity of selection means that individuals one phenotypic standard deviation above the mean were on average more than twice as fit as individuals one deviation below the mean.

These data together with Equation (1) show that natural selection can produce evolutionary rates in excess of $R = 1\%$. By macroevolutionary standards, this rate is simply astounding. If sustained, it would take an animal the size of a mouse to that of an elephant in less than 1,200 generations. In contrast, evolutionary rates seen in the fossil record are orders of magnitude smaller. Among fossil vertebrates, maximum values for R are in the range of 10^{-5} (Gingerich, 1983). (Paleontologists measure rates using Haldane's (1949) unit, the darwin. Assuming an average of one generation per year, a rate measured in darwins can be converted to R simply by multiplying by 10^{-6}.) Thus a rate of $R = 1\%$ is thousands of times larger than the fastest known from fossils. Turning the argument around: the fastest evolutionary rates known from fossils can be explained by a constant selection gradient on the order of $\bar{\beta} = 0.0005$. This strength of selection is so weak that individuals one phenotypic standard deviation above the mean would enjoy fitness only 0.1% higher than those one deviation below the mean.

Molecular evolution also shows that potential rates of evolutionary change are far greater than the usual tempo. Comparisons between coding DNA sequences of humans and rodents reveal codon substitution rates that have averaged around 10^{-9} per year over the last 80 million years (Li and Graur, 1991). But codons for coat proteins in some viruses from living populations are currently evolving at rates up to 8 orders of magnitude faster (Domingo and Holland, 1994), presumably in response to selection to evade host immune systems.

So short-term adaptation has the potential to be remarkably quick, many times faster than the rates seen over geological time. Why the discrepancy? Equation (1) suggests two main possibilities: adaptation stops if selection stops or if variation is exhausted. Which of these two possibilities is likely in the long run? Important morphological traits such as body size usually show no evidence for either. The explanation for their modest rates over long time scales likely involves evolutionary reversals caused by changes in the direction of selection (Gingerich, 1983). Selection on other types of traits, even if consistent in direction, is chronically very weak. For example, theoretical analyses show that some of the behaviors of interest to evolutionary biologists may have vanishingly small fitness effects (Mangel and Ludwig, 1992).

A second possible brake on evolutionary rates is the exhaustion of variation. Some types of changes are evolutionarily forbidden by constraints, as I will discuss later. But even the evolution of traits for which there is now substantial genetic variation will grind to a halt if this fuel is exhausted under strong and consistent directional selection. While this argument is plausible, theory suggests that this kind of limit to adaptation may not be very common. The long-term response to directional selection proceeds in two phases. The first depends on genetic variation that is originally present in the population. For a quantitative trait controlled by many genes, the length of this phase is proportional to N, the effective population size (Robertson, 1960). This observation implies that even moderately large populations can evolve for a substantial time before the original variation becomes depleted. At that point, the population then enters a second phase in which evolutionary change is fueled by the new genetic variation introduced by mutations that appeared since selection began. While evolution may proceed somewhat more slowly during the second phase than the first, the rate can still be substantial (Hill, 1982). The population size N is also critical in determining this ultimate rate. Larger populations in effect present larger antennas to capture new favorable mutations. Thus the ultimate evolutionary rate attained under continued directional selection is proportional to N under a variety of assumptions about mutation and selection (Hill and Keightley, 1987). The accumulation of new favorable mutations may allow polygenic traits to evolve indefinitely in moderate- to large-sized populations. The tremendous rates of progress under long-term artificial selection seen in domesticated animals and plants (some with quite small population sizes) seem to corroborate this view.

Since most natural populations are far larger in size than laboratory populations or agricultural selected stocks, leaving aside rare endangered species, their potential to respond to long-continued selection should be still greater. In sum, the exhaustion of genetic variation by continued directional selection may not be a problem for many kinds of traits in natural populations.

The comparison between the evolutionary rates that genetic theory shows are possible and those actually realized over macroevolutionary times reveals an embarrassment of adaptive potential. Adaptation may be delayed while waiting for the production of a rare evolutionary novelty by mutation or recombination, or as a minor phenotypic feature responds to weak selection. Major quantitative traits that have important fitness effects, however, will often show a very quick and sustained response to a changed environment that favors a new optimal phenotype.

II. How Sensitive Is Adaptation?

The wonders of natural history make it easy to imagine that selection can operate on infinitesimal fitness differences. In some cases, that impression is probably accurate. Biases in the use of synonymous tRNA genes in the bacterium *E. coli* suggest that selection pressures as small as 10^{-5} guide their evolution (Bulmer, 1991). Despite such triumphs of selection, theory shows that there are limits to the sensitivity of adaptation.

Random events can have an important impact on adaptation, even when the population size is infinite and the environment constant. An advantageous mutation is represented by a small number of copies for the first few generations after it appears. During this period, its fate is largely governed by chance. In fact, unless its fitness advantage is very large, there is a considerable chance that a favorable new gene will be lost by accident.

The probability that a new mutation that increases fitness by a fraction s will become established in a large population is only

$$u(s) \approx 2\,s \tag{2}$$

(Fisher, 1958). So a mutation with a 0.1% fitness advantage (which is quite large) has only a 0.2% chance of fixation—very poor odds indeed. Roughly speaking, $1/s$ copies of the gene must accumulate mainly by chance before selection will become sufficiently effective to guarantee that it will be established. The duration of this vulnerable period is proportional to $1/s$.

Actually, the prospects for a lone advantageous mutation are even more grim than those results suggest. Selection on other loci generally impedes the spread of a new advantageous allele (Hill and Robertson, 1966; Felsenstein, 1988). From the gene's view, selection elsewhere in the genome appears as random variation in fitness. A consequence is to reduce the effective population size, which enhances the power of random genetic drift relative to selection. The result is to decrease the chance that the new gene will be established. The size of the effect depends on the form of selection experienced by the other loci and their linkage to the gene of interest (Barton, 1995). A single unlinked substitution elsewhere in the genome will decrease the gene's fixation probability by no more than a few percent. Repeated substitutions at closely linked loci, however, can drastically decrease (and even eliminate) the chance that a new advantageous mutation will survive.

These facts may not have major consequences for adaptation by new genes that are incessantly reintroduced to a population by mutation or migration. While any given copy of the gene has a small chance of surviving, sooner or later one of them will be established. On the other hand, some types of mutations are very rare, even unique. Loci with novel functions are a good example. One hypothesis for the origin of complex loci is that they have been assembled by chromosomal rearrangements which brought together fortuitous combinations of exons or "minigenes" (Gilbert, 1978). The chance that a particular combination of pieces has been brought together many times seems remote. From the above discussion of the survival prospects for new advantageous mutations, it becomes clear that adaptations based on unique mutations largely owe their existence to plain good luck. The functions represented by the major gene families we see today, for example, may be the outcome of a biased evolutionary lottery rather than an adaptive deliberation of infinite sensitivity.

III. How Localized Is Adaptation?

No one expects a lizard to evolve a unique camouflage specialized to a single boulder. Movement of individuals obliterates selective advantages operating on so small a spatial scale. But what about the evolution of camouflage that is adapted to a lava field 1 km across? Or a lava field 100 km across?

Gene flow causes a spatial averaging of selection pressures. This process determines the spatial scale or "grain" of adaptation (Slatkin, 1973, 1987). Movement can introduce genotypes into habitats to which they are not adapted. Dramatic cases in natural populations have been documented, for example, in grasses that straddle different soil conditions (McNeilly, 1968) and spiders living near a stream in a dry grassland (Riechert, 1993).

Theory quantifies the limits to adaptation's spatial resolution in a patchy habitat. In the case of a single gene, this limit is set by the allele's *characteristic length*. This is the spatial scale over which selection averages the gene's fitness. To make matters concrete, consider an allele that has a fitness advantage s when inside one kind of patch and an equally large disadvantage, $-s$, when outside. The characteristic length l_c is then

$$l_c = \frac{m}{\sqrt{s}} , \qquad (3)$$

where m is the mean distance between the birth places of an individual and its parents. [I have defined l_c here in terms of m, which can be easily visualized, rather than the standard deviation of the dispersal function, as is usual in the theoretical literature. For a population living in a two-dimensional habitat, this l_c can be converted to the standard one simply by multiplying by 0.8.]

The characteristic length increases as the movement of individuals increases and decreases as the intensity of selection goes up. This is intuitively sensible: when individuals range widely, each gene copy experiences selection over large spatial areas; when selection is strong, gene frequencies will respond to even small spatial features.

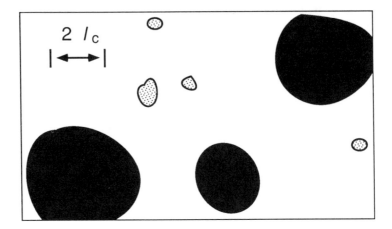

Figure I An allele with a fitness advantage inside habitat patches but a disadvantage outside of them only persists in favorable patches that are larger than a critical minimum size. When the fitness advantage inside the patches equals the fitness loss outside of them, the threshold patch width is about twice l_c, the allele's characteristic length defined by Equation (3). Favorable patches occupied by the allele are black and those where the allele is absent are shaded.

An important fact about the characteristic length is this: l_c determines the minimum size of spatial features in the environment to which the population can adapt. Let us return to the example of lizards under selection for camouflage. Their range includes lava flows that are roughly circular in shape (Fig. 1). The dark color of the rock gives a selective advantage s to a melanic allele in lizards on lava. Dark lizards are conspicuous against the light substrate of the surrounding sandy habitat, however, giving the melanic allele a fitness effect $-s$ there, and heterozygotes have fitness intermediate between the melanic and nonmelanic homozygotes. In this situation, the melanic allele will only survive on lava flows that are larger than about $2\,l_c$ in diameter (Nagylaki, 1975). (The factor of 2 comes from a calculation that accounts for the geometry of this example; other mysterious constants that crop up below have similar origins.) If the lizards move on average 1 km per generation, a melanic allele that has a 1% fitness advantage on a lava flow but a 1% disadvantage elsewhere will only survive in lava flows larger than about 20 km in diameter. In patches much smaller than this, gene flow from the movement of light-colored lizards from the surrounding habitat onto the lava will swamp out the selective advantage that the dark allele has there.

The concept of characteristic length also plays a key role in the evolution of clines. When a species' range straddles a sharp boundary between two habitat types that favor different alleles, a cline in gene frequency results as migration blurs the underlying pattern of selection. Far from the boundary, the alleles favored in each habitat reach fixation. Near the boundary, however, the locus makes a transition from one allele to the other. This transition zone, or cline, is an area where gene flow maintains substantial frequencies of alleles adapted to the other habitat. The shape of this cline is a sigmoid curve centered on the habitat boundary (Fig. 2).

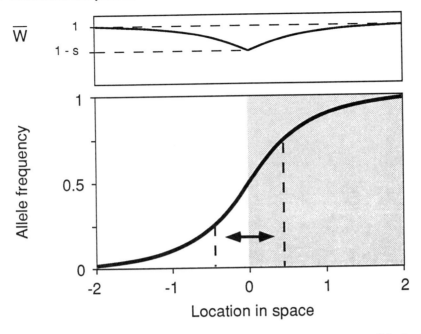

Figure 2 (*Bottom*): Cline in an allele frequency caused by gene flow across an ecotone. Selection is against the allele on the left (unshaded) and in favor of it on the right (shaded). Distance is measured in units of the characteristic length, l_c, defined by Equation (3). The cline's width can be measured by the region where the allele frequency goes from 0.25 to 0.75 (marked by double-headed arrow), which is about 0.9 l_c wide. (*Top*): Mean fitness at different points along the cline.

The cline's width is proportional to the characteristic length. Width is conveniently measured by the region in which the allele frequency changes from 25 to 75%. (This definition is perhaps easier to visualize than the one widely used in the theoretical literature, in which a cline's width is equal to the reciprocal of its greatest slope.) Take the case of our lizard where its range crosses a linear environmental boundary, for example where two large areas of different substrate colors meet. The melanic allele has a selective advantage s on one side of the boundary and a disadvantage s on the other side. This situation produces a cline around the environmental boundary whose width is about 0.9 l_c. Thus large dispersal distances and weak selection yield broad clines, while short dispersal and strong selection do the converse. An allele that enjoys a 1% fitness advantage in one habitat but that incurs a 1% fitness loss in the other will produce a cline about 9 km wide in our lizard. Thus there can be an appreciable region in which an allele appropriate to the other habitat is found at fairly high frequency. The population's mean fitness reaches a low at the habitat boundary, where gene flow decreases it by a factor of s (Fig. 2). Thus the melanism allele with a selection coefficient of $s = 0.01$ will impose a load of 1% on the population's fitness at the ecotone boundary.

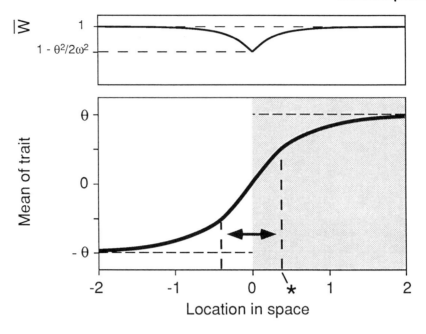

Figure 3 (*Bottom*): Cline in the mean of a quantitative trait caused by gene flow across an ecotone. Survival is maximized by a trait value of $z = -\theta$ to the left of 0 (unshaded) and $z = +\theta$ to the right (shaded). Distance is measured in units of the characteristic length, l_c, defined by Equation (4). The cline's width can be measured by the region where the trait mean goes from 0.25 to 0.75 of the way from $-\theta$ to θ. This region (marked by double-headed arrow) is about $0.8\, l_c$ wide. (*Top*): Mean fitness at different points along the cline. The greatest decrease in mean fitness caused by gene flow occurs at the habitat boundary, and is approximately $\theta^2/2w^2$ (assuming that $\theta^2 \ll 2w^2$).

Clines in quantitative traits follow similar patterns at abrupt habitat boundaries where the optimal value for the trait flips from one value to another (Slatkin, 1978). Instead of an allele frequency changing from 0 to 1, the cline in this case appears as a change in the trait's mean value as it shifts from one optimum to the other (Fig. 3). The equation describing the curve is different than that for a single gene, but the overall shape is very similar (Barton and Gale, 1993). The cline's width can be measured by the region where the change in the mean goes from 25 to 75% of the total difference between the optima. Once again, the concept of a characteristic length is useful. For a quantitative trait whose additive genetic variance stays approximately constant across the cline, the characteristic length is

$$l_c = \frac{m\,\omega}{h},\qquad(4)$$

where h is the square root of the trait's heritability and m is again the mean dispersal distance. The number ω is the width of each fitness peak measured in units of the trait's phenotypic standard deviation (Fig. 4). It reflects the range of trait values around the optimum that have high fitness; smaller values for ω imply a narrower fitness peak and hence more intense stabilizing selection. Comparing Equations (3) and (4) shows that the selection coefficient s has been replaced by

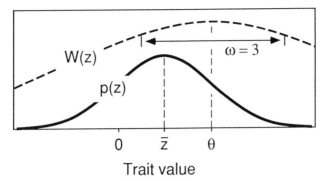

h^2/ω^2. The cline's characteristic length becomes shorter as the heritability of the

Figure 4   Distribution of the trait in the population, $p(z)$, and the individual fitness function, $W(z)$, at the spatial point marked by an " * " in Fig. 3. The discrepancy between the trait mean $\overline{z}$ and the fitness peak θ reflects maladaptation resulting from gene flow. The relative width of the fitness function, ω, represents the range of phenotypes that have high fitness. ω is measured in phenotypic standard deviations; a value of $\omega = 3$ is typical for morphological traits. (The width is defined as $\omega = 1/\sqrt{-c}$, where c is the curvature, or second derivative, of the fitness function at its maximum when the maximum is scaled to 1 and the trait is measured in phenotypic standard deviations. When the fitness function is proportional to a gaussian bell curve, as shown here, ω is equal to the fitness function's standard deviation.)

trait and the intensity of stabilizing selection increase. It is unaffected, though, by the size of the difference between the environmental optima on each side of the habitat boundary.

Measured in these units, the width of a cline in a quantitative trait is about 0.8 l_c, similar to the value for a single locus. We can get some feel for this result by trying out some plausible parameter values. Studies of selection in natural populations often have found values of ω in the range of 3 (reviewed in Turelli, 1984 ; Endler, 1986) (Fig. 4). With this strength of selection and a heritability of h^2 = 0.5, the width of the cline is about 3 times m, the average dispersal distance. Thus a quantitative trait in our lizard will show a cline 3 km wide. Apparently gene flow can produce maladaptation in a quantitative trait over observable spatial scales, just as it can for a single locus. The maximum decrease in mean fitness again occurs at the habitat boundary and is approximately $\theta/2\omega^2$ when the optimal trait values favored in the two habitats differ by 2θ phenotypic standard deviations (Fig. 3). Thus if the trait values favored on each side of the habitat boundary differ by 2 phenotypic standard deviations, the loss in mean fitness caused by gene flow at the boundary will be about 6%.

Migration, like mutation, can be an important source of genetic variation that fuels adaptive change. But in many situations it introduces noise, dulling adaptation's spatial acuity. The spatial patterns of phenotypes we see in nature represent a compromise between what selection favors locally and what migration introduces from afar.

IV. How Far Is Adaptation from Perfection?

In adaptation's utopia, there are no genetic constraints and all traits have had sufficient evolutionary time to reach their fitness peaks. But there is more: individuals are uniformly optimal, and random genetic drift is banished. The real world, of course, includes both nonadaptive variation and drift. How large are these effects on adaptation?

We can measure a population's distance from adaptive perfection using the concept of a fitness load. The idea is to compare the fitness of a population suffering the effects of deleterious mutation, for example, against a hypothetical counterpart in which mutation has been turned off. The mutation load is then simply the proportional decrease in the population's mean fitness caused by mutation. A load near zero means that mutation causes only a small decrease in the fitness of most individuals; a load near one means that the fitness of an imaginary mutation-free individual would be many times greater than that of the average individual.

The next section reviews the fitness load imposed by mutation and phenotypic variation; the section following it quantifies the impact of random genetic drift in similar terms.

A. The Cost of Variation

Variation is a mixed evolutionary blessing. Without variation, of course, there is no evolution. On the other hand, phenotypic variation can represent scatter around fitness peaks. Even though selection is weeding it out, this variation is constantly reintroduced by mutation, migration, and nongenetic noise in developmental processes. A population sitting at a fitness peak would have higher mean fitness if individuals did not vary.

Variation around the optimum thus imposes a load (Haldane, 1937; Muller, 1950; Kondrashov, 1988). Mutation to a deleterious allele at a rate μ per locus decreases the population's mean fitness by an amount μ. That is, the mutation load equals the mutation rate but is independent of the strength of selection. [Why is it independent of selection? Strongly deleterious alleles are kept at low frequency by selection, mitigating their impact, while mildly deleterious alleles are driven to higher frequency by mutation. The effects on mean fitness of a mutant's equilibrium frequency and its selective impact cancel each other.] This rule applies when the mutation rate is small relative to the selection coefficient s, as is true for many types of deleterious mutations.

The load from a single locus is small: the mutation rates at enzyme-coding loci are typically in the range 10^{-4} to 10^{-7}. But the number of loci n in a eukaryotic genome is large, perhaps 10^4 to 10^5. To calculate the genome's total load from mutation, we must commit to an assumption about how the fitness effects of different loci combine. Unfortunately, we have little information about epistasis for fitness effects at the genetic level because selection is typically so weak that it cannot be studied directly. The simplest assumption is that the loci have independent effects so that fitnesses are multiplicative. In that case, the load is

$$L_{\text{mutation}} \approx 1 - e^{-n\mu} . \tag{5}$$

There is a simple rule of thumb for expressions that have the form of (5): their value is about 0.1 when the number in the exponent is 0.1. This means that the load becomes noticeable (that is, it reaches 10%) when the number of loci n is 10% of $1/\mu$, for example when there are $n = 10^4$ loci that mutate at a rate of $\mu = 10^{-5}$. The load climbs to 63% (quite substantial) when there are 100,000 loci.

The cost of variation can also be viewed from the phenotypic level. Here the load is caused by the scatter of phenotypes around a fitness peak. With n quantitative traits under independent selection, the load is

$$L_{\text{variation}} \approx 1 - e^{-n/2\omega^2}, \qquad (6)$$

where ω is again width of the fitness peak (Kirkpatrick and Sanderson, unpublished results). Again using the rule of thumb, the load becomes nontrivial when there are 2 or more traits under selection with $\omega = 3$. The load climbs to 50% with 12 traits, and is over 95% if there are 60 independent quantitative traits under stabilizing selection of this intensity. The number of quantitative traits under appreciable selection has never been estimated, but a minimum number might be established using the methods of Lande and Arnold (1983) or Schluter and Nychka (1994).

Clearly variation can impose major costs on a population's mean fitness. The impact could be smaller if selection operates in ways other than those we have assumed here. Some extreme forms of epistasis can lead to much smaller loads (Kondrashov, 1995). But studies of selection on multiple quantitative traits (see Schluter and Nychka, 1994) imply that the fitness load from phenotypic variation is not trivial, and it seems plausible that a similar conclusion will emerge from the molecular level when enough data accumulate to draw generalizations. The bottom line holds a touch of irony: the variation on which adaptation depends can exact a substantial cost of its own.

B. The Burden of Chance

Random genetic drift introduces another type of load. Drift weakens the efficiency of selection, occasionally allowing a deleterious mutation to spread through a population to fixation. A new mutation that decreases fitness by a fraction s has a chance

$$u(s) \approx \frac{1}{2N} - 2s \qquad (7)$$

to become fixed in a population of size N. This approximation holds so long as $s < 1/4N$; the probability is negligible for mutations that are even more deleterious (Kimura, 1957; Hill and Keightley, 1987).

Fixation of these mildly deleterious mutations results in a fitness load. The implications are particularly dramatic if we consider the DNA in the entire genome since the number of bases is so large (Kondrashov, 1995). When the selection coefficient against an inappropriate base is s, the drift load contributed by a single site is simply $2s$. (This conclusion holds when the selection coefficient and mutation rate are much smaller than $1/N$, which is probably the case for most DNA sites in all but extremely large populations.) If sites have independent fitness effects, then the total load caused by n sites is

$$L_{\text{drift}} \approx 1 - e^{-2ns}. \tag{8}$$

The drift load in this situation is independent of the population size because drift is so strong relative to selection (we assumed that $s \ll 1/N$).

Again using the rule of thumb, the drift load will be appreciable if $2n$ (= twice the number of sites) is 10% of $1/s$ or larger. The number of selected DNA sites in higher eukaryotes probably falls in the range $m = 10^7$ to 10^9. With $n = 10^8$ and $s = 10^{-9}$, the drift load is 18%—population mean fitness is slightly decreased by drift. But with $s = 10^{-7}$, the drift load is 99.9999998%—the population is far, far away from perfect adaptation. Unfortunately, existing data do not let us clearly discriminate where along this range of loads nature typically lies.

Quantitative traits suffer drift as well. When a trait is under stabilizing selection, drift causes the mean to wander from the fitness peak. How far it wanders is determined by the strength of stabilizing selection and the population size. (It is independent, however, of the genetic variance: more variance makes the trait more susceptible to drift but also more responsive to selection, and these factors cancel.) On average, random genetic drift will cause the trait mean to deviate

$$\overline{D} \approx \frac{\omega}{2\sqrt{N}} \tag{9}$$

phenotypic standard deviations from the optimum, on average, where again ω is the width of the fitness peak and N the population size (based on the results of Lande, 1976).

What does this mean in concrete terms? Again taking $\omega = 3$ as reasonable for many traits in natural populations, drift will cause even a small population of size $N = 1000$ to deviate on average only 5% of the trait's phenotypic standard deviation from the fitness optimum—a discrepancy so tiny that it would be impossible to detect in the field. For $\overline{D}$ to be as large as 1 phenotypic standard deviation with a population this size, the fitness peak's width must be larger than $\omega = 60$—extremely weak stabilizing selection. With larger populations, of course, the effects of drift will be even smaller. The impression is that drift generally will not cause the mean to wander far from a fitness peak most of the time. Excursions far from the fitness peak will, however, happen occasionally.

What is the drift load with quantitative characters? Deviation of a trait's mean from its selective optimum lowers the average fitness of the population. The load caused by n quantitative traits in a population of size N is

$$L_{\text{drift}} = 1 - e^{-n/4N} \tag{10}$$

(Lande, 1980). It is interesting that the load is unaffected by either the strength of stabilizing selection or the amount of genetic variation, a fact that also holds for individual loci under strong overdominant selection (Robertson, 1970).

The drift load is small when a small number of traits is under selection in a large population. With 10 traits and 10^6 individuals, drift reduces the mean fitness by only 0.00025%. Alternatively, with 1000 traits and 1000 individuals, $L_{\text{drift}} = 22\%$. Comparing these numbers with results from Equation (6) suggests that the drift load will generally be much smaller than the variation load caused by quantitative traits.

The concept of fitness loads has had a checkered history in evolutionary biology. The reasons are varied: loads do not give predictions for evolutionary trajectories, their calculation requires uncertain assumptions about fitness interactions, and the standard of a genetically perfect individual is a remote abstraction. Nevertheless, the drift and variation loads provide quantitative measures of how far a population is from flawless adaptation. By that standard, at least, ours seems to be far from a perfect world.

V. Is Adaptation Limited by Opportunity?

A major preoccupation of evolutionary theoreticians from Darwin's day to the present involves explaining how evolution might drive a population from one adaptive peak to another, higher fitness peak when the peaks are separated by a maladaptive intermediate. Two major alternatives have been discussed. Wright (1932, 1977 Ch. 3) suggested that genetic drift pushes populations off fitness peaks, occasionally by chance forcing them across adaptive valleys and into the domain of attraction of a higher peak. This is the first phase of Wright's famous "shifting balance" theory. The second possibility is that conditions change in a way that removes the selective barrier between the peaks. The prospects for these mechanisms are discussed in the next two sections.

A. Peak Shifts by Drift

The key issue about peak shifts initiated by random genetic drift is the amount of time that it will take for a population to drift away from the first peak and across the adaptive valley. This period can be extremely long (Lande, 1976, 1985; Barton and Charlesworth, 1984). The average number of generations T is quite insensitive to the genetic details: the basic patterns are the same for quantitative characters and single loci. The waiting time is approximately

$$T \approx k\,e^{2Nd}, \tag{11}$$

where N is the effective population size and d is depth of the fitness valley relative to the height of the original peak. The constant factor k depends on the amount of genetic variation and the details of selection; it may be typically on the order of 100 for quantitative traits (Lande, 1985). Equation (11) shows that the average time it takes for a peak shift to occur is extremely sensitive to the depth of the valley and the population size, but is independent of the height of the new fitness peak or the distance between the peaks. This approximation for T gives the order of magnitude of the average waiting time. While rough, it is sufficiently accurate to draw some interesting conclusions.

Consider a modest adaptive valley in which the population's mean fitness is 5% lower than it is on the peak at which the population currently resides (Fig. 5). Even with population size as small as 100, on the order of one million generations will pass on average before the population discovers a peak on the other side of the valley. The waiting time until a peak shift occurs increases extremely quickly for larger populations. With a population of 500 that reproduces annually, the waiting time will be longer than the present age of the universe. Even when a species is subdivided into a very large number of demes, a long time may pass

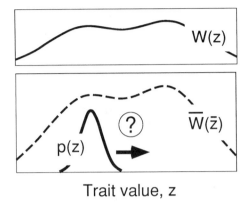

Trait value, z

Figure 5 Fitness landscape with two fitness peaks. The dashed line shows the fitness of individuals as a function of their trait value. The bottom panel shows that $\overline{W(z)}$, the mean fitness of the population as a function of its mean trait value, has an adaptive valley between the two peaks. Mean fitness in the valley is 5% lower than it is at the left hand peak. This prevents selection from taking the population [whose distribution is shown as $p(z)$] from the lower fitness peak on the left to the higher one on the right. The valley might be crossed by random genetic drift. The average waiting time for this event, however, can be extremely long: $T = 8.5$ million generations for the example shown, assuming a heritability of $h^2 = 0.5$ and population of size $N = 100$ [from Equation (7) of Lande (1985) and additional numerical analysis].

before one of them finds a new peak by drift alone. Furthermore, migration between demes during this phase of the process must be limited. If the number of migrants entering each deme from other populations is more than one per generation, gene flow will inhibit differentiation among the demes, making it unlikely that any one of them will cross an adaptive valley (Lande, 1979; Barton and Rouhani, 1993).

Drift could be important in triggering peak shifts if the peaks are separated by very shallow valleys, where "shallow" means that the product $N\,d$ is not much larger than 1. Peaks can be widely separated in morphological space even if the valley between them is slight. Since T is not much affected by the distance between the peaks, this situation sets the stage for drift to trigger a shift of the population to a very different phenotype.

Another hope for peak shifts by drift appears when a species is distributed with very even densities across a large range, rather than being divided into semi-isolated demes (Rouhani and Barton, 1987). Then neighborhoods within the range will be partly isolated if individuals do not disperse widely. Under these conditions, the probability of a peak shift is less sensitive to the depth of the adaptive valley and is more sensitive to the height of the new peak. Still, the rate at which peak shifts occur is low unless the valley is shallow and the "neighborhood size" is small. Here, neighborhood size is measured by $N = 8\,\rho\,m^2$, where ρ is the population density and m is again the average dispersal distance. Consider, for example, the transition to another peak that increases the population's mean fitness by 7% across an extremely shallow adaptive valley that decreases mean fitness by only 0.4%. With a population inhabiting a region that is 80 m by 80 m in size, the average time to a peak shift will be 100,000 generations when the neighborhood size is $N = 37$. When the neighborhood size is $N = 145$, peak shifts are expected only once every 100 million generations.

The general picture seems to be that unusual conditions are required for drift to move populations through maladaptive intermediates. In particular, it has a reasonable chance only when the fitness valleys separating peaks are shallow. If correct, this conclusion has implications for several issues in evolutionary biology. Some verbal theories of speciation and diversification (e.g., Mayr, 1963; Templeton, 1981) involve peak shifts triggered by drift. These theories have been questioned on the grounds that drift may not be a sufficiently powerful force (Barton and Charlesworth, 1984). On the other hand, there are examples in which drift seems to have caused peak shifts. Perhaps the best examples involve the establishment of underdominant chromosomal rearrangements. Meiotic incompatibilities between the chromosomal forms are thought to make the heterozygotes unfit in all ecological conditions. Yet comparisons of populations show that evolution from one form to the other has occurred. This event requires that an adaptive valley be traversed, and so drift (or meiotic drive or some other nonadaptive force) must have been involved.

B. Peak Shifts by Selection

If not by drift, how can evolution find new adaptive peaks? The simplest mechanism occurs when environmental conditions change so that the valley separating them disappears. Then selection is free to carry the population deterministically to the higher peak. Under some conditions, even very subtle changes in the fitness function will trigger a large and rapid evolutionary shift (Fig. 6). Changing ecological conditions, for example, the introduction of a new prey type intermediate in size between two old ones, are one way this happens. Shifts in the adaptive landscape caused by changing ecological conditions are likely to be the most common cause of diversification during evolutionary radiations.

There are other possibilities as well. In Wright's adaptive landscape, a population is represented by a single point. Its coordinates are the mean values of the traits (or gene frequencies), and the altitude of the landscape is the population's mean fitness. In a constant environment, selection drives the population uphill towards higher mean fitness (Wright, 1932; Lande, 1976). Mean fitness, however, depends on the variances of the traits as well as their means. As the phenotypic variance for a trait increases, the fitness landscape becomes smoother. The smoothing results because the population's mean fitness is determined by an average taken over a broader distribution of phenotypes.

The smoothing that results from increased phenotypic variance can actually eliminate a valley separating two adaptive peaks (Kirkpatrick, 1982b; Whitlock, 1995). Several events can cause the variance to increase. For example, the genetic component of the phenotypic variance can increase with an influx of migrants from another population. Alternatively, population bottleneck events occasionally cause an increase in the genetic variance (Bryant et al., 1986). Even small increases in the phenotypic variance can be sufficient to obliterate an adaptive valley in some situations. It is difficult to make quantitative generalizations for this situation, however, because the outcome depends on the specific shapes of the

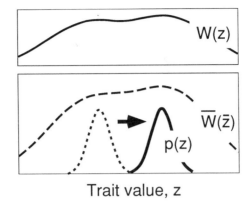

Trait value, z

Figure 6 An adaptive valley can be eliminated by a subtle change in the environment. The individual fitness function $W(z)$, shown in the top panel, is similar to that shown in Fig. 5, but the minimum between the peaks is less deep. This change eliminates the adaptive valley in the mean fitness function $\overline{W(\bar{z})}$, shown in the bottom panel, allowing selection to drive the population from the lower fitness peak to the higher one without the aid of drift.

fitness function and phenotypic distribution.

Peak shifts by selection can be triggered in yet another way. Whenever two traits are genetically correlated (via pleiotropy or linkage disequilibrium), selection on one will cause the other to evolve as a side effect. Under the right conditions, this correlated response can cause a peak shift. When one trait experiences sudden directional selection, the correlated response experienced by its partner can be large enough to pull it off of one adaptive peak, through a valley, and over to another peak (Lande and Kirkpatrick, 1988; Price et al., 1993).

Peak shifts by selection are conceptually more mundane than those caused by drift, and so have gotten less attention from theoreticians. But this mode is theoretically sound and seems likely to be more important in the evolution of most traits.

VI. Is Adaptation Limited by Genetic Resources?

Without inherited variation, evolution goes nowhere. Does a lack of appropriate genetic variation halt adaptation? It does, of course, for at least some traits: no species has yet evolved a life history in which individuals mature instantly, reproduce at an infinite rate, and live forever. While population genetic theory cannot predict which traits are constrained, it does tell us about the consequences of constraints.

Some evolutionary changes that selection favors seem to be out of the question for evolution. At the genetic level, these constraints are reflected by an absence of genetic variation: the heritability for X-ray vision is 0 for all known organisms (on Earth, at least). But constraints can also appear in a more subtle form. An important fact is that the response of a set of traits to selection can be prevented even when genetic variation is present for each of the traits individually.

Genetic correlations between traits restrict combinations of evolutionary changes that are possible. To take a very simple example, consider a set of n quantitative traits, each with the same additive genetic variance. Pleiotropic effects cause a genetic correlation of size r between every pair of traits. If $r = 0$, then every trait is free to respond independently to selection. If r is either positive or negative, however, there will be more genetic variation to evolve towards some combinations of traits than others. The genetic variation available for combinations of changes can therefore be much smaller than the amount suggested by variation for each of the component traits. This situation can dramatically reduce the response to selection. In the case of negative correlations, the constraint becomes absolute when

$$r = \frac{-1}{n} \tag{12}$$

(Dickerson, 1955). At this point, the correlations unite the traits into what is effectively a single character. Thus genetic correlations of only -0.05 between 20 traits will prevent their independent evolution.

Of course the genetic correlations in real populations vary, with both positive and negative values. In this case the formula for when constraints will appear is less simple. But the qualitative pattern is likely to be the same: constraints become increasingly likely as the number of traits increase. This intuition is corroborated by the analysis of additive genetic covariance matrices (Cheverud, 1984; Wagner, 1984). The eigenvalues measure the amount of variation available for evolutionary changes involving different combinations of the traits. Analyses show that the eigenvalues tend to decline exponentially fast. There is typically abundant genetic variation for a small number (say 2 to 5) of combinations of changes. But other combinations have vanishingly small variation. When the number of traits is moderate to large, there are many more evolutionarily forbidden changes than there are evolutionarily permitted ones. This conclusion holds even when none of the pairwise genetic correlations is -1 or $+1$.

Constraints may be particularly acute for certain types of traits. These include growth trajectories, reaction norms, and morphological shapes. Consider a growth trajectory that describes the size of an individual at every age. Since the size at each age can be viewed as a separate trait, and since every individual passes through an infinite number of ages, a growth trajectory is an "infinite-dimensional" trait (Kirkpatrick and Heckman, 1989). Reaction norms and morphological shapes share this infinite-dimensional quality.

Two factors contribute to constraints in infinite-dimensional traits. First, they involve an infinite number of genetic correlations for the simple reason that there is a correlation between every pair of the (infinite number of) ages. Second, the correlations between similar ages will be high: an individual that is large now will be large in the immediate future. Both factors decrease the number of independent evolutionary dimensions or "degrees of freedom" on which adaptation can operate.

These intuitive arguments are fully endorsed by analyses of genetic data on growth trajectories (Kirkpatrick and Lofsvold, 1992). Analysis shows that there is abundant genetic variation for some evolutionary changes, such as increasing or decreasing size at all ages. There is little if any variation, however, available for most changes in the shape of the trajectory. Qualitatively similar conclusions apply to the reaction norms that govern phenotypic plasticity (Gomulkiewicz and Kirkpatrick, 1992).

The picture from these analyses is, if anything, conservative about the ubiquity of constraints. Any empirical study deals with only a small fraction of an organism's traits. The unseen genetic correlations between the characters that are measured in any study and those that are not can only impose further constraints, not alleviate the ones that were identified (Pease and Bull, 1988; Charlesworth, 1990).

What do these constraints mean for adaptation? They frustrate selection from optimizing all traits simultaneously. Evolution cannot respond to immediate selection pressures that favor particular suites of traits. The evolution of one group of traits to their fitness peaks will drag other correlated traits away from their optima. Generally, we expect evolution to strike a compromise in which all of the correlated traits deviate to some extent from their optima. The size of these discrepancies depends on the arrangement of the genetic correlations, the evolutionary history of the population (what combinations of traits it is starting from), and where the fitness peaks are located (that is, what combinations of traits confer high fitness) (Kirkpatrick and Lofsvold, 1992). Depending on these factors, the final equilibrium can lie very close to or very far from a fitness peak.

VII. How Adaptive Is Selection?

The dominant metaphor for adaptation is the fitness landscape introduced by Sewell Wright (1932). A lesson taught in university biology classes everywhere is that selection drives populations uphill, towards a fitness peak. This view of evolution is incomplete. It is sometimes forgotten that selection is not the only evolutionary force acting on populations. Other evolutionary forces (including random genetic drift, mutation, migration, and recombination) can oppose a population's climb towards a fitness peak. An example comes from the evolution of genetic life cycles. Diploids enjoy a fitness advantage over haploids from the masking of deleterious recessive mutations, and this can favor the evolution of diploid-dominated life cycles (Muller, 1932). Yet as individual selection lengthens the diploid phase, the population's mean fitness generally declines in the long run because diploids accumulate more mutations (Jenkins and Kirkpatrick, 1995).

But neglect of forces other than selection is not the only flaw with the simple interpretation of the adaptive landscape. Even in a world governed entirely by selection, evolution does not always go uphill on the mean fitness surface. When the relative fitnesses of phenotypes change in time, there is no guarantee that selection will improve a population's lot. An obvious (if contrived) case is a situation where the climate alternates between wet and dry years. The evolutionary response of an annual plant will leave it worse prepared for the next year than if it had not evolved at all.

More subtle are cases in which the fitness changes are caused not by the abiotic environment but by the population itself. Frequency-dependent selection occurs whenever fitnesses depend on the relative proportions of different phenotypes in the population. While it is treated as a special case by every text on evolution, frequency-dependence acts on many of the traits involved with, for example, foraging and reproduction. It is possible that frequency dependent selection is the rule rather than the exception.

Wright (1969 Ch. 5) recognized that the notion of an adaptive landscape can evaporate completely when there is frequency dependence. A clear example comes from overtopping growth in plants. Taller trees get more sunlight for themselves and shade their neighbors. As selection causes the average tree height to increase, fecundity declines because more of the energy budget is diverted from seed production to wood production; it also may take longer to reach maturity. Arborescent growth is an evolutionary response to selection for competitive ability. Selection favoring the most fit individuals causes the average fitness of the group to decline.

A second case where selection can cause a population's undoing is sexual selection in animals. Mating preferences can establish a trait even if it has negative side effects on survival. A gene for a preferred trait that is expressed in both sexes will spread if its fitness gain through male mating success more than offsets its survival cost averaged over males and females (Kirkpatrick, 1982a). Thus selection can establish traits that have negative effects on female fitness, and hence the population's reproductive output. We expect, and in fact often see, the evolution of modifiers that suppress in females the expression of genes that give a fitness advantage only to males. But the point is that in principle selection itself can cause the evolution of traits that decrease a population's mean fitness.

These two examples illustrate the general point that there is no law preventing selection from causing extinction. Whether or not we view selection as adaptive depends partly on our outlook. Even when selection drives a population extinct, it is the individuals with highest fitness within each generation that push evolution along its trajectory. If being adapted means simply having high relative fitness, then killing all your contemporaries is the pinnacle of adaptation. On the other hand, we might prefer to think of adaptation in terms of the health of the population or species. In this event, selection that causes mean fitness to decline is distinctly not adaptive. Selection in natural populations may well improve their average fitness the vast majority of the time. But if so, it is because of the natural history of the organisms rather than some fundamental law of evolution.

VIII. A Conclusion

Consideration of these seven questions leads to the conclusion that evolution involves substantial adaptive impedimenta. Population genetic theory shows that the adaptive potential of selection is immense, but also that there are consequences of evolutionary forces—gene flow, drift, mutation, and even selection itself—that decrease a population's fitness. The same message emerges when measures of adaptedness other than mean fitness are used. For example, evolution does not always maximize the chance that a population will escape extinction.

It is then remarkable that so many of the organisms around us are so well designed for their lives. The cunning ways by which a malarial parasite evades its host's defenses are beyond the ingenuity of mere human engineering; the intricate behaviors used by leafcutter ants to cultivate their symbiotic fungi tax one's credulity. There are many reasons why such complex life forms might not persist, and yet they flourish. Perhaps it is serendipity that allows adaptation to have the upper hand so often in our world.

Acknowledgments

I thank C. Austin, J. Bull, L. Higgins, B. Mahler, and M. Servedio for comments on the manuscript, and am grateful for support from the N.S.F. (grant DEB-9407969) and the N.I.H. (grant GM-45226-01).

References

Barton, N. H., and Charlesworth, B. (1984). Genetic revolutions, founder effects, and speciation. *Annu. Rev. Ecol. Syst.* 15, 133-164.

Barton, N. H. (1995). Linkage and the limits to natural selection. *Genetics* 140, 821-841.

Barton, N. H., and Gale, K. S. (1993). Genetic analysis of hybrid zones. *In* "Hybrid Zones and the Evolutionary Process" (R. G. Harrison, ed.), Pp. 13-45. Oxford University Press, Oxford.

Barton, N. H., and Rouhani, S. (1993). Adaptation and the "shifting balance." *Genet. Res.* 61, 57-74.

Bulmer, M. J. (1991). The selection-mutation-drift theory of synonymous codon usage. *Genetics* 129, 897-907.

Bryant, E. H., McCommas, S. A., and Combs, L. M. (1986). The effect of an experimental bottleneck upon quantitative genetic variation in the housefly. *Genetics* 114, 1191-1211.

Charlesworth, B. (1990). Optimization models, quantitative genetics, and mutation. *Evolution* 44, 520-538.

Cheverud, J. M. (1984). Quantitative genetics and developmental constraints on evolution by selection. *J. Theor. Biol.* 110, 155-171.

Dickerson, G. E. (1955). Genetic slippage in response to selection for multiple objectives. *Cold Spring Harb. Symp. Quant. Biol.* 20, 25-32.

Domingo, E., and Holland, J. J. (1994). Mutation rates and rapid evolution of RNA viruses. *In* "The Evolutionary Biology of Viruses" (S. S. Morse, ed.), pp. 161-184. Raven Press, New York.

Endler, J. A. (1986). "Natural Selection in the Wild." Princeton University Press, Princeton.

Falconer, D. S. (1981). "Introduction to Quantitative Genetics." Second ed. Longman, London.

Felsenstein, J. (1988). Sex and the evolution of recombination. *In* "The Evolution of Sex: An Examination of Ideas" (R. E. Michod and B. R. Levin, eds.), pp. 74-86. Sinauer, Sunderland, MA.

Fisher, R. A. (1958). "The Genetical Theory of Natural Selection." Second ed. Dover, New York.

Gilbert, W. (1978). Why genes in pieces? *Nature* 271, 501.

Gingerich, P. D. (1983). Rates of evolution: effects of time and temporal scaling. *Science* 222, 159-161.

Gomulkiewicz, R., and Kirkpatrick, M. (1992). Quantitative genetics and the evolution of reaction norms. *Evolution* 46, 390-411.

Haldane, J. B. S. (1937). The effect of variation on fitness. *Am. Nat.* 71, 337-349.

Haldane, J. B. S. (1949). Suggestions as to the quantitative measurement of rates of evolution. *Evolution* 3, 51-56.

Haldane, J. B. S. (1956). The estimation of viabilities. *J. Genet.* 54, 294-296.

Hill, W. G. (1982). Predictions of response to artificial selection from new mutations. *Genet. Res.* 40, 255-278.

Hill, W. G., and Keightley, P. D. (1987). Interrelations of mutation, population size, artificial and natural selection. *In* "Proceedings of the Second International Conference on Quantitative Genetics" (B. S. Weir, E. J. Eisen, M. M. Goodman, and G. Namkoong, eds.), pp. 57-70. Sineaur, Sunderland, MA.

Hill, W. G., and Robertson, A. (1966). The effect of linkage on limits to artificial selection. *Genet. Res.* 8, 269-294.

Jenkins, C. J., and Kirkpatrick, M. (1995). Deleterious mutation and the evolution of genetic life cycles. *Evolution* 49, 512-520.

Kimura, M. (1957). Some problems of stochastic processes in genetics. *Ann. Math. Stat.* 28, 882-901.

Kirkpatrick, M. (1982a). Sexual selection and the evolution of female mating preferences. *Evolution* 36, 1-12.

Kirkpatrick, M. (1982b). Quantum evolution and punctuated equilibrium in continuous genetic characters. *Am. Nat.* 119, 833-848.

Kirkpatrick, M., and Heckman, N. (1989). A quantitative-genetic model for growth, shape, reaction norms, and other infinite-dimensional characters. *J. Math. Biol.* 27, 429-450.

Kirkpatrick, M., and Lofsvold, D. (1992). Measuring selection and constraint in the evolution of growth. *Evolution* 46, 954-971.

Kondrashov, A. S. (1988). Deleterious mutations and the evolution of sexual reproduction. *Nature* 336, 435-440.

Kondrashov, A. S. (1995). Contamination of the genome by very slightly deleterious mutations: why have we not died 100 times over? *J. Theor. Biol.* 175, 583-594.

Lande, R. (1976). Natural selection and random genetic drift in phenotypic evolution. *Evolution* 30, 314-334.

Lande, R. (1979). Effective deme sizes during long-term evolution estimated from rates of chromosomal rearrangement. *Evolution* 33, 234-251.

Lande, R. (1980). Genetic variation and phenotypic evolution during allopatric speciation. *Am. Nat.* 116, 463-477.

Lande, R. (1985). Expected time for random genetic drift of a population between stable phenotypic states. *Proc. Nat. 1. Acad. Sci. USA* 82, 7641-7645.

Lande, R., and Arnold, S. J. (1983). The measurement of selection on correlated characters. *Evolution* 37, 1210-1226.

Lande, R., and Kirkpatrick, M. (1988). Ecological speciation by sexual selection. *J. Theor. Biol.* 133, 85-98.

Li, W.-H., and Graur, D. (1991). "Fundamentals of Molecular Evolution." Sinauer, Sunderland, MA.

Mangel, M., and Ludwig, D. (1992). Definition and evaluation of the fitness of behavioral and developmental programs. *Annu. Rev. Ecol. Syst.* 23, 507-536.

Mayr, E. (1963). "Animal Species and Evolution." Harvard Univ. Press, Cambridge.

McNeilly, T. (1968). Evolution in closely adjacent populations. III. *Agrostis tenuis* on a small copper mine. *Heredity* 23, 99-108.

Mousseau, T. A., and Roff, D. A. (1987). Natural selection and the heritability of fitness components. *Heredity* 59, 181-197.

Muller, H. J. (1932). Some genetic aspects of sex. *Am. Nat.* 66, 118-138.

Muller, H. J. (1950). Our load of mutations. *Am. J. Hum. Genet.* 2, 111-176.

Nagylaki, T. (1975). Conditions for the existence of clines. *Genetics* 80, 595-615.

Pease, C. M., and Bull, J.J. (1988). A critique of methods for measuring life history trade-offs. *J. Evol. Biol.* 1, 293-303.

Price, T. D., Grant, P. R., Gibbs, H. L., and Boag, P. T. (1984). Recurrent patterns of natural selection in a population of Darwin's finches. *Nature* 309, 787-789.

Price, T., Turelli, M., and Slatkin, M. (1993). Peak shifts produced by correlated response to selection. *Evolution* 47, 280-290.

Riechert, S. E. (1993). The evolution of behavioral phenotypes: Lessons learned from divergent spider populations. *In* "Advances in the Study of Behavior" (P. J. B. Slater, J. S. Rosenblatt, C. T. Snowdon, and M. Milinski, eds.), Vol. 22., pp. 103-134. Academic Press, San Diego.

Robertson, A. (1960). A theory of limits in artificial selection. *Proc. Royal Soc. London B* 153, 234-249.

Robertson, A. (1970). The reduction in fitness from genetic drift at heterotic loci in small populations. *Genet. Res.* 15, 257-259.

Rouhani, S., and Barton, N. (1987). Speciation and the "shifting balance" in a continuous population. *Theor. Pop. Biol.* 31, 465-492.

Schluter, D., and Nychka, D. (1994). Exploring fitness surfaces. *Am. Nat.* 143, 597-616.

Slatkin, M. (1973). Gene flow and selection in a cline. *Genetics* 75, 733-756.

Slatkin, M. (1978). Spatial patterns in the distributions of polygenic characters. *J. Theor. Biol.* 70, 213-228.

Slatkin, M. (1987). Gene flow and the geographical structure of natural populations. *Science* 236, 787-792.

Templeton, A. R. (1981). Mechanisms of speciation—a population genetic approach. *Annu. Rev. Ecol. Syst.* 12, 23-48.

Turelli, M. (1984). Heritable genetic variation via mutation–selection balance: Lerch's zeta meets the abdominal bristle. *Theor. Pop. Biol.* 25, 138-193.

Wagner, G. P. (1984). On the eigenvalue distribution of genetic and phenotypic dispersion matrices: evidence for a nonrandom organization of quantitative character variation. *J. Math. Biol.* 21, 77-95.

Whitlock, M. (1995). Variance-induced peak shifts. *Evolution* 49, 252-259.

Wright, S. (1932). The roles of mutation, inbreeding, crossbreeding and selection in evolution. *Proc. VI Internat. Congr. Genet.* 1, 356-366.

Wright, S. (1969). "Evolution and the Genetics of Populations." Vol. 2. Univ. of Chicago Press, Chicago.

Wright, S. (1977). "Evolution and the Genetics of Populations." Vol. 3. Univ. of Chicago Press, Chicago.

Yablokov, A. V. (1974). "Variability of Mammals." Amerind, New Dehli.

PART II

Empirical Methods for

Studying Adaptation

Even evolutionary biologists, with their vast capacity to propose alternative interpretations and invent theories, have to submit to the data eventually. With respect to adaptation, fortunately, there are many different ways in which hypotheses can be confronted with evidence. Less fortunately, there remains a great deal of "wriggle" room. But then Darwin himself was a great wriggler, so perhaps this is only a respectable tradition in evolutionary biology.

There are a variety of elements that cut across the different chapters assembled together in this section. One element is the comparison of different species, which features in Chapters 5, 6, 9, and 10. A revolution in such comparisons has taken place recently in the development of better statistical models and tools for estimating the amount of information in phylogenetic comparisons. On the other hand, there is still a considerable tension concerning the relationship between pattern and causation, a leitmotif throughout these chapters. Are particular comparative patterns or evolutionary dynamics produced by the inferred evolutionary mechanisms or instead by some other arcane confounding mechanism?

An additional important strain running through these chapters is that between comparisons that cut across a variety of species or populations at one time, and the collection of data on evolutionary trajectories through time. Either approach is subject to issues of phylogeny and causation, but they have some radically different features for empiricists. Moreover, different concepts of adaptation place very different weights on these two empirical endeavors. If adaptation refers to whether or not a particular character state enhances fitness, regardless of its origin, then comparative information may be essential. On the other hand, if adaptation refers to the mechanistic source of a character, particularly whether or not that source is natural selection, then evolutionary trajectories may provide the essential information.

We would particularly encourage our readers to balance the contrasting views of the authors carefully; there is an implicit dialogue among these chapters, with areas of considerable disagreement.

CHAPTER 5

Testing Adaptation Using Phenotypic Manipulations

BARRY SINERVO

ALEXANDRA L. BASOLO

I. Introduction

Traditionally, correlational studies have been used to assess the adaptive significance of organismal traits. Correlations between the presence or the absence of a trait in various selective regimes have been used to infer that the trait is an adaptation for the new selective regime. The "correlational approach" (Endler, 1995) has been used extensively to assess adaptations. Another major paradigm for assessing whether or not a trait is an adaptation is the optimality approach in which a trait is assessed in terms of its impact on natural and sexual selection. If the mean of the population is in close proximity to an optimum, the trait would be considered adaptive (Arnold, 1983; Feder, 1987). Additional traits have been used as proxies for fitness. For example, the link between morphology and performance has been advocated as a first step in assessing the selective importance of traits (Arnold, 1983). By their very nature, optimality arguments concerning adaptation and links to natural selection are restricted to the present (Orzack and Sober, 1994). In principle, it is impossible to directly measure natural selection in the past.

Both the correlational approach and the optimality approach implicitly assume that traits were strongly shaped by natural selection and that historical or phylogenetic constraints have played a minor role in adaptive evolution. The comparative method should be integrated into studies of adaptation because many of our notions of adaptation are grounded in the comparative method (Pagel and Harvey, 1989; Harvey and Pagel, 1991). The appearance of traits in a lineage are often correlated with a shift or change in habitat type or new selective regime. Gould and Lewontin (1979) and subsequently Gould and

Vrba (1982) employed stringent historical criteria to define adaptation. The trait must have originated as an adaptation to be considered an adaptation. That is, natural selection must have been an important agent in shaping the trait as an adaptation when the trait arose as a new character. If the trait originated for some other selective and/or historical reason, unrelated to current function, the trait is considered an exaptation. Moreover, when considering adaptive explanations for a character, nonadaptive explanations must necessarily also be considered (Gould and Lewontin, 1979). Gould and Lewontin consider a trait that is demonstrably optimal in the present-day context, but for which there is no information regarding origin, to be an aptation. Additional paleontological or phylogenetic information is necessary to determine the adaptive nature of a trait. We will use the single term adaptation when describing studies investigating the origin and maintenance of traits, but we are explicit with regards to the kind of information different studies present (historically based reconstructions or optimality arguments).

Adaptation as a concept has experienced little opposition, but the testability of adaptation continues to be debated. One experimental tool that has been particularly useful in addressing whether a specific trait has evolved in part by the process of adaptation is manipulation of the phenotype. If our aim is to understand adaptation in natural populations in the present-day context, understanding the ecological causes of natural selection is essential. Experimental manipulations not only provide us with the means of detecting natural selection, but they also provide us with causal links between traits and fitness in the wild. A comprehensive analysis of adaptation should attempt to satisfy at least four criteria. (1) Individual or geographic variation in a trait must have a genetic basis (Falconer, 1981). (2) A trait must influence fitness or a component of fitness (Lande and Arnold, 1983; Endler, 1986). (3) The mechanistic links underlying correlations between the trait and the fitness in the wild (Ferguson and Fox, 1984; Grant, 1986; Jayne and Bennett, 1990) must be elucidated (Arnold, 1983; Levins and Lewontin, 1985). The hypothesis of adaptive variation (point 2) should be independently validated by experimentally manipulating the (4) selective environment or the (5) phenotypic trait itself (Endler, 1986; Mitchell-Olds and Shaw, 1987; Schluter, 1988; Wade and Kalisz, 1990).

Experimental manipulations provide additional insight into the adaptive nature of traits because we can often separate the effects of natural selection from stochastic effects. Manipulations have great utility in (1) assessing the adaptive nature of traits in the present-day context (e.g., whether a trait is optimal), (2) allowing for a direct test, of the adaptive nature of traits, (3) reconstructing ancestral character states that predated the origins of the adaptation, and (4) using simultaneous manipulations of the selective environment and of the phenotype in order to reconstruct historical selective regimes that led to the evolution of the adaptation. In addition, by manipulating the phenotype, one can attribute fitness differences to a specific variable of the phenotype and address questions that are not testable using present-day variation. Experimental manipulations provide additional evidence that can be used to determine whether a trait arose as an adaptation or exaptation, and at the very least provide stringent means for assessing whether a trait is being maintained by current selection. Thus, manipulations complement emergent phylogenetic approaches for assessing the adaptive nature of traits.

In this chapter, we partition the types of phenotypic manipulations into (1) changes in a single trait that do not appear to have an effect on other traits of the individual, (2) changes in a single trait that affects other characters expressed by the individual, (3) alterations of the extended phenotype of an individual, and (4) alterations of the selective environment and the phenotype of the individual. While there is an enormous literature on the effect of the environment on the expression of phenotypic traits, we do not discuss such purely environmental manipulations. We consider such manipulations to be distinct from direct manipulations of the phenotype or extended phenotype of the individual. Such environmental alterations of the phenotype may occur through mechanisms that are unrelated to the genetic mechanisms that underlie the adaptive evolution of the trait. A goal of (1) and (3) is to change a single variable and determine what effect it has on individual fitness; in practice, however, it may be difficult to manipulate a single trait without affecting other characters. The goal of (2) is to manipulate a single mechanism that controls two or more traits to determine the effect of a suite of traits on individual fitness. The goal of (4) is to determine at a population level how changes in the selective regime (i.e., environment) result in a shift in the mean phenotype of a "manipulated" population. In addition, by manipulating the phenotype at the individual level (1 - 4, above), one can assess effects of the manipulation on the individual as well as on other individuals (the biotic environment). These types of manipulations can be important in behavioral studies, including sexual selection, predator-prey interactions, host-parasite interactions, and intra- and interspecific frequency-dependent selection.

First, we discuss the relationship of phenotypic manipulations to the internal constraints of an organism, genetic correlations, that govern the evolution of suites of traits involved in the process of adaptive evolution. Second, we discuss how manipulations are useful in detecting natural selection and the qualitative form of selection (e.g., directional, stabilizing, or disruptive selection) or the performance consequences of adaptations. Third, we illustrate how questions concerning adaptations are also amenable to experimental analysis by reference to studies of sexual selection. Fourth, we discuss how manipulations of the selective environment identify the causal agents of selection. We end the chapter with a discussion of the problems that might arise from manipulation of the phenotype. We do not intend to provide an exhaustive review of studies using phenotypic manipulations, but rather we highlight select studies that illustrate how such manipulations are useful in elucidating both the origins and the maintenance of adaptations.

II. Assessing Constraints on Adaptation and the Utility of Manipulations

In recent years, students of evolution have come to the realization that the process of adaptation is subject to various constraints. One form that these constraints takes is a trade-off between two or more traits. Such trade-offs or constraints should also be reflected as genetic correlations among traits

(Chevrud, 1984) or as negative genetic correlations between the expression of these traits on fitness (Falconer, 1981). Genetic correlations result from either pleiotropy (Rose and Charlesworth, 1981) in which one gene controls the expression of two or more traits or through genetic linkage, e.g., by chromosome linkage or assortative mating (Lande, 1983). Genetic correlations that arise from pleiotropy should be reflected in development or physiological couplings among traits. Manipulation of the mechanisms underlying the physiological or developmental regulation of traits should elucidate the mechanistic causes of genetic correlations (Sinervo, 1993; Sinervo and DeNardo, 1996; Sinervo and Doughty, 1996).

Reznick (1985) provided a useful partitioning of methods for assessing life history trade-offs in which he considered the utility of experiments in studying the process of life history adaptation. Although Reznick's arguments were originally constructed for understanding genetically based trade-offs that constrain the process of life history adaptations involving costs of reproduction, the partitioning he made applies equally well to adaptation involving a range of morphological and physiological traits. The four categories he used were: (1) phenotypic correlations between an index of reproductive effort and some cost such as decreased growth, survival, or fecundity; (2) experimental manipulations of a trait and observing effects on other traits; (3) genetic correlations between traits as detected from correlations among relatives (e.g., half-sib designs, etc.); and (4) genetic correlations as detected in artificial selection studies in which selection on a single trait will also result in a genetic response to selection in other traits if the selected trait is genetically correlated with other traits (Lande, 1979).

Reznick (1985, 1992) concluded that genetic correlations (3) and selection experiments (4) are the only methods that provide the essential information necessary to establish the existence of life history trade-offs. Life history trade-offs that are reflected in phenotypic correlations (1) are confounded by environmental effects and genotype by environment effects that may mask genetically based trade-offs (Reznick, 1985, 1992). Reznick (1985) also concluded that results from experimental manipulations (2) measure plasticity in the life history traits and thus may be more related to genotype by environment interactions than genetically based trade-offs. These views have received explicit experimental attention in studies by Leroi et al. (1994a) and Chippindale et al. (1993) in which the environment of *Drosophila melanogaster* was manipulated in an attempt to detect life history trade-offs among fecundity and energy reserves. Leroi et al. (1994a) and Chippindale et al. (1993) both found that such environmental manipulations cannot be reliably used to predict the direction of evolutionary change because of the existence of genetic correlations among traits that remain unaltered by the environmental manipulation. As such, an estimate of evolutionary response to selection based on environmental manipulations will not include all relevant genetic correlations that govern the life history trade-off.

While we agree in large part with the views presented by Reznick (1985, 1992), we would further partition experiments into the following categories: (i) environmental manipulations, (ii) single-trait manipulations that only affect the trait per se or traits that are involved in the extended phenotype or the organism, and (iii) multitrait manipulations in which manipulation of a mechanism responsible for genetic correlations is used to effect changes in a suite of traits. These manipulations of phenotype vary in their usefulness for understanding constraints arising from genetic correlations.

(i) One can experimentally manipulate the environment to determine whether an environmentally induced change in one trait results in correlated changes in other traits. For example, Hirshfield (1980) manipulated food availability and temperature (three levels of each) in growing rice paddy fish and estimated correlated effects on reproductive effort and survival. Clearly, these results could be influenced by genotype x environment effects and thus are susceptible to Reznick's (1985) basic criticism. While such plasticity resulting from genotype by environment effects is an interesting life history phenomenon, such plasticity does not necessarily demonstrate how genetically correlated traits might respond to selection. Likewise, manipulating food availability in *Drosophila melanogaster* may alter the phenotype in a genotype specific manner that reflects evolved changes in plasticity (Leroi et al., 1994a, Chippindale et al., 1993). The fecundity in lines of *D. melanogaster* that have been previously selected for starvation resistance responds differently when the environment is altered (e.g., lower food) than the fecundity in lines that have not been selected for starvation resistance (controls). The starvation resistant lines incur a smaller cost in fecundity per unit of starvation resistance at different levels of yeast. Such genetically based plasticity obscures the estimation of a genetically based trade-off between fecundity and starvation resistance from a purely environmental manipulation of food availability.

(ii) One can hold the environment constant and instead experimentally manipulate the mean or variance of a single phenotypic trait under investigation (Schluter, 1988). Examples of this approach involve adjusting clutch size in birds (adding or removing eggs), tail length in widow birds (Andersson, 1982), leaf area of plants (Willson and Price, 1980), or offspring size by yolk removal (Sinervo and McEdward, 1988; Sinervo, 1990; Sinervo and Huey, 1990). Single-trait manipulations might have *cascading* effects on other traits. For example, increasing clutch size in birds could result in either decreased fledging success or perhaps reduced survival of the overworked parent. Alterations of the numbers of eggs in a clutch by removing or adding eggs is more akin to an environmental manipulation of the nest environment -- e.g., a manipulation of the "extended phenotype." The removal of eggs does not necessarily capture relevant trade-offs associated with egg production. While such manipulations have great utility in establishing causal relationships between a single trait and fitness (Endler, 1986; Mitchell-Olds and Shaw, 1987; Wade and Kalisz, 1990), they do not necessarily lead us to valid inferences about causal relations between correlated traits involved in genetically based trade-offs because these manipulations do not involve two traits with common mechanisms of regulation.

(iii) One can experimentally manipulate a single physiological mechanism that links two or more traits and study correlated effects on other traits (Marler and Moore, 1988; 1991; Ketterson et al., 1991; Sinervo and Licht, 1991a,b; Ketterson and Nolan, 1992; Nolan et al., 1992; Landwer, 1994; Sinervo and DeNardo, 1996). This type of manipulation maintains the essence of any physiologically based genetic correlations. In the case of clutch size manipulations in birds, the single-trait manipulation of clutch size (i.e., add or remove eggs from the nest) may give dramatically different results compared to hormonal manipulations of clutch size that simultaneously affect the suite of traits governing allocation to reproduction (Fig. 1) (Partridge and Harvey, 1985; Landwer, 1994; Sinervo and DeNardo, 1996). However, multitrait "mechanistic" phenotypic manipulations must be designed carefully and with adequate controls to avoid the introduction of spurious "pleiotropic" effects that result from pharmacological effects of hormones. In addition, one multitrait manipulation may not capture all of the genetically based trade-offs involved in a life history trade-off, and several levels of regulation (e.g., follicle stimulation, follicle growth, and follicle atresia) must be investigated to gain a comprehensive understanding of the entire trade-off involving reproductive allocation.

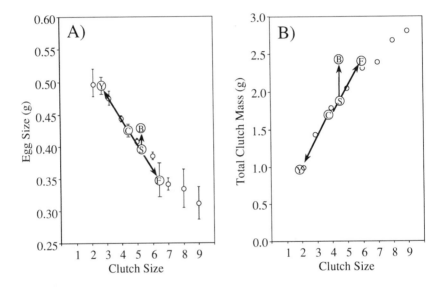

Figure 1 Summary of published data on experimental manipulations relative to natural variation in egg size, clutch size, and total clutch mass. Variation in mean (± S. E.) (A) egg size (g) and clutch size and (B) total clutch mass among female *Uta stansburiana* from three populations (O) from the inner coast range [adapted from Sinervo and Licht (1991a,b) and Sinervo (1994)]. The results from two complementary experimental manipulations of clutch size are also plotted. Females receiving exogenous gonadotropin ovine FSH (F) produce larger clutches of small eggs relative to either sham or control females (C). Females with surgically reduced clutch size achieved by follicle "yolkectomy" (Y) produce a smaller clutch of large eggs. Results for the third manipulation, exogenous corticosterone (B), do not affect clutch size per se, but increase the amount of energy invested in eggs relative to females receiving a saline implant (S). Moreover, this energy is selectively channeled into female offspring (Sinervo and DeNardo, 1996). While the calculated per egg change in offspring size is small, this occurs in half of the eggs (e.g., female offspring) and the resultant effects of corticosterone manipulation on total clutch mass are comparable to the effects of FSH. Differences between control and saline-treated lizards are not biologically significant in that experiments on follicle ablation and follicle-stimulating hormone were conducted in the laboratory using females from a population which naturally produces smaller clutches (see Sinervo, 1984). Data on reproductive traits of saline-treated and corticosterone treated females are from the 1991 experiment. [Redrawn from Sinervo and Licht (1991b); Sinervo and DeNardo (1996)].

III. Experimental Manipulations and Life History Adaptation Involving Offspring Size and Costs of Reproduction

The optimality analysis of adaptation within the context of natural selection only seeks to show that a trait is an adaptation in the present-day context. It has one fundamental premise: if the trait is optimal or adapted to its selective environment, then an individual with a value close to the mean value of the trait in a population should have higher fitness compared to individuals with higher or lower values of the trait. This condition can arise by two mechanisms: (i) stabilizing selection on a single trait, or (ii) conflicting selection pressures on a set of genetically correlated traits. In the first case, individuals in a population with extreme values for a single trait are selected against (Lande and Arnold, 1983). The situation is more complex in the case of a trait that is genetically correlated with multiple fitness traits. For example, a single trait can be negatively correlated with one fitness trait but positively correlated with another fitness trait (see Fig. 1). These two effects would act multiplicatively and be resolved as a fitness optimum or disruptive selection depending on the precise form of the two directional selection fitness surfaces (Arnold and Wade, 1984). Phenotypic manipulations are useful in analysis of the form of the fitness surface.

Many authors have advocated the use of phenotypic manipulations in understanding whether traits are under directional, optimizing or stabilizing selection (Endler, 1986, Mitchell-Olds and Shaw, 1987, Schluter, 1988, Wade and Kalisz, 1990). Manipulating phenotypic traits can greatly expand the range of variation and thus the power to detect selection. The number of individuals in the tails of the frequency distribution can be inflated and these are the individuals that provide the "leverage" to detect selection (Fig. 2). More importantly, comparable results from analyses of natural and experimentally induced variations (Sinervo et al., 1992) illustrate the causal relationship between natural variation and selection.

There is a rich history of "phenotypic manipulations" of life history traits in avian studies in which eggs are added or removed from a clutch and cascading effects on parental survival of offspring fledging success are measured (Lack, 1947; Hogstedt, 1980; Nur, 1984a,b, 1986, 1989; Lima, 1987; Gustafsson and Sutherland, 1988). Such studies address adaptive variation in clutch size observed within and among species. However, as discussed above, alterations of the numbers of eggs in a clutch by removing or adding eggs are more akin to an environmental manipulation of the nest environment. Manipulation of clutch size by direct hormonal manipulation of the female's ovary should better capture the essence of genetically based correlations in the female parent per se. Such manipulations of mechanisms use the built-in regulatory machinery of development and physiology.

Phenotypic manipulations have recently been used to study allocation to reproduction in lizards (Sinervo and Licht, 1991a,b; Sinervo et al., 1992; Landwer, 1994; Sinervo and DeNardo, 1996). Life history theory predicts that offspring size observed in natural populations should be under stringent natural selection arising from two counterbalancing selective forces: (1) fecundity selection arising from the egg size and egg number trade-off, and (2) offspring survival selection arising from the advantages of large neonate size in enhancing survival of offspring to maturity (Fig. 2). Thus, offspring size should be under stabilizing selection in which the optimum offspring size in the population reflects that phenotype that produces the most offspring that survive to maturity.

The predictions of life history theory are in large measure reflected in experimental measurements of natural selection in *Uta stansburiana*. Experimental measurements of fecundity selection, the first component of selection, was achieved by manipulating ovarian development (Sinervo and Licht, 1991a,b; Sinervo et al., 1992). Direct surgical manipulation of the number of follicles on the ovary lowers clutch size and, in a cascading fashion, enhances egg size and thus offspring size. Conversely, stimulating the recruitment of follicles on the ovary using exogenous gonadotropin (follicle-stimulating hormone) results in a larger clutch and, in a cascading fashion, reduces egg size (Sinervo and Licht, 1991a,b; Sinervo et al., 1992; Sinervo, 1993). Experimental measurement of offspring survival selection as a function of offspring size, the second component of selection, was assessed by measuring the survival of free-ranging giant hatchlings, miniaturized hatchlings, and control hatchlings (Sinervo, 1990; Sinervo and Huey, 1990; Sinervo et al., 1992). Giants were obtained from eggs that were produced by ablating follicles on the female's ovary. Miniature hatchlings were not obtained by stimulating the production of more smaller offspring, but by directly removing yolk from freshly laid eggs (aspirating a known volume of yolk). By studying the survival of a large number of such control and experimentally manipulated offspring to maturity, Sinervo et al. (1992) tested whether the seasonal changes in egg size observed in lizards were adaptive (Nussbaum, 1981; DeMarco, 1989). Females typically produce larger offspring in later clutches, and this is in accord with measurements of natural selection on offspring size (Fig. 2). The optimal egg size of later clutches is larger than the optimal clutch size on the first clutch.

However, there is a discrepancy between the "optimal" egg size and the egg size that females produce in natural populations. This discrepancy reflects additional constraints on the production of very large eggs. Females that either produce small clutches of large eggs or were induced to produce small clutches of large eggs experience a higher frequency of burst eggs or oviductally-bound eggs. These difficulties at oviposition may reflect a constraint arising from the diameter of the pelvic girdle, which is thought to be a limiting factor in the evolution of offspring size in vertebrates (see below). Additional "selective constraints" that limit the allocation of large amounts of energy to offspring also arise from costs of reproduction in female parent.

Landwer (1994) and Sinervo and DeNardo (1996) further tested the consequences of such adaptive allocation to offspring size and number on the costs of reproduction incurred by the female parent. Follicle ablation on the ovaries of field active females lowers investment on the first clutch (total clutch mass) and this enhances survival to the next season in *Urosaurus ornatus*

A) First Clutch

B) Later Clutches

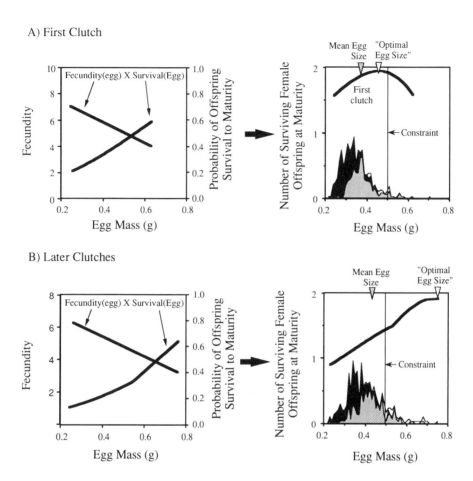

Figure 2 Fecundity selection in natural populations of *Uta stansburiana* in the coast range of California results from the trade-off between egg size and egg number. Females that produce small eggs produce more offspring. Fecundity selection is balanced, however, by the effects of selection on probability of survival to maturity as a function of egg size. Only data on the survival of female offspring to maturity are presented. The fitness surface describing survival of male offspring to maturity is qualitatively different from female offspring (see Sinervo et al., 1992). The number of offspring that survive to maturity is determined by fecundity as a function of egg size times survival of offspring as a function of egg size [fecundity(egg)xsurvival(egg)]. The detection of survival selection is facilitated by augmenting the distribution of control hatchlings (gray histogram) with miniaturized (black histogram) and giant hatchlings (white histogram). Note that the survival curve for later clutch offspring is shifted to larger egg sizes relative to the survival curve for the first clutch offspring. The effect of this seasonal change in offspring survival is a shift in the optimum to larger egg sizes. The optimum egg size (∇) is the reproductive strategy that leaves the most surviving offspring at maturity. The observed shift in average egg size does not, however, match the optimum due to additional selective constraints on egg size (see text). [Redrawn from Sinervo et al. (1992)].

(Landwer, 1994) and to subsequent clutches in *Uta stansburiana* (Sinervo and DeNardo, 1996) (Fig. 3). Conversely, two ovarian manipulations enhance effort and lower survival to the first and second clutches in *U. stansburiana*. Administering follicle stimulating hormone increases clutch size and also total clutch mass, whereas administering the hormone corticosterone increases allocation per offspring (e.g., egg size) and also total clutch mass without altering clutch size (Fig. 1). The results from such experimental manipulations of costs of reproduction in natural populations corroborate observed patterns of selection arising from natural variation (Sinervo and DeNardo, 1996). However, the fitness surfaces involving costs of reproduction shifts among years in *Uta stansburiana* , depending on the end of a long-term seven year drought in California. Prior to the drought (e.g., 1991, Fig. 2), females that were induced to lay large clutches on the first clutch of the reproductive season had the highest survival to subsequent clutches. In the drought-breaking year (1992), females induced to lay a large clutch did not differ in survival to the first clutch compared to sham-manipulated females. After the drought, the pattern of selection reversed and females that were induced to lay large clutches had lower survival to the first clutch compared to sham-manipulated females. The pattern of natural selection observed in females with experimentally-altered clutch mass is similar to comparative data on the effects of natural variation in clutch mass on survival to subsequent clutches (data not shown, but see path analyses of natural selection in Sinervo and DeNardo, 1996). Moreover, the patterns of natural selection measured in experimental and comparative data are consistent with year to year changes in total clutch mass that reflect an evolutionary response to selection on "costs of reproduction". For example, both egg mass and total clutch mass increased from 1991 to 1992. The observed survival selection that favors females producing a large clutch mass would provide the mechanism underlying such presumed genetic changes (Fig. 3). Conversely, both egg mass and total clutch mass decreased significantly in from 1992 to 1993 and the observed survival selection that favors females producing small clutches would provide the mechanism underlying an evolutionary response to selection. The heritability of egg size measured on field active females during the last three drought years (1989-1991) is 0.61, and with such a high heritability the population would be expected to show a rapid evolutionary response to selection (Sinervo and Doughty 1996). The agent of selection is likely to be predatory snakes. We observed a four-fold increase in snake abundance within one year of the lifting of the drought (1993).

The correspondence between inferences from ovarian manipulations and genetically-based correlations is reflected in a comparison of results from experimental manipulation of the life history trade-offs involving energy allocation and heritability estimates of egg size of free-ranging lizards (Sinervo and Doughty, 1996, Sinervo and DeNardo, 1996). Moreover, the correlation between egg size of the dam and clutch size of her offspring suggests that the genetic correlation between egg size and clutch size is also large and negative. Experimental manipulation of clutch size probably uses analogs of the natural mechanisms for regulating clutch size (follicular atresia and follicular recruitment). Thus, manipulation of clutch size probably co-opts physiological mechanisms that underlie the genetic correlation between these traits. As such, experimental manipulation of clutch size and egg size by follicle ablation,

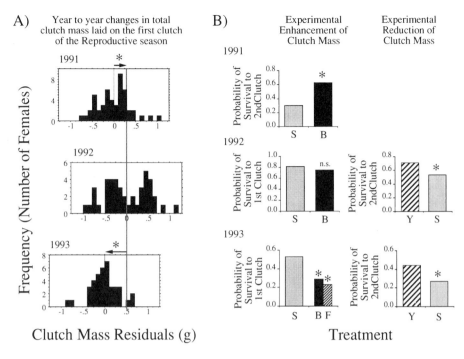

Figure 3 (A) Results of analysis of covariance in total clutch mass in which the effects of postlaying mass were removed by ANCOVA. Data from years 1991 and 1993 were compared to 1992 in a pairwise ANCOVA comparison. Significant differences among years [e.g., arrow describes magnitude of change in mean between the mean in 1992 (heavy shaded vertical line and the light shaded line is the mean in each year relative to 1992)] are denoted by an asterisk ($P<0.05$ 1991 vs. 1992; $P<0.001$ 1992 vs. 1993). The residuals for 1991 and 1993 are offset from the reference year (1992) by the difference in total clutch mass as detected by analysis of covariance. Note that data for 1992 are decidedly nonnormal and appear to be bimodal. Note also that the variance in total clutch mass drops abruptly from 1992 to 1993. These differences in total clutch mass are largely due to significant changes in egg size among years [see Sinervo and DeNardo, (1996)], not changes in clutch size. (B) The effects of enhanced clutch mass (corticosterone implants [B] and FSH implants [F]), and reduced clutch mass (follicle ablation via a yolkectomy [Y]) on survival of adult females during 1991, 1992, and 1993. Survival for females receiving corticosterone implants in 1991 was higher than sham-manipulated females; survival is measured from early March to mid-May or the production of the second clutch (8-week interval). Significant differences in survival between treatments were tested by Chi-square analysis and are denoted by an asterisk ($P<0.05$). Females receiving corticosterone in 1991 also had higher lifetime reproductive success (e.g., survival to a second year of reproduction; number of clutches produced on average for corticosterone females is 2.68 ± 0.50 [S. E.] versus 1.15 ± 0.36 for shams). Survival for females undergoing follicle ablation (and controls) is from early-mid vitellogenesis (mid-April) on the first clutch to production of the second clutch (early May, 6-week interval). Survival for females receiving implants in 1992 and 1993 (corticosterone, FSH, sham, unmanipulated) is from the earliest stages of vitellogenesis (early April) to 1 month later (production of the first clutch). [Redrawn from Sinervo and DeNardo (1996)].

follicle stimulating hormone, and corticosterone are likely to involve mechanisms that are the same as those underlying the life history trade-offs (Sinervo, 1994a,b).

These manipulations have also been used to explore proximate constraints that might set limits on the process of life history adaptation of clutch size in lizards (Sinervo, 1994a,b). For example, all 260+ species of Anoline lizards lay sequential single-egged clutches and this is presumably derived from an ancestral iguanid that laid sequential multi-egged clutches (Andrews and Rand, 1974, Etheridge and De Queiroz, 1988). Sinervo and Licht (1991b) investigated proximate constraints on this evolutionary transition by turning a typical multi-egged iguanid lizard, *Uta stansburiana*, into the single-egged morphology. *Uta stansburiana* never lays single-egged clutches and rarely lays two-egged clutches (2-9 egged clutches are the range in this species). However, Sinervo and Licht (1991b) ablated follicles on the ovary of *U. stansburiana* and produced one-, two-, and three-egged clutches. Females producing experimentally-altered clutches experience a greater number of problems during oviposition in the form of burst eggs and oviductally-bound eggs (dystocia). Moreover, the problems seen in experimentally-altered clutches are comparable in frequency to the rare females that naturally lay two- and three-egged clutches. Sinervo and Licht (1991a,b) hypothesize that proximate constraints arising from the pelvic girdle (Congdon and Gibbons, 1987) may limit the allometric relationships involving egg size, clutch size, and body size. Support for this idea is provided by the similarity of egg size between experimentally-induced clutches of one egg that are produced in *U. stansburiana* and the natural one-egged clutches of an *Anolis* with a comparable body size. Apparently, natural selection has not produced an *Anolis* lizard with an egg that is any larger than the egg size that a modern-day multi-egged species is capable of producing by experimental perturbation (Sinervo, 1994a,b). Such higher-level taxonomic comparisons and experimental manipulations (e.g., *Anolis* versus *Uta*), provide strong inferences regarding the constraints on life history adaptation particularly when such studies are compared to results from within population analyses of natural selection (Fig. 2) within an evolving lineage such as *Uta stansburiana* (Sinervo, 1994a, b)

Manipulations of offspring size and parental effort partitions the impact of the selection on ovarian regulation within the context of three life history trade-offs: (1) the consequences of the egg size and number trade-off in the female parent (fecundity selection), (2) the impact on offspring fitness (survival selection), and (3) costs of reproduction incurred by the female parent (survival selection). In light of the complexity of established patterns of selection on egg size (Sinervo et al., 1992; Sinervo and Doughty, 1996; Sinervo and DeNardo, 1996), it is unlikely that purely comparative approaches will yield fruitful answers to theoretical predictions concerning the evolution of egg size. Conversely, purely experimental approaches are unlikely to provide answers to such questions because the frame of reference, natural variation in the wild, is not explicitly addressed. We strongly advocate a mixture of experimental ("manipulative" and "mechanistic") and comparative approaches to understanding the evolution of reproductive allocation in natural populations.

IV. Manipulations of Morphology and Performance

The experimental manipulation of phenotypic traits within the context of an analysis of natural selection validates the premise that life history adaptations are typically under strong stabilizing selection. Thus, we can determine whether life history traits are a result of the process of adaptation to the environment. The goal of future manipulations should be to extend such observations to traits that are less directly related to fitness. Complex morphological structures undoubtedly play a role in survival and reproduction (Bock, 1977, 1980; Arnold, 1983), but detection of natural selection on such structures using phenotypic manipulations is rare. In principle, such manipulations of morphology must take place during ontogeny. By manipulating ontogenetic trajectories, one could take advantage of the regulative nature of development to achieve a fully functional, albeit experimentally altered, morphology that can be enhanced or reduced in structure.

For example, Sinervo and McEdward (1988) used manipulations of egg size in sea urchins to assess developmental consequences of an evolutionary change in egg size on larval development and morphology. Two congeneric species of sea urchins, *Strongylocentrotus droebachiensis* and *S. purpuratus*, have marked differences in egg size, larval morphology, and development rate. *Strongylocentrotus droebachiensis* begins feeding as a larva with a more elaborate feeding structure (the ciliated band) and develops faster to metamorphosis of the adult morphology. Moreover, egg size was correlated with larval morphology and development rate in a larger species comparison (McEdward, 1986a,b). Sinervo and McEdward (1988) reduced the egg size of the larger-egged *S. droebachiensis* and created half-sized "twins" and one-quarter-sized "quadruplets" by separating blastomeres of two- and four-cell embryos. Differences in morphology and development rate in experimentally altered larvae phenocopy the between-species differences. Thus, an important avenue for the elaboration and evolution of a complex morphological adaptation, the larval ciliated band, can occur through evolved changes in egg size. Because the complexity of the larval feeding structure, the ciliated band, is directly related to the clearance rate of algae from the water (Strathmann, 1971; Hart and Strathmann, 1994), the correlations between egg size and larval morphology have important functional implications. Hart (personal communication) has used the experimental manipulation of larval morphology by egg size manipulation (Sinervo and McEdward, 1988) to directly test the performance consequences of an evolved change in egg size. Hart found that experimentally altered larvae have reduced feeding capability compared to control morphologies.

Experimental embryology provides evolutionary biologists with a rich, although underutilized, "toolbox" for investigating the evolution of other complex morphological adaptations (Sinervo, 1993). The goal of future studies should be to incorporate such manipulations into comparative and phylogenetic tests of adaptations. Despite the paucity of such "ontogenetic manipulations," ablation of a structure can be used as a surrogate to alter morphology, particularly in studies where "performance" is used as a proxy for fitness. The strength of such ablation approaches is enhanced when a structure is ablated early in development and then natural processes of regulation allow a partial compensation of the size of the ablated structure. In this way, a range of phenotypes can be created, spanning the fully developed structure and rudimentary structures. For example, Reilly and Lauder (1991) ablated a specific muscle in larval salamanders and measured the feeding performance of adult salamanders after the larvae metamorphosed. Individuals varied in the degree of muscle regeneration and this allowed a test of the muscle function in the projection of the tongue.

Another manipulation of morphology and performance in which the structure can grow back involves the feeding of crossbills on pine cones. Benkman (1988) and Benkman and Lindholm (1991), studied the feeding performance of crossbills, *Loxia curvirostra,* by filing the length of the upper and lower mandibles down to what is presumably the ancestral condition, a very short beak with no overlap between upper and lower mandibles. Not surprisingly, crossbills with experimentally altered morphology have poor feeding performance. However, as the bills grew back, the gains in performance arising from more bill overlap and greater amounts of bill curvature could be measured. This type of experimental manipulation assesses the improvements in performance (and perhaps fitness) that would be accrued as crossbills evolved from the ancestral condition. For example, the crossbill condition is essential for opening closed cones, but not partially open cones, and thus the evolution of the crossbill trait presumably expanded the temporal opportunity for exploiting cones.

Even if regeneration following ablation is not possible in an organism, analysis of function and thus adaptive value can still be gleaned from such ablation experiments. For example, Carothers (1986) analyzed the adaptive value of the elongated scales of fringe-toed lizards by sprinting fringe-toed lizards with fringes intact and without fringes (fringes cut) down a sand race track. Fringe-toed lizards without fringes sprint slower on sand than fringe-toed lizards with intact fringes, suggesting that this trait has adaptive value in sand environments with which the trait fringe-toed trait is strongly correlated. Given that the fringe-toed trait has evolved as many as 26 times in a variety of lizard taxa that have occupied sand dune habitats (Luke, 1986), fringes not only have adaptive value in the present-day context, but also were likely to be adaptive during the origins of the trait.

Another example of phenotypic manipulation in which the adaptive value of the trait was assessed by using locomotor performance as a proxy for fitness is the relationship between tail length and crawling speed in the snake *Thamnophis sirtalis* (Jayne and Bennett, 1989). Correlational analyses indicated that snakes with an intermediate tail length (postcloacal tail length) relative to bodies (snout-vent length) have the highest crawling speed. Apparently there is an optimal tail length for crawling fast. Jayne and Bennett (1989) tested the adaptive value of tail length as a locomotor structure in snakes and found that removing a portion of the tail had no impact on speed. The negative results of this experiment are illuminating in that tail length has no apparent role in locomotion contrary to results from correlational studies. Thus other adaptive explanations must be sought for the correlational patterns or perhaps nonadaptive explanations that are unrelated to performance.

A final example of a manipulation of morphology and performance deals with a larger macroevolutionary adaptation involving the evolution of the insect wing. Marden and Kramer (1994) used the comparative method and phenotypic manipulations to address whether surface skimming in stoneflies could be an intermediate stage between swimming and flight. That is, that larval gills evolved into wings to accommodate flight. By manipulating wing area in extant stoneflies, they found that even with reduced wing size, test subjects can move faster by skimming than by swimming. While suggestive, these results are difficult to interpret in historical terms because of the lack of phylogenetic information on flight and skimming. Nevertheless, Marden and Kramer's (1994) results provide an alternative explanation to origin of insect flight that is contrary to other exaptational hypotheses regarding the thermoregulatory value of insect wings (Kingsolver and Koehl, 1994). Biophysical modeling suggests an adaptive thermoregulatory value to small insect wings that could ultimately be an exaptation for flying performance. However, Marden and Kramer's (1994) results suggest that rudimentary wings have aerodynamic value in locomotion involved in surface skimming and thus the trait may have originated as an aerodynamic adaptation.

V. Phenotypic Manipulations and Sexual Selection

Phenotypic manipulations can be particularly useful in testing the importance of sexual selection on trait evolution and maintenance. There are two types of sexual selection: intrasexual and intersexual (Darwin 1871). Intrasexual selection involves competition between members of one sex for mating access to members of the opposite sex. Intersexual selection involves mating preferences in one sex for traits in the opposite sex. One important characteristic of both types of sexual selection is that they involve signaler-receiver systems: one individual produces a signal that is transmitted through the environment and received by another individual (Arnold, 1994). In intrasexual selection, the signaler and receiver are both of the same sex, and thus the interaction often involves reciprocal signals. In intersexual selection, the signaler and receiver are of different sexes. In signaler-receiver systems one can examine the adaptive significance of a signal by manipulating the signal and then examining effects of signal variants on receivers, sometimes revealing the adaptive significance of the behavioral response to the signal.

Manipulations of the phenotype are particularly useful in the study of signal-receiver relationships because a single character of an individual can be altered in order to determine whether that one character alone is a target of sexual selection and thus affects fitness. While we discuss intrasexual selection here, we concentrate primarily on intersexual selection. In these latter cases, we are concerned with mating preferences and the evolution of preferred traits (by convention, intersexual selection is discussed in terms of female preferences and male traits, but the ideas are interchangeable regardless of sex). By definition, any male trait preferred by females is adaptive if females have the opportunity to exercise mate choice for that trait (Basolo, 1996). Unless countered by natural selection or sexual selection in another context, preferred male traits should evolve. Although testing adaptive significance of male traits has been successfully accomplished by phenotypic manipulations, establishing that evolution of female preference was adaptive has proven to be problematic (Kirkpatrick, 1987; Andersson, 1994; Basolo, 1996).

One important experimental advantage that manipulation of a male signal has is that phenotypic and genetic correlations between the trait of interest and other male traits can be broken. In many cases, phenotypic manipulations have allowed hypotheses to be tested that would otherwise be difficult to evaluate. For example, when male traits covary and it is not possible to identify all traits that might be important to females, statistical association between female responses to males and male characteristics can lead to misleading inferences about which traits are targets of female choice.

Manipulation of male traits usually involves the increase, decrease, or removal of a simple or complex trait. The types of communication that can be investigated by manipulating male signals include acoustic, visual, chemical (including olfactory and taste), tactile, and electric modalities. Of the numerous ways in which manipulations can be used, we present a small subsample of the following types of experimental manipulations: (1) isolated presentation of variants in a male trait (synthesized signals, disassociated from the male) to females, (2) males or females with manipulated traits presented to the opposite sex, (3) manipulation of the extended male phenotype presented to females, (4) modification of the interface between the signaler and the receiver resulting in an alteration in perception of the signal to the female, and (5) manipulation of the female to alter signal perception. In addition, we discuss some potential pitfalls and drawbacks to using phenotypic manipulations to address questions concerning adaptive evolution.

VI. Manipulations of Signaler Biology that Result in Signal Change

The most common type of phenotypic manipulation in studies of sexual selection is the manipulation of male signals. Manipulation of male signals can be done in two ways. First, the signal can be manipulated indirectly through changes in physiology or environmental conditions. Second, the signal can be manipulated directly through the production of isolated synthetic signals or through direct changes in male morphology. A large body of work has been produced concerning the importance of song or call variation in mate attraction and intersexual competition in crickets (e.g., Popov and Shuvalu, 1977; Stout and McGhee, 1988; Wagner et al., 1995), frogs (e.g., Ryan, 1980; Sullivan, 1983; Wells and Schwartz, 1984; Gerhardt, 1992; Wagner, 1992) and birds (e.g., Searcy, 1984, 1988; Searcy and Marler, 1984; Searcy and Yasukawa, 1990; also see Kroodsma et al., 1982 for review). These studies investigate male-male interactions and female preferences by recording calls or songs, using these natural calls as stimuli, or using these as models to synthesize new signals with altered call components. In a study of male-male interactions in the frog *Acris crepitans blanchardi*, males were found to lower their call's dominant frequency during contests over calling sites with other males (Wagner, 1989). By playing calls to males that varied only in how dominant frequency was changed during a contest, Wagner (1992) showed that males that lower call dominant frequency obtain an advantage relative to males that do not lower their call dominant frequency: they repel opponents more frequently.

Hormonal implants have been used in a number of studies to investigate whether hormone level affects various components of fitness, including intrasexual and intersexual selection (Wittenberger, 1981; Fox, 1983; Crews, 1985; Wingefield and Ramenofsky, 1985; Marler and Moore, 1988; Moore, 1991;

Ketterson et al., 1991; Ketterson and Nolan, 1992; Nolan et al., 1992). This is a type of indirect manipulation of the signal. For example, if hormones are administered and a specific male trait is altered, it may not be known what the specific mechanistic base of change is and/or what other traits have also been affected by the manipulation. If traits additional to the target trait have been affected, the results may be confounded, making it difficult to assign an adaptive function solely to the target trait.

In male *Sceloporus jarrovi* lizards, increased aggression is thought to improve success in intrasexual competition; if so, then why aren't natural levels of aggression higher? Manipulation of the phenotype by implanting testosterone has been used to investigate the effect of increased testosterone levels on male aggression (Moore and Marler, 1987) and survivorship (Marler and Moore, 1988). The results of these manipulations suggest that while implants increase male aggression, more aggressive males have a lower survivorship. Thus, while males may be more successful in obtaining territories and in male-male encounters, this is counterbalanced by a decrease in survivorship. This may explain why testosterone levels are highest during the fall, when they are breeding.

A second type of indirect manipulation of the phenotype is to manipulate food intake. Food can vary in quantity, nutritional value, spatially, or temporally. In a study manipulating food quantity, male green swordtails, *Xiphophorus helleri*, were placed alternately on ad libitum or restricted food regimes at sexual maturity. Males on the restricted regime were found to invest significantly more in sword growth versus body growth relative to ad libitum males (Basolo, in prep.). In this species, female preferences for both larger body size (standard length) and greater sword length have been established. The results of this study suggest that males shift allocation from body growth to sword growth under food-stressed conditions. In addition, preliminary results indicate that the strength of the preference for body size is stronger than that for sword length (Basolo, unpublished). As an incremental increase in sword length is likely less costly than an incremental increase in body growth, it appears that males shift allocation from a more expensive to a less expensive sexually selected trait when food stressed. Like hormone manipulations, food manipulation cannot be used directly to assess whether females select mates based on a trait. However, given an independent assessment of female preferences, they can be used to examine if food-stressed males make adaptive changes in trait expression.

Numerous studies have directly manipulated visual signals by additions of or reductions of male traits. In a classic study, Andersson (1982) manipulated tail length in the widow bird *Euplectes progne* and demonstrated that longer-tailed males are preferred by females and had higher mating success than controls and males with experimentally reduced tails. However, male tail-length reduction was not found to affect the ability of males to retain territories. These results suggest that selection via female choice likely contributed to the evolution of elongated tails and that intrasexual selection likely did not contribute to the evolution of the impressive tails present in male widow birds.

Jones and Hunter (1993) manipulated the forehead crest, which develops during the breeding season in both sexes in the crested auklet *Aethia cristatella*. By using stuffed models with a variation in crest length, these researchers controlled the effects of intrasexual selection, as well as any behavioral changes that may have resulted from direct manipulations of the live birds. Models were presented with either shortened or lengthened crest feathers; both females and males preferred models with longer crests. Thus, it appears then that crests in both males and females are under sexual selection and that crests in females are likely not just a by-product of sexual selection on crests in males, but have evolved at least in part due to male mate choice.

Based on Berry's (1974) observation that male chukawallas appeared to slip to a lower social rank in the field when their tails were lost, Fox and Rostker (1982) quantified male-male social rank in laboratory populations of the lizard *Uta stansburiana* and then decreased the tail length of dominant males. They found that males in which tails were reduced became subordinate to nonmanipulated individuals after the tail had been reduced by two-thirds. In addition, males who decreased in social status with a one-third reduction in their tails regained dominance when their partner in encounters had a tail reduced by two-thirds. This lowering of social status probably reduces a male's ability to secure and defend a territory, as well as to attract a mate.

Through phenotypic manipulation of signals, it is also possible to test hypotheses about the evolution of mating preferences. Houde and Torio (1992) manipulated carotinoid color and saturation in the guppy *Poecilia reticulata* by transferring ectoparasites to males and allowing parasitic infection for nine days. Infection resulted in a significant reduction in carotinoid expression. Following elimination of the parasites, males were tested for female preference. Males exposed to parasites were less preferred by females than they had been prior to infection. Whether females were using carotinoid expression, some other male indicator of past infection, or a combination of the two is not clear. Although females were not using the presence of parasites to avoid males (as models for the avoidance of contracting sexually transmitted diseases would suggest), recently infected males were discriminated against. These results suggest that males who have parasite resistance or who have physiological mechanisms to quickly recover from the effects of parasitic infection likely have an advantage over infected males via mate choice. However, it is not clear from this work whether females were discriminating against less vigorous males, which has nothing specifically to do with parasite resistance or recovery.

In another study investigating the evolution of mating preferences, the effect of visible parasites in male mate choice was investigated. Male pipefish *Syngnathus typhle* are the primary caregivers to young; females court and deposit eggs in male brood pouches where they are fertilized and the embryos are nourished. Pipefish are parasitized by a trematode that penetrates the skin and encysts; heavy infestation can lead to death. Infected pipefish react by secreting melanin around the encysted trematode. Rosenqvist and Johansson (1995) injected black tattoo ink into females to simulate the presence of parasites; for a control, the ink pigment vehicle was injected into a second set of males. The results from male mate choice experiments suggest that males discriminated against females with heavy "parasite" (ink injected) infestations and spent more time with the females with lower apparent infestation intensities. Females with lower natural infestation intensities transfer more eggs to males, which allows the males (who accept eggs from numerous females) to fill their pouches faster. This may diminish the amount of time necessary to participate in the potentially risky (in terms of predation) behavior of courtship. By manipulating the female phenotype, the researchers were able to limit the apparent presence of parasites to one visual cue and thus were able to eliminate characters correlated with parasite infestation as factors in male mate choice.

One area of mate choice evolution in which manipulating phenotypes promises to be particularly informative is in the evolution of female preferences and fluctuating asymmetry. Thornhill (1992) manipulated both forewing symmetry and pheromone release (by covering dispersal gland with glue) in male scorpionflies, *Panorpa japonica*, in a set of mate choice experiments. He found that (a) females showed a preference for males with low asymmetry levels; (b) when males were prevented from expressing pheromone, females no longer preferred more symmetric males; (c) when females were given a choice between a highly asymmetric, pheromone-releasing male with a mealworm and a mealworm alone, females preferred males with mealworms; and (d) when pheromone was not blocked, but the degree of asymmetry was experimentally manipulated, females did not prefer the more symmetric males. Manipulations in these experiments demonstrated that it was not the visual or mechanical aspects of forewing asymmetry that affected female preference, but that females based their preference on the pheromone and that males that had asymmetrical forewings apparently also had a deficiency in pheromone production. The author suggests that the degree of forewing asymmetry may affect foraging ability and therefore that less symmetric males were unable to produce sufficient attractive pheromones.

VII. Manipulations of the Extended Phenotype Resulting in Signal Change

One time-honored method for manipulating the phenotype is indirectly by manipulating "resources," i.e., food, space, etc. Some of these manipulations fall into the category of the extended phenotype. Manipulating the extended phenotype is an important method in studying life history trade-offs. If done properly, manipulation of the extended phenotype can be used to change the least number of variables.

For both intrasexual and intersexual selection, male territory quality can be important. Males who obtain and successfully defend a resource-rich territory may benefit directly by having more resources to utilize themselves, as well as benefit indirectly by attracting mates and by providing resources to both his offspring and the mother of his offspring. Hews (1993) manipulated male-territory quality in the lizard *Uta palmeri* by supplementing available food to determine whether territory quality affects male-mating success. She found that females shift their home range to better quality territories and that there is an increase in courtship rates for males with improved territories. This study suggests that males that obtain and defend high-quality territories likely have access to more potential mates. With this manipulation of the extended phenotype, it appears that the acquisition and retention of a high-quality territory likely increases a male's mating success in relation to males with lower quality territories. It is not clear, however, whether females shift to supplemented territories because they obtain more food for themselves and their offspring, and/or whether the male increases his food intake which affects his behavior and morphology.

Male great gray shrikes, *Lanius excubitor*, attract females to their territories by impaling prey, as well as nonedible materials, on thorns or other projections. By supplementation of the territories of some male great gray shrikes and removing naturally impaled caches in others, Yosef (1991, 1992) demonstrated that impalement appears to be adaptive in the context of sexual selection. Those males whose territories were supplemented were seen with females much earlier in the breeding season than control males, and those whose territories had been manipulated so as to lose a portion of their cache had difficulty attracting females and eventually abandoned their territories. In addition, mated pairs on supplemented territories nested three times during the breeding season, while controls nested only once or twice.

VIII. Modifications of the Environmental Interface between Signaler and Receiver

Beyond the extended phenotype, parameters of the biotic and abiotic environment shared by the signaler and receiver can be manipulated. Based on the perception of the male guppy color pattern in a natural light environment, female choice tests suggest that female guppies use orange spots in mate choice and prefer males with greater amounts of orange (Long and Houde, 1989). However, these tests could not distinguish whether female were basing their choice on the actual amount of orange present or on some character correlated with the amount of orange. Therefore, Long and Houde (1992) altered the perceived color patterns of males by manipulating the spectrum of light present. Wavelengths of incident light were manipulated such that the detection of orange spots was diminished; this reduced or eliminated female discrimination of differences in orange spots on males. In the presence of light transmitted through orange filters (orange light), males with higher amounts of actual orange coloration were preferred to the same degree as males with little or no orange under natural light. The results from these two studies suggest that female guppies prefer males with more orange to males with less orange coloration and that this preference is based on the orange color specifically and not some character correlated with orange coloration. These results also suggest that in nature, water color (clear versus tannin-stained) may affect female perception of male color patterns and possibly conspicuousness of these patterns to unintended receivers, i.e., predatory fish.

IX. Modification of Receiver Biology Resulting in Perceptual Difference in Signal

Receiver biology can be manipulated to change receiver perception of signals. This can be accomplished in a number of ways, including hormone implants, surgical alteration of sensory receptors or other tissues, and food intake control, to name a few. When it is unclear which sensory modality is involved in signal perception, one potentially interesting avenue of pursuit is to alter receiver biology to limit receiver perception to only one modality. For instance, in the Thornhill work with scorpionflies discussed earlier, females could potentially have been fitted with eye cups or other devices to eliminate visual cues, thus limiting perceived cues to chemical ones. Likewise, females could be manipulated such that they could not receive chemical cues (block receptors) and determine whether visual cues play any role in mate choice. By eliminating the reception of signals through two or more modalities, one can investigate the relative contribution of each modality to the mating decision. It is thereby possible to evaluate the relative strength of selection on each component of the signal.

X. Manipulation of the Phenotype, Phylogenetic Information, and Preexisting Biases

As we have illustrated, phenotypic manipulations can be used to determine if a male trait is preferred by females; this, however, will only indicate whether the trait is currently favored by sexual selection, not whether the trait evolved due to sexual selection. To demonstrate that a trait evolved at least in part for its current function, it is necessary to show that the source of selection currently favoring the trait was present at the time the trait evolved. This is a historical question that requires phylogenetic information to test. First, we must know when the trait evolved. Second, we must make inferences about selection at the time the trait evolved.

Phenotypic manipulations can be useful in inferring historical patterns of sexual selection. Consider a male trait currently preferred by females. If a male trait evolved because of female preferences, we would predict either that the trait arose after the evolution of the preference or that the preference and the trait coevolved. For this second possibility, the trait and preference could have arose and coevolved together or the trait could have preceded the origin of the preference. Resolution provided by phylogenetic inference in distinguishing between these two scenarios is low. If we have a good phylogeny for a group, we can infer the evolutionary history of a male trait. We can identify closely related species with the primitive state of trait absence (e.g., species in Fig. 4). If we add the trait to males of a species with the primitive state, we can then examine the phylogenetic distribution of the preference. If the preference evolved prior to or together with the male trait, this would suggest that the female preference not only maintains the male trait in the species in which it is present, but also that female preferences are at least partially responsible for the initial evolution of the male trait (Basolo, 1996).

The most unambiguous evidence for the evolution of a male trait as an adaptation in the context of sexual selection is provided when a preference favoring the trait is present prior to the origin of the trait (Fig. 4b). Evolution of a preference prior to the appearance of the male trait it favors has been described in several ways and with several different terms: (1) sensory trap, West-Eberhard (1984); (2) preexisting bias, Kirkpatrick (1987), and Basolo (1990a); (3) sensory drive, Endler and McLellan (1988); and (4) sensory exploitation, Ryan (1990), and Ryan and Rand (1993). Here we use the term

(a)

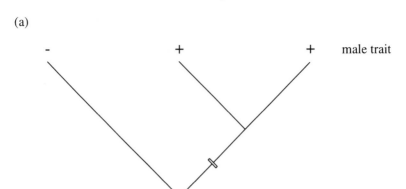

(b)

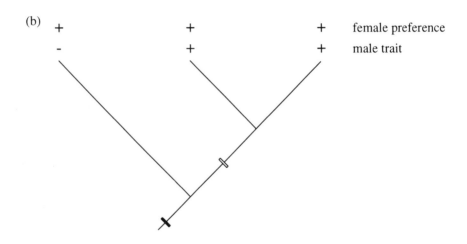

(c)

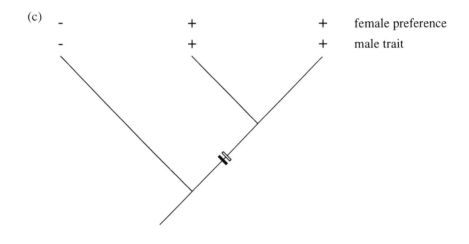

preexisting bias to refer to the condition in which a bias resulting in selection on a male trait arises prior to the origin of the male trait it favors. The bias may be specific for a single trait or may be more general in context with the potential for selecting on a suite of male traits.

By manipulating the phenotype, preexisting biases have been implicated in a growing number of species. This technique has been used to present novel male signals to females. In common grackles, *Quiscalus quiscula*, Searcy (1992) found that females prefer larger repertoires compared to an equal number of repetitions of single song types, yet males sing only one song type each. He suggests that female preference for larger repertoires may predate the appearance of male song repertoires. Clark and Uetz (1992) found that female jumping spiders, *Maevia inclemens*, may orient towards potential mates who make movements like prey. They used video playback and computer animation to control for differences in male courtship behavior and conducted female choice tests. When a simultaneous presentation of the same televised courting male was presented to a female, females chose the moving male. Female jumping spiders have a tendency to orient towards moving objects in the peripheral field of vision; this preexisting bias in turn appears to affect mating success in males. In the American goldfinch *Carduelis tristis*, Johnson et al. (1993) demonstrated that when females were given a choice between males wearing an orange band and males without a band, females preferred orange-banded males. They suggest that because male bills that are naturally brighter are preferred by females, females may be attracted to orange in general, regardless of where it is expressed on males.

These studies suggest that a preexisting bias may have played a selective role in the evolution of male morphological and behavioral traits. However, the historical relationship between the putative female bias and the male trait it favors is not known. To rigorously test the preexisting bias model, the comparative method should be used. As contended by Pagel and Harvey (1989) and Harvey and Pagel (1991), the concept of adaptation is based on the comparative method; adaptive traits can only be designated so by using the comparative method. Adaptive phenotypes are those that have the highest inclusive fitness in relation to other phenotypes in the population. They also stress the importance of considering phylogenetic relationships when utilizing the comparative method.

To determine that a male trait originated as an adaptation due to a preexisting bias, two criteria must be met. First, it must be shown that the trait presently exists along with the preference in a species (or clade). This indicates that intersexual selection on the trait is at least partially responsible for maintenance of the trait. Satisfaction of the first criterion alone is sufficient to

---

Figure 4   Hypothetical relationships between a male trait and a female preference. (a) Phylogenetic distribution of the male trait. (b) Phylogenetic distribution of the female preference which would suggest that the preference evolved prior to the male trait. (c) Phylogenetic distribution of the female preference which cannot distinguish among the hypotheses of coevolution of the trait and preference, evolution of the preference prior to the trait, or evolution of the trait prior to the preference. Thin hatch mark = origin of male trait, thick hatch mark = origin of female preference.

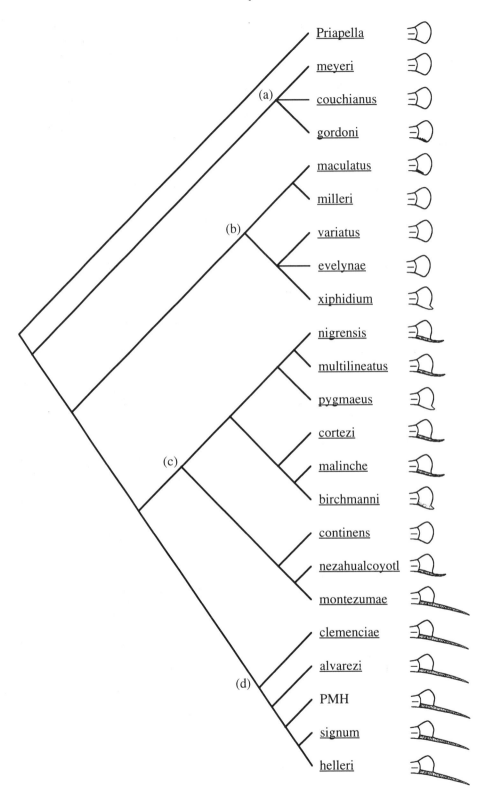

suggest that the trait has an adaptive function presently. [Though this does not distinguish whether it is an aptation or an adaptation, according to Gould and Vrba (1982).] Second, it must be shown that the absence of the trait and the presence of the preference are primitive states. Satisfaction of these two criteria demonstrates both that the trait likely arose as an adaptation and that the trait is presently maintained as an adaptation (Basolo, 1996). These criteria can be satisfied by combining the use of phenotypic manipulations with the comparative method.

Information concerning the evolutionary history of a male trait combined with phenotypic manipulations have been used to investigate the evolution of female preferences and male traits. Ryan et al. (1990) found that female frogs, *Physalaemus coloradorum* and *P. pustulosus*, have the best excitatory frequencies for the basilar papilla, the hearing organ important in call detection, at 2200 Hz. Male *P. pustulosus* produce calls with components at this frequency, while *P. coloradorum* express the primitive state for this character, the lack of a call component at this frequency. They suggested that this primitive sensory bias was exploited by *P. pustulosus* males. In 1992, Ryan and Rand (reported in Kirkpatrick and Ryan, 1992) found that female *P. coloradorum* prefer synthesized calls with components added at 2200 Hz. Therefore, female morphology, combined with phylogenetic information, was predictive of a female preference. These results suggest that a call component at 2200 Hz in *P. coloradorum* will be adaptive if it arises. Thus by combining the use of phenotypic manipulations with the comparative method, it can be shown that a male trait was likely adaptive in the context of sexual selection at its origin (Basolo, 1996).

XI. The Evolutionary History of a Mating Preference in Swordtail Fish

Phylogenetic information and phenotypic manipulations have been combined in a second study addressing the evolutionary history of a female preference and the male trait favored by the preference. In swordtail fish (*Xiphophorus* spp.), males possess a colored extension of the caudal fin called a sword. The most elaborate swords are tri-colored rays at the base of the caudal fin originating at the proximal ray insertion points and extending well beyond the distal end of the caudal fin (a composite trait consisting of at least four components; Figs. 5c and 5d). Darwin (1871) suggested that the sword evolved due to sexual selection, and recent experiments show that female green swordtails, *X. helleri*, prefer males with longer swords (Basolo, 1990a). The sword thus appears to be a target of female mate choice and sexual selection appears to contribute in part to the maintenance of the trait.

The genus *Xiphophorus* contains not only swordtails, but also platyfish, which

Figure 5 Composite phylogeny using morphological, allozyme, and hybridization data (from Rosen, 1979; Rauchenberger et al., 1990). Normal caudal fin condition is depicted; species that are polymorphic for the presence of a character state are represented as possessing the trait. (a) the Rio Grande platyfish, (b) the southeastern platyfish, (c) the northern swordtails, and (d) the southern swordtails. Note: The phylogenetic relationship for *X. andersi* is not represented (from Basolo, 1995a).

are unsworded (Figs. 5a and 5b). To determine if the preference favoring a sword is a consequence of a preexisting bias that was present prior to the origin of the sword, preferences in platyfish were examined. Previous work manipulated male sword length through the surgical attachment of artificial swords (Nelson, 1976). Using an extension of this technique, Basolo (1990a,c) surgically attached artificial swords to male platyfish. Swords were cut from clear plastic film and were hand-colored to resemble natural swords. A microfine suture and nylon thread were used to secure the artificial swords to the musculature in the caudal peduncle of anesthetized males. In addition, clear plastic extensions were surgically attached to a second set of anesthetized males. Males were monitored for a postoperative recuperative period (never less than 3 days) and then used in female choice tests. Females were given the choice of a male with an artificial sword and a male with a clear extension (simulating the visual absence of a sword, but controlling for the affects of the attachment on male behavior). The time a female spent in association with each male was recorded. Females of the two platyfish species (*X. maculatus* and *X. variatus*) demonstrated a preference for conspecific males with surgically attached swords over conspecific males with clear plastic extensions (Basolo, 1990b,c).

Two phylogenies (Rauchenberger et al., 1990; Haas, 1993) suggest that the complete sword is a derived character of swordtails, while a third suggests that it may be primitive for the genus (Meyer et al., 1994). If the sword is a synapomorphy of swordtails, the presence of a preference favoring a sword in

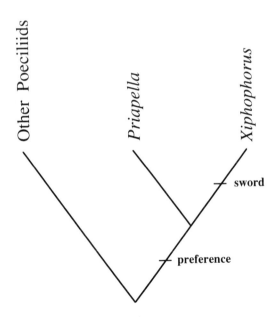

Figure 6 Phylogenetic relationship for the sister taxa *Priapella* and *Xiphophorus* with appearance of the preference favoring the sword and the sword indicated. A preference favoring a sword is a shared, primitive trait, while the sword is a uniquely derived trait within *Xiphophorus* (from Basolo, 1995b).

platyfish suggests that the preference arose prior to the male traits. However, if the sword is a sympleisiomorphy, the preference in platyfish may be a residual preference (Basolo, 1995a). The sister genus *Priapella* lacks a sword, as do other close relatives of *Xiphophorus*. Therefore, it appears that the sword arose after the divergence of *Priapella* and *Xiphophorus*. Thus, the state of swordlessness is the primitive state for the *Priapella/Xiphophorus* complex; this conclusion is in agreement with all three phylogenies (Fig. 6). Female *P. olmecae* were tested for sword preference by attaching swords as described for platyfish and were found to share the preference favoring a sword. These results suggest that a preexisting bias favoring the sword is a primitive state for this group and that it arose prior to the appearance of the sword (Fig. 6). By combining the use of phenotypic manipulations with phylogenetic information, it appears that the sword was adaptive at its origin (Basolo, 1996).

Using manipulations and phylogenetic information, these types of studies can be extended. First, the strength of the preference favoring a sword in *Priapella* can be compared to that of *X. helleri* to determine if the preference functions are different. In *P. olmeceae*, the strength of the preference favoring sword is an increasing function of sword length (Fig. 7). If the strength of the preference in the swordtail is stronger than in *Priapella*, this would suggest that there has been coevolution between the preference and the trait. The strength of the preference in *X. helleri* is currently unknown. Second, the history of the preference can be examined further to determine when it arose; more distantly related outgroups could be tested to determine when the bias favoring a sword

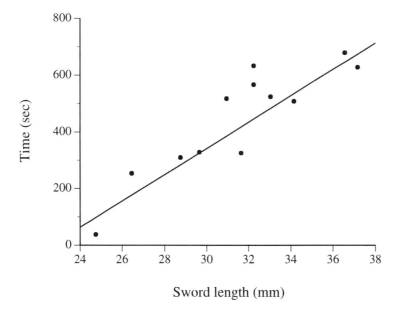

Figure 7 Strength of female preference for sword length; relationship between the amount of time females spent in courtship with sworded males and male sword length (from Basolo, 1995b).

evolved. Third, the base of the preference favoring the sword could be investigated to determine if the bias arose for some other function. For instance, it may be that there is a general preference for conspicuous male traits in the family Poeciliidae (live-bearing, freshwater fish). This could have resulted in different types of conspicuous traits being selected in different lineages and could explain high dorsal fins in mollies, bright coloration in platyfish and guppies, etc. This line of investigation may produce information concerning whether the bias had an adaptive function in another selective context (sexual or natural selection).

In both the *Physalaemus* frogs and the swordtails, it appears that male trait evolution was adaptive, as the selective agent, female preference, was present prior to the appearance of the trait. Although the use of the comparative method and phylogenetic information in the investigation of adaptation has been called into question (Reeve and Sherman, 1993; Frumhoff and Reeve, 1994; Leroi et al., 1994b), arguments in its favor have been made (Leroi et al., 1994b; Basolo, 1996). With proper consideration of phylogenetic methodology, an understanding of what the limitations of the methods are (e.g., resolution and parsimony), and an understanding of the trait under consideration, the use of phenotypic manipulations promises to be a useful method in determining whether male traits likely evolved, at least in part, as adaptations for mate attraction.

XII. Experimental Manipulation of the Agents of Natural Selection

This section discusses how the detection of natural selection is used as an analytic tool to test adaptive hypotheses. The objective of many previous studies of selection has been to merely detect its presence in the wild (Endler, 1986). The success of these ventures is amply documented in Endler's (1986) review of studies of selection in nature. In most cases, however, the agents of selection are largely unknown, and Endler declared that the challenge for future studies is to understand how it is that selection shapes traits. This approach reflects a growing sensitivity to understanding the causal basis of patterns of selection in natural populations. Instead of choosing an arbitrary trait and then attempting to elucidate how that trait is selected, one might choose specific traits for which adaptive explanations have been proposed and test whether those traits are under selection as predicted by theory (e.g., life history allocation theories, see above; ecological theories of character displacement; or phenotypic plasticity in morphology).

The evolutionary divergence of species' morphologies in the presence of competitors was originally offered as an adaptive explanation of a common pattern in geographic variation: morphologies of species that use similar resources are more dissimilar when their populations are sympatric than when they are allopatric (Brown and Wilson, 1956). Satisfactory demonstrations of character displacement have been difficult to achieve because many factors differ between areas of sympatry and allopatry. However, experimental studies of the adaptive phenomenon of character displacement which involve manipulations of the competitive regime in combination with phenotypic manipulations (generating hybrids) provide strong support for character displacement (Schluter, 1994).

Allopatric populations of sticklebacks typically have an intermediate number of gill rakers that are important in the feeding mechanism of sticklebacks (Schluter and McPhail, 1992, 1993; Schluter, 1994, 1995). However, sympatric limnetic sticklebacks are characterized by a high gill raker number; conversely, sympatric benthic sticklebacks are characterized by a low gill raker number. Schluter (1994) created manipulated phenotypes of sticklebacks by hybridizing an allopatric species (Cranby Lake) with limnetic and benthic specialists from a nearby lake that has sympatric populations of sticklebacks (Paxton Lake). Hybridized populations of sticklebacks ("probe" for selection) with an intermediate gill raker number and a high variance in gill raker number were placed in a competitive arena (large experimental ponds) with individuals derived from the allopatric Cranby Lake sticklebacks ("target" of selection). Schluter observed directional "selection" (growth was used as a proxy for fitness) in the target species in response to the presence of the "probe" species that is consistent with the character displacement hypothesis of divergence in gill raker number under competition.

Brown and Wilson (1956) broadly defined character displacement to include any evolutionary response to the presence of competitors, but it is more commonly defined as evolutionary changes in characters restricted to resource exploitation that result from interspecific competition (e.g., Arthur, 1982). Grant's (1972) historically based definition discriminates between changes in morphology resulting from a transition from allopatric populations to sympatric populations as character displacement. Grant (1972) considers the transition from sympatric populations to allopatric populations the phenomenon of "character release." Schluter's (1994) experimental manipulation of morphology and selective regime documents the importance of competition in shaping trophic characters as well as elucidating the historical origins of the pattern of character displacement. By generating hybrid phenotypes, Schluter reconstructs the process of adaptation in the earliest stages when the species has a less derived morphology. Moreover, these manipulations of the selective environment and morphology document the importance of frequency-dependent selection in shaping trophic morphology. Selection on morphology is contingent on the presence or absence of competitors and the kind of morphology present in the competitor. Selection is more intense if the competitors are more similar in morphology than if they are dissimilar in morphology. Information from such manipulations concerning the origins of the adaptation could be further verified by phylogenetic reconstructions of the species involved in the community (Losos and Miles, 1994).

Pfennig's (1990) manipulation of food availability to determine the effects on subsequent morphology also involves a manipulation of the phenotype and a manipulation of the selective environment. In the toad *Scaphiopus multiplicatus*, if a tadpole consumes a critical quantity of fairy shrimp, it becomes a carnivore. If it doesn't, it is an omnivore. He was able to affect tadpole phenotype indirectly by manipulating fairy shrimp density. Higher densities of fairy shrimp occur in more ephemeral ponds and carnivorous morphs have faster developmental rates. Thereby, by consuming fairy shrimp, carnivores have a faster developmental rate which results in them being more likely to metamorphose into froglets before the pond dries up. These results suggest that phenotypic plasticity for these strategies likely evolved in part as an adaptation. Proximate control of the morphs involves the potent amphibian metamorphic hormone thyroxine. Elevation of thyroxine levels of the omnivore morph results in transformation to the carnivore morph in a matter of days (Pfennig, 1992a). More importantly, Pfennig's manipulations of food availability and the carnivore and omnivore morphs provide important evidence regarding the role of frequency-dependent selection in the evolution of these adaptations (Pfennig, 1990, 1992b). If the numbers of shrimp in a pond are not above a critical level to sustain the growth of the carnivore morph, then the omnivore morph will metamorphose in better condition (greater fat reserves). The optimal numbers of carnivores and omnivores depend on the frequency of both morphs and the densities of shrimp. Manipulation of tadpole morphology provides insights into the selective environment that leads to the adaptation. At very low frequencies (e.g., during the origin of the carnivore form), even low numbers of fairy shrimp which the carnivore can exploit more efficiently would allow the carnivore to have a selective advantage over the omnivore; thus a mutation would likely be incorporated into a population because of strong frequency-dependent selection.

The two examples discussed above involve a combination of phenotypic manipulations and environmental manipulations. Such double manipulations are currently underutilized in evolutionary studies of adaptation. For example, the process of natural or sexual selection may be inherently frequency dependent. In principle, one could manipulate the frequencies of morphologies or the frequencies of males that receive hormonal supplements (e.g., testosterone implants) to test the adaptive value of traits under various frequencies. Such simultaneous manipulations of phenotype and selective environment (e.g., frequency of male types) can be directed at understanding the origin of such sexually selected characters and testing ESS ideas concerning the "invasion of mutant phenotypes" into a population (Maynard Smith, 1982).

XIII. Potential Complications and the Future of Phenotypic Manipulations

The goal of many experimental phenotypic manipulations in addressing adaptive hypotheses for the evolution of traits is to alter only the trait under consideration in order to determine the effect of the trait by itself on fitness. In practice, limiting the effects of the manipulation to the trait under study may be very difficult. Careful controls for the manipulations and random assignment of animals to treatment groups are critical. However, in studies of natural selection, it may be very important to use a manipulation that affects a suite of traits so that the full impact (e.g., life history trade-offs) of the trait on natural selection and adaptation can be captured.

When investigating whether a male trait evolved due to intersexual selection, the manipulation may increase a male's attractiveness, but may reduce a male's survivorship. Therefore, correlated consequences of the manipulation need to be considered. For example, Moller (1992, 1994) added tail feathers to the swallow *Hirundo rustica* to investigate whether females preferred longer-tailed males to shorter-tailed males. Females were found to prefer males with longer tails, suggesting that mate choice was at least partially responsible for the evolution of tail length in this animal. Subsequently, Balmford and Thomas (1993) estimated the aerodynamic costs of various degrees of tail elongation. They suggested that for some birds, manipulations of tail feathers would have a negative affect on aerodynamics, but not for the type of tails that barn swallows express. Therefore, by considering flight costs, it appeared that the long tail feathers (streamers) which occur naturally in barn swallows are a result of sexual rather than natural selection. But Norberg (1994) extended the studies of aerodynamics in swallows and concluded that the streamers may indeed have a function in aerodynamics. Thus, experimental manipulations of tail feather may negatively affect male flight.

Some signal-receiver interactions are one-way, but many involve interactions between signalers and receivers. In these cases, responses may escalate, which artificial manipulations may not account for. In other cases, the male trait may be facultative, and prolonged or inappropriate manipulation of such traits may give spurious results. For example, in many fish and frogs, a type of melanophore coloration is facultative. This coloration may be used in aggressive interactions with males or/and it may function in mate choice. A male may be given a manipulated melanophore signal that is inappropriate for a given situation; it may indicate a higher level of challenge than he would naturally indicate to the male with which he is paired. Alternatively, the artificial signal may not indicate the degree to which he is willing to escalate. Either possibility may result in behavior that bears little resemblance to a normal situation. For a static morphological trait, this may not be as much of a problem, but one must still strive to ensure that the manipulated trait functions "normally" in all aspects, other than the one under consideration.

Throughout this chapter, we have argued that phenotypic manipulations are useful in elucidating not only the maintenance, but also the origins of adaptations. In understanding the origins of adaptations, the manipulations must be carried out in a phylogenetic context. Manipulation of a group with a synapomorphy must be referenced to a group with a sympleisiomorphy. If the manipulation involves a mechanistic basis, then additional insights are provided regarding the developmental or mechanistic evolutionary origins of the adaptation. A simple, yet famous, example will suffice to illustrate this point. The axolotl is a derived condition in the salamanders, in that most ambystomatid salamanders have a terrestrial adult phase. An aquatic adult morphology is presumably an adaptation to a low predation aquatic environment or a stable aquatic environment. Metamorphosis of the axolotl to a terrestrial adult is readily achieved with the hormone thyroxine, suggesting that the axolotl's condition was derived by loss of the thyroxine production mechanism associated with metamorphic climax. Such mechanistic manipulations within a phylogenetic context (Shaffer, 1993) provide clear evidence concerning the mechanistic origins of the paedomorphic condition. Recent genetic complementation studies (Shaffer, personal communication) in which different populations of axolotls are hybridized to determine whether "rescue" of the axolotl to the adult occurs will provide evidence regarding the uniqueness of genetic transitions in different populations and thus whether or not the "axolotl" trait has arisen repeatedly in evolutionary time from ancestral *Ambystoma mexicanum* populations.

Future studies involving manipulations should incorporate "ontogenetic manipulations" of phenotype in which specific structures are targeted for change. Endocrinological manipulations of phenotype have great promise, in that implants can be designed to maintain hormones within the "normal range" and not at pharmacological levels. However, what is currently lacking from many tests of adaptations that involve hormonal manipulations is a comparative or phylogenetic context for the manipulations. Such manipulations should be framed within the context of a comparison of closely related species with ancestral and derived conditions.

"Ablation" or "paste" manipulations have great utility in restricting the manipulation to the trait of interest. However, in complex morphologies, such manipulations have limited utility. Manipulations of complex morphologies should occur through ontogenetic manipulations (Sinervo and McEdward, 1988; Sinervo, 1993; Reilly and Lauder, 1991) and there are many tools already developed by classical and modern experimental biology for altering ontogeny. However, such manipulations should also be placed in an explicit comparative test that involves a phylogenetically based comparison of trait evolution. The goal of future phenotypic manipulations should be to synthesize experiments with phylogenetic analysis of trait evolution in order to reveal the developmental and physiological mechanisms underlying adaptive evolution.

Acknowledgments

We thank W. E. Wagner, Jr. for his input and for his constructive comments on the manuscript and R. R. Repasky for engaging discussions on the subject of adaptation.

References

Andersson, M. (1982). Female choice selects for extreme tail length in a widowbird. *Nature* 299, 818-820.

Andersson, M. (1994). "Sexual Selection." Princeton Press, N. J.

Andrews, R., and Rand, A. S. (1974). Reproductive effort in anoline lizards. *Ecology* 55: 1317-1327.

Arnold, S. J. (1983). Morphology, performance and fitness. *Am. Zool.* 23, 347-361.

Arnold, S. J., and Wade, M. J. (1984). On the measurement of natural and sexual selection: theory. *Evolution* 38, 709-719.

Arnold, S. J. (1994). Is there a unifying concept of sexual selection that applies to both plants and animals. *Amer. Nat.* 144, S1-S12.

Arthur, W. (1982). The evolutionary consequences of interspecific competition. *Adv. Ecol. Res.* 12, 127-187.

Balmford, A., Thomas, A. L. R., and Jones, I. L. (1993). Aerodynamics and the evolution of long tails in birds. *Nature* 361: 628-631.

Basolo, A. L. (1990a). Female preference predates the evolution of the sword in swordtail fish. *Science* 250, 808-810.

Basolo, A. L. (1990b). Female preference for male sword length in the green swordtail, *Xiphophorus helleri* (Pisces: Poeciliidae). *Anim. Behav.* 40, 332-338.

Basolo, A. L. (1990c). Preexisting mating biases and the evolution of the sword in the genus *Xiphophorus*. *Amer. Zool.* 30, 80A.

Basolo, A. L. (1995a). A further examination of a preexisting bias favouring a sword in the genus

Xiphophorus. *Anim. Behav.* 50, 365-375.

Basolo, A. L. (1995b). Phylogenetic evidence for the role of a pre-existing bias in sexual selection. *Proc. Roy. Soc. Lond. B* 307-311.

Basolo, A. L. (1996). The evolutionary history of a female preference. Syst. Biol. (in Press).

Benkman, C. W. (1988). On the advantages of crossed mandibles: An experimental approach. *Ibis* 130, 288-293.

Benkman, C. W., and Lindholm, A. K. (1991). The advantages and evolution of a morphological novelty. *Nature* 349, 519-520.

Berry, K. H. (1974). Univ. Calif. Berkeley Publ. Zool. 101.

Bock, W. J. (1977). Toward an ecological morphology. *Vogelwarte* 29, 127-135.

Bock, W. J. (1980). The definition and recognition of biological adaptation. *Amer. Zool.* 20, 217-227.

Brown, W., Jr, and Wilson, E. O. (1956). Character displacement. *Syst. Zool.* 5, 49-64.

Carothers, J. H. (1986). An experimental confirmation of morphological adaptation: toe fringes in the sand-dwelling lizard *Uma scoparia. Evolution* 40, 871-874.

Chevrud, J. M. (1984). Quantitative genetics and developmental constraints on evolution by selection. *J. Theor. Biol.* 110, 155-171.

Chippindale, A. K., Leroi, A. M., Kim, S. B., and Rose, M. R. (1993). Phenotypic plasticity and selection in *Drosophila* life-history evolution. I. Nutrition and the costs of reproduction. *J. Evol. Biol.* 6: 171-193.

Clark, D. L., and Uetz, G. W. (1992). Morph-independent mate selection in a dimorphic jumping spider: demonstration of movement bias in female choice using video-controlled courtship behavior. *Anim. Behav.* 43, 247-254.

Congdon, J. D., and Gibbons, J. W. (1987). Morphological constraint on egg size: a challenge to optimal egg size theory? *Proc. Natl. Acad. Sci. USA* 84: 4145-4147.

Crews, D. (1985). Effects of early sex steroid hormone treatments on courtship behavior and sexual attractivity in the red-sided garter snake, *Thamnophis sirtalis parientalis. Physiol. Behav.* 35, 560-576.

Darwin, C. (1871). "The Descent of Man and Selection in Relation to Sex." Murray, London.

DeMarco, V. G. (1989). Annual variation in the seasonal shift in egg size and clutch size in *Sceloporus woodi. Oecologia* 80, 525-532.

Endler, J. A. (1986). "Natural selection in the wild." Princeton University Press, Princeton, N. J.

Endler, J. A. (1995). Multiple-trait coevolution and environmental gradients in guppies. *Trends Ecol. Evol.* 10, 22-29.

Endler, J. A., and McLellan, T. (1988). The process of evolution: Towards a newer synthesis. *Annu. Rev. Ecol. Syst.* 19, 395-421.

Etheridge, R., and De Queiroz, K. (1988). A phylogeny of Iguanidae. *In* "Phylogenetic Relationships of the Lizard Families" (R. Etheridge, and R. Estes, eds.). pp. 283-367. Stanford University Press, Stanford, CA.

Falconer, D. S. (1981). "Introduction to Quantitative Genetics." Longman, New York, N. Y.

Feder, M. E. (1987). The analysis of physiological diversity: the prospects for pattern documentation and general questions in ecological physiology. *In* "New Directions in Ecological Physiology" (M. E. Feder, A. F. Bennett, W. Burggren, and R. B. Huey, eds.), pp. 38-70. Cambridge University Press, Cambridge

Ferguson, G. W., and Fox, S. F. (1984). Annual variation of survival advantage of large juvenile side-blotched lizards, *Uta stansburiana*: its causes and evolutionary significance. *Evolution* 38, 342-349.

Fox, S. F. (1983). Fitness, home-range quality, and aggression in *Uta stansburiana. In* "Lizard Ecology, Studies of a Model Organism" (R. B. Huey, E. R. Pianka, and T. W. Schoener, eds.), pp. 149-168. Harvard University Press, Cambridge, MA.

Fox, S., and Rostker, M. A. (1982). Social cost of tail loss in *Uta stansburiana. Science* 218, 692-693.

Frumhoff, P. C., and Reeve, H. K. (1994). Using phylogenies to test hypotheses of adaptation, a critique of some current proposals. *Evolution* 48, 172-180.

Gerhardt, H. C. (1992). Female mate choice in tree frogs: static and dynamic acoustic criteria. *Anim. Behav.* 43, 615-635.

Gould, S. J., and Lewontin, R. C. (1979). The spandrels of San Marco and the Panglossian paradigm: a critique of the adaptationist programme. *Proc. R. Soc. Lond. B* 205, 581-598.

Gould, S. J., and Vrba, E. S. (1982). Exaptation – a missing term in the science of form. *Paleobiology*

8, 4-15.

Grant, P. R. (1972). Convergent and divergent character displacement. *Biol. J. Linnean Soc.* 4, 39-68.

Grant, P. R. (1986). "Ecology and Evolution of Darwin's Finches." Princeton University Press, Princeton, N. J.

Gustafsson, L., and Sutherland, W. J. (1988). The costs of reproduction in collared flycatcher *Ficedula albicollis. Nature* 335, 813-815.

Haas, V. (1993). *Xiphophorus* phylogeny, reviewed on the basis of the courtship behaviour. *In* "Trends in Ichthyology" (J. H. Schroder, J. Bauer, and M. Schartl, eds.), pp. 279-288. Blackwell, London.

Harvey, P. H., and Pagel, M. D. (1991). "The comparative method in evolutionary biology." Oxford University Press, Oxford, N. Y.

Hews, D. K. (1993). Food resources affect female distribution and male mating opportunities in the iguanian lizard *Uta palmeri. Anim. Behav.* 46, 279-291.

Hart, M. W., and Strathmann, R. R. (1994). Functional consequences of phenotypic plasticity in echinoid larvae. *Biol. Bull.* 186, 291-299.

Houde, A. E., and Torio, A. J. (1992). Effect of parasitic infection on male color pattern and female choice in guppies. *Behav. Ecol.* 3, 346-351.

Hirshfield, M. F. (1980). An experimental analysis of reproductive effort and cost in the Japanese Medaka *Oryzias latipes. Ecology* 61, 282-292.

Hogstedt, G. (1980). Evolution of clutch size in birds: an adaptive variation in relation to territory quality. *Science* 210, 1148-1150.

Jayne, B. C., and Bennett, A. F. (1989). The effect of tail morphology on locomotor performance of snakes: A comparison of experimental and correlative methods. *J. Exp. Zool.* 252, 126-133.

Jayne, B. C., and Bennett, A. F. (1990). Selection on locomotor performance capacity in natural populations of garter snakes. *Evolution* 44, 1204-1229.

Johnson, K., Dalton, R., and Burley, N. (1993). Preferences of female American goldfinches (*Carduelis tristis*) for natural and artificial male traits. *Behav. Ecol.* 4, 138-143.

Jones, I. A., and Hunter, F. M. (1993). Mutual sexual selection in a monogamous seabird. *Nature* 362, 238-239.

Ketterson, E. D., Nolan, V., Jr., Wolf, L., Ziegenfus, C., Dofty, A. M., Jr., Ball, G. F., and Johnsen, T. S. (1991). Testosterone and avian life histories: the effect of experimentally elevated testosterone on corticosterone and body mass in dark-eyed juncos. *Horm. Behav.* 25, 489-503.

Ketterson, E. D., and Nolan, V., Jr. (1992). Hormones and life histories: An integrative approach. *Amer. Nat.* 140, S33-S62.

Kingsolver, J. G., and Koehl, M. A. R. (Eds.). (1994). "Selective factors in the evolution of insect wings." Annual Review of Entomology. Annual Reviews Inc., Palo Alto, CA.

Kirkpatrick, M. (1987). Sexual selection by female choice in polygynous animals. *Annu. Rev. Ecol. Syst.* 18, 43-70.

Kirkpatrick, M., and Ryan, M. J. (1992) The evolution of mating preferences and the paradox of the lek. *Nature* 350, 33-38.

Kroodsma, E., Miller, E. H., and Oullet, H. (eds.). (1982). "Acoustic Communication in Birds." Academic Press, New York.

Lack, D. (1947). The significance of clutch size. *Ibis* 89, 302-352.

Lande, R. (1979). Quantitative genetic analysis of multivariate evolution, applied to brain: body size allometry. *Evolution* 33, 402-416.

Lande, R. (1983). A quantitative genetic theory of life history evolution. *Ecology* 63, 607-615.

Lande, R., and Arnold, S. J. (1983). The measurement of selection on correlated characters. *Evolution* 37, 1210-1226.

Landwer, A. J. (1994). Manipulation of egg production reveals costs of reproduction in the tree lizard (*Urosaurus ornatus*). *Oeocologia* 100, 243-249.

Leroi, A. M., Kim, S. B., and Rose, M. R. (1994a). The evolution of phenotypic life-history trade-offs: an experimental study using *Drosophila melanogaster. Amer. Nat.* 144, 661-676.

Leroi, A. M., Rose, M. R., and Lauder, G. V. (1994b). What does the comparative method reveal about adaptation? *Amer. Nat.* 143, 381-402.

Levins, R., and Lewontin, R. C. (1985). "The Dialectical Biologist." Harvard Univ. Press, Cambridge, MA.

Lima, S. L. (1987). Clutch size in birds: a predation perspective. *Ecology* 68, 1062-1070.

Long, K. D., and Houde, A. E. (1989). Orange spots as a visual cue for female mate choice on the guppy, *Poecilia reticulata. Ethology* 82, 316-324.

Long, K. D., and Houde, A. E. (1992). Color as a visual cue for female choice in the guppy (*Poecilia reticulata*). Ethology 82, 316-324.

Losos, J. B., and Miles, D. B. (1994). Adaptation, constraint, and the comparative method: phylogenetic issues and methods. *In* "Ecological Morphology: Integrative Organismal Biology" (P. C. Wainwright and S. M. Reilly, eds.). University of Chicago Press, Chicago

Luke, C. (1986). Convergent evolution of lizard toe fringes. *Biol. J. Linn. Soc.* 27, 1-16.

Marden, J. H., and Kramer, M. G. (1994). Surface-skimming stoneflies: a possible intermediate stage in insect flight evolution. *Science* 266, 427-430.

Marler, C. A., and Moore, M. C. (1988). Evolutionary costs of aggression revealed by testosterone manipulations in free-living male lizards. *Behav. Ecol. Sociobiol.* 23, 21-26.

Marler, C. A., and Moore, M. C. (1991). Supplementary feeding compensates for testosterone-induced costs of aggression in male mountain spiny lizards, *Sceloporus jarrovi. Anim. Behav.* 42, 209-219.

Maynard Smith, J. (1982). "Evolution and the Theory of Games." Cambridge University Press, Cambridge

McEdward, L. R. (1986a). Comparative morphometrics of echinoderm larvae. I. Some relationships between egg size and initial larval form in echinoids. *J. Exp. Mar. Biol.* 93, 169-181.

McEdward, L. R. (1986b). Comparative morphometrics of echinoderm larvae. II. Larval size, shape, growth, and the scaling of feeding and metabolism in echinoplutei. *J. Exp. Mar. Biol.* 96, 267-286.

Meyer, A., Morrissey, J., and Schartl, M. (1994). Molecular phylogeny of fishes of the genus *Xiphophorus* suggests repeated evolution of a sexually selected trait. *Nature London* 368, 539-542.

Mitchell-Olds, T., and Shaw, R. G. (1987). Regression analysis of natural selection: statistical and biological interpretation. *Evolution* 41, 1149-1161.

Moller, A. P. (1992). Female swallow preference for symmetrical male sexual ornaments. *Nature* 357, 238-240.

Moller, A. P. (1994). "Sexual selection and the barn swallow." Oxford University Press, Oxford.

Moore, M. C. (1991). Application of organization-activation theory to alternative male reproductive strategies: a review. *Horm. and Behav.* 25, 154-179.

Moore, M. C., and Marler, C. A. (1987). Effects of testosterone manipulations on non-breeding season territorial aggression in free-living lizards, *Sceloperus jarrovi. Gen. Comp. Endocrinol.* 65, 225-232.

Nelson, J. L. (1976). "Sexual Selection and the Swortail Fish, *Xiphophorus helleri*." Ph.D. Dissertation: University of California, Santa Cruz.

Nolan, V. J., Ketterson, E. D., Ziegenfus, C., and Cullen, D. P. (1992). Testosterone and avian life histories: effects of experimentally elevated testosterone on prebasic molt and survival in male dark-eyed juncos. *Condor* 94, 364-370.

Norberg, R. A. (1994). Swallow tail streamer is a mechanical device for self-deflection of tailleading edge, enhancing aerodynamic efficiency and flight maneuverability. *Proc. Roy. Soc. Lond. B.* 257, 227-233.

Nur, N. (1984a). The consequences of brood size for breeding blue tits. I. Adult survival, weight change and the cost of reproduction. *J. Anim. Ecol.* 53, 479-496.

Nur, N. (1984b). The consequences of brood size for breeding blue tits. II. Nestling weight, offspring survival and optimal brood size. *J. Anim. Ecol.* 53, 497-517.

Nur, N. (1986). Is clutch size variation in the blue tit (*Parus caeruleus*) adaptive? An experimental study. *J. Anim. Ecol.* 55, 983-999.

Nur, N. (1989). The cost of reproduction in birds: an examination of the evidence. *Ardea* 45: 17-29.

Nussbaum, R. A. (1981). Seasonal shifts in clutch size and egg size in the side-blotched lizard, *Uta stansburiana* Baird and Girard. *Oecologia* 49, 8-13.

Orzack, S. H., and Sober, E. (1994). Optimality models and the test of adaptationism. *Amer. Nat.* 143, 361-380.

Pagel, M.D., and Harvey, P.H. (1989). Comparative methods for examining adaptation depend on evolutionary models. Folia Primatol. 53, 203-220.

Partridge, L., and Harvey, P. H. (1985). Costs of reproduction. *Nature* 316, 20-21.

Pfennig, D. (1990). The adaptive significance of an environmentally-cued developmental switch in an anuran tadpole. *Oecologia* 85, 101-107.

Pfennig, D. W. (1992a). Polyphenism in spadefoot toad tadpoles as a locally adjusted evolutionarily stable strategy. *Evolution* 46, 1408-1420.

Pfennig, D. W. (1992b). Proximate and functional causes of polyphenism in an anuran tadpole. *Funct. Ecol.* 6, 167-174.

Popov A. V., and Shuvalov, V. F. (1977). Phonotactic behavior of crickets. *J. Comp. Physiol.* 119, 111-126.

Rauchenberger, M., Kallman, K. D. and Morizot, D. C. (1990). Monophyly and geography of the Panuco Basin swordtails (Genus *Xiphophorus*) with descriptions of four new species. *Am. Mus. Nat. Hist. Novit.* 2974, 1-41.

Reeve, H. K., and Sherman, P. W. (1993). Adaptation and the goals of evolutionary research. *Quart. Rev. Biol.* 68, 1-32.

Reilly, S. M., and Lauder, G. V. (1991). Experimental morphology of the feeding mechanism in salamanders. *J. Morph.* 210, 33-44.

Reznick, D. (1985). Costs of reproduction: an evaluation of the empirical evidence. *Oikos* 44, 257-267.

Reznick, D. (1992). Measuring costs of reproduction. *Trends Ecol. Evol.* 7, 42-45.

Rose, M., and Charlesworth, B. (1981). Genetics of life history in *Drosophila melanogaster*. II. exploratory selection experiments. *Genetics* 97, 187-196.

Rosen, D. E. (1979). Fishes from the uplands and intermontane basins of Guatemala: Revisionary studies and comparative geography. *Bull. Am. Mus. Nat. Hist.* 162, 267-376.

Rosenqvist, G., and Johansson, K. (1995). Male avoidance of parasitized females explained by direct benefits in pipefish. *Anim. Behav.* 49, 1039-1045.

Ryan, M. R. (1980). Female mate choice in a Neotropical frog. *Science* 209, 523-525.

Ryan, M. J. (1990). Sexual selection, sensory systems, and sensory exploitation. *Oxf. Surv. Evol. Bio.* 7, 156-195.

Ryan, M. J., Fox, J. H., Wilczynski, W., and Rand, A. S. (1990). Sexual selection for sensory exploitation in the frog *Physalemus pustulosus*. *Nature London* 343, 66-67.

Ryan, M. J., and Rand, A. S. (1993). Sexual selection and signal evolution-the ghost of biases past. *Phil. Trans. R. Soc. Lond.* B 340, 187-195.

Schluter, D. (1988). Estimating the form of natural selection on a quantitative trait. *Evolution* 42, 849-861.

Schluter, D. (1994). Experimental evidence that competition promotes divergence in adaptive radiation. *Science* 266, 798-801.

Schluter, D. (1995). Adaptive radiation in sticklebacks: trade-offs in feeding performance and growth. *Ecology* 76, 82-90.

Schluter, D., and McPhail, J. D. (1992). Ecological character displacement and speciation in sticklebacks. *Amer. Nat.* 140, 85-108.

Schluter, D. and McPhail, J. D. (1993). Character displacement and replicate adaptive radiation. *Trends Ecol. Evol.* 8, 197-200.

Searcy, W. A. (1984). Song repertoire size and female preferences in song sparrows. *Behav. Ecol. Sociobiol.* 14, 281-284.

Searcy, W. A. (1988). Dual intersexual and intrasexual functions of song in red-winged black-birds. *Proc. XIX Int. Congr. Ornithol.* 1, 1373-1381.

Searcy, W. A., and Marler, P. (1984). Interspecific differences in the response of female song birds to song repertoires. *Z. Tierpsychol.* 66, 128-142.

Searcy, W. A., and Yasukawa. (1990). Use of song repertoire and intersexual and intrasexual contexts by male red-winged blackbirds. *Behav. Ecol. Sociobiol.* 27, 123-128.

Searcy, W. A. (1992). Song repertoire and mate choice in birds. *Am. Zool.* 32, 71-80.

Shaffer, H. B. (1993). Phylogenetics of model organisms: The laboratory axolotl, *Ambystoma mexicanum*. *Syst. Biol.* 42, 508-522.

Sinervo, B. (1990). The evolution of maternal investment in lizards: an experimental and comparative analysis of egg size and its effects on offspring performance. *Evolution* 44, 279-294.

Sinervo, B. (1993). The effect of offspring size on physiology and life history: manipulation of size using allometric engineering. *Bioscience* 43, 210-218:

Sinervo, B. (1994a). Experimental manipulations of clutch and egg size of lizards: mechanistic, evolutionary, and conservation aspects. *In* "Captive Management and Conservation of Amphibians and Reptiles" (J. B. Murphy, K. Adler and J. T. Collins, eds.), Society for the Study of Amphibians and Reptiles, Ithaca, N.Y.

Sinervo, B. (1994b). Experimental tests of allocation paradigms. *In* "Lizard Ecology III" (E. R.

Pianka, and L. J Vitt, eds.), Princeton Univ. Press, Princeton, NJ.

Sinervo, B., and DeNardo, D. F. (1996). Costs of reproduction in the wild: path analysis of natural selection and experimental tests of causation. *Evolution* 50, 1299-1313.

Sinervo, B., and Doughty, P. (1996). Interactive effects of offspring size and timing of reproduction on offspring reproduction: experimental, maternal, and quantitative genetic aspects. *Evolution* 50, 1314-1327.

Sinervo, B., Doughty, P., Huey, R. B., and Zamudio, K. (1992). Allometric engineering: a causal analysis of natural selection on offspring size. *Science* 258, 1927-1930.

Sinervo, B., and Huey, R. B. (1990). Allometric engineering: an experimental test of the causes of interpopulational differences in locomotor performance. *Science* 248, 1106-1109.

Sinervo, B., and Licht, P. (1991a). The physiological and hormonal control of clutch size, egg size, and egg shape in *Uta stansburiana*: constraints on the evolution of lizard life histories. *J. Exp. Zool.* 257, 252-264.

Sinervo, B., and Licht, P. (1991b). Proximate constraints on the evolution of egg size, egg number, and total clutch mass in lizards. *Science* 252, 1300-1302.

Sinervo, B., and McEdward, L. R. (1988). Developmental consequences of an evolutionary change in egg size: an experimental test. *Evolution* 42, 885-899.

Strathmann, R. R. (1971). The feeding behavior of planktotrophic echinoderm larvae: mechanisms, regulation, and rates of suspension-feeding. *J. Exp. Mar. Biol. Ecol.* 6, 109-160.

Stout, J. F., and McGhee, R. (1988). Attractiveness of the male *Acheta domestica* calling song to females. II. The relative importance of syllable period, internsity and chirp rate. *J. Comp. Physiol.* 164, 277-287.

Sullivan, B. (1983). Sexual selection in Woodhouse's toad (*Bufo woodhousei*). *Anim. Behav.* 31, 1011-1017.

Thornhill, R. (1992). Female preference for the pheromone of males with low fluctuating asymmetry in the Japanese scorpionfly (*Panorpa japonica*: Mecoptera). *Behav. Ecol.* 3, 277-283.

Wade, M. J., and Kalisz, S. (1990). The causes of natural selection. *Evolution* 44, 1947-1955.

Willson, M. F., and Price, P. W. (1980). Resource limitation of fruit and seed production in some *Aclepias* species. *Can. J. Biol.* 58, 2229-2233.

Wagner, W. E., Jr. (1989). Fighting, assessment, and frequency alteration in Blanchard's cricket frog. *Behav. Ecol. Sociobiol.* 25, 429-436.

Wagner, W. E., Jr. (1992). Deceptive or honest signalling of fighting ability? A test of alternative hypotheses for the function of changes in call dominant frequency by male cricket frogs. Anim. Behav. 44, 449-462.

Wagner, W. E., Jr., Murray, A. M., and Cade, W. H. (1995). Phenotypic variation in the mating preferences of female field crickets, *Gryllus integer*. *Anim. Behav.* 49, 1269-1281.

Wells, K. D., and Schwartz, J. J. (1984). Vocal communication in a neotropical treefrog, *Hyla ebraccata*: aggressive call. *Behavior* 91, 128-145.

West-Eberhard, M. (1984). Sexual selection, competitive communication and species-specific signals in insects. *In* "Insect Communication" (T. Lewis, ed.), pp. 283-324. Academic Press, Toronto.

Wingfield, J. C., and Ramenofsky, M. (1985). Testosterone and aggressive behavior during the reproductive cycle of male birds. *In* "Neurobiology" (R. Gilles, and J. Balthazart, eds.), pp. 92-102. Springer, Berlin.

Wittenberger, J. (1981). "Animal Social Behavior." Duxbury Press, Boston.

Yosef, R. (1991). Females seek males with ready cache. *Nat. Hist.* 6, 37-37.

Yosef, R. (1992). From nest building to fledging of young in great grey shrikes (*Lanius-excubitor*) at Sede Boqer, Israel. *J. ornithol.* 133, 279-285.

CHAPTER 6

Phylogenetic Systematics of Adaptation

ALLAN LARSON

JONATHAN B. LOSOS

Darwin's theory of natural selection was proposed to explain adaptation, the phenomenon that many organismal characteristics appear to have been designed to perform particular biological functions. Because natural selection is the only process within Darwinian evolutionary theory by which a character might be molded specifically to enhance organismal functions and survival, the evolutionary definition of adaptation now specifies that an organismal character constitutes an adaptation if it performs a function that is of utility to the organisms possessing it and if the character evolved by natural selection for that particular function (Gould and Vrba, 1982; Baum and Larson, 1991). The hypothesis that a character constitutes an adaptation makes specific predictions regarding the utility of the character, its phylogenetic origin, and how it may confer to its possessors an advantage for survival or reproductive success not provided by alternative characters. Testing these predictions requires multidisciplinary approaches that incorporate functional morphology, behavior, ecology, phylogenetic systematics, genetics, and natural history.

Hypotheses of adaptation and their tests are fundamentally comparative. To propose that a particular character is adaptive implies that the character confers an advantage which promotes the survival or reproductive success of its carriers relative to organisms lacking the trait. The character is to be compared specifically to phylogenetically antecedent conditions that occur as alternative variants within populations or in related evolutionary lineages. Adaptation is a meaningful hypothesis only if alternative possible explanations exist for the evolutionary origin and maintenance of organismal characters and their variation. We emphasize here the use of deductive methodology (see Mayr, 1982) and the importance of explicit alternative hypotheses because nonrigorous use of adaptive explanations has hindered evolutionary research (Lewontin, 1977). Several kinds of observations potentially can falsify the hypothesis of adaptation. For example, the hypothetically adaptive character may not enhance performance or survival relative to alternative variants that are its phylogenetic antecedents, or the character may originate in an ancestral lineage for which its current biological role would have been irrelevant (Greene, 1986a).

Studies of adaptation have followed two complementary evolutionary traditions, the first using systematic methodology to address macroevolutionary questions and the second using population genetic methodology to address microevolutionary ones. The macroevolutionary tradition uses phylogenetic analysis to test historical hypotheses of adaptation as explanations for the origins of the characters of species and higher taxa. The microevolutionary, or ecological genetic, tradition examines the action of natural selection on populational variation and is considered elsewhere in this book. The macroevolutionary aspects of adaptation constitute a major issue to which Darwin's theory of natural selection was addressed and it is the primary focus of this chapter.

The importance of using a phylogenetic perspective for macroevolutionary studies of adaptation is evident from the criticisms of Lewontin (1977) and Gould and Lewontin (1979). These authors discredit adaptationist studies conducted before such studies incorporated the rigorous concepts and methods of phylogenetic systematics. Their criticisms led directly to the reformulation of adaptationist studies in an explicitly phylogenetic perspective. We present the phylogenetic study of adaptation in the context of these criticisms. We then summarize in the form of a general protocol the procedures needed for testing macroevolutionary hypotheses of adaptation and discriminating adaptation from alternative explanations of character evolution. Statistical approaches useful for this procedure are reviewed. The procedures are illustrated using examples from the adaptive radiation of lizards of the genus *Anolis* on Caribbean islands.

I. Previous Criticism and Phylogenetic Revision of the Adaptationist Program

Lewontin's (1977) critique of adaptationist studies makes three principal criticisms: (1) Adaptationist studies often partition the organism into traits and the environment into problems that have no well-founded biological basis. Lewontin (1977) illustrates the arbitrariness of many such decisions using examples from sociobiological explanations of human behavior in which traits such as "indoctrinability" and "blind faith" have been considered adaptive biological characters. (2) Characters are studied in isolation from each other with the questionable assumption that there are no significant effects of interaction among characters in determining their utility to the organism. (3) All characters are assumed to be adaptive, with the main goal of adaptationist studies being to discover how they are adaptive rather than potentially to reject the adaptive explanation in favor of alternatives.

These criticisms reveal the importance of the initial stages of an adaptationist study, in which the investigator chooses the characters to be studied and generates hypotheses regarding their contributions to organismal survival or reproduction. Phylogenetic solutions to these three problems are now considered sequentially.

A. Homology and Hypotheses of Adaptation

The principles of homology and phylogenetic analysis are indispensable for answering Lewontin's (1977) first criticism that adaptationist studies arbitrarily atomize the organism into traits and the environment into problems that the traits are designed to solve. The study of homology addresses this criticism by providing objective criteria for identifying characters as nonarbitrary components of organismal phenotype. The concept of homology used here combines elements of several published concepts (Patterson, 1982; Wagner, 1989; Hall, 1994). The characters to be studied as potential adaptations should exhibit three fundamental properties of homology: conservation, individualization, and uniqueness (Wagner, 1989). Conservation is the evolutionary persistence or stability of attributes of a character among lineages. Individualization denotes the separate developmental pathway of the character relative to the remainder of the organism. Uniqueness specifies that the distribution of the character defines a monophyletic group. We regard characters that satisfy these criteria as historically individualized components of the phenotype appropriate for study as hypothetical adaptations.

Patterson (1982) describes three empirical tests (similarity, congruence, and conjunction) by which one can judge whether a hypothetically adaptive character meets these basic criteria of homology. Although Patterson's (1982) tests were formulated within the conceptual framework of pattern cladism, we apply them using the principles of phylogenetic systematics (see de Queiroz, 1985).

Patterson's (1982) test of similarity examines shared attributes of a character among lineages. The form of a character, its developmental origin, and its position within the organism may reveal conserved patterns of similarity indicative of homology. These criteria may reveal that shared attributes among lineages are only superficial and not truly indicative of homology. For example, elongation of the trunk in the bolitoglossine salamanders *Lineatriton* and *Oedipina* fails the similarity test of homology because elongation occurs by lengthening individual vertebrae in *Lineatriton* and by increasing the number of vertebrae in *Oedipina* (Wake, 1966). Kaplan (1984) shows how intermediate forms observed in ontogeny, paleontological series, or interspecific comparisons among extant species may reveal similarities of morphological characters. Characters that meet rigorous criteria of similarity are then to be subjected to the tests of congruence and conjunction.

The test of congruence uses comparisons to additional taxonomic characters to ask whether the character being examined constitutes a synapomorphy of a monophyletic group. To the phylogenetic systematist, this test is equivalent to testing the hypothesis that the character had a single evolutionary origin. The test of congruence is failed if monophyly of the species sharing the character being examined is contradicted by the variation of other taxonomic characters (Patterson, 1982). The vertebrate "wing" (present only in bats and birds) fails the congruence test of homology because other synapomorphic characters (including presence of feathers, hair, and mammary glands) identify monophyletic groups (birds, mammals) that contradict a group containing only the winged vertebrates (bats + birds). The test of congruence is most effective when many congruent taxonomic characters are available for the taxon being studied and the phylogeny of the taxon, therefore, is well resolved.

If alternative characters occur as a stable polymorphism within lineages that undergo successive events of evolutionary branching, the congruence test of homology may be failed by a character that is nonetheless homologous in the sense of having had a single evolutionary origin. Technical problems that result from this phenomenon, called lineage sorting, are considered in detail

Figure 1 Toepads are composed of expanded scales that are termed lamellae. Lamellae are covered by microscopic setae; those of a geckonid lizard are illustrated (from Hildebrand, 1988; Reprinted by permission of John Wiley & Sons, Inc.).

by Roth (1991), who notes that the problem potentially occurs with characters that are emergent at the molecular or organismal levels (including morphological and behavioral characters).

The test of conjunction guards against inappropriate taxonomic comparisons of a character to a serial homologue that was produced by evolutionary duplication and divergence of characters within a lineage. The test is failed if two hypothetically homologous characters are observed together in the same organism (Patterson, 1982). For example, the hypothesis that the halteres of flies are homologous to hindwings of other insects would be falsified if both halteres and hindwings were observed in the same individuals.

We illustrate the tests of homology using the toepads present in almost all anoline lizards (Peterson, 1983; Fig. 1). These structures are composed of expanded subdigital scales, termed lamellae. The lamellae are covered with millions of microscopic setae. Electrons on the surface of the setae form bonds with electrons on the substrate, and the force of these intermolecular bonds allows adhesion to smooth surfaces (Cartmill, 1985). Electron microscopy establishes detailed similarity of structure among toepads of different anoles

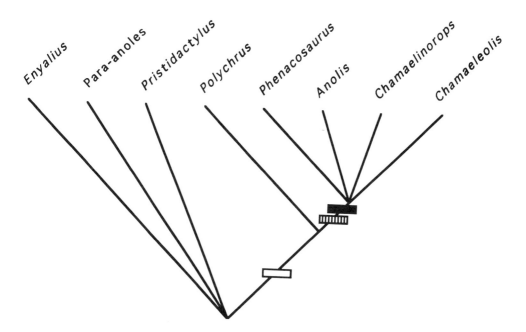

Figure 2 Evolution of toepads in anoline lizards. The black bar indicates origin of toepads, the hatched bar indicates origin of the ability to cling to smooth surfaces, and the open bar indicates origin of the use of arboreal habitats. The phylogeny is from Frost and Etheridge (1989). Relationships among the anoline genera are uncertain: *Chamaeleolis, Chamaelinorops,* and *Phenacosaurus* all possibly arose within *Anolis* (Hass et al., 1993; Jackman et al., unpublished). More detailed information on the natural history and/or phylogeny of *Enyalius, Pristidactylus,* and the para-anoles may produce different conclusions about when arboreality evolved, but will not alter the fundamental conclusion that it preceded the evolution of expanded toepads in the anolines.

and reveals no profound differences that would falsify similarity (Ruibal and Ernst, 1965; Peterson, 1983). Monophyly of the anoline lizards is established by several synapomorphies (Frost and Etheridge, 1989). The distribution of toepads is congruent with these characters (Fig. 2), except for secondary absence in several taxa (Peterson and Williams, 1981). Toepads are present on all digits and therefore constitute serially repeated structures within the organism. The presence of toepads on all digits, rather than specific attributes of the toepads of individual digits, constitutes the character being studied as a hypothetical adaptation. The character conceived in this way passes the conjunction test of homology because only a single set of toepads is present in any animal. Thus, toepads of anoline lizards meet the criteria of homology and constitute individualized components of the phenotype appropriate for testing as hypothetical adaptations.

The terms nonhomology and homoplasy describe characters that fail one or more of the three tests described above. Patterson (1982) categorizes the different kinds of nonhomology according to which of the three tests are passed or failed. We discuss below the use of homoplastic patterns of evolution to test the predictions of what we call general hypotheses of adaptation. Two categories of nonhomology are important in this context, parallelism and convergence. A character exhibits parallelism if it fails the congruence test, but passes similarity and conjunction. Parallelism involves more than one evolutionary origin of a characteristic that arises by similar developmental processes from similar ancestral conditions. Parallelism has been used to test hypotheses of developmental constraints on evolutionary change, as well as hypotheses of adaptation (Wake, 1991). A character exhibits convergence if it passes the test of conjunction, but fails both similarity and congruence. Convergence denotes the independent evolutionary origin of superficially similar characters and also has been invoked as a test of hypotheses of adaptation (Harvey and Pagel, 1991; Wake, 1991).

Toepads superficially similar to those of anoline lizards occur in two other groups of lizards, a monophyletic group within skinks (genera *Prasinohaema* and *Lipinia*) and the Gekkonidae. When compared to the toepads of anoles, toepads of these taxa fail the similarity test because they have conspicuous differences in the setae (Williams and Peterson, 1982). The congruence test is failed because of conflict with numerous characters used to examine the family-level relationships of lizards (Estes et al., 1988). The conjunction test is passed because only a single set of toepads is found in any animal. The toepads of gekkonids and skinks therefore are convergent with those of anoles and do not collectively constitute homologous characters.

Phylogenetic analysis also answers Lewontin's (1977) criticism of the arbitrary partitioning of the environment into problems that the characters must solve. Lewontin (1977) notes that organisms determine to some degree their effective environments through their active choice and utilization of resources. Physical and biological components of environments utilized by related species can be examined comparatively, and phylogenetic analysis can be used to identify the evolutionary succession of environmental factors that characterize the history of an evolving lineage. Utilization of environmental factors by populations can be examined analytically using the same principles and tests described above.

B. Hypotheses of Adaptation and Interactions Among Characters

Principles of homology are important also for answering Lewontin's (1977) second criticism, that adaptive studies of individual characters must not disregard their interactions with other characters in their contributions to organismal survival. Phylogenetic trees enable one to examine the historical associations of characters to identify cases of evolutionary nonindependence. Two different characters may characterize exactly the same taxa (Fig. 3a), in which case their phylogenetic histories coincide and tests of the adaptive status

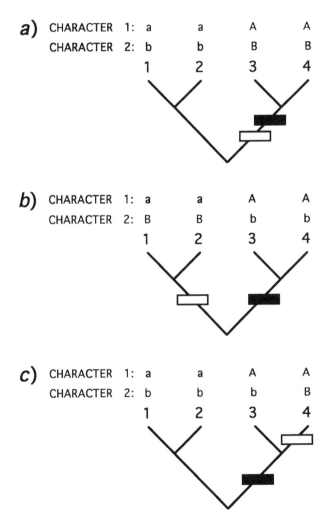

Figure 3   Possible relationships between two phylogenetically congruent characters: (a) the characters specify the same clade; (b) the characters specify mutually exclusive clades; or (c) the clade specified by one character is nested within the clade specified by the other.

of either character are done entirely in the context of the other one. Alternatively, the characters may originate on different lineages. The characters may occur on mutually exclusive lineages (Fig. 3b), in which case evolutionary associations among them are absent. If one character (1) arose on a lineage ancestral to the one on which a second character (2) arose (Fig. 3c), character 2 is studied in the context of character 1. To investigate the origins of character 1, however, it could be misleading to examine its utility only in taxa that also possess character 2, because character 1 arose in the absence of character 2. The use of appropriate phylogenetic comparisons can reveal whether the presence or absence of character 2 affects the biological role of character 1.

An example comes from communication in anoles. Anoles have a large extensible fan on the throat, termed a dewlap, that is used in social contexts by males and, in some species, by females (Fig. 4). Anoles also usually use stereotyped patterns of head-bobbing in such displays. Widespread occurrence of head-bobbing in iguanian lizards indicates that head-bobbing arose much earlier than the anoline dewlap in the evolutionary history of lizards (see Jenssen, 1977; Carpenter, 1986). The hypothesis that head-bobbing evolved initially as a biomechanical necessity for unfolding the dewlap or as an adaptation for displaying the dewlap thus would be inappropriate. A more appropriate hypothesis is that the dewlap evolved as an adaptation for communication and that head-bobbing may have been a prerequisite for the evolution of the dewlap.

The phylogenetic context will identify situations in which a hypothetical character is simply a structural consequence of organismal architecture, development, and allometry (Gould and Lewontin, 1979); such characters will

Figure 4 *Anolis grahami* using its dewlap during a display.

Figure 5 *Anolis luteogularis,* one of the largest anoles, illustrating a large and rugose head. Compare it to the smaller, presumably more ancestral, head structure illustrated by *Anolis grahami* (Fig. 4).

originate as part of a complex of related features of organismal design that are phylogenetically inseparable from each other. Concordant evolution of two characters suggests that purely structural or developmental explanations for their coupling be investigated prior to testing hypotheses of adaptation. For example, large and rugose heads have evolved independently several times within *Anolis* (Fig. 5). In each case, the evolution of these features occurred simultaneously with an increase in body size (see Cannatella and de Queiroz, 1989). The hypothesis that large heads are simply an allometric consequence of an increase in body size, and not an independent character, should be tested and rejected before postulating adaptive explanations specifically for the large heads. Even if these characters are not necessarily coupled developmentally, adaptive explanations for large heads must be formulated within the context of large body size because of the phylogenetic association of these characters; large heads may be adaptive only in animals having large body size. The phylogenetic context therefore restricts nonarbitrarily the combinations of characters that should be examined to test biologically meaningful hypotheses of interaction among characters.

C. Alternatives to Adaptation

Lewontin's (1977) third major criticism of adaptationist studies is that all characters are assumed adaptive, causing the investigator to search only for the means by which a character is adaptive rather than subjecting the hypothesis of adaptation to potential falsification. The revised definition of adaptation given by Gould and Vrba (1982), in which the concepts of adaptation, exaptation, and nonaptation are logically separated, provided the first step in answering this criticism. The hypothesis of adaptation makes several predictions, one of which is that a character confers an advantage relative to its phylogenetic antecedents. If the character cannot be distinguished from its phylogenetic antecedents with respect to conferring an advantage to the organisms possessing it, it is termed a nonaptation (Fig. 6; Baum and Larson, 1991).

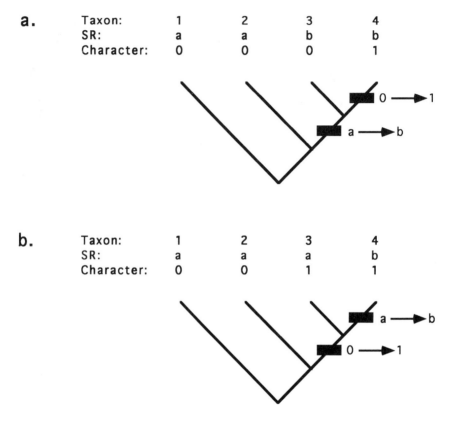

Figure 6 Illustration of the phylogenetic predictions of hypotheses of adaptation, exaptation, nonaptation, and disaptation using the selective regime (SR; after Baum and Larson, 1991). Hypotheses are illustrated with reference to taxon 4. (a) A clade having a derived character (1) is phylogenetically nested within a clade having a derived selective regime (*b*). Adaptation predicts that utility or performance of the derived character (1) will exceed that of the ancestral condition (0) in selective regime *b*. Primary nonaptation predicts that characters 0 and 1 have equivalent utility under selective regime *b*. Primary disaptation predicts that character 0 has greater utility than character 1 in selective regime *b*. (b) A clade having a derived selective regime (*b*) is nested within a clade having a derived character (1). Exaptation predicts that the utility of the derived character (1) exceeds that of the ancestral character (0) in selective regime *b*. Secondary nonaptation predicts that characters 0 and 1 are equivalent in utility within selective regime *b*. Secondary disaptation predicts that character 0 has greater utility than character 1 in selective regime *b*.

If the character is found to confer an advantage, the hypothesis of adaptation requires further that the character's evolutionary origin occurred by natural selection for its currently observed biological role. This criterion makes several predictions potentially falsifiable by phylogenetic analysis. The biological role currently observed for the character must have been present at its origin. This prediction is tested by examining the character and its utility in different species in a phylogenetic context (Fig. 6). Thorough testing requires comparative studies of functional morphology and selective factors affecting the character (Baum and Larson, 1991) and detailed studies of the natural history of the species being studied (Greene, 1986b, 1994). The hypothesis that a character evolved as an adaptation for a biological role is falsified if that role is associated only with a subset of taxa nested within the clade possessing the character; in this case, the current utility of the character evolved after the character itself (Fig. 6). Using the terminology of Gould and Vrba (1982), such a character is termed an exaptation for the biological role being examined. The history of character evolution in a group of species often may comprise a nested hierarchy of exaptations and adaptations. A character may acquire biological roles other than ones for which it arose by natural selection, and its newly acquired role may lead to selectively driven modification of that character and others. These modifications represent adaptations arranged phylogenetically in a nested hierarchy. The broader goal of studies of adaptation, therefore, can be seen as that of examining the interaction of adaptive, exaptive, and nonaptive influences on evolutionary diversification (see Arnold, 1994).

The nested hierarchy of adaptations and exaptations is illustrated by the evolution of the dewlap in anoles. The dewlap is expanded using the second ceratobranchial bone, a component of the hyoid apparatus. The hyoid apparatus is homologous to the gill arches of aquatic vertebrates; the second ceratobranchial of tetrapods is a homology nested within the more inclusive homology that includes the corresponding gill arches of aquatic vertebrates. The gill arch to which the second ceratobranchial is homologous presumably arose by selection for a respiratory function and constitutes an adaptation for respiration. The gill arches were coopted for feeding in tetrapods, making gill arches an exaptation for feeding, but the derived characteristics of the hyoid apparatus that arose in the context of feeding, including the second ceratobranchial bone, constitute adaptations for feeding. The second ceratobranchial bone then was coopted in anoles for expanding the dewlap and is an exaptation for communication. Derived characteristics of the anoline second ceratobranchial bone evolved for expanding the dewlap constitute adaptations for communication.

One should consider the alternative hypotheses of character evolution to be investigated, especially when hypotheses of adaptation are falsified. Vrba and Gould (1986) present a taxonomy of the deterministic and stochastic processes that may underlie the evolution of characters. They emphasize that evolutionary studies in the past often have emphasized adaptive interpretations because alternative hypotheses involving nonrandom processes were largely absent from traditional Darwinian theory, which invoked random drift as the main alternative to natural selection. They emphasize the importance of a hierarchical expansion of traditional Darwinism to include deterministic processes acting at the species level (species selection) and the genomic level (selfish DNA) whose consequences often include evolutionary sorting of the organismal-level characters that are usually the focus of adaptationist studies. Organismal characters generated through the indirect action of species selection or selfish DNA are not necessarily expected to have a biological role conferring advantages at the organismal level. The evolutionary study of the effects of selfish DNA and species selection on evolution of organismal characters is a new field. We anticipate that as phylogenetic approaches are used more extensively to test hypotheses of adaptation, rejection of adaptive explanations will lead to greater investigation of these alternative processes as factors affecting the evolution of organismal characters. Vrba and Gould (1986) offer a more detailed discussion of the predictions of hypotheses of species selection and selfish DNA.

So far we have emphasized single processes (natural selection for a particular biological role, exaptation, species selection, selfish DNA, stochastic processes) as the primary explanations of character evolution. The actual evolutionary history of a group undoubtedly features numerous processes acting simultaneously or in sequence to influence the evolution of organismal characters. Different factors may act synergistically or in opposition to each other. Empirical discrimination of alternative causal factors will be most effective when they act in opposition to each other. As the concept of exaptation implies, different biological roles may be performed by a single character and the different roles may exert conflicting selective influences on the differential survival and reproduction of alternative characters. Likewise, species selection and natural selection may have opposite effects on the evolutionary sorting of character variation.

This chapter emphasizes the initial stages of testing hypotheses of adaptation and should be viewed as the beginning of an iterative, deductive process. Hypotheses of evolutionary process often make specific predictions regarding evolutionary pattern that permit rigorous testing and potential falsification. Complex interactions among factors influencing organismal evolution, such as those discussed by Vrba and Gould (1986), may obscure the role of natural selection. Initial results therefore will reveal whether selective factors prevail over any conflicting influences on the evolution of organismal characters. If predictions of natural selection are falsified, alternative causal hypotheses of character evolution should be investigated (Baum and Larson, 1991). If the predictions of natural selection are upheld, the hypothesis of adaptation is subjected to a more refined and detailed investigation (Baum and Larson, 1991; Coddington, 1994).

Because the study of adaptation requires the use of many evolutionary concepts whose contexts and meanings have varied, and because precise definitions are needed for employing phylogenetic methodology, we provide specific definitions and explanations of key concepts in Table 1.

II. A General Protocol for Testing Hypotheses of Adaptation

The hypothesis that a character evolved as an adaptation makes two predictions: (1) that the character evolved in the context of a particular

TABLE I

Phylogenetic Definitions

Adaptation	An organismal character produced by natural selection for a particular biological role (modified from Gould and Vrba, 1982; Baum and Larson, 1991).
Aptation	An organismal character that confers utility to the organism regardless of whether its phylogenetic origin featured natural selection (from Gould and Vrba, 1982).
Biological role	An action or use of a character by the organism during the course of its life history (from Bock and von Wahlert, 1965; Bock, 1979).
Character	A component of organismal phenotype subject to cladistic analysis and to Patterson's (1982) tests of homology. A character that passes Patterson's (1982) tests may be termed a homology or homologous character.
Disaptation	An organismal character whose utility to the organism is demonstrably inferior to that of a phylogenetically antecedent character (from Baum and Larson, 1991).
Effect	The biological role of an exaptation, a use for which the character was coopted, but which was not a factor in the character's evolutionary origin (from Gould and Vrba, 1982).
Exaptation	An organismal character that has been coopted for a use unrelated to its origin. An exaptive character originally may have been an adaptation for a different use or a nonaptation (from Gould and Vrba, 1982).
Function	The biological role of an adaptation, the use for which the character evolved by natural selection (from Gould and Vrba, 1982)
Homology	A component of the organismal phenotype exhibiting the evolutionary properties of conservation, individualization, and uniqueness (Wagner, 1989; see text); empirically identifiable using Patterson's (1982) tests of similarity, congruence, and conjunction.
Homoplasy	Possession by two or more species of a shared attribute that was not derived from their most recent common ancestor; embraces convergence, parallelism, and evolutionary reversal (modified from Futuyma, 1986).
Natural selection	Those interactions between heritable organismal character variation and the environment that cause differences in rates of birth or death among varying organisms in a population (from Vrba and Gould, 1986).
Nonaptation	An organismal character that confers no utility for organismal survival or reproduction relative to phylogenetically antecedent characters. A primary nonaptation is one whose origin cannot be ascribed to the direct action of natural selection. Nonaptation also may arise secondarily by loss of the utility of a character (modified from Gould and Vrba, 1982; Baum and Larson, 1991).
Parsimony	Resolution of homoplasy among phylogenetic characters by choosing as the best working hypothesis the phylogenetic topology that minimizes homoplasy.
Performance advantage	A result of comparative functional analysis showing that a character has enhanced utility not associated with an alternative character (from Greene, 1986a).
Selective regime	Critical aspects of organismal/environmental interaction that are postulated by a hypothesis of adaptation to be significant factors influencing natural selection of the characters being studied. It provides predictions of how natural selection would direct evolution of a character under study as contrasted with other potential influences including stochastic factors, genetic or developmental correlations with other selected characters, or incidental effects of species selection or selfish DNA (from Baum and Larson, 1991).
Sorting	Differential rates of birth or death among varying organisms in populations resulting from any deterministic or stochastic causes (from Vrba and Gould, 1986).

selective regime; and (2) that the character is more advantageous than phylogenetic antecedents in that context. Generally applicable protocols for testing these predictions have been formulated by Greene (1986a), Coddington (1988, 1990), and Baum and Larson (1991). These protocols are compared and contrasted by Baum and Larson (1991) and, although their details differ, they share common goals. A synopsis of the protocol described by Baum and Larson (1991), which incorporates many features of the earlier protocols and adds some new procedures, is given below. Some minor modifications have been incorporated. We present this protocol as a series of six steps to be applied iteratively in testing hypotheses of adaptation.

Step 1: Formulation of Hypotheses of Adaptation - A hypothesis of adaptation must be formulated by making specific statements regarding (1) the character hypothesized to be adaptive, (2) the taxa in which the character is observed and to which the hypothesis of adaptation pertains, and (3) the postulated biological role of the character. The phylogenetic methodology is potentially applicable to taxa of any rank, but we agree with Cracraft (1990) that the most effective resolution of the evolutionary history of a character occurs when comparisons involve closely related species (Cracraft, 1989). The characters studied usually will be fixed or nearly fixed in the species studied except for the effects of recurring mutation. Some cases of intraspecific polymorphism may be appropriate for study if outgroup comparison provides an unambiguous inference of evolutionary polarity of the character states.

Step 2: Reconstruction of Phylogeny - Any methodology that produces a rooted tree relating the taxa containing the character being studied and their most closely related outgroups can be used for this study (see Swofford et al., 1996). Phylogenetic inferences should be most robust when based upon a large number of characters drawn from diverse morphological, molecular, and behavioral systems, and examined for any systematic conflicts among characters (see Larson, 1994). We agree with de Queiroz (1989) that it is not necessarily circular to consider the same characters as sources of phylogenetic information and subject to analysis as potential adaptations. When a phylogenetic hypothesis is based on a large number of characters, however, the phylogenetic hypothesis generally will not depend heavily on the effects of the particular characters being investigated as potential adaptations.

Step 3: Scoring and Phylogenetic Partitioning of Characters - As noted above, every effort must be made to establish homology relations among the characters studied as potential examples of adaptation. The procedures noted here for assessing homology of the characters being studied as potential adaptations are the same procedures applied to all morphological or behavioral characters used to generate the phylogeny in the previous step. Patterson's (1982) tests of similarity, conjunction, and congruence are applied, with congruence assessed using the phylogenetic analysis described in the previous step. A detailed discussion of criteria used for testing similarity of characters is given by Kaplan (1984). Developmental information can be particularly important for determining whether similarity among characters is indicative of homology or only a superficial similarity produced by convergence. Although the importance of developmental information is uncontroversial, the ways in which the information is used to assess homology are highly controversial (Patterson, 1982; de Queiroz, 1985; de Pinna, 1991). A particularly controversial issue is whether a character should be conceived and measured at only a single stage (usually the adult) in the organism's life cycle or whether the character should be conceived as an ontogenetic transformation transcending different stages of the organism's life history (de Queiroz, 1985; de Pinna, 1991).

Incongruence among characters is usually resolved using the criterion of maximum parsimony, although other methods have been proposed. In many cases, more than one equally parsimonious optimization of character changes on the phylogenetic tree will be possible, precluding unambiguous assessment of homology. The implications of the alternative optimizations for testing hypotheses of adaptation must be examined (Maddison and Maddison, 1992; Losos and Miles, 1994).

Step 4: Scoring and Phylogenetic Partitioning of Selective Regimes - The critical factors comprising the selective regime depend on the specific hypothesis of adaptation being tested. The selective regime can incorporate abiotic climatic factors, biotic environmental factors, organismal features, or any combination of these factors. A detailed study of the natural history of the species being studied is generally required for the realistic characterization of selective regimes (Greene, 1986b). Because selective regimes comprise organismal/environmental interactions, they are subject to the same criteria of homology and phylogenetic analysis that apply to morphological characters as discussed in the previous step. Comparative studies of selection acting within populations of related extant species provide potentially the most effective test of the macroevolutionary stability of selective regimes.

Step 5: Assessing the Biological Role or Utility of Characters - Results of the previous two steps are combined to infer selective regimes for the lineages on which a character arises. Of particular interest are the selective regimes under which a character evolved and the selective regimes of extant species that possess the character. The selective regime forms the framework for testing hypotheses of the biological role or utility of the hypothetically adaptive character. The hypothesis of adaptation postulates that the character performs a specific function and that this performance enhances the organism's survival or reproductive success. Predictions of the hypothesis of selective origin of a character therefore can be tested in two ways: (1) functional morphological studies to test the prediction that possession of the character enhances performance, and (2) studies to test the prediction that possession of the character increases rates of survival or reproductive success. Both approaches encompass numerous specific tests and provide valid tests of the selective hypothesis. The particular approaches chosen will depend on the characteristics of the organisms being studied and the ease by which appropriate manipulations and observations can be made.

The most direct approach for evaluating the utility of a character is to study its use in a task, such as the ability to escape predators, that is important for survival or reproductive success. Performance is measured to test the hypothesis that a character provides greater utility than the antecedent character in the relevant selective regime. Several different kinds of comparative and experimental approaches can be used to obtain the relevant measurements. Whenever possible, naturally occurring variation should be used to obtain the alternative characters whose performance is to be compared. When this option is not possible, experimental manipulation can be used to simulate the alternative characters (e.g., Carothers, 1986; Emerson and Koehl, 1990). Physical models also may be useful for examining how alternative characters affect performance (Rudwick, 1964; Fisher, 1985). Phylogenetic methods are used to infer the utilities of characters as they occurred in ancestral lineages. Detailed coverage of the testing of utility of characters is provided by Fisher (1985).

Measurement of performance is usually the most useful way to test predictions of hypotheses of adaptation because the predictions made are very specific. The question that the character's performance actually enhances survival or reproductive components of fitness (Lauder et al., 1993) can be examined by measuring within populations the rates of survival or reproduction of individuals possessing the hypothetically adaptive character versus its evolutionary antecedent. As with the measurement of performance, use of naturally occurring polymorphism is desirable, although appropriate character variation could be induced using genetic or phenotypic manipulations. Studies of this sort frequently look at the effect of character manipulation on survival (e.g., Kettlewell, 1973; Schluter, 1994), but less frequently examine reproductive aspects of fitness.

Step 6: Classifying Traits into Categories of Utility/Historical Genesis - The data gathered above permit characters to be categorized as adaptations, exaptations, nonaptations, or disaptations (Baum and Larson, 1991; Arnold, 1994). A character qualifies as an aptation if it demonstrates significantly greater utility than its phylogenetic antecedent for performing a task that promotes organismal survival or reproductive fitness in a particular selective regime. If the selective regime for which the character is advantageous characterizes the lineage on which it arose, the character is termed an adaptation. The character retains the status of adaptation in all lineages that retain the selective regime present at its origin. A subsequent evolutionary change of selective regime may cause the character to be exaptive, nonaptive, or even disaptive on more recent lineages. A character is termed an exaptation where it is advantageous for a selective regime that arose subsequent to its evolutionary origin. Prior to the change of selective regime, the character may have occupied any of the alternative categories.

If a character is found not to differ significantly in utility from its phylogenetic antecedent, nonaptation constitutes the working hypothesis for further investigation of its evolution. A subsequent study of the natural history of the species possessing the character may reveal undiscovered components of the selective regime or biological roles that may lead ultimately to rejection of the hypothesis of nonaptation; until nonaptation is rejected, however, the character cannot be considered an adaptation or exaptation. Primary nonaptation denotes a character that arose by means other than natural selection, whereas secondary nonaptation denotes a formerly aptive character whose utility was lost by an evolutionary change in the selective regime (Baum and Larson, 1991).

Disaptation describes characters whose utility in a particular selective regime is significantly less than that of their phylogenetic antecedents. Formerly aptive or nonaptive characters can become disaptations through an evolutionary change of selective regime. Primary disaptation implies that the character replaced one of superior utility at its origin, which runs counter to the theory of natural selection. Empirical results suggesting primary disaptation probably indicate an erroneous assessment of selective regime or homology of characters, particularly lack of individualization from other features subject to selection (see Baum and Larson, 1991). Note that primary nonaptations and primary disaptations are the only categories for which origin of a character by natural selection is explicitly rejected.

As this discussion illustrates, the adaptive status of a character is not necessarily stable during its evolutionary history and may be categorized differently at different points in its evolutionary history. Characters that qualify as adaptations must arise as such; characters whose evolutionary origin is nonadaptive do not become adaptations secondarily by subsequent evolutionary change. Characters never originate as exaptations, but may become exaptive only through evolutionary change in the selective regime and utility of the character (Fig. 6). Characters may be nonaptive or disaptive either by their evolutionary origin or secondarily by evolutionary changes in selective regime.

Rigorous discrimination of the categories described above occurs only when phylogenetic resolution is precise. An ambiguous situation occurs, for example, when evolutionary changes in both the character and selective regime occur on the same branch of the phylogenetic tree. In such cases, one cannot determine whether the character arose before or after the new selective regime, precluding discrimination of adaptation and exaptation. If appropriate variation occurs, one would expect response to a new selective regime to occur rapidly; entry into the new selective regime and character change then would not be separated by phylogenetic branching and would be reconstructed on the same branch of the evolutionary tree. Similarly, exaptation could occur rapidly if a new character leads to modification of the selective regime which would be expected, for example, for characters termed key innovations (Baum and Larson, 1991). However, in many cases of exaptation, the new selective regime will not be immediately available for a variety of reasons (e.g., the new regime occurs in other biogeographic areas or results from subsequent evolution of other taxa). In these cases, changes in the character and selective regime are more likely to be separated by a branching of lineages. For this reason, Arnold (1994) argues that if one observes many instances of simultaneous evolution of a character and a selective regime, adaptation is the preferred working hypothesis.

The evolution of toepads in anoles, mentioned previously, can be used to illustrate this protocol for testing hypotheses of adaptation. Toepads are widely considered adaptive for arboreality, but this hypothesis has not been tested rigorously. The manner in which this hypothesis can be tested is straightforward. The hypothesis states that the evolution of toepads conferred increased clinging capability and occurred in the context of a selective regime favoring enhanced arboreality. Cartmill (1985) identifies two ways in which arboreality may be achieved, by grasping and by adhesion. The presence of claws in lizards provides some grasping ability, but toepads potentially provide both improved grasping ability and adhesion. The phylogenetically antecedent condition of anoline digits is to possess claws, but lack toepads; the presence of toepads in addition to claws is the derived condition. Arboreality in anoles, therefore, encompasses at least two selective regimes potentially affecting the evolution of toepads for clinging. Toepads and their setae may enhance the grasping of narrow supports and provide adhesion on smooth surfaces. The phylogenetic hypothesis necessary to examine this question is presented in Fig. 2. We already have discussed the evidence that toepads pass empirical tests of homology. Almost all anoles are arboreal; the few that are not have become terrestrial secondarily (Peterson and Williams, 1981). Phylogenetic reconstruction indicates that arboreality evolved before the evolution of toepads. Functional studies demonstrate that all 17 species of *Anolis* examined can cling to a smooth surface (Fig. 7; Losos, 1990a; Irschick et al., in press); lizard species without toepads are unable to cling to such surfaces (Losos, unpublished; *Polychrus*, however, has not been examined). Functional studies are not yet available to examine the potential utility of toepads for grasping narrow branches. Thus, we can reject the null hypothesis of functional equivalence for comparing digits that contain toepads versus the phylogenetically antecedent condition in which they are absent. The evolution of toepads is associated functionally and phylogenetically with increased clinging ability, consistent with the hypothesis that toepads constitute an adaptation to adhesion on smooth arboreal surfaces. Further tests of this hypothesis might include experimental removal of the setae to examine their effect on clinging in the lab or survival in the field, detailed behavioral studies (Greene, 1986b, 1994) to refine understanding of the utility of the pad, and measurements of grasping ability on rough or narrow surfaces.

Figure 7 Measurement of clinging ability in a tokay gecko *(Gekko gecko)*. Lizards are placed on a force plate and pulled downward, permitting measurement of the adhesive force generated (see Irschick et al., in press). Photograph by D. J. Irschick.

III. Using Homoplasy to Test General Hypotheses of Adaptation

Parallelism and convergence produce common attributes in nonhomologous characters of different taxa. Common attributes produced by parallelism may constitute genuine similarity whereas convergence produces only superficial resemblances among characters. The different lineages in which parallel or convergent evolution are observed may share a selective regime through common ancestry or may have evolved comparable selective regimes via parallel or convergent evolution. Parallelism permits a nearly exact replication of tests of adaptive hypotheses, and convergence permits repeated testing of adaptive hypotheses for the common attributes of convergently evolved characters. Each of the characters related by parallel or convergent origin should individually pass the tests of homology and be subjected to the phylogenetic tests of adaptation described in the protocol outlined above.

Parallelism and convergence also permit testing what we call general hypotheses of adaptation. If similar attributes have evolved multiple times in different lineages, one can test the general hypothesis that evolution occurred in each case under similar conditions of selective regime. This comparative approach, which has been used in evolutionary biology since the field's inception, has been reviewed by Pagel (1994a). Common attributes of different characters are hypothesized to be functionally analogous (McLennan and Brooks, 1993; Wenzel and Carpenter, 1994); this hypothesis should be tested by incorporating analyses of performance when investigating a correlation between the evolution of common attributes and selective regime (e.g., Losos, 1990a; Arnold, 1994; Coddington, 1994).

A general hypothesis of adaptation is strongest if one observes many independent evolutionary origins of similar attributes occurring in the contexts of comparable selective environments. Statistical methodology can be employed to test the null hypothesis that an observed association between shared attributes of characters, functional performance, and selective regime occurred by chance. The wide variety of statistical methods that has been proposed in the past few years is too great to review here; interested readers may consult reviews by Maddison and Maddison (1992) and Losos and Miles (1994).

One commonly used statistical approach is the concentrated changes test (Maddison, 1990; Maddison and Maddison, 1992; Sillén-Tullberg, 1993) [see Pagel (1994b) for a recent method that relies on maximum likelihood rather than parsimony]. Given two attributes (one of which could be a selective regime), #1 and #2, each of which has two states, A and a and B and b, this test asks whether the evolution of $a \rightarrow A$ in character #1 occurs more often when the state of character #2 is B than would be expected by chance. For studies of adaptation, this translates to asking whether a particular character evolved in taxa occupying a specific selective regime more often than expected by chance. This test is necessary because if most members of a lineage occupy a particular selective regime, then one would expect most instances of character evolution to occur in taxa occupying that selective regime by chance alone. For small numbers of taxa and instances of character evolution, exact calculations of probability are possible, whereas for larger numbers, simulation is necessary. Either type of analysis can be performed using MacClade (Maddison and

Maddison, 1992).

The evolution of lizard toepads is useful for illustrating this approach. As mentioned earlier, toepads evolved convergently in three lineages of lizards. We therefore can test the general hypothesis that lizard toepads represent an adaptation to arboreality. Within geckos, homology of the toepads is clear (Ruibal and Ernst, 1965), but within skinks, microstructure of the toepad varies (Williams and Peterson, 1982), which suggests that toepads of different species might not be homologous. For this analysis, assumption of homology within skinks is statistically conservative because it minimizes the number of times that toepads have evolved in arboreal selective regimes. Measurements of performance are scant for geckos, toepad-bearing skinks, and their outgroups, but existing data suggest that all toepad-bearing lizards have enhanced clinging ability (Irschick et al., in press).

Consequently, the first hypothesis to test is that the evolution of toepads and clinging ability is phylogenetically associated in lizards. Testing this hypothesis requires a fully resolved phylogeny for lizards (including snakes that are phylogenetically part of the lizard clade), which is currently unavailable. For heuristic purposes, we present a phylogeny (Fig. 8) that follows Estes et al. (1988) for interfamilial relationships among limb-bearing lizards (i.e., excluding pygopodids, snakes, and other legless, and thus padless, squamates).

Given the phylogeny in Fig. 8, expanded toepads and enhanced clinging capability are inferred to have evolved simultaneously three times and no times separately. Using MacClade, we calculate that the probability of three events of toepad evolution occurring simultaneously with the evolution of clinging ability is $P = 0.000008$. Each instance of evolution of toepad morphology and enhanced clinging ability occurs on an arboreal lineage. The probability of this occurring by chance is $P = 0.028$. Thus, these analyses uphold the hypothesis that expanded toepads in lizards have evolved as an adaptation to increase the clinging ability in arboreal situations.

The relationship between general hypotheses of adaptation and hypotheses specific for a particular character in a particular taxon is complex. A statistically significant association between the shared attributes of characters and selective regime has been interpreted as strengthening the hypothesis that each character is individually adaptive as proposed. However, a particular character may depart from the general trend and fail to constitute adaptation even if other characters sharing attributes with it are adaptive. Furthermore, a particular character may be adaptive as proposed even if the general hypothesis of adaptation is not upheld.

IV. Adaptation and Quantitative Characters

The protocol just outlined emphasizes discrete (or categorical) characters. Essentially the same conceptual approaches can be used to study adaptive evolution in quantitative (or continuous) characters, although details of the phylogenetic analysis will differ. A number of methods based on parsimony can be used to reconstruct the evolution of continuous characters; one of these is mathematically essentially identical to the one used for categorical characters (Swofford and Maddison, 1987). For continuous characters, most extant taxa will differ in their character states as a result of sampling, genetic drift, and other processes. Consequently, reconstructions often will reveal evolutionary change occurring on many branches of the tree. Thus, rather than testing the hypothesis that the character evolved within a particular selective regime, as one would do with a categorical character, one might test the hypothesis that character evolution has been greater in lineages experiencing the selective regime than in lineages not exposed to it. Furthermore, because populations often contain substantial genetic variation for continuous characters (Falconer, 1981), one might expect characters to evolve quickly after the lineage enters a new selective regime. Using this assumption, one can test the more restrictive hypothesis that evolutionary change of a character is greater on branches experiencing a change of selective regime.

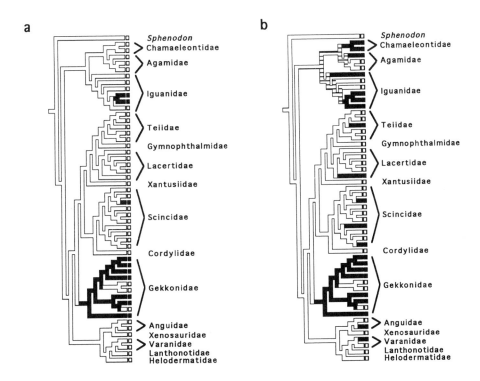

Figure 8 (a) Toepads and clinging ability are distributed identically in lizards and are shown together. Black bars represent lineages characterized by expanded pads and clinging ability. Interfamilial relationships follow Estes et al. (1988). The number of lineages illustrated within each family is roughly proportional to the number of species within each family, but is not meant to depict specific relationships; rather, the tree is presented for heuristic purposes. Figure prepared using MacClade (Maddison and Maddison, 1992). (b) Evolution of use of arboreal habitats. Hatched lineages indicate ambiguous reconstruction of ancestral habitats. In this case, arboreality could have arisen once in the Acrodonta (=Chamaeleontidae, Agamidae, and Iguanidae) followed by three separate losses of arboreality for a total of four evolutionary changes. However, an equally parsimonious reconstruction would postulate four independent origins of arboreality with no reversals; other equally parsimonious resolutions also exist.

This approach can be applied to cases in which comparable selective regimes have arisen multiple times. For example, evolution of body size in Lesser Antillean anoles is hypothesized to be greater on the lineages that moved from a one-species to a two-species island than on the remaining branches showing no change in the number of species. In this case, the selective regime refers to size of prey; in sympatry, anoles diverge in body size to utilize different sizes of prey (Schoener, 1970). The hypothesis was upheld for anoles of the northern Lesser Antilles, but not for those in the southern Lesser Antilles (Losos, 1990b).

In some cases, the focal character, performance, and selective regime all will be continuous variables. Several examples of continuous selective regimes are temperature, dimensions of structural habitat, and size of prey. For studies in which all three parameters are continuous variables, the hypothesis of adaptation would predict that all three variables are correlated in their evolution. Numerous methods have been advanced to investigate the correlated evolution of continuous characters. They are reviewed elsewhere (e.g., Harvey and Pagel, 1991; Gittleman and Luh, 1992, 1994; Maddison and Maddison, 1992; Miles and Dunham, 1993; Losos and Miles, 1994; Purvis et al., 1994; Westneat, 1995) and the field is changing rapidly with new developments. Broadly speaking, however, three explicitly phylogenetic approaches exist:

1. Evolutionary reconstructional methods ("Directional Methods" of Harvey and Pagel, 1991) use one of several algorithms based on parsimony, or other criteria (e.g., Harvey and Purvis, 1991; Lynch, 1991), to reconstruct evolutionary change of each variable. Then, the amounts of change occurring on all lineages are calculated for all variables and examined for correlated evolution.

2. Autocorrelational methods (Cheverud et al., 1985; Gittleman and Kot, 1990; see also Lynch, 1991) were developed to examine the extent to which character values among extant taxa reflect their phylogenetic relationships; for example, is the similarity between pairs of taxa a function of their relatedness? This approach uses a phylogeny to generate a matrix of phylogenetic relatedness among all taxa. This matrix is then used in a maximum likelihood-based regression in which species' values are estimated using the weighted average of related species, the weights being specified by the matrix of phylogenetic relationships. If, in fact, no relationship exists between observed and estimated values for each species, then the autocorrelation coefficient will not differ significantly from zero and phylogenetic information may not be informative in studying character evolution (see Martins, 1996). Presumably, this result occurs when characters evolve so rapidly that closely related species are not necessarily similar phenotypically. These methods can be used also to calculate how different each species' phenotypic value is from that predicted based on its phylogenetic relationships. This method has been used to examine whether a correlation exists between such "specific values" calculated for several variables (reviewed in Purvis et al., 1994; Martins, 1996), although the underlying evolutionary basis for this approach is not clear (Losos and Miles, 1994).

3. The independent contrasts method (Felsenstein, 1985) is the most widely used method for studying correlated evolution. The rationale underlying

this method is that each ancestral node in a phylogenetic tree gives rise to two descendants (either extant species or other ancestral nodes). The amount of difference in value of a continuous character between these two descendants, termed a contrast, represents the amount of evolutionary change that has accrued since the descendants diverged from their common ancestor; however, contrary to directional methods, independent contrasts do not specify on which of the two branches the evolutionary change occurred. For a fully resolved phylogenetic tree with n taxa, $n - 1$ contrasts exist. Statistical methods are used to calculate contrasts using a model of character evolution and the lengths of branches in units of expected evolutionary change (proportional to time in a model of gradual evolution). Contrasts calculated for several variables are then examined for correlation (see Garland et al., 1992 for details). In the absence of a fully resolved phylogeny, one can calculate contrasts for phylogenetically nonoverlapping pairs of extant species (Felsenstein, 1985; Burt, 1989).

We illustrate the study of adaptation in quantitative characters using independent contrasts (data are presented in Table 2). Distantly related species of *Anolis* that occupy similar structural habitats often are morphologically similar (Williams, 1972, 1983). One character that often differs among species occupying different habitats is relative length of the hindlimb (Moermond, 1979; Williams, 1983; Pounds, 1988; Losos, 1990a), which is related to running ability (Losos, 1990c; Losos and Irschick, 1996). Analysis of principal components of several variables of the structural habitat indicates that the second axis reflects a trade-off between low, broad supports (high values on PC II) versus high, narrow supports (low values on PC II). One might hypothesize that this variation in habitat use is associated with variation in length of limbs; lizards having high values on PC II might have evolved long legs to run quickly on the ground in pursuit of prey and to confront conspecifics, whereas lizards having low scores on PC II have shorter limbs to enhance locomotion on narrower supports. The following hypothesis of adaptation is suggested: species of *Anolis* have adapted to changes in structural habitat by modifying the relative length of the hindlimb to maximize capability for effective locomotion.

Figure 9 presents the phylogeny used in this analysis and indicates 13 independent contrasts. Given the lack of knowledge concerning modes of character evolution, the most conservative approach is to try several maximally distinctive models to examine whether different assumptions qualitatively alter the results of the analysis. Consequently, each analysis was run twice, once assuming a model of gradual evolution using the branch lengths indicated in the figure, which were derived from molecular systematic studies, and once assuming that change occurred only at speciation events; hence lengths of all branches were set equal [see Garland et al. (1992) for further details on use of independent contrasts].

We first tested the hypothesis that structural habitat is an important selective regime affecting locomotion. If this hypothesis is correct, then locomotor behavior should have evolved concordantly with structural habitat. In support of this hypothesis, evolutionary increases in the use of low habitats were associated with evolutionary increases in the frequency of running and decreases in the frequency of walking (i.e., contrasts that were large and positive for PC II were correlated with large positive contrasts for change in running frequency and large negative contrasts for change in walking frequency;

TABLE 2

Data for Anolis Species in Independent Contrasts Example[1]

	Snout-vent length (mm)	Hindlimb length (mm)	Sprint speed (m/s)	Run frequency (% all moves)	Walk frequency (% all moves)	PC II score
A. *cristatellus*	4.156	3.985	-2.154	0.869	0.548	0.783
A. *cuvieri*	4.844	4.559	-2.172	---	---	-1.676
A. *evermanni*	4.132	3.869	-1.988	0.600	0.821	-0.118
A. *gundlachi*	4.176	4.054	-2.154	0.714	0.621	1.124
A. *krugi*	3.884	3.660	-1.966	0.533	0.781	0.456
A. *poncensis*	3.781	3.440	-1.952	0.660	0.720	0.471
A. *pulchellus*	3.774	3.482	-1.917	0.452	0.896	-0.170
A. *stratulus*	3.794	3.489	-1.784	0.838	0.631	-1.517
A. *garmani*	4.681	4.367	-2.293	0.739	0.736	-0.381
A. *grahami*	4.112	3.816	-2.087	0.778	0.596	-0.707
A. *lineatopus*	4.035	3.830	-2.079	0.861	0.402	0.730
A. *opalinus*	3.867	3.528	-1.981	0.987	0.461	0.424
A. *sagrei*	3.892	3.589	-1.988	0.844	0.445	1.707
A. *valencienni*	4.286	3.656	-1.897	0.420	1.012	-1.125

[1] Data from Losos (1990a); no locomotor behavioral data are available for *cuvieri*. Morphological and sprint data are ln transformed; locomotor behavioral data are arcsine square-root transformed.

Fig. 10a; Table 3). Hence, the assumption that this aspect of structural habitat may be an important selective regime for locomotion appears reasonable.

We then examined whether the evolutionary changes of morphology, locomotor capabilities, and habitat are associated. We removed the effect of body size on hindlimb length and sprinting capability, but not structural habitat (see Losos, 1990a), by calculating residuals of each variable regressed upon snout-vent length, a common proxy for body size. These residuals were calculated using the contrasts for each variable. With the effect of size removed, evolution in limb length is positively associated with evolutionary change in sprinting capabilities (Fig. 10b; Table 3). In turn, an evolutionary change in structural habitat is associated with a change in sprinting capability (Fig. 10c; Table 3). Hence, we suggest that relative hindlimb length in these lizards has evolved as an adaptation to changes in structural habitat. Further tests of this hypothesis would include a more detailed examination of anoline behavior and use of habitats, and the effect of limb length on survival and reproduction in natural populations.

V. Criticisms of Phylogenetic Approaches to the Study of Adaptation

Adaptation and phylogenetic analyses are inherently controversial subjects; inevitably, criticisms will be made regarding the phylogenetic methodology for studying adaptation advocated here. Two critiques of these methods have

TABLE 3
Results of Independent Contrasts Analyses[1]

	Gradual model		Speciational model	
	r^2	P	r^2	P
Run frequency vs. habitat use (PC II)	0.28	0.062	0.27	0.070
Walk frequency vs. habitat use	0.43	0.016	0.45	0.013
Relative sprint speed vs. relative hindlimb length	0.66	0.001	0.68	0.001
Habitat use vs. relative hindlimb length	0.50	0.005	0.56	0.002

[1]All analyses are regressions through the origin. Successful standardization
of contrasts was verified prior to analyses. See Garland et al. (1992) for details.

appeared (Leroi et al., 1994; Frumhoff and Reeve, 1994). Here we discuss these
critiques and conclude that their arguments do not provide a compelling
reason to abandon phylogenetic methods for the study of adaptation.

Leroi et al. (1994) acknowledge that the phylogenetic methodology that we
describe will identify adaptations accurately in some instances, such as the
repeated evolution of "crown of thorns" bristles in flea lineages whose hosts live
in hazardous habitats (Traub, 1980) or fringes on the toes of phylogenetically
diverse lizards that utilize sandy substrates (Luke, 1986). Nonetheless, Leroi et

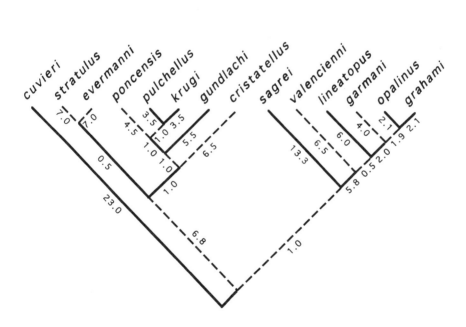

Figure 9 Phylogeny of *Anolis* lizards used in the example of independent contrasts. Thirteen
contrasts are indicated by the black and hatched lines. Each contrast represents the two branches
diverging from an ancestral node. Numbers represent the lengths of branches, which are not drawn
to scale. Sources of phylogenetic information are discussed in Losos (1990a, 1992).

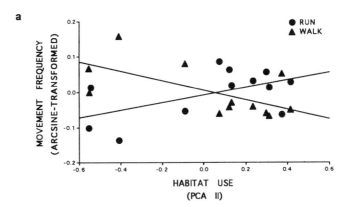

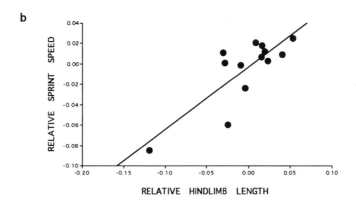

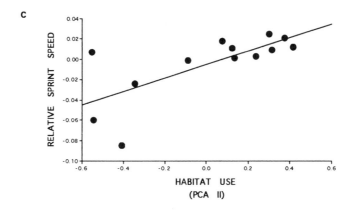

al. (1994) consider such cases exceptions, arguing (p. 383) that "phylogenetic patterns will often suggest that a trait is an adaptation when in fact it is not and suggest that it is not, when in fact it is." This conclusion stems from three major sources: consideration of the genetics of character evolution, a critique of the concept of the selective regime, and a reliance on inductive methods of scientific inquiry. We discuss these criticisms together with criticisms concerning the use of parsimony (Frumhoff and Reeve, 1994).

A. Genetics of Character Evolution: Microevolutionary and Macroevolutionary Perspectives

The critique of Leroi et al. (1994) contains an extended argument that genetic information is critical for evaluating the role of adaptive explanations in character evolution. In essence, Leroi et al. (1994) view the phylogenetic study of adaptation as if its goal were to identify the genetic details of the response to natural selection occurring in a single population in the evolutionary past. This is a misrepresentation of the goals of macroevolutionary studies of adaptation. Although we agree that genetic information is important for understanding the responses to selection occurring within a population on a microevolutionary time scale, we do not agree that lack of such information compromises phylogenetic tests of the role of adaptation occurring on a macroevolutionary time scale. In a microevolutionary study, the evolutionary response of a population to selection on character variation depends critically on the specific genetic system or background underlying this variation. In contrast, for a macroevolutionary study, we begin with the observation that a character has evolved and ask whether natural selection can explain its origin.

Leroi et al. (1994) note that a character may evolve for reasons other than natural selection, such as pleiotropy or genetic correlation. The purpose of the methodology outlined in this chapter is to identify and test predictions that are specific to a hypothesis of character evolution by natural selection, but would not be made by alternative hypotheses. In their criticisms of phylogenetic approaches, Leroi et al. (1994) overlook the emphasis on functional studies (e.g., Greene, 1986a; Coddington, 1990; Baum and Larson, 1991) which provide a means of testing predictions specific to the hypothesis of natural selection. Other potential explanations, including developmental or genetic correlations of the character being studied with other selected characters, do not predict an increase in utility or a specific biological role for the character being studied.

Leroi et al. (1994) proposed an alternative method by which phylogenetic information could be used to study character evolution. They suggest (p. 397)

Figure 10 Analysis of locomotor adaptations in *Anolis*. (a) The relationship between habitat use and frequencies of walking and running. Habitat use is represented by the second axis in an analysis of principle components (Losos, unpublished). Each point represents one contrast in a phylogenetic analysis of independent contrasts using the gradual model from Table 3. Regression lines are for heuristic purposes; analyses using contrasts require regression through the origin (Garland et al., 1992). (b) The relationship between relative limb length and relative sprint speed. Variables are residuals calculated from regressions against snout-vent length. Residuals are calculated using values of contrasts. (c) Relationship between relative sprint speed and habitat use.

that "patterns of selection, genetic variation, covariation, and interaction might be measured for several species belonging to a single monophyletic group" and this information could then be optimized onto a phylogenetic tree to infer historical events. They illustrate this method as a means of testing whether a genetic correlation was responsible for correlated evolution of two traits. Certainly, there can be long-term patterns of correlation among characters and such correlations will be of interest in evaluating hypotheses of character evolution. Epigenetic or developmental factors producing correlation of characters can be stable on a macroevolutionary time scale and may be manifested in patterns of pleiotropy and epistasis observed within populations at any given period of evolutionary history. Rather than an alternative to our approach, this methodology is an appropriate extension of it because it tests phylogenetic predictions specific to other processes that could be responsible for character evolution. Hypotheses of developmental constraint, for example, can be tested phylogenetically using a protocol comparable to the one that we describe for studying adaptation; this approach, however, is in no manner a replacement for the study of natural selection.

B. Critique of the Use of Selective Regime

Leroi et al. (1994) criticized the concept of selective regime as being defined so broadly as to be useless. As with any methodology, the use of selective regimes can be abused. The concept is meaningful only if one can identify with precision important components of organismal/environmental interaction that may influence the action of selection on character variation. Oddly, Leroi et al. (1994) accept the selective regime postulated by Luke (1986), which they state to be "type of substrate," but are critical of the use of "scansorial/arboreal vs. terrestrial" as different selective regimes in plethodontid salamanders (Baum and Larson, 1991).

In both cases, as well as those discussed in this chapter, it is important that the selective regime be described in sufficient detail to predict whether selection would favor individuals with some capabilities over individuals with others. Certainly, the hypothesis that an attribute increases locomotor performance on sand, as fringes do (Carothers, 1986), would evolve in taxa living on sandy substrates is reasonable and worth testing, as is the hypothesis stated earlier that a feature that enhances the ability to hold onto a rough surface would evolve in arboreal and scansorial salamanders. Thus, the concept of the selective regime has utility when properly employed.

In this vein, the criticisms of Leroi et al. (1994) regarding the analysis of arboreal vs. terrestrial selective regimes in plethodonid salamanders represent a superficial reading and misrepresentation of Baum and Larson (1991). Contrary to the statements of Leroi et al. (1994), Baum and Larson provide an extensive discussion of the relationships among morphology, function, and arboreality in salamanders, as evidenced by the comparison of several different plethodontid lineages that occupy different arboreal selective regimes. Leroi et al. (1994) also misrepresent Baum and Larson (1991) by stating that the evolution of derived attributes studied as adaptations in *Aneides* did not occur in other salamanders, when, in fact, several similar instances of such evolution were explicitly discussed (pp. 14 and 15), allowing a test of a general hypothesis of adaptation.

C. Use of Parsimony for Reconstructing the Evolution of Characters and Selective Regimes

Frumhoff and Reeve (1994) identify an important assumption of phylogenetic approaches to the study of adaptation, namely, that the evolution of a character can be reconstructed accurately. If reconstructions of character changes are incorrect, mistaken conclusions may follow. Frumhoff and Reeve (1994, p. 173) identify assumptions implicit in parsimony: "Character state optimization will accurately reveal the timing and direction of historical transitions between character states only if the rate of character change within lineages is low relative to the rate of ... cladogenesis. If, in particular, transitions between character states are sufficiently frequent within a species lineage, the phylogenetic 'memory' of these traits will decay such that their mapping onto a cladogram may bear little relation to the actual sequence of character state transitions between ancestral and derived taxa...." We also agree with Frumhoff and Reeve (1994, p. 174) that "parsimonious inferences of the ancestral states of a given character may often not be robust and should be interpreted with caution." Schultz et al. (1996) provide a thorough reanalysis of this argument, however, and show that the error rates for reconstruction of ancestral characters are considerably lower than the estimates of Frumhoff and Reeve (1994).

One must be aware of parsimony's underlying assumptions and take care to assess whether they are likely to be violated. For example, characters that differ greatly among even closely related species may experience rates of evolution too great to be studied using older phylogenetic divergences. As discussed earlier, phylogenetic autocorrelation may be an appropriate means of identifying such situations. Parallel evolution in closely related taxa will be more difficult to detect. However, if it occurs as a result of parallel evolution in selective regimes, as Frumhoff and Reeves speculate, then, in fact, the procedure will be conservative by identifying one instance of evolution of a trait in a particular selective regime when, in fact, multiple instances occurred.

Frumhoff and Reeve (1994) also doubt that the evolution of selective regimes will be conserved enough to permit accurate reconstruction via parsimony. As with character evolution, some selective regimes may evolve too rapidly for these methods to be useful. Nonetheless, the similarity in environmental circumstances that often characterize large, old clades (e.g., the arboreal habitats of *Anolis,* the aquatic lifestyle of whales) indicates that selective regimes often evolve slowly enough to be useful in phylogenetic studies of adaptation.

Any phylogenetic study of adaptation must examine characters and selective regimes at a phylogenetic level for which evolutionary changes are observed, but not extensively superimposed on single lineages. The problems associated with the uses of parsimony discussed above are diminished in importance if the most appropriate phylogenetic level is established; closely related lineages are used to investigate rapidly evolving characters and more distantly related lineages are compared to study more slowly evolving characters. Undoubtedly, certain characters and taxa will be more amenable than others to studies of adaptation because of historical details regarding rates of evolution, branching of lineages, and extinction. Although these factors are beyond the control of the investigator, choice of taxa and characters amenable to thorough phylogenetic analysis can ensure that studies of adaptation can be productive. As illustrated by our examples from anoles, characters of interest to evolutionary ecologists often are amenable to phylogenetic analysis. If the criticisms of Frumhoff and Reeve (1994) are taken to their logical extreme, the concepts of homology and phylogeny would be meaningless and functional properties of characters would not be generalizable beyond the species level. This conclusion is equivalent to discrediting the entire field of phylogenetic systematics, as well as comparative anatomy, paleontology, and, in fact, all of comparative biology.

D. Hypothetico-deductive vs Inductive Approaches to Scientific Inquiry

Criticisms of phylogenetic approaches to studying adaptation invariably emphasize the inability to demonstrate in a positive sense the occurrence of natural selection in the evolutionary past. Expectation of positive demonstration or proof reflects an inductive approach to science that we consider inappropriate, not only to the study of historical processes, but to studies of any natural phenomenon. Rather than attempting to show by induction that natural selection has acted in a particular population, we derive predictions from the hypothesis that natural selection has acted and seek empirical data that potentially would falsify that hypothesis. We argue above that the historical hypothesis of adaptive origin of a character makes specific, testable predictions that are not made by alternative hypotheses explaining the evolutionary origin of a character. These hypotheses, even for explaining unique evolutionary events, make numerous predictions whose testing requires studies of morphology, function, development, phylogeny, genetics, ecology, and behavior. Empirical results in all of these fields provide potential falsifiers of hypotheses of adaptation and sources for refining those hypotheses.

Associated with the inductive arguments used to criticize phylogenetic studies of adaptation is the statement that results of the proposed phylogenetic study of adaptation may be in error. There is no methodology in any aspect of evolutionary biology (or any science) that is guaranteed to produce a completely correct result in a single study. The usefulness of a general methodology, such as the one presented here, is to provide a basis for repeated empirical testing of hypotheses. The methodology, if applied in an iterative manner, should lead to correction of errors (Baum and Larson, 1991; Coddington, 1994). The diverse and specific predictions of hypotheses of adaptation ensure repeated testing of predictions and potential falsification of the hypotheses. Hypotheses of adaptation, or of any of the alternative explanations for the evolutionary origin of a character, are viewed from the deductive framework as working hypotheses whose successive testing leads to improvement of those hypotheses. By correcting errors, application of the deductive method approximates the truth, although we do not reach a point in the investigation at which absence of error is assured.

The testing of historical hypotheses in evolutionary biology does not require a compromise of the principles of hypothetico-deductive science. [See Mayr (1982) for further discussion of this issue.] We reject in principle the inductive arguments used to discredit the testing of historical hypotheses of adaptation. Further criticism and improvement of the testing of historical hypotheses of adaptation should be made using an explicitly deductive framework.

Acknowledgments

We acknowledge David Baum, Thore Bergman, James Cheverud, Doug Creer, Kevin de Queiroz, Harry Greene, Tomas Hrbek, Duncan Irschick, Todd Jackman, Susan Jacobs, George Lauder, David Pepin, Robert Robertson, and Mike Veith for assistance and helpful comments on issues discussed here. Support was provided by NSF grants DEB-9318642 and BSR 9106898.

References

Arnold, E. N. (1994). Investigating the origins of performance advantage: Adaptation, exaptation and lineage effects. *In* "Phylogenetics and Ecology" (P. Eggleton and R. Vane-Wright, eds.), pp. 123-168. Academic Press, London.

Baum, D. A., and Larson, A. (1991). Adaptation reviewed: A phylogenetic methodology for studying character macroevolution. *Syst. Zool.* 40, 1-18.

Bock, W. J. (1979). The synthetic explanation of macroevolutionary change - A reductionist approach. *Bull. Carnegie Mus. Nat. Hist.* 13, 20-69.

Bock, W. J., and von Wahlert, G. (1965). Adaptation and the form-function complex. *Evolution* 19, 269-299.

Burt, A. (1989). Comparative methods using phylogenetically independent contrasts. *Oxf. Surv. Evol. Biol.* 8, 33-53.

Cannatella, D. C., and de Queiroz, K. (1989). Phylogenetic systematics of anoles: Is a new taxonomy warranted? *Syst. Zool.* 38, 57-69.

Carothers, J. H. (1986). An experimental confirmation of morphological adaptation: toe fringes in the sand-dwelling lizard *Uma scoparia. Evolution* 40, 871-874.

Carpenter, C. C. (1986). An inventory of the display-action-patterns in lizards. *Smith. Herpetol. Info. Serv.* 68, 1-18.

Cartmill, M. (1985). Climbing. *In* "Functional Vertebrate Morphology" (M. Hildebrand, D. M. Bramble, K. F. Liem, and D. B. Wake, eds.), pp. 73-88. Harvard University Press, Cambridge.

Cheverud, J.M., Dow, M.M., and Leutenegger, W. (1985). The quantitative assessment of phylogenetic constraints in comparative analyses: Sexual dimorphism in body weight among primates. *Evolution* 39, 1335-1351.

Coddington, J. A. (1988). Cladistic tests of adaptational hypotheses. *Cladistics* 4, 3-22.

Coddington, J. A. (1990). Bridges between evolutionary pattern and process. *Cladistics* 6, 379-386.

Coddington, J. A. (1994). The roles of homology and convergence in studies of adaptation. *In* "Phylogenetics and Ecology" (P. Eggleton and R.I. Vane-Wright, eds.), pp. 53-78. Academic Press, London.

Cracraft, J. (1989). Speciation and its ontology: The empirical consequences of alternative species concepts for understanding patterns and processes of differentiation. *In* "Speciation and Its Consequences" (D. Otte and J. A. Endler, eds.), pp. 28-59. Sinauer Assoc., Sunderland, MA.

Cracraft, J. (1990). The origin of evolutionary novelties: Pattern and process at different hierarchical levels. *In* "Evolutionary Innovations" (M. H. Nitecki, ed.), pp. 21-44. University of Chicago Press, Chicago.

de Pinna, M. C. C. (1991). Concepts and tests of homology in the cladistic paradigm. *Cladistics* 7, 367-394.

de Queiroz, K. (1985). The ontogenetic method for determining character polarity and its relevance to phylogenetic systematics. *Syst. Zool.* 34, 280-299.

de Queiroz, K. (1989). Morphological and Biochemical Evolution in the Sand Lizards. Ph.D. Dissertation, Univ. California, Berkeley.

Emerson, S. B., and Koehl, M. A. R. (1990). The interaction of behavioral and morphological change in the evolution of a novel locomotor type: "flying" frogs. *Evolution* 44, 1931-1946.

Estes, R., de Queiroz, K., and Gauthier, J. (1988). Phylogenetic relationships within Squamata. *In* "Phylogenetic Relationships of the Lizard Families: Essays Commemorating Charles L. Camp" (R. Estes and G. Pregill, eds.), pp. 119-281. Stanford Univ. Press, Stanford, CA.

Falconer, D. S. (1981). "Introduction to Quantitative Genetics." 2nd Ed. Longman Group Ltd., London.

Felsenstein, J. (1985). Phylogenies and the comparative method. *Amer. Nat.* 125, 1-15.

Fisher, D. C. (1985). Evolutionary morphology: Beyond the analogous, the anecdotal, and the ad hoc. *Paleobiology* 11, 120-138.

Frost, D. R., and Etheridge, R. (1989). A phylogenetic analysis and taxonomy of iguanian lizards (Reptilia: Squamata). *Univ. Kans. Mus. Nat. Hist. Misc. Publ.* 81, 1-65.

Frumhoff, P. C., and Reeve, H. K. (1994). Using phylogenies to test hypotheses of adaptation: A critique of some current proposals. *Evolution* 48, 172-180.

Futuyma, D. J. (1986). "Evolutionary Biology." 2nd Edition. Sinauer Assoc., Sunderland, MA.

Garland, T., Jr., Harvey, P.H., and Ives, A.R. (1992). Procedures for the analysis of comparative data using phylogenetically independent contrasts. *Syst. Biol.* 41, 18-32.

Gittleman, J.L., and Kot, M. (1990). Adaptation: Statistics and a null model for estimating phylogenetic effects. *Syst. Zool.* 39, 227-241.

Gittleman, J.L., and Luh, H.-K. (1992). On comparing comparative methods. *Annu. Rev. Ecol. Syst.* 23, 383-404.

Gittleman, J.L., and Luh, H.-K. (1994). Phylogeny, evolutionary models, and comparative methods: A simulation study. *In* "Phylogenetics and Ecology" (P. Eggleton and R.I. Vane-Wright, eds.), pp. 103-122. Academic Press, London.

Gould, S. J., and Lewontin, R. (1979). The spandrels of San Marcos and the panglossian paradigm: A critique of the adaptationist programme. *Proc. Royal Soc. London B* 205, 581-598.

Gould, S. J., and Vrba, E. S. (1982). Exaptation–a missing term in the science of form. *Paleobiology* 8, 4-15.

Greene, H. W. (1986a). Diet and arboreality in the emerald monitor, *Varanus prasinus*, with comments on the study of adaptation. *Fieldiana Zool. New Ser.* 31, 1-12.

Greene, H. W. (1986b). Natural history and evolutionary biology. *In* "Predator Prey Relationships" (M. E. Feder and G. V. Lauder, eds.), pp. 99-108. Univ. of Chicago Press, Chicago.

Greene, H. W. (1994). Homology and behavioral repertoires. *In* "Homology: The Hierarchical Basis of Comparative Biology" (B. K. Hall, ed.), pp. 369-391. Academic Press, San Diego.

Hall, B. K. (ed.) (1994). "Homology: The Hierarchical Basis of Comparative Biology." Academic Press, San Diego.

Harvey, P. H., and Pagel, M. D. (1991). "The Comparative Method in Evolutionary Biology." Oxford Univ. Press, Oxford.

Harvey, P. H., and Purvis, A. (1991). Comparative methods for explaining adaptations. *Nature* 351, 619-624.

Hass, C. A., Hedges, S. B., and Maxson, L. R. (1993). Molecular insights into the relationships and biogeography of West Indian anoline lizards. *Biochem. Syst. Ecol.* 21, 97-114.

Hildebrand, M. (1988). "Analysis of Vertebrate Structure." 3rd Ed. Wiley and Sons, New York.

Irschick, D. J., Austin, C. C., Petren, K., Fisher, R. N., Losos, J. B., and Ellers, O. (1996). A comparative analysis of clinging ability among pad-bearing lizards. *Biol. J. Linn. Soc.* (in press).

Jenssen, T. A. (1977). Evolution of anoline lizard display behavior. *Amer. Zool.* 17, 203-215.

Kaplan, D. R. (1984). The concept of homology and its central role in the elucidation of plant systematic relationships. *In* "Cladistics: Perspectives on the Reconstruction of Evolutionary History" (T. Duncan and T. F. Stuessy, eds.), pp. 51-70. Columbia Univ. Press, New York.

Kettlewell, H. B. D. (1973). "The Evolution of Melanism." Clarendon, Oxford.

Larson, A. (1994). The comparison of morphological and molecular data in phylogenetic systematics. *In* "Molecular Ecology and Evolution: Approaches and Applications" (B.

Schierwater, B. Streit, G. P. Wagner, and R. DeSalle, eds.), pp. 371-390. Birkhäuser Verlag, Basel.

Lauder, G. V., Leroi, A. M., and Rose, M. R. (1993). Adaptations and history. *Trends Ecol. Evol.* 8, 294-297.

Leroi, A. M., Rose, M. R., and Lauder, G. V. (1994). What does the comparative method reveal about adaptation? *Amer. Nat.* 143, 381-402.

Lewontin, R. (1977). Adattamento. *In* "Enciclopedia Einaudi, Vol 1" (G. Einaudi, ed.), Turin, Italy. Reprinted in English in R. Levins and R. Lewontin. 1985. "The Dialectical Biologist" Harvard Univ. Press, Cambridge, Massachusetts.

Losos, J. B. (1990a). Ecomorphology, performance capability, and scaling of West Indian *Anolis* lizards: An evolutionary analysis. *Ecol. Monogr.* 60, 369-388.

Losos, J. B. (1990b). The evolution of form and function: Morphology and locomotor performance in West Indian *Anolis* lizards. *Evolution* 44, 1189-1203.

Losos, J. B. (1990c). A phylogenetic analysis of character displacement in Caribbean *Anolis* lizards. *Evolution* 44, 558-569.

Losos, J. B. (1992). The evolution of convergent structure in Caribbean Anolis communities. *Syst. Biol.* 41, 403-420.

Losos, J. B., and Irschick, D. J. (1996). The effect of perch diameter on escape behaviour of Anolis lizards: Laboratory predictions and field tests. *Anim. Behav.* 51,593-602.

Losos, J. B., and Miles, D. B. (1994). Adaptation, constraint, and the comparative method: Phylogenetic issues and methods. In "Ecological Morphology: Integrative Organismal Biology" (P. C. Wainwright and S. M. Reilly, eds.), pp. 60-98. Univ. Chicago Press, Chicago.

Luke, C. (1986). Convergent evolution of lizard toe fringes. *Biol. J. Linnean Soc.* 27, 1-16.

Lynch, M. (1991). Methods for the analysis of comparative data in evolutionary biology. *Evolution* 45, 1065-1080.

Maddison, W. P. (1990). A method for testing the correlated evolution of two binary characters: Are gains or losses concentrated on certain branches of a phylogenetic tree? *Evolution* 44, 539-557.

Maddison, W. P., and Maddison, D. R. (1992). "MacClade, Version 3." Sinauer Publishers, Sunderland, MA.

Martins, E. P. (1996). Conducting phylogenetic comparative studies when the phylogeny is not known. *Evolution* 50, 12-22.

Mayr, E. (1982). "The Growth of Biological Thought." Belknap Press, Cambridge, MA.

McLennan, D. A. and Brooks, D. R. (1993). The phylogenetic component of cooperative breeding in perching birds: A commentary. *Amer. Nat.* 141, 790-795.

Miles, D. B., and Dunham, A. E. (1993). Historical perspectives in ecology and evolution. *Annu. Rev. Ecol. Syst.* 24, 587-619.

Moermond, T. (1979). Habitat constraints on the behavior, morphology, and community structure of *Anolis* lizards. *Ecology* 60, 152-164.

Pagel, M. (1994a). The adaptationist wager. *In* "Phylogenetics and Ecology" (P. Eggleton and R. Vane-Wright, eds.), pp. 29-52. Academic Press, London.

Pagel, M. (1994b). Detecting correlated evolution on phylogenies: A general method for the comparative analysis of discrete characters. *Proc. Roy. Soc. London B* 255, 37-45.

Patterson, C. (1982). Morphological characters and homology. *In* "Problems of Phylogenetic Reconstruction" (K. A. Joysey and A. E. Friday, eds.), pp. 21-74. Academic Press, New York.

Peterson, J. A. (1983). The evolution of the subdigital pad in *Anolis*. I. Comparisons among the anoline genera. *In* "Advances in Herpetology and Evolutionary Biology: Essays in Honor of Ernest E. Williams" (A.G.J. Rhodin and K. Miyata, eds.), pp. 245-283. Museum of Comparative Zoology, Harvard University, Cambridge, MA.

Peterson, J. A., and Williams, E. E. (1981). A case study in retrograde evolution: The *onca* lineage in anoline lizards. II. Subdigital fine structure. *Bull. Mus. Comp. Zool.* 149, 215-268.

Pounds, J. A. (1988). Ecomorphology, locomotion, and microhabitat structure: Patterns in a tropical mainland *Anolis* community. *Ecol. Monogr.* 58, 299-320.

Purvis, A., Gittleman, J. L., and Luh, H.-K. (1994). Truth or consequences: Effects of phylogenetic accuracy on two comparative methods. *J. Theor. Biol.* 167, 293-300.

Roth, V. L. (1991). Homology and hierarchies: Problems solved and unresolved. *J. Evol. Biol.* 4, 167-194.

Rudwick, M. J. S. (1964). The inference of function from structure in fossils. *Br. J. Philos. Sci.* 15,

27-40.

Ruibal, R., and V. Ernst. (1965). The structure of the digital setae of lizards. *J. Morphol.* 117, 271-294.

Schluter, D. (1994). Experimental evidence that competition promotes divergence in adaptive radiation. *Science* 266, 798-801.

Schoener, T. W. (1970). Size patterns in West Indian *Anolis* lizards. II. Correlations with the sizes of particular sympatric species - displacement and convergence. *Amer. Nat.* 104, 155-174.

Schultz, T. R., Cocroft, R. B., and Churchill, G. A. (1996). The reconstruction of ancestral character states. *Evolution* 50, 504-511.

Sillén-Tullberg, B. (1993). The effect of biased inclusion of taxa on the correlation between discrete characters in phylogenetic trees. *Evolution* 47, 1182-1191.

Swofford, D. L. and Maddison, W. P. (1987). Reconstructing ancestral character states under Wagner parsimony. *Math. Biosci.* 87, 199-229.

Swofford, D. L., Olson, G. J., Waddell, P. J., and Hillis, D. M. (1996). Phylogenetic inference. *In* "Molecular Systematics" second edition (D. M. Hillis, C. Moritz, and B. K. Mable, eds.), pp. 407-514. Sinauer Associates, Sunderland, MA.

Traub, R. (1980). Some adaptive modifications in fleas. *In* "Proceedings of the International Conference on Fleas" Ashton Wold, U.K, June, 1977. (R. Traub and H. Starcke, eds.), pp. 33-67, Balkema, Rotterdam.

Vrba, E. S., and Gould, S. J. (1986). The hierarchical expansion of sorting and selection: Sorting and selection cannot be equated. *Paleobiology* 12, 217-228.

Wagner, G. P. (1989). The origin of morphological characters and the biological basis of homology. *Evolution* 43, 1157-1171.

Wake, D. B. (1966). Comparative osteology and evolution of the lungless salamanders, family Plethodontidae. *Mem. So. California Acad. Sci.* 4, 1-111.

Wake, D. B. (1991). Homoplasy: The result of natural selection, or evidence of design limitations? *Amer. Nat.* 138, 543-567.

Wenzel, J. W., and Carpenter, J. M. (1994). Comparing methods: adaptive traits and tests of adaptation. *In* "Phylogenetics and Ecology" (P. Eggleton and R. Vane-Wright, eds.), pp. 79-101. Academic Press, London.

Westneat, M. W. (1995). Phylogenetic systematics and biomechanics in ecomorphology. *Environ. Biol. Fishes* 44, 263-283.

Williams, E. E. (1972). The origin of faunas: evolution of lizard congeners in a complex island fauna: a trial analysis. *Evol. Biol.* 6, 47-89.

Williams, E. E. (1983). Ecomorphs, faunas, island size, and diverse end points in island radiations of *Anolis. In* "Lizard Ecology: Studies of a Model Organism" (R. B Huey, E. R. Pianka, and T. W. Schoe ner, eds.), pp. 326-370. Harvard Univ. Press, Cambridge, MA.

Williams, E. E., and Peterson, J. A. (1982). Convergent and alternative designs in the digital adhesive pads of scincid lizards. *Science* 215, 1509-1511.

CHAPTER 7

Laboratory Evolution: The Experimental Wonderland and the Cheshire Cat Syndrome

MICHAEL R. ROSE

THEODORE J. NUSBAUM

ADAM K. CHIPPINDALE

I. Introduction: Victorian Schoolgirls Revisited

On an earlier occasion, we proposed that evolutionary theories "are like sheltered, upper-class, Victorian schoolgirls, dressed up entirely in white and severely scrubbed" (Rose et al., 1987, p. 95). This simile arose from our view that, while theories may be pretty to look at, they may not be that robust after all when confronted with the dangers of the real, empirical world. Indeed, we went further and proposed that the confrontation of theories with data could be likened to taking these schoolgirls on a trip through a jungle. There is a problem, however, with this situation. Many schoolgirls will die for reasons that are more accidental than a test of their true fitness. It will just happen that Prudence will be bitten by a venomous snake, while Cicely fortuitously escaped death by running ahead to talk with Abigail. To test the schoolgirls more appropriately, an environment combining danger with some type of order is preferable.

Our simile can be continued at this point by recalling *Alice's Adventures in Wonderland* by Lewis Carroll. In this charming tale, a classic Victorian schoolgirl has a fantastical adventure in which she is stringently tested, but also somehow free from mortal danger. For us, the experimental realm of laboratory evolution provides the comparable Wonderland to that of Alice. It is possible to test theories with greater fairness in the laboratory, in that the number of confounded, distracting, but unknown factors is reduced. On the other hand, the laboratory introduces artifactual problems that are every bit as perverse as the fantastical creatures that Alice met in Wonderland. Like Alice, the experimenter may become entirely confused as to what is going on.

As scientists, the Wonderland of laboratory evolution is our world of choice. Indeed, for the detailed study of adaptation as a process, which is to say the response to natural selection, there may be no empirical approach more powerful than laboratory evolution. Only with laboratory evolution experiments is it possible to perform experiments in which evolving populations are extensively replicated. This allows great statistical power of inference in a field, evolutionary biology, in which the population is the true unit of observation. Only with laboratory evolution is it possible to start evolutionary trajectories again and again, deliberately varying the conditions or ancestry of the evolving populations. This makes possible the repeated testing of general evolutionary hypotheses, especially those concerning the effects of specific selection regimes. Only with laboratory evolution can the products of evolution be kept for repeated comparison under standardized and appropriate conditions, allowing a deeper and deeper characterization of both adaptation as a process and the products of adaptation. That is, laboratory evolution is an outstanding system for investigating the operation of selection, adaptation as a process, and the end products of selection, adaptation as a product. In our discussion, both of these types of adaptation will be considered; what they have in common is the action of natural selection. Evidently, therefore, our working definition of adaptation is couched in terms of natural selection. Beneficial structures that have not been shaped by natural selection would not, on this interpretation, be considered adaptations.

In what follows, we will try to provide a brief guidebook to the Wonderland of laboratory evolution, sketching a map of the terrain, describing some instructive examples of research in this area, and discussing the paradoxical beasts that afflict the unwary traveler. So, "Down the rabbithole!"

II. Topography of the Wonderland

A. Definition of Laboratory Evolution

"What do you mean by that?" said the Caterpillar sternly. "Explain yourself!"

By "laboratory evolution" we mean experiments in which populations are cultured in the laboratory so that selection proceeds under defined, reproducible conditions, but without the direct choice of reproducing individuals by the experimenter. This would subsume experiments that have been described as "laboratory natural selection" and experiments that have been called "laboratory culling," such as those of Crow (1954) and Rose et al. (1990, 1992), which involve killing large numbers of flies in selected populations using a lethal stress. In these designs, the individual phenotypes are not measured for selection and neither involves the explicit selection of reproductive individuals by the experimenter. An important feature of laboratory evolution as an experimental strategy is that the experimenter does not endeavor to control, or forestall, all forms of natural selection other than those chosen for manipulation.

This definition of laboratory evolution explicitly excludes the "artificial selection" of quantitative genetics (e.g., Pollak et al., 1977; Falconer, 1981; Weir et al., 1988; Hill and Mackay, 1989). In artificial selection, the phenotypes undergoing selection are measured by the experimenter, and the reproducing adults that produce the next generation are explicitly chosen. At least ideally, experimenters using artificial selection strive to prevent any form of natural selection from affecting the response to artificial selection.

B. Principles of Experimental Design

To take advantage of the potential power offered by laboratory evolution experiments, it is necessary to give careful consideration to issues of experimental design before initiating experiments, most importantly: (i) founding laboratory populations, (ii) population size, (iii) replication, and (iv) controls.

Founding Laboratory Populations: All laboratory populations ultimately descend from wild sources and, therefore, at some point in their cultivation, must be subject to a change in the conditions and selective forces that they encounter. Even with matched controls, when populations undergo rapid adaptation to the laboratory, the study of adaptation to specific treatments deliberately imposed by the experimenter may be compromised by interactions between lab adaptation and selection treatment adaptation. Like Alice growing larger and smaller, it may be impossible for the experimenter to understand what is going on under these conditions.

"Dear, dear! How queer everything is to-day! And yesterday things went on just as usual. I wonder if I've been changed in the night? Let me think: was I the same when I got up this morning?"

Such problems will be exacerbated when selection treatments are not carefully run in parallel with controls. If, for example, two selection treatments are handled asynchronously, as when their treatments differ in demographic features, the opportunity for adaptation to laboratory conditions will be greater in the treatment having more generations of laboratory culture. Furthermore, if common elements of the culture regimes of different treatments are not strictly standardized, laboratory adaptation may also proceed at different rates with respect to these elements of culture. Given the foregoing arguments, the ideal experimental material will be populations that have ceased to undergo rapid adaptation to general laboratory conditions.

Population Size: Natural populations tend to be large relative to laboratory populations, raising a number of important considerations for the design of experiments. In the study of adaptation in the laboratory, large population sizes are almost always desirable. If small populations are employed, the potential arises for population genetic processes other than selection, like genetic drift and inbreeding depression, to play a major role in the differentiation of populations, confounding the inference of selective differentiation.

Replication: The problem of replicate number is intrinsically statistical. In particular, the replication of experimental populations is necessary because populations tend to differentiate with respect to one another in the absence of any intervention by the experimenter. Genetic drift is one population genetic mechanism that can produce this effect, but divergent responses to selection can, in principle, as well. Given that this type of replicate differentiation is expected *a priori*, experimental designs must allow for it. Furthermore, since the basic unit of observation is the population, in most applications, each population's trajectory is a single datum. For all these reasons, laboratory evolution experiments must be replicated. The quantitative degree of replication required will depend on the system under study. In systems that have little initial genetic variation in common among the experimental populations, such as bacterial clones, the response to selection will depend on the supply of mutations. The response to selection may then depend on the particular mutations that arise in each line and these mutations may be "private" to such asexual lines. This type of experiment may require more replication than systems where all replicates exploit a large sample of the same pool of genetic variation, such as highly outbred insect populations.

The ultimate goal of studies of adaptation is to establish causal associations between selective conditions and selection response. To this end, the degree of replication required will depend on the scale of resolution desired. Here, a trade-off may naturally arise between the number of replicate populations and the number of repeat measures made from each population; the compromise reached will depend upon the scale of change of the traits under selection and the level of error associated with their measurement. Ultimately, however, the population trait-mean will be a crucial datum in the analysis, the degrees of freedom of which cannot legitimately be inflated by measuring more individuals within a population. For example, in a paired design, a *t*-test comparing five paired replicate population means from each of two treatments would have four *df.* and, in an unpaired design, 8 *df.*, irrespective of the number of individuals measured within each population. With respect to statistical power, comparisons of paired replicates between laboratory selection treatments have the following critical *t*-test values, at $p < 0.05$, for the difference between treatments: 12.7 with two replicates per treatment, 4.3 with three replicates, 3.2 with four, and so on. Therefore, it will usually be desirable to replicate selection experiments at least threefold for statistical purposes given the onerous requirements for statistical significance with only two replicates per treatment.

Controls: The perfect control may be the unattainable ideal of scientific research, but the absence of any type of control in laboratory evolution experiments will doom them to irrelevance. In the simplest laboratory adaptation experiment, the best control may be perfectly preserved specimens from the founding population. Such a control will rarely be available to the researcher working on the live properties of the organism, although some *E. coli* work has come close to solving these problems (see Lenski et al., 1991). The next best thing is controls that are run exactly in parallel with selection lines, leaving aside only the form of selection imposed by the experimenter. In general, controls may be inadvertently subject to novel types of selection rather than undergoing mere release from selection. When this occurs the experimental results may be bizarre and unpredictable; data analysis will approach the confusion of the trial presided over by the King and Queen of Hearts at the end of *Alice's Adventures in Wonderland.*

III. Studies of Adaptation Using Laboratory Evolution: Adventures in Wonderland

An experiment on laboratory evolution is almost inherently a study of adaptation because populations evolving in the laboratory are necessarily subject to some pattern of selection. Their response to the selection regime then instantiates adaptation to such conditions. However, some of this adaptation may be trivial, obscure, or irrelevant. There is some historic literature in evolutionary biology, dating back to the 1930s, that is based on laboratory evolution experiments, as reviewed by Wright (1977, Ch. 9). Many of the historic experiments used mutant stocks and other genetic contrivances that undermine much of the value of these experiments for the understanding of adaptation in normal, outbred organisms. A few experiments involved laboratory adaptation to resistance to toxic substances, such as DDT (e.g., Crow, 1954), but these were unreplicated, and in any case are of no clear relevance to adaptation in nonagricultural environments. Evidently, we cannot discuss the full range of the experimental literature on laboratory evolution. The studies that remain have been chosen for their completeness, pertinence to adapatation, and familiarity to the authors.

It should also be noted that, in characterizing any one selection regime, we do not suppose that the adaptive response to these regimes is entirely specific to the features of the selection regime that were chosen by the experimenter. Naturally, it is inevitable that diverse forms of natural selection will remain in effect, giving results that may be entirely unexpected by the experimenter. To some extent, the experimenter just has to drink from the bottle labeled "DRINK ME" and proceed.

A. *Laboratory Evolution in* Escherichia coli

The most elementary of all laboratory evolution experiments is the mere culture of populations in a laboratory environment. For most populations, excepting only those that have undergone long-term cultivation in the same laboratory environment, any particular laboratory environment will be a novel environment to some extent. Typically, this "experimental design" will be inadvertent in that organisms are frequently sampled from the wild and then haphazardly cultured in the laboratory.

An experiment of more appropriate design was performed by Lenski et al. (1991), using *Escherichia coli*. Three features of this experiment were notable: (i) Lenski et al. compared the fitness of the laboratory-evolving lines with controls obtained by thawing ancestral cells, frozen at the beginning of the experiment; (ii) Lenski et al. replicated the experiment 12-fold, handling the replicates in parallel, but without cross-contamination; and (iii) the experiment was conducted with samples every 100 generations over a total of 2000 generations. These three points exemplify general principles of good experimental design in laboratory evolution experiments. First, these experiments have the ideal control: largely unmodified derivatives of the ancestral population. Comparison with this control gives a relatively fixed point of reference for characterizing the evolutionary trajectory. Evidently, the control was a reasonable representative of the initial state of the evolving populations. Most experiments with multicellular organisms will not have

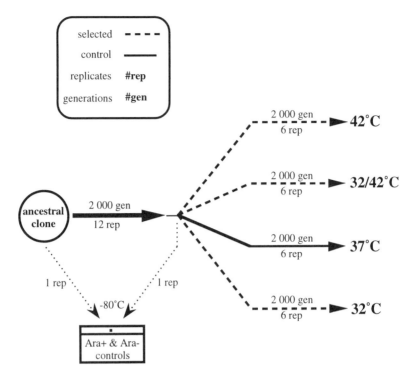

Figure I Two large selection experiments performed by Lenski, Bennett, and colleagues on *E. coli*. In the first experiment, all 12 replicate populations were exposed to selection for a standard set of laboratory conditions (Lenski et al., 1991). In the second experiment, one of the lines from the earlier, 2000 generation experiment was used to found another set of lines; these were selected for four different temperature conditions.

this feature because it will not usually be possible to freeze and recover live specimens representative of the founding generation for a laboratory evolution experiment.

Lenski et al. (1991) also extensively replicated their experiment, 12-fold. The fundamental significance of this design decision is that the laboratory evolution experiments have the trajectory of evolution for individual populations as their elementary unit of observation, as discussed above. The *E. coli* work handily illustrates this principle because in any one culture the total number of cells, and thus individuals, is in the range 10^6-10^9. But in the comparison of the 12 populations with their controls, the actual degrees of freedom are not more than 11 (there were some complications involving marker phenotypes) instead of the astronomical number of degrees of freedom that might be imagined from the number of individuals involved. If, for example, there had been only one evolving lineage undergoing comparison with the ancestor, there would have been only one degree of freedom for a statistical comparison since it is the difference between ancestor and selected population that is the raw datum, despite the trillions of cells involved.

Finally, it is important that Lenski et al. continued culturing their lines for 10,000 generations, with frequent sampling, as reported in Lenski and Travisano

(1994). This enabled them to differentiate distinct epochs in the evolutionary process. However, it should be noted that this is probably particularly important in bacterial experiments starting from small populations because these will usually lack much standing genetic variation. Instead, these populations must "wait" for the occurrence of selectable mutations before the evolutionary process can begin.

The results of this experiment were dramatic. After an initial lack of response to selection, over the first few hundred generations, the bacterial populations responded rapidly to selection. Interestingly, this response to selection decelerated over time so that the selected populations appeared, on average, to be approaching some asymptotic limit on adaptation. The total increase in fitness measured over the first 2000 generations of laboratory evolution was about 30%. Over 10,000 generations of selection, the total increase in fitness averaged about 50% (Lenski and Travisano, 1994).

Bennett et al. (1992) and Bennett and Lenski (1993) have also imposed four temperature regimes on populations of E. coli, with 6-fold replication. Interestingly, these experiments were started with the lines adapted to the laboratory for 2000 generations, described in Lenski et al. (1991), as shown in Fig. 1. These temperature-selected lines were propagated for an additional 2000 generations at their temperature level. Fitness continued to increase in these lines by up to 50%. But under temperature-specific selection it had become temperature dependent: lines from specific temperature treatments increased in fitness more under the conditions of that temperature treatment. Sometimes there were apparent "trade-offs" between adaptation to different temperature regimes, but it was not the rule. In particular, the range of the overall "thermal niche," within which the bacterial populations could maintain themselves, did not appear to change as a result of temperature-specific laboratory evolution.

B. Density-dependent Laboratory Evolution in Drosophila

A type specimen of laboratory natural selection on a sexual organism was studied by Mueller and Ayala (1981). This study is particularly exemplary in that it was directed at general theoretical questions. Mueller and Ayala were interested in testing the popular dichotomy of r and K selection, particularly propounded by MacArthur and Wilson (1967) and Pianka (1972). In this dichotomy, "r" selection refers to selection for rapid exploitation of newly available habitat (particularly via rapid population growth rate), whereas "K" refers to selection under relatively stable conditions, close to the carrying capacity of the environment. Mueller and Ayala devised two sets of threefold-replicated D. melanogaster populations; the first set of three was reproduced in uncrowded vials using discrete generations, while the second set of three was cultured in population cages at high densities with overlapping generations, as shown in Fig. 2. These were intended to represent r and K selection, respectively. The scientific question was whether or not laboratory evolution would result in increasing r under the r regime, and increasing K in the K regime. This was indeed the effect observed; in the broadest sense of the original r and K theory, the results were corroborative.

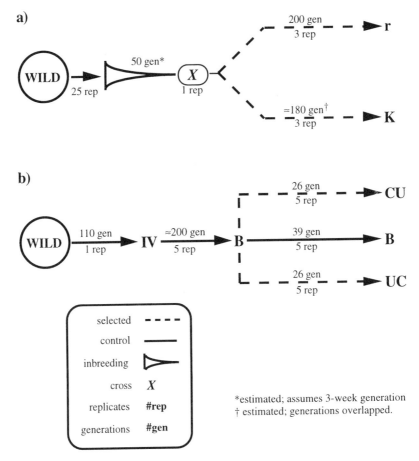

Figure 2 Density-dependent selection on *D. melanogaster* by Mueller and colleagues. (a) Selection for adaptation to "r" and "K" environments is diagrammed. The "K" environment in this experiment allowed crowding of both larvae and adults, along with overlapping generations. (b) Selection for crowded and uncrowded conditions is shown in which the condition of crowding was standardized separately for both larval and adult periods. Therefore, selection for crowding of one life stage was not confounded with the other. Thus, "UC" populations were uncrowded as larvae but crowded as adults.

But Mueller studied these populations further, with a view to examining some of the more specific ideas that have been associated with the r and K dichotomy. The interesting thing about his experimental strategy is that the existence of the r and K populations provided him with established material for testing more specific hypotheses. Mueller failed to confirm many assumptions about the expected differentiation between r and K populations, such as the notion that K individuals should be more efficient competitors (Joshi and Mueller, 1988; Mueller, 1988, 1990). Indeed, K individuals proved to be significantly less efficient.

An additional development in the work with these populations has been the realization that the r and K selection regimes extensively confound multiple differences in selection pressures, making them unsuitable for the specific resolution of the effects of density-dependent selection mechanisms. With this in mind, Mueller et al. (1993) went on to create stocks of *D. melanogaster* selected specifically at high or low densities during the larval or adult phases of life. With such stocks, Mueller was able to distinguish the specific selection mechanisms that differentiated particular characters in his original r and K stocks. This progression illustrates two different goals of work in laboratory evolution. The first is that the experimenter often seeks to test general theories using selection conditions that constitute good tests of those theories. The second goal is to determine the specific selection mechanisms that produce observed patterns of laboratory evolution. The ideal experimental designs for these two purposes may not be the same.

C. Age-dependent Laboratory Evolution in Drosophila

Similar experimental designs have been employed to study the efficacy of age-specific selection in *Drosophila*. One of the key expectations of age-structured

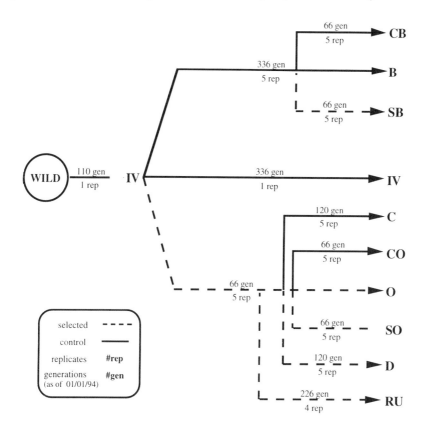

Figure 3 The *Drosophila* selection regimes of Rose and colleagues. In this system, "B" corresponds to early fertility selection, "O" to late fertility selection, "RU" to reverse selection (O->B conditions), "S" to adult starvation resistance selection, "D" to adult desiccation resistance selection, and "C" to controls for these respective treatments.

population genetics theory is that populations cultured at later ages exclusively are expected to evolve increased longevities compared to controls reproduced at earlier ages exclusively (Charlesworth, 1980; Rose, 1991). This approach has been used by Rose and Charlesworth (1980) on a pair of unreplicated populations, Luckinbill et al. (1984) with threefold replication, Rose (1984a) and Hutchinson and Rose (1991) with fivefold replication, and Partridge and Fowler (1992) in two experiments involving threefold replication each. [There are earlier experiments of related design, but varied intentions and interpretations (see Rose, 1991).] One example of such a design is given in Fig. 3. This is one of the more commonly performed laboratory evolution experiments. It has been found that increased longevities do indeed evolve with reproduction at late ages, as expected by population genetics theory. One of the features of these experiments that has remained somewhat uncertain is whether or not the evolution of increased longevity is associated with decreased early fecundity. This result was found by Rose and Charlesworth (1980), Rose (1984a), and Luckinbill et al. (1984), but not by Partridge and Fowler (1992) or Chippindale et al. (1993). (See the discussion of the Cheshire Cat, below.)

One experimental strategy that has a lot of potential in the study of laboratory evolution is reversion to the culture regime that antedated the

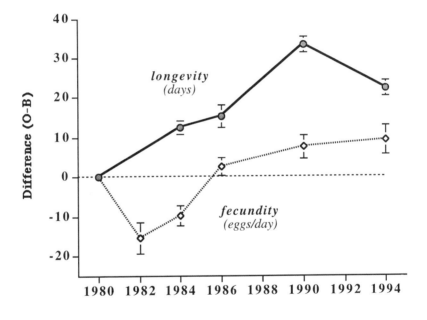

Year

Figure 4 Long-term evolution of a life history trade-off in *D. melanogaster*. (Bars indicate standard errors.) In the early evolution of postponed senescence populations, a decline in early fecundity of the long-lived O's suggested antagonistic pleiotropy between early and late life fitness characters. As shown here, subsequent selection reversed this result, such that, in the "standard" experimental environment, O's produced significantly more eggs early in life (as well as later). Like the Cheshire Cat, however, the trade-off reappears when the flies are conditioned under more "B-like" conditions (see Fig. 5).

imposition of a particular selection regime. This is also called "relaxed selection," although it should be noted that natural selection will normally also occur in the "relaxed" regime. For this reason, we prefer the term "reversed selection." Characters may or may not respond to reversed selection. "Reversing" responses are usually interpreted in terms of direct selection or an evolutionary trade-off between fitness under the different selection regimes. However, some subtle interaction between reversed selection and inbreeding depression may be involved. It is possible, for example, that selection for increased values of fitness components is acting to oppose the fixation of deleterious alleles by inbreeding in experimental populations of small effective sizes. This problem may be revealed by changes in characters other than those undergoing selection if these characters are sensitive to inbreeding depression. But in any case it will always be dubious to merely assume that there will be no response to the relaxation of selection, although that case can of course arise.

One hypothesis by which the absence of a response to reversed selection can be explained is the fixation of relevant genetic variation. An alternative hypothesis is that differentiation has occurred under the selective regime where the alleles underlying this differentiation are neutral under "relaxed" or reverse selection conditions. The first hypothesis can be tested using additional selection in the same direction, a significant response indicating the continued existence of selectable genetic variation. When this occurs, an absence of response to reversed selection could reflect neutrality under the reversed conditions. An additional possibility is the achievement of a new selective equilibrium, which is actively maintained by natural selection, at the point reached by the populations subject to the experimentally imposed selection regime.

One of the more extensive studies of reversed selection was performed by Service et al. (1988) using *Drosophila* that had been selected for later reproduction. Service et al. derived five reversed selection lines that were reverted to the ancestral, and control, early reproduction regime. They studied the response of four characters to reversed selection: early fecundity, starvation resistance, ethanol vapor resistance, and desiccation resistance. These characters were monitored every other generation for 20 generations. In addition, two out of four of these characters were later subjected to selection beyond the limits of late reproduction lines; this was a check for the exhaustion of genetic variability in the lines undergoing reversed selection (Rose et al., 1992). Finally, the reversed lines were also studied more than 100 generations after their initiation, which allowed the evaluation of whether or not the selected lines from which they were derived, the late reproduction stocks, had achieved a novel, selectively maintained equilibrium (Graves et al., 1992). The idea is that if the reverse selection stocks did not initially change for some character, then the role of selection in maintaining a new equilibrium for that character would be revealed by its continued maintenance some generations later. On the other hand, if a character of the late reproduction stocks had undergone differentiation that was merely neutral, under the original selection regime, recurrent mutations should cause the character to evolve over long periods of laboratory maintenance. The findings from these reversed selection experiments were relatively simple. First, complete fixation of genetic variation for fitness-related characters does not seem to have occurred. These characters readily respond to further laboratory evolution (e.g., Rose et al., 1992). Second, some characters exhibit evolutionary trade-offs. In particular, early fecundity appears to trade off with starvation resistance (Service et al., 1988). Third, other characters appear to have either less pronounced trade-offs or none at all. Specifically, there is no evidence for such trade-offs affecting desiccation resistance and ethanol resistance (Service et al., 1988).

IV. The Cheshire Cat and Other Syndromes

Any Wonderland has strange things in it, from bread and butterflies to gryphons. A problem is that these bizarre denizens may lurk behind the data instead of jumping out from behind the bushes. In interpreting the results of laboratory selection, one would like to assume that the observed evolution arises because of the particular selection regimes chosen by the experimenter, but in fact these responses may arise because of other functional or genetic causes, which may be unknown. Here we discuss four such factors: cryptic selection, genotype-by-environment interaction, inbreeding, and linkage disequilibrium. We do not wish to argue that this is an exhaustive list of potential artifacts, but we will argue that these four factors will usually merit examination, in principle if not in actual experiments. Many syndromes result from these artifacts, among them that of the Cheshire Cat.

A. Cryptic Selection

Like the rule of thumb in political economy that it is impossible for legislation to change only one part of the economy at a time, so it will be virtually impossible to select on only one character at a time. Inevitably, some aspect of the methods used during laboratory evolution will give rise to selection differentials that are unintended. Note that these methods may have nothing to do with the intended selection and may instead arise because of some quirk of routine culture, such as humidity, size of enclosure, or accumulation of waste products. However, lest this caution suggest the untenability of laboratory evolution experiments, it should be realized that in the laboratory there is a reasonable chance of sorting out such cryptic selection. In field evolution experiments, to say nothing of the comparative method (cf. Leroi et al., 1994a), there will be far less prospect of discovering this type of artifact because there are necessarily even more uncontrollable, confounding variables outside of the laboratory. Nonetheless, the relative superiority of laboratory evolution does not exonerate its practioners from the need to address the problem of cryptic selection.

Here we discuss two examples of cryptic selection in *Drosophila* laboratory selection, one from Partridge and Fowler (1992) and one from our own work (Chippindale et al., 1994). Perhaps the most common oversight in laboratory selection experiments is the failure to control conditions in life cycle stages that are not directly targeted by the selection procedures. For example, in *Drosophila* lab evolution, the discrete life cycle stages of the insect may lead to the perception that these discrete stages are wholly independent. In this class of problems, larval culture density is perhaps the most commonly neglected variable. In the case of the Partridge and Fowler (1992) study, selection was targeted at adult demographic parameters, specifically the timing of reproduction. These authors employed two sets of selection treatments, each made up of three replicate populations. Selection consisted of holding the adults of populations selected for postponed senescence for longer and longer periods of time before reproduction. These experiments were done ostensibly in the fashion of Rose (1984a) and Luckinbill et al. (1984), with adequate replication, and with large initial population sizes.

Considerable difficulties in the interpretation of the Partridge and Fowler (1992) results arise, however, because of what might seem a fairly small oversight: larval culture densities were controlled by the duration allowed for the adults to lay rather than by any direct standardization procedure. Aged flies will not produce as many eggs as younger flies, and flies that have died of old age will lay no eggs at all. As a result, culture densities varied substantially between the selection treatments, with the cultures reproduced earlier having much greater densities. The consequences of this systematic difference in handling were several: First, strong inadvertent selection for density-related traits may have led to the observed strong differentiation among treatments in larval competitive ability, development rate, and size of the imago. These results were interpreted as correlated responses to selection for late-life fertility. Later results from their own laboratory (Roper et al., 1993) indicate that these responses arose as a result of cryptic selection due to larval crowding; results from other laboratories also support this conclusion (e.g., Zwaan, 1993; Chippindale et al., 1994). Another problem generated by the lack of control over larval density comes in the interpretation of the direct responses to selection. Here, the effect of differential larval crowding on adult body size alone could severely complicate attempts to untangle the relationships among characters. It is conspicuous that the Partridge and Fowler study failed to find a trade-off between early- and late-life fecundity, or between early-fecundity and longevity, unlike previous studies (e.g., Luckinbill et al., 1984). While the experiments of Partridge and Fowler are not completely vitiated by the problem of confounded density-specific selection, their interpretation is at least greatly complicated.

In a similar experiment, Rose and colleagues (Rose, 1984a; Hutchinson and Rose, 1991; Leroi et al., 1994b) have been selecting for later reproduction, as discussed above. It has been found that the longer-lived stocks also have greater development times and higher larval viability (Chippindale et al., 1994). This has arisen despite the identical handling of control and selected lines for the first 14 days of life, a period that includes the entire period of larval and pupal development. This might lead an experimenter to explain the increased developmental period as a functional contributor to the increased life span exhibited by the selected lines. But it has also been found that stocks that are not selected for such late reproduction, yet experience an absence of selection for reproduction at 14 days of age, have evolved an increased development time and an increased viability as well. The interpretation offered by Chippindale et al. (1994) is that the age-specific selection regime has been confounded with a relaxation of selection for early development, the type of selection normally imposed on the controls, because the controls reproduce immediately, within a few days of adult eclosion. Thus these alterations to larval life history are essentially spurious to age-specific adaptation.

B. Genotype-by-environment Interaction

Genotype-by-environment interaction (GxE) is a long-known (e.g., Falconer, 1981) and now popular (e.g., Via, 1987) area of study in evolutionary quantitative genetics. Laboratory evolution experiments can reveal the importance of GxE in bold relief. In the work of Bennett and Lenski (1993), their *E. coli* exhibited remarkably specific differential adaptation to the culture temperatures at which they were cultivated. For example, lines adapted to 42°C have significantly enhanced fitness at 40-42°C, but not at lower temperatures. This rise in fitness is remarkably steep. While at 41°C the relative fitness advantage was 1.12, or 12%, at 41.8°C the relative advantage was 45%. This is an example of a set of experiments in which GxE was anticipated, and therefore not a problem for interpretation.

But GxE may arise in cases in which it is not anticipated, thereby wreaking havoc with the use of laboratory evolution experiments to test evolutionary hypotheses. Chippindale et al. (1993) did not find the trade-off between longevity and early fecundity that had been found earlier (e.g., Rose, 1984a; Luckinbill et al., 1984). Indeed, the longer-lived stocks had the same or greater fecundity than controls at every adult age. Like the Cheshire Cat, the trade-off between longevity and early fecundity had first appeared in the study of Rose (1984a) and disappeared in the study of Chippindale et al. (1993). This was curious indeed.

Leroi et al. (1994b, c) endeavored to sort out this anomaly. They found that there is no general or universal reduction in early fecundity as a result of the laboratory evolution of increased longevity. But there is a reduction in early fecundity in the original stock culture conditions; there has indeed been an evolutionary trade-off between fecundity and longevity under the conditions that

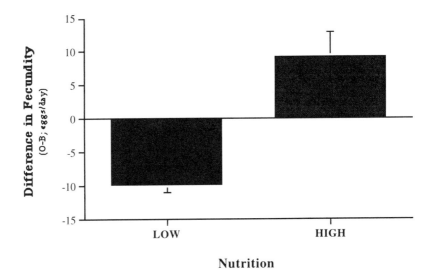

Figure 5 The Cheshire Cat in action: The trade-off between early fecundity and longevity appears and disappears from one environment to another. (The bars indicate standard errors.) These data from 1994 show that the superiority of the O (late-reproduced) populations, found in the high nutrition environment, disappears at lower food levels. Lower yeast levels are more representative of the B's egg-laying environment under selection, while high yeast levels are more like the O's egg-laying regime.

are most relevant to the experiment. That is, they could get the Cheshire Cat to reappear again, under the right conditions. Specifically, the longer-lived stocks evolve in such a way that they lose early fecundity under the conditions used to maintain their ancestral shorter-lived stocks, as shown in Fig. 5. This was not an easy conclusion to obtain: Leroi et al. had to sort through many types of potential genotype-by-environment interactions involving the use of carbon dioxide, timing and frequency of transfer, nutrition, accumulation of waste products, and so on. While the GxE involved was large in its effects on the differentiation of stocks, quite subtle, multiple, environmental differences were involved. Sorting out GxE when it inadvertently affects experimental results will be difficult in many cases. When it is not sorted out, trade-offs may appear and disappear and reappear, as inexplicably as the Cheshire Cat.

C. Inbreeding

Inbreeding is extremely commonplace in artificial selection because of the small size of the groups usually selected for the next generation, sometimes fewer than 20. This may not be a large problem in selection on characters like bristle number in *Drosophila* or coat color in domestic breeds of mammal. These characters appear to be resistant to ill effects from inbreeding, but this is not true of fitness characters for most outbreeding organisms (cf. Johnson and Wood, 1982). As Provine (1971) described in the case of Raymond Pearl's historic selection on egg laying in chickens, such inbreeding may cause selection to fail to make any progress in increasing fitness-related characters. On the other hand, this same effect might accelerate the progress of selection downward on fitness-related characters (Falconer, 1981). In the study of Clarke et al. (1961), selection for increased and decreased development time made progress only in the direction of increased development time, the direction of lowered fitness. This kind of effect is also present in the work of Rose (1984b). Selection for greater fecundity in *Drosophila* was first halted and then, in some lines, was actually reversed by inbreeding depression. Clearly, inbreeding depression can be an unmitigated disaster for laboratory evolution experiments, addling the fitness components of the organism under study, like the mind of the Mad Hatter was addled by the heavy metals used in haberdashery. For laboratory evolution, inbreeding depression is the veritable Mad Hatter Syndrome: everything goes to hell.

The study of Partridge and Fowler (1992) provides a more recent example of inbreeding depression. In the original paper, some indication of inbreeding depression of the late-reproduced lines was apparent in their very low preadult survivorship and in the variability among replicate populations for this character. Perhaps as a result of their failure to standardize larval culture densities, effective population sizes in their late-reproduced selection treatments were lower than in the early reproduced treatment. A later study by Roper et al. (1993) confirmed that inbreeding depression had, indeed, afflicted the late-reproduced lines. Some attempt was made by these authors to recover the populations from the effects of inbreeding by performing crosses among replicates. This sort of procedure, however, does not completely rescue an experiment from the problem of inbreeding since the hybrids will consist of unusual genotypes that would not be present if selection had been allowed to proceed without inbreeding.

D. Linkage Disequilibrium

One of the most attractive applications of laboratory evolution, at least in the mind's eye, is to create a set of differentiated populations, relative to matched controls, for some adaptation of interest and then compare the allele frequencies at various loci for associations between adaptation and genetic differentiation. For example, Tyler et al. (1993) compared the allele frequencies of Cu, Zn superoxide dismutase among the lines of *D. melanogaster* created by Rose (1984a). They found that the slower electromorph was more common in long-lived populations. This allele is also more active in free-radical scavenging than its faster moving variant. Does this result show that improvements in free-radical scavenging tend to increase life span?

Not necessarily. The response of the alleles of superoxide dismutase could arise entirely from linkage disequilibrium between the slow allele and an allele at a nearby locus, where it is the latter locus, of unknown function, that is in fact physiologically contributing to the longer life span. Solutions to this problem are not easy. One is to test for the effects of alternative alleles at a locus directly by transformation experiments. This has in fact been done for superoxide dismutase (e.g., Reveillaud et al., 1991; Orr and Sohal, 1994). More active alleles at this locus do indeed confer increased longevity, corroborating the finding of Tyler et al. (1993). Another possible solution would be parallel laboratory evolution across a range of species, especially nonsibling species. Common responses of alleles to standardized selection across species would reveal that linkage disequilibrium is not a likely source of an artifactual correlation between a particular form of laboratory evolution and a particular genetic response.

V. Comparing Looking Glasses

What you see depends on the glass in which you see it, or through which you see it. All empirical approaches have substantial limitations for the study of adaptation. An important issue is the severity of these limitations in each case. We would argue that the above case studies and general arguments constitute a compelling, though not overwhelming, case for the use of laboratory evolution to unravel adaptation. In laboratory evolution experiments, selection is imposed by the experimenter, with varying degrees of precision, and the effects of this selection on a wide range of genotypic and phenotypic attributes can be measured with ease. For this reason, there are few better contexts for observing the relationship between selection regime and the response to selection, and thus adaptation, despite the intrusion of Cheshire Cats and the occasional Mad Hatter. Nonetheless, it is worth considering whether or not other methods might, at least in some cases, yield insights as, or even more, readily than laboratory evolution.

A. Laboratory Versus Field Evolution

To the extent to which the experiments of laboratory evolution can be performed using populations in the wild, but with all other aspects of experimental design and control the same, then the experiments in the wild are of course to be preferred. The *Poecilia* transplantation experiments of Reznick (see Reznick and Travis, this volume) approach this ideal. But it is doubtful that there will be many other systems of comparable manipulability and replicability in the wild. Most other species are difficult to confine without imposing artifacts that are analogous to those in the laboratory. Extremely few species will be amenable to observation in the wild, due to their size and lack of crypsis, and yet be small enough to be handled as entire populations. (However, it might be said that there are analogous problems with the number of species that are amenable to laboratory evolution experimentation.) We emphatically predict that Cheshire Cats will be common in field studies of evolution (though not Mad Hatters), leaving aside still other creatures too fantastic for us to imagine at this point. Finally, there is the profound problem of achieving replication, where the different replicates are close enough to be handled together and yet sufficiently isolated from each other genetically.

B. Laboratory Evolution vs Phenotypic Manipulation

Phenotypic manipulation, in which organisms are subjected to environmental perturbation, has been offered as one of the highways to knowledge of the architecture of biological adaptation (e.g., Partridge and Harvey, 1988; but see Reznick, 1985, 1992; Sinervo and Basolo, this volume). This would be wonderful, if it were true. The problem is that the response to phenotypic manipulation may be a poor guide to the response to selection. One example of this is provided by the study by Chippindale et al. (1993) of the relationship among longevity, fecundity, starvation resistance, and nutrition. They found that the evolutionary interrelationship among longevity, fecundity, and starvation resistance was broadly similar to the interrelationship that arose with phenotypic manipulation. But in important respects, the response to phenotypic manipulation was disanalogous. Phenotypic manipulation may or may not reveal the foundations of a particular biological adaptation. It appears that it is not possible to find out which situation obtains without doing an evolutionary experiment of some kind. Therefore, this method seems decidedly inferior to laboratory evolution for the study of adaptation.

C. Laboratory Evolution vs Correlations Among Relatives

There is a close relationship among the matrix of additive genetic variances and covariances, the matrix of phenotypic variances and covariances, and the immediate response to selection (see Falconer, 1981; Lande, 1979; but see Turelli, 1988). Furthermore, these matrices can be estimated using the classical correlations between relatives designs, such as sib analysis and parent-offspring regression (Kempthorne, 1957; Falconer, 1981). If one assumes that these matrices are stable, then one can extrapolate evolution over millions of years, as in Lande's (1979) study of mammalian brain-body allometry. No experiment in laboratory evolution can compete with this kind of explanatory power.

Of course, this entire inferential strategy depends on the assumption that the genetic and phenotypic variance-covariance matrices remain stable throughout the period of extrapolation. This may not be true. Gene frequencies may change in such a way as to change genetic variance components. Of perhaps still greater importance, the genetic or phenotypic variance matrix terms may change in response to environmental change due to genotype-by-environment interaction, as discussed earlier (see also Service and Rose, 1985).

But even if all these caveats are not relevant, there are still some purely practical considerations to bear in mind when comparing selection with studies of the correlations between relatives. The most important of these is the statistical power of particular experimental designs. Sib analyses and the like are notoriously weak in statistical power (Klein et al., 1973). Furthermore, a sib analysis of a particular population has a large, and often unknown, sampling variance *among populations* that arises from differentiation among populations. That is, the parameter estimates obtained for a particular adaptation in a particular population might be relatively accurate *for that population*, but not accurate over the ensemble of similar populations within a species. Selection experiments can resolve both of these problems. First, selection experiments accumulate direct and indirect responses to selection over each generation of selection so that the statistical power of the experiment can be increased over time. Second, multiple replicates within selection treatments can be used to provide estimates of among population variance, leading to a more reliable accounting for among-population heterogeneity. For these reasons, a laboratory evolution experiment may be a better guide to the evolutionary basis of an adaptation, leaving aside broader assumptions, such as the constancy of variance-covariance matrices.

VI. The Mock Turtle's Story: What Did We Learn about Adaptation from Laboratory Evolution?

> *Alice did not feel encouraged to ask any more questions about it, so she turned to the Mock Turtle, and said, "What else had you to learn?"*
>
> *"Well, there was Mystery,"* the Mock Turtle replied, *counting off the subjects on his flappers,—"Mystery, ancient and modern, with Seaography: then drawling – the Drawling-master was an old conger-eel, that used to come once a week:* **he** *taught us Drawling, Stretching, and Fainting in Coils."*

Adaptation is characteristically a matter of specific adjustments to particular environments. The ability of the experimenter to impose such environments in the course of laboratory evolution gives the method the power to unravel many of these specific adaptations, be they adaptations to temperature in *E. coli* (e.g., Bennett et al., 1992), adaptations to timing of reproduction with age-specific selection in *Drosophila* (Rose, 1991), adaptations to crowding in *Drosophila* (e.g., Mueller et al., 1993), and these examples are of course not exhaustive. Nonetheless, if the study of adaptation is to be more than the proverbial "stamp-collecting" of natural history, there must be some generalities that can be gleaned about adaptation from laboratory evolution experiments. Here we provide a few examples of such generalities. We offer them to illustrate what may develop from studies of laboratory evolution where our understanding of adaptation is concerned.

(i) *Adaptation rarely seems to be limited, at least not for long, by a complete lack of genetic variation in large populations.* We know this because it is almost always possible to obtain a response to selection in laboratory evolution experiments, even when selection acts on fitness or fitness-related characters. In particular, this suggests that most simpleminded notions of evolutionary stasis cannot appeal to a lack of available, relevant, genetic variation.

(ii) *Evolutionary genetic trade-offs are common, but not universal.* Adaptations appear to be constrained in some cases by antagonistic pleiotropy between fitness-related characters. But this is by no means always the case.

(iii) *The evolution of multiple populations facing the same selection pressures is often qualitatively uniform.* One thing that has been surprising for us is the extent to which multiple lines and multiple laboratories can yield largely parallel results when they use the same basic experimental designs. This has not been a universal or uncomplicated parallelism, and we have discussed reasons why this might be so above. But it is more common than we expected.

(iv) *Adaptation may be startlingly specific to the selection regime imposed, with strong genotype-by-environment interaction.* This pattern arises when it is sought and, unfortunately, when it has not been anticipated. The observation of adaptations may depend stringently on the environment in which observation takes place. In particular, genotype-by-environment interaction may obscure adaptations whose expression depends on such interactions. This is the Cheshire Cat Syndrome. Often, the ways in which organisms adapt exhibit a **type** of specificity that will be entirely unexpected by the experimenter. Adaptation can be entirely nonintuitive *a priori*, whatever claims are made *post hoc*.

VII. Goodbye to Wonderland

So what is the value of traveling to Wonderland? Evolutionary theories, like Victorian schoolgirls, need to be improved. In the 19th century, this would have entailed visiting severe maiden aunts and going to uplifting cultural events. In visiting Wonderland with Alice, in reading the book, our young girl from 1866 is experiencing a range of intellectual and emotional puzzles. In overcoming them, absorbing them, or going beyond them, she develops into a young lady.

When we go to the laboratory and set up an experiment in which the intensity of age-specific selection is varied among replicated laboratory populations, we are taking theoretical concepts and applying them in an artificial world. This world is full of artifacts and problems of interpretation, associated here with the Cheshire Cat and the Mad Hatter. And many of these artifacts make interpreting experiments difficult. But from our encounter with this often confusing and unfair world, we can learn about our theories and improve them. Like generations of children educated by Lewis Carroll, our understanding of evolution in general will be treated to few experiences better than an Adventure in the Wonderland of laboratory evolution.

Acknowledgments

We are grateful to G. Benford, A. F. Bennett, F. Cohan, J. F. Crow, B. Hawkins, G. V. Lauder, R. E. Lenski, A. M. Leroi, L. D. Mueller, P. M. Service, and K. Weber for their comments on an earlier version. Our research has been supported by US-PHS grants AG06346 and AG09970 and NSF grant DEB-9207757 to M. R. R., as well as through the generosity of the Robert H. Tyler estate.

References

Bennett, A. F., and Lenski, R. E. (1993). Evolutionary adaptation to temperature. II. Thermal niches of experimental lines of *Escherichia coli*. *Evolution* 47, 1-12.

Bennett, A. F., Lenski, R. E., and Mittler, J. E. (1992). Evolutionary adaptation to temperature. I. Fitness responses of *Escherichia coli* to changes in its thermal environment. *Evolution* 46, 16-30.

Charlesworth, B., (1980). "Evolution in Age-Structured Populations." Cambridge University Press, Cambridge.

Chippindale, A. K., Leroi, A. M., Kim, S. B., and Rose, M. R. (1993). Phenotypic plasticity and selection in *Drosophila* life-history evolution. I. Nutrition and the cost of reproduction. *J. Evol. Biol.* 6, 171-193.

Chippindale, A. K., Hoang, D. T., Service, P. M., and Rose, M. R. (1994). The evolution of development in *Drosophila melanogaster* selected for postponed senescence. *Evolution* 48, 1880-1899.

Clarke, J. M., Maynard Smith, J., and Sondhi, K. C. (1961). Asymmetrical response to selection for rate of development in *Drosophila subobscura*. *Genet. Res.* 2, 70-81.

Crow, J. F., (1954). Analysis of DDT-resistant strains of *Drosophila*. *J. Econ. Entomol.* 47, 393-398.

Falconer, D. S., (1981). "Introduction to Quantitative Genetics." 2nd ed. Longman, London.

Graves, J. L., Toolson, E. C., Jeong, C., Vu, L. N., and Rose, M. R. (1992). Desiccation, flight, glycogen, and postponed senescence in *Drosophila melanogaster*. *Physiol. Zool.* 65, 268-286.

Hill, W. G., and Mackay, T. F. C., eds., (1989). "Evolution and Animal Breeding." C•A•B International, Wallingford, U.K.

Hutchinson, E. W., and Rose, M. R. (1991). Quantitative genetics of postponed aging in *Drosophila melanogaster*. I. Analysis of outbred populations. *Genetics* 127, 719-727.

Johnson, T. E., and Wood, W. B. (1982). Genetic analysis of life-span in *Caenorhabditis elegans*. *Proc. Natl. Acad. Sci. U.S.A.* 79, 6603-6607.

Joshi, A., and Mueller, L. D. (1988). Directional and stabilizing density-dependent natural selection for pupation height in *Drosophila melanogaster*. *Evolution* 47, 176-184.

Kempthorne, O., (1957). "An Introduction to Genetic Statistics." Wiley, New York.

Klein, T. W., DeFries, J. C., and Finkbeiner, C. T. (1973). Heritability and genetic correlation: Standard errors of estimates and sample size. *Behav. Genet.* 3, 355-364.

Lande, R., (1979). Quantitative genetic analysis of brain: body size allometry. *Evolution* 33, 402-416.

Lenski, R. E., Rose, M. R., Simpson, S. C., and Tadler, S. C. (1991). Long-term experimental evolution in *Escherichia coli*. I. Adaptation and divergence during 2,000 generations. *Amer. Nat.* 138, 1315-1341.

Lenski, R. E., and Travisano, M. (1994). Dynamics of adaptation and diversification: A 10,000-generation experiment with bacterial populations. *Proc. Natl. Acad. Sci.* 91, 6808-6814.

Leroi, A. M., Rose, M. R., and Lauder, G. V. (1994a). What does the comparative method reveal about adaptation? *Amer. Nat.* 143, 381-402.

Leroi, A. M., Chippindale, A. K., and Rose, M. R. (1994b). Long-term laboratory evolution of a genetic life-history trade-off in *Drosophila melanogaster*. 2. The role of genotype x environment interaction. *Evolution* 48, 1244-1257.

Leroi, A. M., Chen, W. R., and Rose, M. R. (1994c). Long-term laboratory evolution of a genetic life-history trade-off in *Drosophila melanogaster*. 2. Stability of genetic correlations. *Evolution* 48, 1258-1268.

Luckinbill L. S., Arking, R., Clare, M. J., Cirocco, W. C., and Buck, S. A. (1984). Selection for delayed senescence in *Drosophila melanogaster*. *Evolution* 38, 996-1003.

MacArthur, R. H., and Wilson, E. O. (1967). "The Theory of Island Biogeography." Princeton University Press, Princeton.

Mueller, L. D., (1988). Density-dependent population growth and natural selection in food limited environments: The *Drosophila* model. *Am. Nat.* 132, 786-809.

Mueller, L. D., (1990). Density-dependent natural selection does not increase efficiency. *Evol. Ecol.* 4, 290-297.

Mueller, L. D., and Ayala, F. J. (1981). Trade-off between r-selection and K-selection in *Drosophila* populations. *Proc. Natl. Acad. Sci. U.S.A.* 78, 1303-1305.

Mueller L. D., Graves, J. L., and Rose, M. R. (1993). Interactions between density-dependent and age-specific selection in *Drosophila melanogaster*. *Func. Ecol.* 7, 469-479.

Orr, W. C., and Sohal, R. S. (1994). Extension of life-span by overexpression of superoxide dismutase and catalase in *Drosophila melanogaster*. *Science* 263, 11-28-1130.

Partridge, L., and Harvey, P. H. (1988). The ecological context of life history evolution. *Science* 241,

1449-1455.

Partridge, L., and Fowler, K. (1992). Direct and correlated responses to selection on age at reproduction in *Drosophila melanogaster. Evolution* 46, 76-91.

Pianka, E., (1972). r- and K-selection or *b* and *d* selection? *Am. Nat.* 106, 581-588.

Pollak, E., Kempthorne, O., and Bailey, T. B., Jr., eds., (1977). "Proceedings of the International Conference on Quantitative Genetics." Iowa State University Press, Ames, IA.

Provine, W. B., (1971). "The Origin of Theoretical Population Genetics." University of Chicago Press, Chicago.

Reveillaud, I., Niedzwiecki, A., and Fleming, J. E. (1991). Expression of bovine superoxide dismutase in *Drosophila melanogaster* augments resistance to oxidative stress. *Mol. Cell. Biol.* 11, 632-640.

Reznick, D. (1985). Costs of reproduction: An evaluation of the empirical evidence. *Oikos* 44, 257-267.

Reznick, D. (1992). Measuring the costs of reproduction. *Trends Ecol. Evolut.* 7, 42-45.

Roper, C., Pignatelli, P., and Partridge, L. (1993). Evolutionary effects of selection on age at reproduction in larval and adult *Drosophila melanogaster. Evolution* 47, 445-455.

Rose, M. R., (1984a). Laboratory evolution of postponed senescence in *Drosophila melanogaster. Evolution* 38, 1004-1010.

Rose, M. R., (1984b). Artificial selection on a fitness-component in *Drosophila melanogaster. Evolution* 38, 516-526.

Rose, M. R., (1991). "Evolutionary Biology of Aging." Oxford University Press, New York.

Rose, M. R., and Charlesworth, B. (1980). A test of evolutionary theories of senescence. *Nature* 287, 141-142.

Rose, M. R., Service, P. M., and Hutchinson, E. W. (1987). Three approaches to trade-offs in life-history evolution. *In* "Genetic Constraints on Adaptive Evolution" (V. Loeschcke, ed.), Springer-Verlag, Berlin.

Rose, M. R., Graves, J. L., and Hutchinson, E. W. (1990). The use of selection to probe patterns of pleiotropy in fitness characters. *In* "Insect Life Cycles" (F. Gilbert, ed.),Springer-Verlag, New York.

Rose, M. R., Vu, L. N., Park, S. U., and Graves, J. L. (1992). Selection on stress resistance increases longevity in *Drosophila melanogaster. Exp. Gerontol.* 27, 241-250.

Service, P. M., and Rose, M. R. (1985). Genetic covariation among life-history components: The effect of novel environments. *Evolution* 39, 943-945.

Service, P. M., Hutchinson, E. W., and Rose, M. R. (1988). Multiple genetic mechanisms for the evolution of senescence in *Drosophila melanogaster. Evolution* 42, 708-716.

Turelli, M., (1988). Phenotypic evolution, constant covariances, and the maintenance of additive genetic variance. *Evolution* 42, 1342-1347.

Tyler, R. H., Brar, H., Singh, M., Latorre, A., Graves, J. L., Mueller, L. D., Rose, M. R., and Ayala, F. J. (1993). The effect of superoxide dismutase alleles on aging in *Drosophila. Genetica* 91, 143-149.

Via, S., (1987). Genetic constraints on the evolution of phenotypic plasticity. *In* "Genetic Constraints on Adaptive Evolution" (Loeschcke, V., ed.), pp. 46-71. Springer-Verlag, Berlin.

Weir, B. S., Eisen, E. J., Goodman, M. M., and Namkoong, G., eds. (1988). "Proceedings of the Second International Conference on Quantitative Genetics." Sinauer Associates, Sunderland, MA.

Wright, S., (1977). "Evolution and Genetics of Populations. Vol. 3, Experimental Results and Evolutionary Deduction." University of Chicago Press, Chicago.

Zwaan, B. J., (1993). "Genetical and Environmental Aspects of Ageing in Drosophila melanogaster: An Evolutionary Perspective." Diss. University of Groningen

CHAPTER 8

The Empirical Study of Adaptation in Natural Populations

DAVID REZNICK

JOSEPH TRAVIS

I. General Introduction: The Controversy Concerning "Adaptation"

Among all of Darwin's (1859) original ideas in *The Origin of Species*, the principle of adaptation, or evolution by natural selection, was the most central to his theory. It was also the most controversial. The propositions that organisms could evolve and that evolution could result in the formation of new species were not original to Darwin. Both were readily accepted by much of the scientific community after the publication of *The Origin of Species*. But Darwin's proposed mechanism for the process of evolution was met with considerable skepticism (Provine, 1971). The initial debates about the efficacy of his concept of natural selection as the mechanism for evolution were fueled largely by an ignorance of the mode of inheritance (Provine, 1971); however, the controversies over natural selection and adaptation have long outlived the discovery of Mendelian inheritance and its incorporation into evolutionary theory.

The current focus of the debate about natural selection lies in defining the cause and effect relationship between a trait that appears to be adaptive and the factors that selected for the evolution of the trait. One problem is identifying traits that evolved in direct response to natural selection, as opposed to some other cause. For example, Gould and Lewontin (1979) argued that traits could arise as a consequence of design constraints rather than as adaptations to a specific form of selection. Gould and Vrba (1982) made a formal distinction between traits that evolved in direct response to a specific form of selection from traits that existed in close to their current form and were then coopted for a new function. A popular approach, taken by Coddington (1988), Greene (1986), and Brooks (Brooks and McLennan, 1991), is to distinguish between the novel appearance of a trait within a phylogeny from its being inherited from an ancestor. All of these attempts to recognize the patterns that might be the signature of adaptation hinge on an acceptable definition of the term. The current literature also includes considerable debate about the definition of adaptation (e.g., Reeve and Sherman, 1993).

Our goal is not to add to the debate about what an adaptation is or the relative importance of natural selection in evolution, but rather to summarize progress in the empirical study of adaptation in natural or "wild" populations. To do so, we will first summarize Darwin's definition of adaptation and natural selection since we think that he got it right the first time. We will then briefly summarize the history of empirical research on adaptation in natural populations. Finally we will present what we consider to be a modern program of research.

To Darwin (1859), an adaptation is any feature of an organism that arose as a consequence of natural selection and hence enhances the fitness of the individual. The concept of "adaptedness," or the fact that organisms seemed so well suited to their lifestyles, was an old one and was a popular subject of discussion for the natural theologians of Darwin's time. His contribution was to provide a naturalistic explanation and mechanism for the origin of the phenomenon. From the beginning, Darwin also recognized that one character might be functionally "tied" to another, so that the response to selection on one trait might cause other changes in morphology that were not themselves necessarily adaptive. Darwin also recognized that organisms had various forms of constraints that would limit their ability to adapt or change in response to natural selection. Finally, he recognized that the morphology of an organism was laden with its history and not necessarily representative of perfect adaptation to current circumstances. His first description of the process thus foreshadowed many of the current areas of debate.

One feature of Darwin's proposal that attracted criticism was his belief in natural selection as a continuous process, albeit having a rate that might vary over time. It was this aspect of the proposal that attracted Huxley's famous admonition that "...you have loaded yourself with an unnecessary difficulty in adopting Natura non facit saltum [nature does not make jumps] so unreservedly"(Provine, 1971; p. 12). A second source of criticism was his adherence to continuous variation as the most important material for the process of natural selection. Most scientists associated this form of variation with blending inheritance, which would rapidly eliminate heritable variation. To counter this constant loss of variation, Darwin (1868) proposed in his theory of pangenesis that the inheritance of acquired traits could provide a constant source of new variations upon which natural selection could act.

Much of the early research stimulated by Darwin's proposals focused on the mode of inheritance. Another key area of research was developmental biology, its chief focus being to deduce the relationship between the development of an organism and its evolutionary history (Gould, 1977). The direct study of natural selection and adaptation was a minor component of the early research efforts of evolutionary biologists. The attention devoted to these other topics is one possible explanation for the dearth of empirical investigations concerning the actual process of evolution. We suspect that a second likely explanation for the lack of interest in natural selection is that the importance of the process was seriously questioned (Provine, 1971).

The history of the conflict over the importance of natural selection and its resolution is well presented by Provine (1971), whom we gratefully acknowledge as the original source of our understanding of 19th and early 20th century evolutionary biology. Our chronology below will be similar to his, but the emphasis will be different since we will concentrate on the empirical study of adaptation in natural populations. Our thesis is that even though there was a basic understanding of the term "adaptation" from the beginning, a full appreciation of the inherent subtleties affecting the study of this central aspect of Darwin's theory took over 70 years to develop. The main reason for this sluggish development is that the study of adaptation was tied to the debates about the importance of natural selection in evolution. These debates had to be resolved before much progress could be made. We will argue that much of the newer controversy about adaptation fades in significance if one studies the dynamics of adaptation rather than simply trying to interpret the adaptive significance of a trait.

II. The Empirical Study of Adaptation Prior to the Modern Synthesis

A. The Biometricians

There were some attempts to study adaptation prior to the Modern Synthesis [we mark the beginning of this period with Fisher's (1930) and Haldane's (1932) books]. Work at this time was dominated by what was later labeled as the Panglossian Paradigm (Gould and Lewontin, 1979). Observations from nature were followed by interpretations of their adaptive significance. One famous example, to be discussed in more detail below, was Batesian mimicry.

The earliest coherent program advanced for the empirical study of adaptation was developed by the "Evolution Committee of the Royal Society," headed initially by W. F. R. Weldon and Karl Pearson. These investigators were highly quantitative and remarkably modern in their approach to the problem. Their approach is well summarized by Weldon's following statement (1893; after Provine, p. 31):

"It cannot be too strongly urged that the problem of animal evolution is essentially a statistical problem: that before we can properly estimate the changes at present going on in a race or species we must know accurately (a) the percentage of animals which exhibit a given amount of abnormality with regard to a particular character; (b) the degree of abnormality of other organs which accompanies a given abnormality of one; (c) the difference between the death rate per cent in animals of different degrees of abnormality with respect to any organ; (d) the abnormality of offspring in terms of the abnormality of the parents and vice versa. These are all questions of arithmetic; and when we know the numerical answers to these questions for a number of species, we shall know the deviation and the rate of change in these species at the present day – a knowledge which is the only legitimate basis for speculations as to their past history, and future fate."

By "abnormality," Weldon meant individual deviations from the population mean so, for example, "the abnormality of the offspring in terms of the abnormality of the parents" refers to the degree to which parents and offspring look alike, or deviate similarly from the population mean.

Weldon and Pearson championed the study of evolution through the mathematical description of the distribution of quantitative traits in natural populations in the same way as current investigators, such as in the studies of natural selection in Galapagos finches (e.g., Gibbs and Grant, 1987). They measured the values of a series of traits for each individual and described the population in terms of the mean and variance of these traits, plus the degree of correlation among traits. They emphasized the need to study changes in the distribution of traits as a way of making deductions about differential survival and hence the adaptive significance of a given trait.

One example of this methodology was Weldon's (1895) comparison of the distribution of morphological traits in juvenile versus adult crabs. They found that the variance of the target traits was lower in the adults than in juveniles and interpreted this as being caused by the selective death of extreme phenotypes, or what we now describe as stabilizing selection. Weldon and Pearson estimated the magnitude of selective death that would be required to generate the observed reduction in variance. They also defined directional selection, or a change in the mean value of a trait, as an alternative form of natural selection. They postulated that stabilizing selection was much more likely to be seen in routine study than directional selection. This is because such directional shifts would only occur when there happened to be a corresponding shift in the environment and such shifts would rapidly run their course. One would only see such a directional change if one were lucky enough to look at the right time.

Weldon (1899) reported one possible example of such a rare episode of directional selection as part of his Presidential Address to the Royal Society. He and Herbert Thompson did a longitudinal study of the frontal breadth relative to carapace length in crabs from Plymouth Sound. Between 1893 and 1898 there was a substantial reduction in the relative frontal breadth. This was associated with the completion of a breakwater that prevented the tides from removing silt washed in from the local streams, plus an increase in ship traffic and human waste dumped into the sound. Weldon performed experiments in which he exposed crabs with a range of frontal breadths to a suspension of fine clay sediment, similar to what was accumulating in the sound. He found that the survivors had relatively narrower frontal breadths than the nonsurvivors. He postulated that the crabs' ability to filter the sediment out of water that enters the gill chamber was correlated with the relative frontal breadth. One of the goals of Weldon's address was to respond to the criticism that natural selection appeared to be too slow a process to account for the origin of biological diversity. He concluded that "...we have here a case of Natural Selection acting with great rapidity because of the rapidity with which the conditions of life are changing" (p. 900). His case for natural selection is incomplete in some important ways, specifically in its failure to address the genetic basis of the observed changes, but this gap is no surprise given the understanding of inheritance at the time.

This approach to the study of adaptation was not revisited until the 1970s, beginning with papers by Lande (1976,1979). It is difficult to know why such a promising beginning was stillborn, but it was perhaps a victim of the Biometrician-Mendelian controversies, the alignment of the Biometricians with a rejection of Mendelian inheritance, and the subsequent diversion of attention to the study of genetics.

B. Mimicry

One prominent example of the observational approach to the study of adaptation that predated the biometricians is mimicry. This example is of historical interest because it was one of the most popular sources of studies of adaptation up through the 1930s and has been a source of continuing study and controversy. The controversies have often served as a microcosm of the larger-scale debates about evolution. This brief history also serves as a microcosm of how the empirical study of adaptation developed from before Weldon's time to the present.

Bates (1862) originally proposed that the similarity in appearance between allegedly palatable and unpalatable butterflies is an adaptation because it reduces the probability that the former will be preyed upon by diurnal, visually-oriented predators like birds. It had been proposed earlier that the warning coloration of unpalatable species was an asset because it made them easier to recognize and hence more efficiently trained their would-be killers to avoid them. The mimics gained by being mistaken for the foul-tasting models and hence profiting from this avoidance. By 1915, investigators had demonstrated that some butterflies did indeed vary in how tasty they were; some clearly inspired revulsion on the part of predators. Many additional examples of possible mimicry have been described, and alternative mechanisms for the evolution of mimicry (e.g., Mullerian, Mertensian, and Aggressive mimicry; reviewed by Wickler, 1968) had been proposed.

Punnett's (1915) review of mimicry neatly encapsulates the Mendelian-Biometrician controversy that dominated the arguments about natural selection between 1890 and approximately 1920 (Provine, 1971). He shows that the controversy was still alive and well even after the expansion of Mendelian genetics between 1900 and 1915, yet he also looks forward to the modern synthesis that would follow. Punnett displayed the judgement of a gifted empirical scientist by outlining what needed to be done to build an experimental evaluation of mimicry as an adaptation. He emphasized that to prove that butterfly color patterns are adaptations, one must study the properties of birds as predators. The evolution of aposematic coloration and mimicry must mirror a bird's perceptions and learning abilities. Would birds really learn to avoid brightly colored potential prey that packed a bad experience, what kind of training was required, how long would the memory last, and how readily would the experience be generalized to different species of prey? Punnett's logic neatly foreshadowed a program of study that would not begin until much later.

The positive aspects of Punnett's proposals for an empirical study of mimicry have been overshadowed by his adherence to De Vries' mutation theory. His model for the evolution of mimicry was based on a reasonable interpretation of early studies of the genetics of polymorphic mimicry in *Papilio dardanus*. These early investigations revealed that complex, alternative color patterns were inherited as if they were alternative alleles at a single locus. Punnett reasoned that such alleles must have arisen through a single macromutation. Each allele then either failed or succeeded based on its similarity to available models. This trial and error approach to selection based on mutations with large effects represents

DeVries' mutation theory; it was the main alternative to Darwin's view of adaptations being the result of progressive changes in the phenotype through the modification of what we later recognized as polygenic traits.

Fisher (1930) revisited Punnet's review of the evolution of mimicry and proposed an alternative mechanism for its evolution. His treatment of mimicry was part of Fisher's more general revision of Darwin's proposal of evolution by natural selection because he showed that it was entirely consistent with Mendelian inheritance. A major change in perspective between Fisher and Punnett was the development of an appreciation of polygenic inheritance and quantitative traits. Fisher proposed that mimicry was a polygenic trait and that a partial resemblance between a model and mimic would be sufficient to give the mimic a selective advantage over a nonmimetic conspecific. Partial mimicry could then be progressively improved upon as natural selection acted on loci that modified and perfected the resemblance between the model and the mimic. Fisher's proposals for the evolution of mimicry were similar in principle to his general treatment of evolution by natural selection. He effectively recast Darwin's original model for evolution, but substituted polygenic traits and additive effects of genes for Darwin's continuous variation and blending inheritance.

By 1936, Robson and Richards report that "the number of cases...is now very large, and most of the chief insect and arachnid groups contain typical examples of the phenomenon." Furthermore, "...the large amount and varied nature of the available data, and the fact emphasized by Fisher (1930, p. 146) that if the theory of mimicry is mainly true, then we appear to have a long series of cases in which the characters...are of adaptive value" (p. 251).

The implementation of Punnett's suggested program of study and the discovery and characterization of the supergenes that explained the early results of genetic investigations of polymorphic mimicry did not begin until the 1950s and 1960s (e.g., Van Zandt-Brower, 1958a-c; Clarke and Sheppard, 1960a-c). Ridley (1993) summarized some of this work and notes that, even today, Clarke and Sheppard's case for the evolution of polymorphic mimicry represents an untested hypothesis.

The study of mimicry from the 1860s through the 1930s thus began with natural history observations that were followed by an adaptive interpretation of their significance. While early experiments with predators lent support to the idea that warning coloration and mimicry could be effective, it was difficult for Punnett to accept mimicry as being a product of natural selection because it did not appear that the known mechanisms of inheritance could explain the gradual evolution of such a complex trait. This specific example is representative of the more general rejection of evolution by natural selection by much of the scientific community prior to the Modern Synthesis.

III. The British School of Ecological Genetics and the Influence of the Modern Synthesis

A. The Early Development of the Field

It was the development of population genetics theory during the 1910s and 1920s that rehabilitated Darwin's proposed mechanism for evolution by natural selection. It is no coincidence that the development of this also marks the beginning of the development of the first program for studying adaptation that has since continued through an unbroken line of intellectual descent. The important features of the approach are (after Ford, 1971): (1) it focused on distinct aspects of the phenotype, often discrete morphological polymorphisms, that were variable in natural populations; (2) it usually incorporated an evaluation of the genetic basis for these morphological variations; (3) it evaluated the impact of this variation on components of fitness in natural populations; and (4) it often used comparative studies among populations as a way of identifying environmental factors that might be responsible for natural selection. The emphasis on polymorphisms to study adaptation looks superficially like an application of de Vries' mutation theory, but by then they could understand polymorphisms as being governed by the same rules as any other form of genetic variation. The emphasis on polymorphisms was just a convenience for field studies, because the variation was so easy to recognize. The emphases on genetics and comparative studies are both direct outgrowths of the modern synthesis era. The contribution of genetics is obvious as the beginning of ecological genetics follows directly on the heels of the development of population genetics. The origins of the emphasis on comparisons made over a wide geographical area are less obvious. This was the approach being taken by an increasing number of systematists that provided the data base for Dobzhansky's (1937) and Mayr's (1942) development of the biological species concept, so perhaps it was borrowed from that source. The general approach was thus to begin with morphological variation, then to evaluate the genetic basis of this variation and its contribution to individual fitness. This approach was foreshadowed in part by Weldon, Pearson, and others, but they do not seem to have had much direct influence on the development of the program of research beginning in the 1930s. We feel that these earlier authors did not have much influence on the development of ecological genetics because we rarely see citations of their work, plus we do not see anything similar to their quantitative approach to characterizing morphological variation in natural populations.

Some examples of successful models of this research program and early, key publications include:

1. Industrial melanism in *Biston betularia* (Kettlewell, 1955, 1956, 1958): the demonstration that the rapid increase in the frequency of melanic moths in industrial England was caused by background matching and selective predation by birds.

2. Shell banding polymorphisms in *Cepaea* snails (Cain and Sheppard, 1950, 1954; Sheppard, 1951): the demonstration that the relative frequency of shell color and banding was a function of background color matching and selective predation by birds.

3. Background color matching in *Peromyscus* (Dice, 1947): the demonstration that pelage color in mice, like shell color in snails or wing color in moths, was selected to match background color by avian predators.

4. Heavy metal tolerance in *Agrostis tenuis* (McNeilly, 1968; McNeilly and Antonovics, 1968): the finding that plants growing on mine tailings were less susceptible to the effects of toxic heavy metals than those from adjacent, uncontaminated soils.

5. Inversion polymorphisms in *Drosophila* (Dobzhansky, 1948): the association between the frequency of chromosome inversion patterns and altitude or season.

All of these studies demonstrated that the characters under study had a genetic basis, although the details of the genetic analyses varied from case to case. All of them demonstrated that the characters had some impact on some component of fitness, usually short-term survival. These results varied from seasonal changes in the frequency of types (*Cepaea*) to short-term adult survival (*Cepaea*, Industrial melanism), to more complete analyses of the components of fitness (McNeilly and Antonovics, 1968).

B. Biochemical Applications

The success of the model systems of ecological genetics for characterizing the process of adaptation might be tempered by the criticism that they are not representative of the majority of genetic polymorphisms that are acted upon by selection, and thus not representative case studies in adaptation. Heavy metal tolerance and industrial melanism represent adaptations to intense anthropogenic selection that may not represent the magnitude of natural selection. The classic polymorphisms might be a distinct subset of genetic polymorphisms because of the visual apparency of the phenotypic variants. If the magnitude of selection on such polymorphisms is unusually high, then this apparency would lead to a bias in the models selected for study by biologists.

Some of these criticisms can be arguably refuted by Endler's (1986) compilation of studies of selection, in which the model systems of ecological genetics fall comfortably within the parameter ranges found in hundreds of subsequent studies. The estimated magnitudes of selection on individual loci vary widely among cases. Estimates of large selection coefficients are sufficiently abundant from a sufficient diversity of systems that the classic models of ecological genetics no longer seem idiosyncratic. However, before such data were accumulated, the problem was very real and led two research groups, R. C. Lewontin and colleagues at Chicago, and H. Harris and colleagues at London, to study the variation at genes encoding enzymes (see Lewontin, 1974, for discussion). The rationale for these model systems included the possibility of a truly random sample of the genome, which in turn would lead to a more objective understanding of the frequency with which natural selection acted on individual genes and the magnitude with which it acted.

The resultant explosion of studies of biochemical variants fell squarely in the tradition of ecological genetics. Evolutionary biologists attempted to identify the ecological factors that generated natural selection on these traits; they elucidated how the enzyme functioned in the context of those ecological factors; they examined local dynamics of allelic variants; and they surveyed geographic variation in the putative ecological factor and the allele frequencies at the locus in hopes that a concordance could be found to support the conclusion of biochemical adaptation.

While most studies of biochemical variants attained only a suggestive, inferential conclusion for the importance of natural selection, some research programs built convincing cases for biochemical adaptation. These studies are distinctive because they went beyond establishing an association between changes in allele frequencies and some feature of the environment, although the methods varied from study to study. Some of the best known examples include the studies of leucine aminopeptidase variants in *Mytilus edulis* (Koehn and Hilbish, 1987), the lactic dehydogenase polymorphism in *Fundulus heteroclitus* (Powers et al., 1983), or the variants at the phosphoglucose isomerase locus in Colias butterflies (Watt, 1977, 1983; Watt et al., 1983). Other well-investigated examples include studies of bacterial adaptation to nutrients (Dykhuizen and Dean, 1990), the action of different glutamic-pyruvic transaminase genotypes in *Tigriopus californicus* (Burton and Feldman, 1983), or hemoglobin variants in *Peromyscus* (Chappel and Snyder, 1984; Chappel et al., 1988; Snyder et al., 1988).

The most extensively studied biochemical polymorphism is the alcohol dehydrogenase gene in *Drosophila* (reviews include those by David et al., 1983; Van Delden, 1984; Chambers, 1988). The best-known polymorphism consists of the "Fast" and "Slow" alleles in *D. melanogaster*, but there is additional variation within and among species throughout the group. All of the elements of classical ecological genetics can be found in this literature, from comparative surveys of variation and its environmental correlates to detailed experimental studies on the role of the genetic variants in their ecological context.

Alcohol dehydrogenase is essential for the degradation of environmental ethanol in the larval habitat (Van der Zel et al., 1991). The ethanol content of natural larval habitats varies considerably from wineries to decaying fruits to decaying flowers to decaying leaves and fungi (Gibson et al., 1981; Oakeshott et al., 1982a, b; McKechnie and Morgan, 1982; Capy et al., 1988). There is a striking concordance among levels of ethanol tolerance, *Adh* activity, and larval habitat among species across the genus *Drosophila* and allied genera (Merçot et al., 1994), which strongly suggests that some form of adaptation to environmental ethanol has occurred and that adaptation might involve the alleles at the *Adh* locus. Indeed, there is widespread geographic polymorphism at the *Adh* gene that involves electrophoretic allelic variants (Oakeshott et al., 1982b) and nucleotide variations within electrophoretically detectable alleles (Simmons et al., 1989).

Three lines of experimental evidence support an adaptive interpretation of the variation at the *Adh* locus. First, studies of physiological function of allozymes support an adaptive explanation for their distribution (Heinstra et al., 1988; Freriksen et al., 1994). Second, studies of behavioral patterns of flies with different *Adh* genotypes bolster the credibility of the observed associations of ethanol tolerance, allozyme kinetics, and habitat (Hougoto et al., 1982; Hoffman and Parsons, 1984). Third, an enormous number of cage experiments under a variety of conditions demonstrate the adaptive significance of the allelic variation (Chambers, 1988; Freriksen et al., 1994).

C. What We Have Learned About Adaptation From Ecological Genetics?

Many of the classic studies of ecological genetics were based on phenomena that were well known long before the key experiments were done. Some study systems that we now use as the classic examples of adaptation were initially misinterpreted. For example, Dobzhansky (1937) said of industrial melanism:

> "The possibility that the selection favoring the melanic mutants may be operative is at least not excluded, although the attempts to ascribe to the darker coloration a protective

significance are certainly not convincing. Comparative physiological studies on the dark and light forms of the same species would seem to be the most hopeful source of information on the subject." (p. 160)

Mayr (1942) said of banding patterns on *Cepaea*:

"There is, however, considerable indirect evidence that most of the characters that are involved in polymorphism are completely neutral, as far as survival value is concerned. There is, for example, no reason to believe that the presence or absence of a band on a snail shell would be a noticeable selective advantage or disadvantage." (p. 75)

While neither of these authors would have required any convincing that adaptation was important, these quotes are a telling commentary on the state of the science by the early 1940s. Even at this time there were still very few studies that provided strong evidence for the adaptation. Circumstances had begun to change by the 1950s since the number of good experimental evaluations of possible adaptations had increased substantially. The continued study of these adaptations has begun to flesh out what seems to be lessons or rules of the process of adaptation.

Lesson 1 - *Adaptation is Not Simple:* All of the classics of ecological genetics generally began by testing a simple, unifactorial proposition, such as "visually oriented predators select for background color matching in their prey." In each of these cases above, experimental evidence supports this hypothesis. However, whenever these seemingly simple traits were examined more closely, it turned out that adaptation and selection were far more complicated issues.

Jones et al. (1977) summarize the collective literature on shell color and banding polymorphisms in *Cepaea nemoralis* and conclude that "...at least eight evolutionary forces are now known to affect its shell polymorphism" (p. 136). In addition to the background color matching and predator selection studied by Clarke and Sheppard, there clearly is an overriding effect of the environmental factors, such as temperature. Darker shells heat more rapidly in direct sunlight and there is a progressive increase in the frequency of darker shells along a south to north cline in western Europe. Similar patterns can be found on a more local scale, such as along altitudinal gradients. Because predators often form search images, there is frequency-dependent selection, favoring rare phenotypes. Because predators will prey on the congeners *C. hortensis* and *C. nemoralis,* when the two species are sympatric, morph frequencies in one species can influence morph frequencies in the other. Morph frequencies have also been found to be affected by gene flow from neighboring populations with different color backgrounds, genetic drift, and linkage disequalibria among color and banding genes. The conclusion, then, is that even a trait as simple as color polymorphism is influenced by a diversity of factors. A number of these, such as background matching, thermal biology, and frequency dependence, fall under the rubric of "natural selection."

If the classic examples of ecological genetics were more complicated than the initial euphoric claims indicated, so much more so were the case studies of biochemical polymorphisms. One might argue on *a priori* grounds that it was inevitable that the biochemical polymorphisms would be equally or more complex than the others. Most, if not all, of these studies identified an abiotic factor as the principal agent of selection, and abiotic factors rarely act in ecological independence from one another, and arguably never act independently of biotic factors (Dunson and Travis, 1991). Further, considerations of metabolic control suggest that it might not be possible to examine polymorphisms at individual loci independently of one another (Clark and Koehn, 1992; Whitlock et al., 1995).

The polymorphism at the *Adh* locus illustrates these lessons as well; not all of the variation in ethanol tolerance can be explained by the distribution of the *Adh* alleles, and the distribution of the *Adh* alleles is affected by more than just the distribution of ethanol levels in the larval environment (see Geer et al., 1993). Four lines of evidence lead to these conclusions. First, additional environmental factors such as temperature, carbohydrate level, and the presence of other volatiles in the environment play a significant role in either the success with which the allozymes function or the overall effectiveness of the metabolic pathway in which they are embedded (Starmer et al., 1986; Oudman et al., 1991). This means that the distribution of ethanol alone cannot fully explain the *Adh* variation. Second, while it is tempting, and probably not wholly incorrect, to regard the kinetic properties of the allozymes as the major link between genotype and fitness, it appears that the regulation of cellular *Adh* concentration may be at least as vital a component of that link as the properties of the allozymes, if not more so (McDonald et al., 1977; Laurie-Ahlberg et al., 1980). This means that the *Adh* alleles alone are insufficient to account for all of the variance in ethanol tolerance. Third, a variety of evidence implicates interactions with alleles at other loci as important elements in the determination of fitness (Cavener and Clegg, 1981; Van Delden and Kamping, 1989), which means that ethanol degradation is not the sole force behind the biochemical evolution in the larval flies. Fourth, flies use a variety of methods to cope with the ethanol challenge other than simple detoxification (Geer et al., 1993), which means that in some cases the *Adh* variation may have only minor consequences for fitness.

Jones et al. (1977) note that our scientific training leads us to think of the world in terms of exclusive, alternative hypotheses that are to be either accepted or rejected. Every example of an adaptation that has been studied from different perspectives has revealed that it is influenced by a multiplicity of potentially conflicting influences and hence is "not necessarily explicable in a simple and unitary way" nor can it be evaluated with hypotheses that are mutually exclusive. The empirical study of adaptation thus demands a more flexible application of the scientific method.

Lesson 2 - *Adaptation Requires Compromise:* The more detailed analyses of adaptation also show that adaptation often involves trade-offs or compromises among competing functions. The potentially conflicting demand of background coloration and thermal physiology in *Cepaea* represents one such trade-off. A second is reported by McNeilly (1968) for the evolution of heavy metal tolerance in *Agrostis tenuis*. When seed from copper tolerant and nontolerant plants were compared on copper-contaminated soils, all nontolerant seeds died within 7 weeks. All tolerant plants survived for that same interval. When the two types of seeds were grown together at high densities on uncontaminated soil, the nontolerant plants had a substantial growth advantage. Such a growth advantage generally translates into higher fitness. The implication here is that adaptation to copper-contaminated soils was gained at some cost, although the nature of the cost in this case is not known. Such trade-offs have proven to be a common feature of evolution by natural selection. They might be present because of some form of internal physiological mechanism, as is likely to be true for *Agrostis tenuis,* or because of the conflicting demands of the multiple selective factors that act simultaneously on an organism, such as for shell color in *Cepaea.*

An alternative way of viewing constraints is to realize that natural selection acts on individuals, not on specific features of individuals. When any feature of an individual is modified by the process of natural selection, there may well be correlated changes that will also influence fitness. The product of the process will be a balance of all of these changes. This has been more fully appreciated since the advent of the quantitative genetic approach (see below).

Lesson 3 - *Fitness is Relative and Context Specific:* Two underlying assumptions of population genetics are: (1) fitness can only be defined for a given phenotype relative to other phenotypes in the population and (2) the fitness of a phenotype is defined only for a specific environment. Ecological genetics has provided many concrete examples that illustrate that adaptation involves sorting among alternative phenotypes that differ in relative fitness and that the relative fitness of phenotypes changes with the environment. In some applications of ecological genetics, adaptation is gauged by changes in the frequency of a given morph in a given set of circumstances, such as an increase in the frequency of melanic moths in association with industrialization. Alternatively, adaptation is inferred from comparisons of morph frequencies in different populations, correlated with differences in the environment, such as the higher frequency of melanic morphs in industrialized areas. In all cases, an essential feature of the studies is that the relative fitness of the alternative types changes with the environment so, for example, melanic morphs have a higher relative fitness in industrialized areas, but peppered morphs have a higher relative fitness in nonindustrialized areas. This may seem too obvious to be worth mentioning, but it becomes part of a powerful argument against the claim that "evolution is tautological" (e.g., Peters, 1976).

The substance of this claim of tautology is that fitness is judged in retrospect; the phenotype that increases in frequency at the expense of others is by definition the one with the highest fitness. In ecological genetics, fitness is indeed first inferred from a correlation between the phenotype and the environment; however, such correlations represent only the first step in the investigation. Given the suggestion that a given aspect of the environment has selected for some form of adaptation, this then becomes a hypothesis that can be, and often has been, evaluated experimentally. Differences in morph frequencies are caused by differences in specific components of fitness, such as the relative survival, growth, and reproductive success of different morphs in alternative environments. At least some of these components of fitness have been quantified for all of the better studied cases in ecological genetics. Finally, in some cases it has been possible to manipulate the environment and evaluate the prediction that a presumably more fit phenotype for a given set of circumstances will increase in frequency (see example below for guppies).

IV. The Quantitative Genetic Approach - The Generalization of Ecological Genetics to Continuous Variation

Most of the classic ecological genetic studies began with the observation of a persistent discrete polymorphism in a relatively simple genetic system. Discrete polymorphisms are relatively rare and are not representative of the way morphological variation is distributed for most organisms in most circumstances. Instead, we most often see phenotypic variation that is continuously distributed, whether the traits are morphological, physiological, behavioral, or developmental. The multivariate distributions of such traits, more often than not, define individual species and characterize populations or groups of populations and as such are the very stuff of evolution. An empirical ecological genetics of continuously distributed characters took much longer to coalesce. When it did so, one of its first major elements was the explicit consideration of the correlations among traits and how direct selection on one trait would indirectly influence other traits. It is no coincidence that some of its proponents found their inspiration in the work of the biometricians of a previous era (Lande, 1988; Turelli and Barton, 1990).

A. Background

The quantitative genetic approach to adaptation emerged from the confluence of two intellectual streams. Beginning in the 1960s, a series of analytic models of quantitative genetic evolution were developed that allowed the exploration of phenotypic and genetic dynamics under the joint influence of selection, mutation, migration, drift, and various mating systems (reviewed at various levels of detail by Turelli, 1988; Barton and Turelli, 1989; Bulmer, 1989; and Turelli and Barton, 1990). One might trace this stream to its headwaters in the papers of Griffing (1960), Latter (1960), and Kimura (1965); however, Lande's work perhaps best illustrated how these models could be directed toward several problems in evolutionary biology (review in Lande, 1988; Arnold, 1994).

The second stream might be traced to Crow's (1958) examination of how fitness is distributed continuously within a population and how the variance in fitness might be partitioned into an "opportunity for selection" at various stages of the life cycle. In a series of papers that blended theoretical and empirical material, Wade and Arnold (1980) and Arnold and Wade (1984a, b) extended this idea and combined it with Price's (1970, 1972) covariance approach to selection to formulate a readily applied method for defining and quantifying phenotypic selection and separating it from the genetic response to selection. This work defined fitness as an individual's contribution to the next generation; this definition makes fitness a property of an individual. It defined selection as the covariance between a trait and fitness. While the covariance approach had been used in population genetics (e.g., Crow and Nagylaki, 1976), this application to the phenotypic level created a clear, operational distinction among fitness, selection, and genetic evolution.

This distinction, in conjunction with Sober's (1984) discussion of causation in selection, eliminated the last vestiges of the tautology criticism from the formulation of adaptive evolution. From these treatments it became clear that selection was something other than "the survival of the fittest" because fitness could be defined in the absence of selection. "Selection" was to be taken as the causal covariance between an individual's value of a trait and its fitness; the "causal covariance" was meant to signify that an individual's fitness took on the value it did because of the functional significance of its phenotypic values in a well-defined ecological context. Finally, only when trait variation had a genetic basis could it be said that selection would lead to evolution.

The confluence of these streams occurred with Lande and Arnold's (1983) multivariate extension of the Wade-Arnold conceptualization. Lande and Arnold combined explicit methods for quantifying direct and indirect selection on an arbitrary suite of traits with an analytic genetic model, derived from the larger series of models to which we referred earlier, to examine the consequent phenotypic dynamics. Their success is visible in the subsequent explosion of empirical work that used this approach [see Arnold (1994) and the collections of papers in Boake (1994) and Fritz and Simms, 1992)] and the collateral eruption of methodological studies on how best to detect and quantify selection and how best to interpret the results (e.g., Manly, 1985; Mitchell-Olds and Shaw, 1987; Schluter, 1988; Crespi and Bookstein, 1989; Phillips and Arnold, 1989; Wade and Kalisz, 1990).

B. Examples

Simms and Rauscher (1989) and Rauscher and Simms (1989) used this method to study the coevolutionary interaction between plants and herbivores. The work is exemplary in including a formal quantitative genetic design so that it is possible to look at the selection gradients on the phenotypes and breeding values (genotypes). Their study plant, the morning glory (*Ipomoea purpurea*), has additive genetic variation for resistence to herbivory, possibly in the form of secondary compounds that inhibit herbivory. The plants suffer a phenotypic cost of herbivory in the sense that plants produced an average of 20% more seeds when herbivores were excluded with insecticides. Our normal expectation when there is additive variance for a trait combined with evidence for directional selection is that the trait should show a progressive response to selection. In this case, one would predict a steady increase in plant resistance until all herbivory is eliminated.

Rauscher and Simms (1989) summarize three hypotheses for the observation of persistent herbivory in the face of additive genetic variation for resistance to herbivory. The most commonly assumed mechanism is that there will be a cost to resistance. Such a cost, in combination with other conditions, can result in an equilibrium at some intermediate level of resistance. A second possibility is that the plants are being observed at an intermediate stage during an episode of directional selection so that resistance is still increasing in frequency. A third is that the additive variance for resistance to herbivory is selectively neutral, possibly because low levels of damage do not result in a loss of fitness or because there is no cost associated with the observed variation in resistance.

Lande and Arnold's (1983) method allows one to discriminate in part among these alternatives because it is possible to estimate the magnitude of both directional and stabilizing selection on the different components of resistance to herbivory. The "cost" hypothesis predicts significant stabilizing selection, while the other two hypotheses do not. It turned out in this case that there was no evidence of stabilizing selection, but there was evidence for directional selection for resistance to one of the main herbivores. While there was evidence for a phenotypic cost of herbivory, there was no evidence that plants that were resistant to herbivory sustained a cost in terms of reduced seed production. The net result is that it is not possible at this time to distinguish between the second and the third hypotheses. Some data imply that resistance to herbivory might be an adaptation on the part of a plant, but the results do not resolve the anomaly of persistent genetic variation for a trait that appears to strongly influence fitness (i.e., significantly reduces seed production).

The crucial contribution of the quantitative genetic approach has been to force an explicit consideration of the covariances among characters. The phenotypic covariances, regardless of their origin, can create significant indirect selection on individual traits; the genetic covariances constrain the direction of the multivariate response to selection, at least in the short term. Grant and Grant (1995) illustrate the subtlety of such effects in their evaluation of a selection episode in the medium ground finch, *Geospiza fortis*, during a drought in the Galapogos in 1976-1977. Selection clearly favored an increase in body size, as evidenced by a significant net effect of selection on all individual body measurements. However, the ecological forces acting through food availability were actually selecting for a decrease in bill width to allow more effective harvesting of the available food. Nonetheless, the bill widths of the surviving birds and their offspring represented a net increase over the average values before the episode of selection. This occurred because there was very strong selection for an increase in bill length and bill depth, which are both positively correlated phenotypically and genetically with bill width. The net indirect effect of selection for wider bills, which occurred through the strong correlations with bill length and depth, overwhelmed the direct force of selection for a narrower bill.

C. Limitations

If this approach is employed in a strictly statistical manner, without collateral ecological expertise, it can mislead. While the Lande-Arnold method allows the separation of direct and indirect selection on traits, its accuracy depends on the investigator's initial diagnosis of the trait cluster and on the precision and accuracy of the estimation of fitness (Manly, 1985).

The first of these linchpins, the choice of traits for inclusion, might be best expressed as a repetition of the age-old admonition to "know your organism." Any investigator will rely on preliminary data, prior work, or a knowledge of the ecological role of individual traits and their developmental connections in order to decide which traits to include. Functional constraints intrinsic to the organism can be used as well, although there may not be a one-to-one correspondence between a specifically known constraint and the genetic covariance that is measurable. For example, in most butterflies and moths there will be a functional constraint between a female's reproductive effort for the first brood and that for subsequent broods. This constraint is based on the fact that all of the protein reserves required for reproduction are obtained in the larval stage; more expenditure on the first brood means less on the remaining broods. However, if the major axis of genetic variation is found in the amount of protein harvested as a larva, then there can be a positive covariance between present and future reproduction that does not reflect the functional constraint known to every student of entomology. Thus a functional constraint need not be reflected in an observed genetic covariance, and an observed genetic covariance may not reveal all of the germane functional constraints. This is essentially the argument offered by Van Noordwijk and De Jong (1986) in their graphical model and bolstered by Charnov (1989) in the terms of optimization and Charlesworth (1990) and Houle (1991) in quantitative genetic form. The statistically inclined will recognize that some form of partial correlation would clarify these situations, but the point to be made here is that a thorough knowledge of the organism, its ontogeny, development, and physiology is likely to be the best prerequisite for choosing trait clusters for study.

The second linchpin, the estimation of fitness, is more troublesome because even accomplished field biologists will find fitness difficult to estimate. This is especially true if generations overlap and the populations under scrutiny are not near their demographic equilibria. Various surrogates for fitness may yield the correct rank order of individuals but not the correct variance. The detailed estimation of direct and indirect selection forces depends on correct estimation of the variance in fitness (see Manly, 1985; Travis and Henrich, 1986).

In any statistical method, the practitioner must not confuse correlation and causation. The Lande-Arnold method and its progeny contain no guarantee against this confusion. This issue has sparked debate over the utility of this approach [see Mitchell-Olds and Shaw (1987), plus other papers cited above]. The problem of causation is not a problem inherent in the quantitative genetic approach but in its careless application. Nonetheless, we occasionally take an impression from some presentations that the investigator expects the methods themselves to guard against this miscue, which is incorrect.

The most controversial limitation of this approach lies, ironically, in what is also its greatest virtue, namely the identification of genetic covariances that constrain adaptation. There is no question that these covariances constrain present response, at least in the short term. What is unclear is whether these covariances function as long-term constraints. If they do, then the vectors of selection gradients, in combination with the additive genetic covariances, would allow us to infer the large-scale, long-term history of multivariate adaptation and evolution (Grant, 1986; Arnold, 1988). Such a "genetic uniformitarianism" would link micro- and macroevolutionary studies.

Two classes of problems potentially limit the usefulness of the genetic covariances uncovered by the quantitative genetic approach. First, selection may be highly variable on a time scale longer than that of the research program. This might be true because the population is not at an ecological equilibrium with its abiotic or biotic environment or because short-term studies offer a misleading picture of the long-term nature of selection. For many natural populations, one or the other situation seems quite likely to be common (e.g., Trexler et al., 1992). This is an empirical issue, to be sure; only the accrual of long-term studies will tell us how problematic these considerations might be.

The second set of problems revolves around the constancy of genetic covariances. If these covariances change more slowly than mean genotypic values, then the covariances indeed indicate the larger constraints that channel multivariate evolution. The covariance matrix will function in this way under a suite of conditions: weak to moderate selection, moderate to large numbers of loci governing the trait, large effective population sizes under selection, and particular distributions of pleiotropic effects of alleles (e.g., Turelli and Barton, 1990;

Arnold, 1994). How often these conditions are met in the process of adaptation is unknown. Empirical approaches to this issue range from comparisons of genetic or phenotypic covariances with the major axes of population differentiation (Palmer and Dingle, 1986; Lofsvold, 1986; James, 1991), and examinations of genetic covariances before, during, and after artificial selection (Wilkinson et al., 1990). While some of these studies (among others in the literature) suggest support for the constancy of these covariances, others (e.g., James et al., 1991) do not. The number of studies done to date are too few and too restrictive in types of traits and time scales of change to offer any conclusions; this is an open question.

V. Methods for the Empirical Study of Adaptation In Nature

A. Empirical Perspectives on Adaptation

When faced with explaining why organisms look, function, and act as they do, biologists from different disciplines are prone to shade their explanations with their own predilections. Systematists might look first to an organism's phylogenetic heritage, ecologists to its environmental milieux, and geneticists to its amount and distribution of genetic variation. The major lesson to be drawn from the debate about adaptation over the last 17 years is the danger of a narrow perspective (Gould and Lewontin, 1979; Maynard Smith, 1980; Gould and Vrba, 1982; Reeve and Sherman, 1993). Explanations from each of these different perspectives are correct in part, but each is by itself incomplete (Mayr, 1988). The critical issues are how strong are the influences of each factor and whether one factor overwhelms any of the others.

These critical issues cannot be resolved by considering adaptations only in a static context of which traits are seen in which taxa in which ecological milieux. The reason for this is that static observations cannot define cause and effect relationships. The most effective way to establish cause and effect is to examine the evolutionary dynamics of adaptations. For one reason, it is the only way we will evaluate the usefulness of the larger genetic models of adaptation, whether the roots of those models are in quantitative or population genetics. But more importantly, the study of evolution in action can provide the most convincing line of evidence for adaptation.

Obviously this prescription is more easily written than accomplished because we cannot recreate the prior evolutionary dynamics of wild populations to satisfy our scholarly curiosity. Even if we could do so, we would have no guarantee that we had recreated history as it unfolded the first time; we are prisoners of the very historical phenomena we seek to understand. Our only recourse is to observe patterns in nature and to somehow devise complementary studies of contemporary dynamics that can uncover the extent to which these patterns have been molded by adaptive evolution. This is what Weldon (1899) attempted when he exposed crabs to suspended clay sediments to imitate what he thought caused morphological evolution in natural populations. How best to accomplish this feat is the subject of subsequent sections.

Before considering methodology, it is important to return to the issue of multiple perspectives because the most compelling cases for adaptation combine the observation of patterns (a static approach) with experimental studies (a dynamic approach) from several perspectives. The literature on the alcohol dehydrogenase polymorphism in *Drosophila* illustrates this point. Four lines of "static" evidence implicate biochemical adaptation. First, there is the striking concordance among levels of ethanol tolerance, *Adh* activity, and larval habitat among species to which we referred earlier. However, this concordance is heavily determined by the nonrandom distribution of clades among habitats, which implies a significant role for phylogeny in determining the taxa in which higher ethanol tolerance is observed. Second, there are repeated patterns of geographic variation in allele frequencies at the *Adh* locus on different continents (Oakeshott et al., 1982b). These are suggestive but not conclusive because there is a plausible alternative explanation rooted in a historical effect (Kreitman et al., 1992). Third, the number of fixed amino acid replacements in the *Adh* gene between species exceeds the expectation based on neutral evolution of silent substitutions (McDonald and Kreitman, 1991). Fourth, the levels and locations of nucleotide polymorphisms within species are incompatible with the expectation derived from a neutral evolutionary process (Kreitman and Hudson, 1991).

While some may find these arguments compelling, they remain demonstrations of pattern in nature that are only plausible outcomes of an adaptational process. They are made more compelling by their support in the three lines of experimental evidence that we described earlier. No single line of evidence would convince all readers; yet, given the evidence in total, it is difficult to avoid the conclusion that the *Adh* polymorphisms and their distributions have adaptive significance. The patterns in nature reflect plausible results of the qualitative dynamics observed in laboratory experiments; conversely, the laboratory experiments illustrate the type of dynamics to be expected if the patterns in nature had been developed through a process of natural selection. We argue that this confluence of evidence offers the most compelling empirical argument for adaptation.

Our expertise is developing the lines of evidence for the evaluation of adaptation in animal populations. In the remainder of this section we offer our perspectives on how this is best accomplished.

B. Methods for Studying Adaptation

The development of ecological genetics encouraged the development of new techniques of general utility to the study of evolution in natural populations. We will argue that these methods, if applied hierarchically, can provide progressively stronger arguments that a given trait is an adaptation to specific type(s) of selection. They can also reveal the details of the mechanisms of natural selection. We will highlight aspects of our own research programs to illustrate the application of these techniques.

1. Comparative Studies

The starting point for most ecological genetics studies has been to make comparisons among populations or a temporal series of collections. For industrial melanism, the first comparison was among samples collected at different times, revealing a progressive increase in the frequency of melanic morphs. The second comparison was among contemporaneous samples collected from areas that differed in the degree of industrialization. The associations revealed by these comparisons were the first step in identifying factors that might have selected for changes in morph frequency. More generally, the comparative method represents a powerful technique for establishing a relationship between the phenotype and specific features of the environment, and hence providing clues to the potential adaptive significance of those traits.

2. Genetic Analyses

The second hierarchical component in the study of adaptation is the demonstration that the trait variation under scrutiny has a genetic basis. This follows from the ecological genetic approach, which is to ask whether the phenotypic variation among populations is based on genetic differences that have arisen as divergent responses to different pressures of natural selection. While this protocol may appear straightforward, the term "genetic basis" can have a variety of interpretations. Moreover, several factors can complicate the attempt to discover the genetic basis for phenotypic trait variation. These two sets of issues are not independent of one another and are vital considerations for designing the most appropriate experiment.

The first issue to confront is whether the trait variation is more likely to be controlled by a simple one- or two-locus genetic system or many loci, each of which exerts a small to moderate effect (the paradigmatic quantitative genetic system). The empirical problem is straightforward for discrete, single-locus or simple two-locus traits: perform Mendelian crosses among morphs in order to uncover the inheritance of the trait. In many such cases, the variation among populations is then readily interpreted as variation in the relative frequencies of different alleles. This is obviously the case for most biochemical polymorphisms. One set of experimental crosses, properly executed, allows the levels of genetic variation within and among populations to be quantified simultaneously by the subsequent assessment of allele, genotype, or haplotype frequencies. The most divergent of these populations can be selected for subsequent ecological study.

This simplicity does not extend to quantitative genetic variation. In these systems, levels of within- and among-population genetic variation are potentially independent quantities. To see this, consider a situation in which the age at maturity is affected by 600 loci, 200 of which are polymorphic within a species like the guppy. The age at maturity can vary widely within and among populations. We might find one pair of populations in which each is fixed for alternate alleles at all 200 loci; one population has fixed alleles that induce an age earlier than the species' average and the other has fixed alleles that induce a later age. The genetic component of their divergent phenotypes is very large, but there is no heritability within either population. Conversely, we might find another pair of populations that harbor intermediate allele frequencies at all 200 loci, with one population having a small majority of alleles that induce earlier age at maturity (relative to the species' average) and the other having a small majority of alleles that induce later age. The genetic differences between these populations are slight, as will be their phenotypic differences, but the heritability within each population will be quite large. Thus, in principle, the demonstration of genetic variation for a trait within a single population is not sufficient to demonstrate that the variation among populations has a genetic basis; conversely, the lack of genetic variation within a population need carry no information about the genetic basis of variation among populations. To make a stronger statement, the demonstration of a trait's heritability sensu strictu is neither a necessary nor a sufficient condition for studying the adaptive significance of trait variation among populations.

The simplest interpretation of what a "genetic basis" can mean in this context then is that the phenotypic differences are maintained when individuals from different populations are raised in a common environment (e.g., Law et al., 1977). This method has the advantage of simplicity of execution but it carries three potential disadvantages. First, the maintenance of differences can, in some circumstances, be due to effects other than genetic ones. For example, maternal effects can be very influential for many types of traits. These effects can be induced by the environment experienced by the female parent (Roach and Wulff, 1987; Mousseau and Dingle, 1991) and can persist for one or more generations; if female parents from different populations have different environmental experiences then the maintenance of phenotypic variation in offspring raised in a common environment can reflect these differences and not the "genetic basis" sought by the investigator. Some maternal effects may themselves be genetically based, in which case two generations of hybridizations are necessary to interpret the maintenance of phenotypic differences as indicative of genetic distinctions (Reznick, 1981, 1982a). To preclude either type of maternal effect it is necessary to rear the organisms through at least two generations in the common environment.

The second potential disadvantage of the simple common garden experiment is that the environment of the "common garden" may not be one in which genuine, relevant, genetically based differences are manifested. This can occur in two ways involving phenotypic plasticity in the trait under scrutiny. The first occurs when no differences are manifested; here the common environment is an unfortunate combination of conditions in which different genotypes happen to produce similar phenotypes. The second is when the common environment represents a novel combination of environmental stimuli that induces genetically based phenotypic variation that would not be expressed in any of the natural environments. These two results are extreme cases of the well-documented general principle that the levels of genetic distinctions among populations will depend upon the environment in which they are measured (reviewed by Travis, 1994a).

These concerns lead to the prominence of the reciprocal transplant method, in which individuals from two (e.g., Berven et al., 1979) or more (e.g., Schmidt and Levin, 1985) populations are raised in each location. In this design the phenotypic variation can be partitioned into effects attributable to local environment, population of origin, and the interaction of population and environment. The second and third of these carry information about the genetic basis of trait variation. This approach (presuming that maternal effects have been accommodated) has the advantage of allowing genuine, relevant, genetically based variation to be detected. The chief disadvantage is logistical; this is difficult work. In some cases it may not be possible to perform these experiments in the field; one solution to this problem is to perform common garden studies with at least two "common gardens" that vary in a set of environmental conditions meant to mimic natural variation among environments (Winn and Evans, 1991).

The third potential disadvantage of the common garden experiment is one that it shares with the simplest of reciprocal transplant experiments: neither reveals the underlying nature of the genetic distinctions among populations. While at first glance it may seem esoteric to bother with the question of whether genetic differences are purely additive or not, this information can elucidate a number of additional questions about the nature of adaptation. For example, various combinations of hybridizations and backcrosses can, in principle, suggest the number of independent genes on which the adaptation is based (Cockerham, 1986). In turn, this information can illuminate the debate over the constancy of genetic covariances because constancy is unlikely when a small number of independent loci control trait variation (Turelli and Barton, 1990). Several other issues can be illuminated by data arising from interpopulational crosses of various types (reviewed in Whitlock et al., 1995), provided that the environments in which the offspring from such crosses are raised are appropriate ones for inducing germane phenotypic variation.

This discussion should not be taken to indicate that we are opposed to common garden experiments. All of the approaches we have discussed have advantages and disadvantages; each can provide a different kind of information and each carries its own potential pitfalls. The choice of approach should be based on the level of knowledge about the "genetic basis" that is practical to acquire and after careful consideration of which pitfalls can be avoided and which advantages are most desirable.

Nor do we relegate quantitative genetics of intrapopulation variation to a clearly subsidiary role. The major contribution of this approach is to identify the genetic covariances that constrain current responses to selection. We feel that it is important to identify these constraints, indeed that adaptation cannot be fully understood without such knowledge. Moreover, we feel it is still an open empirical issue (as do others, e.g., Arnold, 1994) whether the adaptive differentiation of populations has in fact occurred along axes that are described by the patterns of genetic covariances among traits. What we are advocating is a clear distinction between the goals of detecting within- and among-population genetic variance and covariance and that methods designed for one goal not be misinterpreted in their role in the search for the other.

3. Mark-recapture

An early innovation of the school of ecological genetics was the refinement of mark-recapture techniques (Fisher and Ford, 1947). These were originally used to estimate population size and to evaluate the movement of individuals. However, being able to know organisms as individuals and to examine them repeatedly also allows us to evaluate mortality rates (LeBreton et al., 1992) and lifetime reproductive success (Clutton-Brock, 1988). When such data are combined with an evaluation of variable and potentially adaptive features of organisms, it is possible to generate detailed studies of the dynamics of natural selection. While such studies are rare, the existing samples are sufficient to illustrate the power of the technique.

Undoubtedly the best example is the extended body of research on the Galapagos finches by Peter and Rosemary Grant and their colleagues (e.g., Grant, 1986; Grant and Grant, 1989). One key result of this work has been to follow the effects of El Niño events on the quantity and quality of food available to finches and on the correlated changes in body size and bill dimensions. Their combination of ecological and behavioral observations with details on the relative reproductive success of individuals as a function of their morphology leaves no doubt that body size and bill dimensions represent adaptations to prevailing food supplies. Their results also emphasize the context-specific nature of relative fitness since it has been possible to follow repeated cycles of increases and decreases in body size, in association with cyclical changes in the environment (e.g., Gibbs and Grant, 1987).

4. Experimental Approaches

One tradition established in some ecological genetics studies involved manipulating a population in a way that allows one to more directly evaluate the contribution of a trait to fitness. For example, Brower and colleagues (1967) manipulated the color pattern of a mimetic species of butterfly by adding and removing characteristic marks on its wings, thus affecting its qualities as a mimic. Improved mimicry was associated with a short-term increase in the probability of recapture and, presumably, an increase in survivorship. In general, such experiments allow us to more precisely interpret variations in the phenotype as adaptations to a given set of circumstances. The multifactorial nature of selection on any single trait, much less the likely importance of multiple traits in the phenotype, has led us to be strong advocates of experimental studies as a necessary (but not sufficient) element in the study of adaptation.

We will illustrate the application of these methods with examples from our own research programs.

VI. Life History Evolution in Guppies

One of us (DR), along with Heather Bryga, Mark Butler, John Endler, and Helen Rodd, has applied all of these methods as part of considering whether or not the life history traits of guppies (*Poecilia reticulata*) are adaptations to predator-induced mortality rates. Guppies are small, live-bearing fish in the family Poeciliidae. I studied populations from Trinidad that had been previously studied over a 25-year period by a series of investigators (Haskins et al., 1961; Liley and Seghers, 1975; Endler, 1978). I focused on guppies from two types of communities that differed in the types of predators with which guppies co-occur. One type had predators that frequently prey on adult size classes of guppies, suggesting a high rate of predation on guppies. We refer to these localities as "*Crenicichla*" localities, so named for the pike cichlid *Crenicichla alta* that is found there and that appears to be a key guppy predator (Haskins et al., 1961; Liley and Seghers, 1975; Endler, 1978). In the alternative type of community, only the killifish *Rivulus harti* co-occurs with guppies; these sites will hence be referred to as "*Rivulus*" localities. *Rivulus* is an omnivore that only occasionally includes guppies in its diet. Because *Rivulus* is the only serious predator at these sites and because of its relatively low predation rate on guppies, we assumed that predator-induced mortality rates would be much lower in this type of environment. Also, *Rivulus* preys predominantly on small, immature size classes of guppies. I therefore predicted that there would also be differences among localities in the susceptibility of different age classes of guppies to predation; adults would suffer higher mortality rates in *Crenicichla* localities, while juveniles would suffer higher mortality rates in *Rivulus* localities.

An important feature of the distribution of these communities is that they are often found in the same river valley and in very similar environments. The distributions of predators and guppies are often punctuated by waterfalls and rapids that exclude some species, creating different fish communities above and below the barrier, with no discernible difference in other features of the environment. These patterns are repeated in a series of drainages, providing many replicates of the contrast between *Rivulus* and *Crenicichla* localities.

Prior to the initiation of this work, three theoretical treatments of life history evolution (Gadgil and Bossert, 1970; Law, 1979; Michod, 1979) considered the potential consequences of changes in age-specific mortality. All three predicted

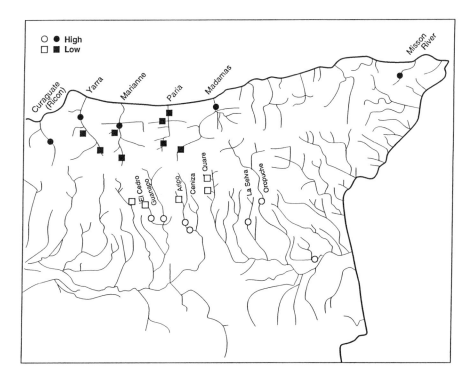

Figure 1 Map of northeastern Trinidad and the main north and south flowing drainages of the northern range. Solid symbols indicate sampling sites included in our survey of north flowing streams. The dominant predators in the high predation sites of the north slope include the goby *Eleotris pisonis* and the mullet *Agonostomus monticola*. The low predation sites include the killifish *Rivulus harti* and prawns in the genus *Macrobrachium*. Open symbols indicate sampling sites from south flowing streams. Dominant predators in high predation sites on the south slope include the cichlid *Crenicichla alta*, plus species in the family Characidae.

that an increase in adult mortality rates would select for a decrease in the age at maturity and an increase in reproductive effort. Subsequent theory (e.g., Charlesworth, 1994) often supported this prediction, but the fabric of life history theory has also become far more complex and now admits alternative predictions (e.g., Orzack and Tuljapurkar, 1989). Nevertheless, the existing theory provided a basis for making predictions about how guppy life histories would evolve in response to these differences in predation.

My first step was a comparative study to evaluate the correlation between guppy life histories and predator community. My colleagues and I initially evaluated the life history phenotypes of wild-caught guppies from 12 field localities (Fig. 1; Reznick and Endler, 1982). I also evaluated the genetic basis of these life history differences in the second generation, laboratory-reared descendants from two *Crenicichla* and two *Rivulus* localities (Reznick, 1982b). I reared the fish

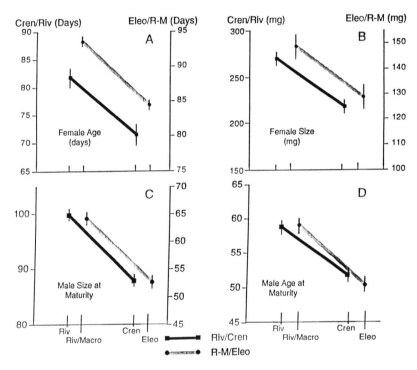

Figure 2 Least square means (+ 1 standard error) for the age and size at maturity and size at first parturition in females from high versus low predation environments. These data are based on the comparison of fish from two high and two low predation sites on the north and south slopes, for a total of eight localities. The stippled lines represent the means for the high (Eleo) versus low (Riv/Macro) predation sites on the north slope (values from Reznick et al., 1996b). The solid lines are the corresponding values for the high (Cren) and low (Riv) predation sites on the south slope (from Reznick, 1982b). Lower levels of food availability were used in the experiment on north slope guppies. A consequence of this is that these fish tended to mature at a later age and smaller size than the south slope guppies. To facilitate comparisons among the two experiments, the y-axes were scaled differently so that the proportional differences between high and low predation fish remained the same. All differences between the high and low predation sites were significant (p<0.05). 2(A). Female age at first parturition (days). 2(B). Female size at first parturition (wet weight in mg). 2(C). Male size at maturity (wet weight in mg). 2(D). Male age at maturity (days).

through two generations in a common environment before making comparisons to control for any possible influences of the environment on the life history phenotype. I assumed that differences among population means that persist after two generations in a common environment have a genetic basis. These initial comparisons revealed both phenotypic and genetic differences in guppy life histories that were consistent with the predictions of life history theory; guppies from *Rivulus* localities were older at maturity than their counterparts from *Crenicichla* localities (Fig. 2). In separate analyses, we also showed that guppies from high predation localities had higher levels of reproductive effort, meaning that they devoted a higher percentage of consumed resources to reproduction (Reznick, 1982b; Reznick and Bryga, 1996). I extended these comparisons to additional dependent variables not covered by the life history models cited above,

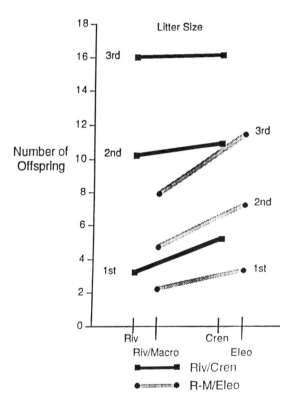

Figure 3 Number of offspring in the first three litters for guppies from high versus low predation sites on both slopes of the northern range. These data are from the same experiments as summarized in Fig. 2. The stippled lines represent the results for the north slope comparisons. In this case, guppies from the high predation localities produced significantly more offspring, after correcting for female size differences, than their counterparts from the low predation localities. The solid lines represent the results from the south slope comparisons. Guppies from high predation localities produced significantly more offspring in the first litter, but not in the second or third litters.

such as the number and size of offspring or the frequency of reproduction (Figs. 3 and 4). Guppies from high predation localities produced more and smaller offspring per litter plus reproduced more frequently. These additional details reveal how the fish could devote different proportions of consumed energy to reproduction. Guppies from *Crenicichla* communities devoted more to reproduction primarily through timing; they began to reproduce sooner and reproduced more frequently. They also augmented their reproductive efforts by producing litters that were a larger proportion of their body weight.

The phenotypic comparisons have since been extended to 40 new localities in Trinidad, Tobago, and Venezuela and to new types of predator committees

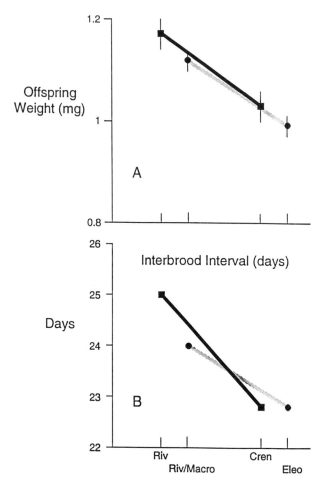

Figure 4 Least square means (+ 1 S. E.) for high versus low predation sites for (A) offspring weight (third litter only) and (B) interbrood interval These data are from the same experiments as summarized in Fig. 2. The stippled lines represent results from the north slope comparisons, while the solid lines represent results from the south slope comparisons. Guppies from high predation localities produced smaller offspring and had shorter intervals between broods than their counterparts from low predation localities.

(Reznick, 1989; Reznick et al., 1996a; Reznick, in preparation). Some of these communities, found on the north slope of the Northern Range of Trinidad and on Tobago, are similar to the south slope in having a contrast between high and low predation; however, they contain a very different suite of predators. *Crenicichla* and all of the other predators associated with the south slope of the Northern Range are absent. They are replaced with species of gobies and mullets, derived from a marine environment. Nevertheless, the life history patterns are the same as on the south slope. The genetic comparisons have been extended to 10 new localities, including the types of fish communities found on the north slope (Figs. 2-4; Reznick and Bryga, 1996; Reznick, unpublished). All of these new results are consistent with the original observations. They strengthen the idea that it is mortality rates rather than other variables that matter because these patterns persist in spite of large differences in the nature of the habitat and in spite of there being an entirely different suite of predators in the high predation communities.

New data on the genetic affinities of guppies from these different rivers demonstrate that guppies from *Rivulus* communities in a given drainage are most closely related to those from *Crenicichla* communities in the same drainage (Carvalho et al., 1991). This suggests that the differences in life histories evolved independently in each river, again strengthening the argument these life history patterns represent adaptations to local conditions.

Taken together, these data suggest that guppy life histories have evolved in response to these differences in predation and hence their life histories are an adaptation to predator-induced mortality. This amount of evidence simply establishes a pattern, but does not define cause and effect relationships. Specifically, if mortality rates have selected for guppy life history evolution, then it should be possible to document such effects in natural populations.

The proposition that guppies from *Crenicichla* localities experience higher mortality rates was based primarily on laboratory and field observations of predation and stomach content data (Haskins et al., 1961; Seghers, 1973; Endler, 1978; Mattingly and Butler, 1994). I felt that it was most relevant to characterize the mortality rates of guppies in natural populations since these, rather than the indirect characterizations of the predators, are what will determine the course of evolution. I used mark-recapture techniques to do a comparative study of mortality rates of guppies from high versus low predation localities. If the differences in predation caused differences in guppy mortality rates, then the recapture probabilities of guppies from *Crenicichla* localities should be lower. If the *Crenicichla* prey selectively on adults, then the difference in mortality should be more dramatic for the adult age classes. We did, indeed, find that the overall probability of recapture was substantially and significantly lower in *Crenicichla* localities, implying higher mortality rates (Fig. 5; from Reznick et al., 1996b). Such a result reveals a potential mechanism of life history evolution and thus goes a step farther in arguing that the differences in guppy life histories among *Crenicichla* and *Rivulus* localities represent an adaptation to predator induced mortality.

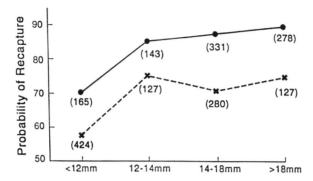

Figure 5 Recapture probabilities of juveniles, mature males, and females from seven high (*Crenicichla*) and seven low (*Rivulus*) predation localities. For purposes of analyses, we divided the fish into four size classes. The smallest two (<12 and 12-14 mm) consist of immature individuals. The 14 to 18-mm size class includes >95% of mature males and all females reproducing for the first time. The >18-mm size class includes females that were producing their second or subsequent litters. The number of individuals marked and released in each size class is indicated in parentheses next to each data point. The probability values are derived from a log-linear analysis with survival, size, and predator as independent variables. The probability of recapture of guppies from *Crenicichla* localities was significantly lower than for *Rivulus* localities (p<0.001). The probability of recapture from both types of localities was significantly less in the <12-mm size class than in the larger size classes (p<0.001).

There are, however, complications in interpreting mortality rates as the sole or direct mechanism that has selected for life history evolution. First, Fig. 5 reveals that, while the mortality rates are higher in *Crenicichla* localities, these higher rates apply equally to all size classes rather than being concentrated on adults. Some treatments of life history theory predict that no differences in life histories will evolve in such circumstances (Law, 1979; Michod, 1979). Others predict the evolution of patterns similar to what we have observed (Kozlowski and Uchmansky, 1987; Charlesworth, 1994). There is some evidence that these differences in predation have also caused differences in the pattern of habitat utilization, population density, and resource availability. Charlesworth (1994) predicts that when density dependence is present, the way the life history evolves will depend on how density effects are manifested. We are currently investigating the potential importance of density dependence and the possible interactions between predation and density. As with shell color and banding polymorphisms in *Cepaea*, the ultimate explanation for the evolution of these patterns is not likely to be a simple, unifactorial one.

A second way to strengthen the argument that life histories are an adaptation to mortality schedules is to manipulate the mortality rates of guppies in nature to evaluate whether or not they will evolve in a way that mimics the comparative pattern. I have exploited the discontinuities imposed by natural barriers on the distribution of guppies and their predators to evaluate more directly the effects of predation. In two experiments (Reznick and Bryga, 1987; Reznick et al., 1990), we worked on streams with barrier waterfalls that stopped the upstream dispersal of all species of fish except *Rivulus*. Guppies were found below the barrier waterfall in *Crenicichla* communities. We introduced these guppies over the barrier waterfall and hence moved them from a high predation to a low predation environment (Fig. 6). If predators are responsible for the evolution of the life history

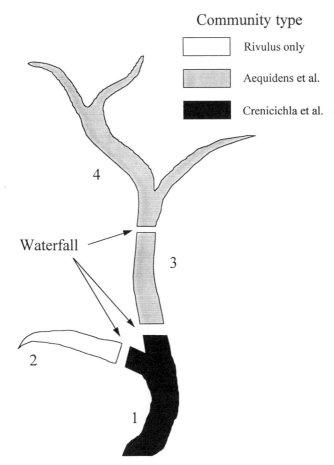

Community type

Rivulus only

Aequidens et al.

Crenicichla et al.

4

Waterfall

3

2

1

Figure 6   Schematic of guppy introduction experiments in Trinidad. In one type of experiment, a barrier waterfall excludes all fish except *Rivulus*. In this case, guppies from the high predation locality below the barrier waterfall (Site 1) were introduced into the low predation locality above the waterfall (Site 3). Two experiments of this sort have been initiated. Some of the results of this work are summarized in the text and in Table 1. In a second type of experiment, the barrier waterfall stops predators like *Crenicichla alta*, but not guppies, *Rivulus* or some other species of fish, including the cichlid *Aequidens pulchur*. In this case, the level of predation on guppies was increased by introducing predators (*C. alta*) from below the barrier waterfall over the waterfall (from Site 1 into Site 3). This presumably causes an increase in guppy mortality rates. Life history evolution is evaluated with a comparison of genetic differences in guppy life histories in the second laboratory generation of fish. For the two experiments reported in the text, guppies from Site 2 (experimental reduction in mortality rate) were compare with those from Site 1 (high predation controls). For the experimental increase in mortality rate (not reported in the text), guppies from Site 3 (predator addition) were compare with those from Sites 1 and 4 (high and low predation controls, respectively).

differences of guppies from the two types of populations, then we predict the evolution of life histories in the introduced guppies to match their new mortality patterns.

We followed the evolution of the guppy life histories and found that they changed in a way that is consistent with the predictions of life history theory and our comparative studies; guppies introduced into a *Rivulus* environment evolved delayed maturity and decreased reproductive effort relative to their progenitors from the *Crenicichla* locality downstream (Table 1). These experiments have thus allowed us to follow the dynamics of the evolution of the trait. It leaves no doubt that something about these differences in life histories, or something highly correlated with them, is adaptive. The mark-recapture methodology demonstrates that the guppies from the introduction sites experience lower mortality rates than those from the high predation controls. It thus appears very likely that predator-induced mortality is important as either a direct or an indirect agent in selecting for these differences, although the entire story is certainly not yet revealed.

All of these results are required to argue that differences in guppy life history patterns among localities are adaptations to predator-induced mortality rates. Comparative studies reveal a strong association between predation and life histories, but do not reveal cause and effect relationships. The studies of a wider range of communities, sometimes with entirely different suites of predators, strengthen the argument for predator-induced mortality as the likely cause rather than other aspects of the environment. Introduction experiments reveal that life histories will evolve in a predictable way in response to local conditions and that the response parallels the patterns revealed by the comparative studies; however, the results of each experiment do not generalize to other localities nor do they specify what caused the life histories to evolve. Mark-recapture studies reveal the expected differences in mortality rates, but not the actual causes of the differences in mortality rates among natural populations. Lab studies and field observations of the predators reveal that they really do differ in their tendency to eat guppies, but do not confirm that they cause differences in mortality rates in nature. Finally, there are the potential effects that are confounded with predation, such as guppy population density. Each result by itself has some weaknesses. When they are all taken together, they make a strong case for guppy life histories being an evolved response to prevailing mortality rates and for predators being a major cause of these differences in mortality rates.

VII. Body Size and Life History in the Sailfin Mollie

The sailfin mollie, *Poecilia latipinna*, is a poeciliid species native to the salt marshes of the southeastern United States and northeastern Mexico and to the freshwater springs and lakes of peninsular Florida. One of us (JT), along with Joel Trexler, James Farr, Michael McManus, and Margaret Ptacek, has examined the enormous variation in the body size of mature males within and among populations throughout the eastern half of the species range [Fig. 7; see Travis (1994b) for a longer review of this work]. Like all poeciliid males, mollies do not grow appreciably after maturity so the size distribution in nature reflects the distribution of size at maturity. The major source of size variation is a Y-linked series of alleles (Fig. 8), so in part the variation among populations reflects distinct distributions of these alleles. A suite of other life historical and behavioral traits covary tightly with male body size within each population; the covariation is not a function of social conditioning or experience and appears to be genetically based (Fig. 9). However, variation among populations in the behavioral traits is not fully dependent on the body size variation (Fig. 10), which means that at least to some extent body size and behavior have been decoupled in evolution. These are the

TABLE I

Results of two introduction experiments: Both of the introduction experiments summarized here follow the model presented in Fig. 6, in which guppies are introduced from a high predation locality (Site 1) into a low predation locality (Site 2). All of these results represent are introduced from a high predation locality (Site 2). The first two columns (from Reznick and Bryga, "common garden" comparisons of the second generation, lab-reared offspring from wild-caught females. The first two columns (from Reznick and Bryga, 1987) represent a comparison of the control and experimental populations for guppies introduced over a barrier waterfall on the El Cedro river. This assay represents a comparison of fish collected 4 years after the introduction, which is equal to approximately 6.9 generations. At this time, only male age and size at maturity had changed. In a subsequent assay, based on fish collected 7.5 years after the introduction or approximately 12.7 generations, both males and females from the introduction site had delayed maturity. The results from Reznick et al. (1990) are based on a duplicate experiment on a tributary to the Aripo river. In this case, the parental generation was collected 11 years, or approximately 18.1 generations, after the introduction. Most aspects of the life history had evolved in the predicted fashion in this study. The Reznick (1982) data are from a comparison of two Rivulus and two Crenicichla localities and are included here to provide a frame of reference. Note that differences in mean values among experiments are largely attributable to differences in food availability.

| | Reznick and Bryga (1987) | | Reznick et al. (1990) | | Reznick (1982) | |
Life history trait[o]	Control (Crenicichla)	Introduction (Rivulus)	Control (Crenicichla)	Introduction (Rivulus)	Crenicichla	Rivulus
Male age at maturity (days)	60.6 (1.8)	72.7 (1.8) •	48.5 (1.2)	58.2 (1.4) •	51.8 (1.1)	58.8 (1.0) •
Male size at maturity (mg-wet)	56.0 (1.4)	62.4 (1.5) •	67.5 (1.2)	76.1(1.9) •	87.7 (2.8)	99.7 (2.5) •
Female age at first parturition (days)	94.1 (1.8)	95.5 (1.8) (NS)	85.7 (2.2)	92.3 (2.6) *	71.5 (2.0)	81.9 (1.9) •
Female size at first perturition (mg-wet)	116.5 (3.7)	118.9 (3.7) (NS)	161.5 (6.4)	185.6 (7.5) •	218.0 (8.4)	270.0 (8.2) •
Brood size, litter 1	2.5 (0.2)	3.0 (0.2) (NS)	4.5 (0.4)	3.3 (0.4) *	5.2 (0.4)	3.2 (0.5) •
Brood size, litter 2	6.3 (0.3)	7.0 (0.3)§	8.1 (0.6)	7.5 (0.7) (NS)	10.9 (0.6)	10.2 (0.8) (NS)
Brood size, litter 3 ∇	–	–	11.4 (0.8)	11.5 (0.9) (NS)	16.1 (0.9)	16.0 (1.1) (NS)
Offspring size (mg-dry), litter 1	0.91 (0.02)	0.87 (0.02) (NS)	0.87 (0.02)	0.95 (0.02)§	0.84 (0.02)	0.99 (0.03) •
Offspring size, litter 2	0.93 (0.02)	0.86 (0.02) *	0.90 (0.03)	1.02 (0.04) •	0.95 (0.02)	1.05 (0.03) *
Offspring size, litter 3 ∇	–	–	1.10 (0.03)	1.17 (0.04) (NS)	1.03 (0.03)	1.17 (0.04) •
Interbrood interval (days)	24.9 (0.4)	24.89 (0.4) (NS)	24.5 (0.3)	25.2 (0.3) (NS)	22.8 (0.3)	25.0 (0.03) •
Reproductive effort (%) #	4.0 (0.1) *	3.9 (0.1) (NS)	22.0 (1.8)	18.5 (2.1) (NS)	25.1 (1.6)	19.2 (1.5) *

NS, not significant; *, P<0.05; •, P<0.01; §, 0.05<P<0.10.

[o]Differences in mean values among experiments are attributable to differences in food availability. Reznick (1982) had the highest levels, Reznick et al. (1990) was intermediate, and Reznick and Bryga (1987) had the lowest levels. ∇ Fish were only kept until they produced two litters of young in Reznick and Bryga (1987).

#Values for reproductive effort in Reznick and Bryga (1987) represent a single estimate made at the end of the experiment; those for the other two studies represent the sum of four consecutive estimates. See Reznick (1982) for details on the latter analysis.

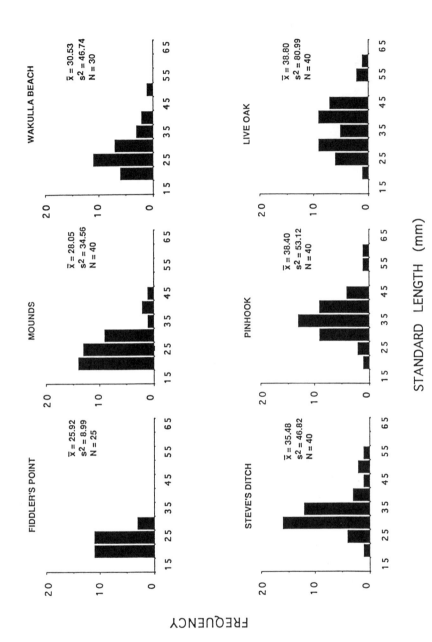

general features of phenotypic variation that must be explained.

Earlier work by other authors indicated that size distributions could vary appreciably among populations, and our initial observations in north Florida confirmed this. Trexler undertook a wide survey of nearly 70 populations in the eastern half of the species range and quantified a number of characteristics of males and females, along with measuring environmental parameters. For a subset of these populations, scattered throughout the initial range, he also examined levels of allozyme variation at 26 loci. The goal was to compare the spatial pattern of allozyme variation with the spatial patterns of differentiation in quantitative traits. The allozyme data, when averaged across loci, should reflect levels of gene flow and the scale of presumptively neutral character differentiation; as such, it would represent a null hypothesis against which to contrast the spatial patterns of quantitative character variation.

The allozyme data revealed a smooth pattern of isolation by distance (Trexler, 1988) that was strikingly at odds with the spatial pattern of quantitative character differentiation. Spatial autocorrelation analysis indicated that the allozyme correlation between any two populations became zero at 57 km separation; for quantitative characters, there is no consistent spatial autocorrelation (Travis and Trexler, 1987). Hierarchical analyses of variance showed that most of the variation in both types of traits was found in a typical local population (69-70%); however, regional variation accounted for 25% of the protein variation but only 6% of the male body size variation. Variation among local populations accounted for about 5% of protein variation but 24% of the male body size variation. The obvious conclusion is that one of these sets of traits must be under selection because dual neutrality should produce qualitatively similar spatial distributions of variation. This comparative approach was facilitated by the fact that we knew the inheritance of body size to be based in part on a Y-linked allelic series. Thus the comparison between allozyme data and body size was a comparison between two sets of traits with comparable mechanisms of inheritance.

Of course it was conceivable that the body size variation among populations, though not concordant with the allozyme variation, was influenced solely by some suite of environmental variables that changed on a spatial scale smaller than the scale of isolation by distance revealed by the allozymes. The fact that female body size and brood sizes, adjusted for female size, varied concordantly with the variation in male size served to bolster this possibility: it is easy to postulate a physiological effect of enhanced food resources or water temperature that would induce larger male sizes as well as larger sizes of females and their broods. However, the measurements of environmental variables in the 3 years surveyed produced an interesting result: there were no consistent associations of any single variable or linear combination of variables with the variation in either male body size or female life history traits (Travis and Trexler, 1987). Of course, this result is inconclusive; it is easy to have environmental effects that cancel each other's direct influence on growth and development (locations that induce higher demand

Figure 7 Histograms of variation in the standard length (linear measure of body size) of mature males in six north Florida populations of *Poecilia latipinna*. The Live Oak population has an average male size that is among the largest observed in the eastern half of the species range (Travis and Trexler, 1987), while the Fiddler's Point population has an average that is about 9 mm greater than the smallest average size observed (from Ptacek and Travis, 1996).

metabolism also are those with higher productivity of food resources so there is no net effect of either the stress due to the demand or the enhanced food).

Trexler and Travis (1990) undertook two extensive studies of environmental effects. One was a reciprocal transplant approach in which half-sib broods from two populations were split between enclosures in two natural habitats. The other study divided half-sib broods among replicated combinations of three salinities, two temperatures, and two food levels in a factorial design in the laboratory (Trexler et al., 1990). The two experiments yielded remarkably complementary results. Male body size was insensitive to environmental effects, and males whose parents were from different populations emerged with different size distributions regardless of the environment of ontogeny. Thus the extensive male body size variation among populations (i) cannot be an environmental effect, (ii) has a genetic basis, and (iii) is incompatible with the spatial pattern of variation expected under a null hypothesis of genetic drift. Moreover, data from the lab experiment were consistent with the hypothesis that selection continually reassorts male size distributions in the face of ongoing gene flow: body size patterns of lab-reared, unselected progeny from different populations were more similar than the recurrent patterns seen in the two natural populations from which they were drawn.

Yet males were not insensitive to environmental effects. Lower temperatures induced longer times to maturity by inducing slower growth; males did not respond to lowered growth rates by maturing at smaller sizes. Indeed, the data

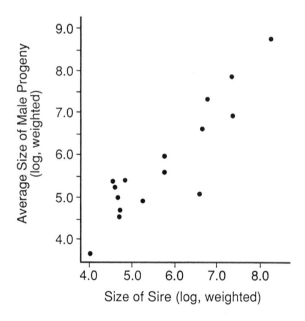

Figure 8 Average body sizes at maturity, measured as standard lengths, of half-sib *Poecilia latipinna* males raised in laboratory crosses (vertical axis) plotted against the body size of their fathers (horizontal axis) on a log-log scale. The regression, weighted by the sample size of each family, has a slope of nearly 1.0, which indicates Y-linkage in the transmission of size between generations (from Travis, 1994c).

show that males sacrifice time to maturity to attain a specific body size. This offers additional circumstantial evidence in favor of the hypothesis that body size variation among populations is adaptive.

Having documented the patterns of variation and investigated, at least broadly, the nature of the genetic variation within and among populations together with the potential for environmental effects, we employed two types of experimental regimes to test directly for the action of natural selection. In the first type, we manipulated the size distributions of males and females exposed to agents of mortality. These experiments showed that larger males were more likely to be preyed upon by wading birds (Trexler et al., 1994) but that they are more likely to survive cold winters, especially in low salinity habitats (Trexler et al., 1992).

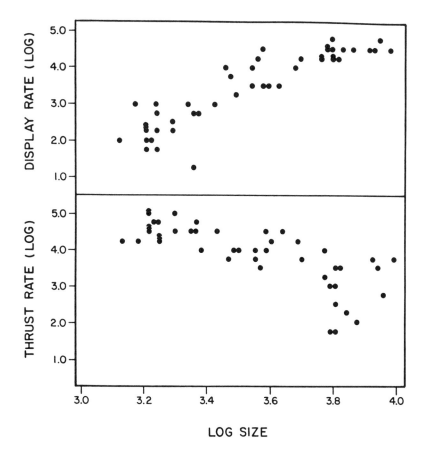

Figure 9 The log of the rate of courtship displays (upper panel) and gonopodial thrusts (lower panel), derived from the numbers of each behavior exhibited by a male in a 20-min observation period, plotted against the log of male body size (measured as standard length) on the horizontal axis. The males were derived from controlled crosses and raised from birth in isolation in laboratory aquaria. Each behavior has a strong relationship with body size. The smallest males rely on gonopodial thrusting, which is an attempt to inseminate a female without first eliciting her cooperation, while the largest males rarely use thrusting and rely heavily on eliciting female cooperation through courtship displays (from Travis, 1994c).

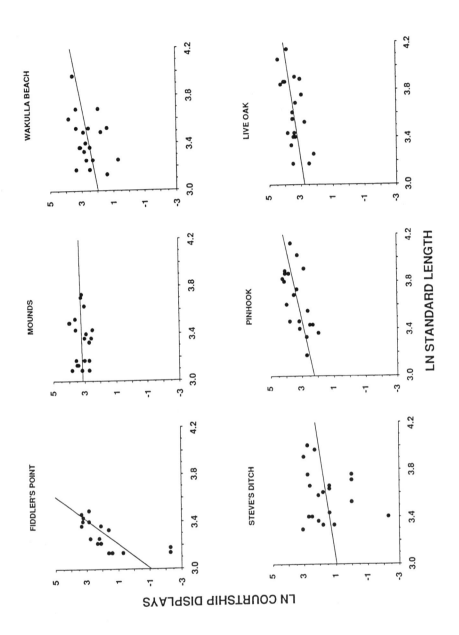

Subsequent experiments have shown that the ability of larger males to withstand colder winters is traceable to their proportionately larger storage depots for triacylglycerol, depots that are larger at the cost of a proportionately smaller testis (McManus, 1994). In the second type of experiment we manipulated the size distribution and social milieu of sets of males and females to uncover the action of the mating system and thereby discovered that, as many authors have postulated, sexual selection favors larger males (Travis, 1994b).

The investigation of the variation among populations in the patterns of male sexual behavior (Fig. 10) is less well developed at this time, but has proceeded in the same fashion: quantifying phenotypic variation, performing crosses to investigate whether the variation has a genetic basis, and developing experiments to ascertain whether there is adaptive significance to the behavioral variation. In principle, behavioral traits are no different than morphological or life history traits in the problems that they present to the student of adaptation. However, the potential effects of ontogenetic experience on behavioral expression and the ease with which individuals adjust behavior to short-term changes in social situations place an additional premium on defining the environmental conditions under which the genetic control of behavior will be measured [Travis (1994c) addresses this issue more fully].

Our studies of sailfin mollies carry many lessons for the study of adaptation, but one particular lesson bears emphasis here. Many of our conclusions are reliable only because we have repeated the same types of observational and experimental studies over several years. Indeed, temporal variation is a major element in the dynamics of this system. For example, Trexler's large-scale surveys of body size were repeated three times, and he made two replicated surveys of protein variation. The reciprocal transplant experiments were performed four times, and the "winter mortality" experiments were executed in three distinct winters and replicated in one growing season. We are now engaged in similar repetitive work on behavioral patterns (e.g., Ptacek and Travis, 1996). The outcome of a single survey or experiment was not necessarily representative of the results suggested by the ensemble of replicated studies. This is no surprise. If local adaptation has occurred, it has had many growing seasons in which to manifest itself, and indeed to be genuine must be a phenomenon that persists despite the temporal variation in environmental conditions. It would be naive to expect all of the answers to our questions to be discoverable in single replications of field studies.

An implicit message in both of our synopses of research is that there is not a discrete list of types of experiments that can be performed to study adaptation; the actual designs and questions posed are often specifically tailored to the study system. The more important message is that experiments can and should be done to either manipulate the environment or the organism to evaluate the

Figure 10   Scatter plots of the rate of courtship displays on a logarithmic scale (vertical axis) as a function of male body size (standard length) on a logarithmic scale (horizontal axis) for males from the six north Florida populations illustrated in Fig. 7. While in most cases there is a positive relationship between the rate of courtship displays and male size variation, males from one populations (Mounds) show no such relationship. Even among those populations with a positive relationship, the rates of courtship displays for males of the same body size vary widely among populations. These data indicate that the strong relationship between these variables often seen within any one population need not be a strong constraint on the level of divergence among populations in behavioral repertoires (from Ptacek and Travis, 1996).

consequences of alternative phenotypes on fitness. A third message, revealed in observations of Galapagos finches or the introduction experiments on guppies, is that when organisms experience an episode of directional selection, the rate of evolution can be very high and observable within a relatively brief interval of time, as originally suggested by Weldon (1899). Such results should encourage future investigators to test their evolutionary hypotheses more directly.

VIII. A Generalized Concept of Fitness: A Challenge for Future Empirical Studies of Adaptation

One important change in the past 65 years has been a broadening of the way we think about fitness. The prevailing theme in Darwin's (1859) *The Origin of Species* was that fitness was defined primarily by individual survival. A consequence of longer life in a given environment was having more opportunities to reproduce and hence the opportunity to make a greater contribution to future generations than shorter-lived phenotypes. This narrow definition of fitness in part inspired the argument that the theory of evolution is tautological (Peters, 1976).

Beginning with Fisher (1930), the concept of fitness was broadened to include all components of the life history, instead of just survival. Fisher formally defined reproductive value, or the expected future reproductive contribution of an individual, as a function of its age, expectation for survival, and expectation for reproduction. The impact of a given mortality factor, such as an increase in age/size specific mortality, would be a function of the reproductive value of the affected individuals. Williams (1957, 1966) further developed the concept of reproductive value by suggesting that there might be a functional relationship between the different components of the life history. In the latter paper, he formally partitioned reproductive value into two components. These are what would be allocated to reproduction now (reproductive effort) versus what is expected to be invested in the future (residual reproductive value). He hypothesized that an increase in reproductive effort would be gained at a cost in residual reproductive value through reduced survival, reduced future reproduction, or both. Finally, he suggested that natural selection would maximize the sum of these two quantities. By this interpretation, fitness is gauged by a composite of birth and death schedules. We are left with competing ways to estimate fitness given such schedules, including the intrinsic rate of increase and the net reproductive rate (Charlesworth, 1994). Nevertheless, it is clear that we must consider more than life span when evaluating the fitness of alternative phenotypes.

The actual implementation of this broader definition of fitness represents an area in which the empirical evaluation of adaptation could be improved. Most often, investigators rely on evaluating just one or a few components of fitness, e.g., most of the vintage ecological genetics of the 1950s and 1960s just considered the short-term mortality rates of alternative phenotypes. McNeilly (1968) carried the analysis further by considering both survival and growth rates and by making comparisons under a greater diversity of environmental conditions, such as different densities and the presence or absence of plants from the alternative type of environment.

Clutton-Brock (1988) and Service and Lenski (1982; Lenski and Service, 1982) have both proposed methods to evaluate fitness more completely in natural populations. Clutton-Brock championed the application of long-term mark recapture studies to evaluate lifetime reproductive success, or the number of surviving offspring produced by an individual. Variation in lifetime reproductive success would presumably represent a more complete estimate of variation in fitness. Lenski and Service instead characterized each individual by its intrinsic rate of increase relative to others in the population. The difference between these two techniques lies in the timing in when offspring are produced. The technique used in the Clutton-Brock analyses ignores everything about when offspring are produced in the lifetime of an individual. The Lenski-Service method is sensitive to timing, including variables such as the age at maturity and age-specific fecundity. The former method will give a good index of fitness if the population size is relatively stable, while the latter will be more appropriate if there are wide fluctuations in population size, particularly if there are long-term trends of change in population size (Charlesworth, 1994).

We have discovered for ourselves why the application of such an extended definition of fitness has been so rare. In our guppy work, it has taken 15 years of field and laboratory study to evaluate all of the components of the life history necessary to construct a life table. These include growth rate, size-specific survival, age at maturity, frequency of reproduction, and the number and size of offspring. These are also what we think of as the components of fitness. Actually demonstrating how each aspect of the life history influences fitness in a given context is one of our future goals, as is the further characterization of the interrelationships among different components of the life history. One approach has been to simulate population growth rates for a given schedule of individual growth rates and reproduction under alternative mortality schedules. A life history typical of guppies in a *Rivulus* locality has a higher population growth rate than one typical of guppies in a *Crenicichla* locality when they are evaluated under a *Rivulus* mortality schedule. The opposite is true when the two life histories are evaluated under a *Crenicichla* mortality schedule (Rodd and Reznick, unpublished). This result again implies that the life history of each type of guppy is an adaptation to its prevailing mortality schedule.

IX. Summary and Conclusions

1.   In developing the concept of adaptation, Darwin argued that, if certain conditions were satisfied, then evolution by natural selection had to follow.

2.   The proposition was so logical and simple that it seems that acceptance and subsequent experimental evaluations of specific adaptations, or characterizations of the process, should have readily followed Darwin's (1859) publication of *The Origin of Species.*

3.   This did not happen. One reason is that the underlying conditions for the process to work, such as the source and maintenance of heritable variations, were not understood at the time and took decades to

characterize.

4. Our programs for the empirical study of adaptation developed along with the modern synthesis, beginning with the work of Ford and Fisher. Their approach involved elements of very different facets of the modern synthesis era, ranging from population genetics to systematics.

5. The influence of population genetics on the empirical study of adaptation is clear in the emphasis on evaluating the genetic basis of trait variation and the influence of traits on fitness.

6. The aspect of the approach that was possibly attributable to systematics is the emphasis on comparative studies and the study of the distribution of phenotypic variation on a wide geographic scale, combined with correlations between the phenotype and the environment.

7. Besides providing us with many empirical examples, variously supported, for the phenomenon of adaptation, these studies have also provided some important lessons about adaptations in general. All of these lessons were articulated before the advent of extensive empirical studies of adaptation, but the work gave them substance. Two such lessons are that adaptations are not simple and that adaptation often involves compromise. Because natural selection acts on whole organisms and not traits, the final product will necessarily represent a compromise among all the different ways that a given trait influences fitness, all the different traits that comprise a complete organismal phenotype, and all the correlations that exist among different components of the phenotype. Adaptation is also a compromise among all the different biotic and abiotic aspects of the environment that are the sources of natural selection. The third lesson is that the fitness of a phenotype is only definable with respect to the alternatives present in the population and that it can change with any change in the abiotic or biotic environment.

8. While the initial focus of ecological genetics was on obvious morphological polymorphisms, the same general methods have now been applied to biochemical polymorphisms and polygenic traits as well. Adaptation has thus been studied at diverse levels of biological organization.

9. Finally, existing programs of study have considerably elaborated on multiple methodologies for the experimental evaluation of adaptations. These include comparative studies, either of a temporal sequence of samples or among geographically distinct populations, population studies including mark-recapture methods, and a diversity of experimental manipulations.

10. Although the scientific community is not fully unified behind this claim, the empirical study of adaptation clearly demonstrates that the theory of evolution by natural selection is not tautological and that adaptation is a prevalent phenomenon in natural populations.

Acknowledgments

We dedicate this paper to Robert Ricklefs for having inspired our interest in this topic when we took his course "Adaptation" in 1974. We are very grateful to Helen Rodd, Daniel Simberloff, and Mary Ruckelshaus for their helpful comments on the manuscript. We were supported by NSF grants DEB92-20849 (JT), DEB91-19432 (DR), and DEB94-19823 (DR) during the preparation of this manuscript.

References

Arnold, S. J. (1988). Quantitative genetics and selection in natural populations: microevolution of vertebral numbers in the garter snake *Thamnophis elegans*. *In* "Proceedings of the Second International Conference on Quantitative Genetics" (B. S. Weir, E. J. Eisen, M. M. Goodman, and G. Namkoong, eds.), pp. 619-636. Sinauer, Sunderland, MA.

Arnold, S. J. (1994). Constraints on phenotypic evolution. *In* "Behavioral Mechanisms in Evolutionary Ecology" (L. A. Real, ed.), pp. 258-278. University of Chicago Press, Chicago.

Arnold, S. J., and Wade, M. J. (1984a). On the measurement of natural and sexual selection. *Theory Evolution* 38, 709-719.

Arnold, S. J., and Wade, M. J. (1984b). On the measurement of natural and sexual selection: applications. *Evolution* 38, 720-734.

Barton, N. H., and Turelli, M. (1989). Evolutionary quantitative genetics: how little do we know? *Annu. Rev. Gen.* 23, 337-370.

Bates, H. W. (1862). Contributions to an insect fauna of the Amazon valley; Lepidoptera: Heliconidae. *Trans. Linn. Soc., London* 23, 495-566.

Berven, K. A., Gill, D. E., and Smith-Gill, S. J. (1979). Countergradient selection in the green frog, *Rana clamitans*. *Evolution* 33, 609-623.

Boake, C. R. B., ed. (1994). "Quantitative Genetic Studies of Behavioral Evolution." University of Chicago Press, Chicago.

Brooks, D. R., and McLennan, D. A. (1991). "Phylogeny, Ecology, and Behavior: A Research Program in Comparative Biology." University of Chicago Press, Chicago.

Brower, L. P., Cook, L. M., and Croze, H. J. (1967). Predator responses to artificial Batesian mimics released in a neotropical environment. *Evolution* 21, 11-23.

Bulmer, M. G. (1989). Maintenance of genetic variability by mutation-selection balance: a child's guide through the jungle. *Genome* 31, 761-767.

Burton, R. S., and Feldman, M. W. (1983). Physiological effects of an allozyme polymorphism: glutamate-pyruvate transaminase and response to hyperosmotic stress in the copepod *Tigriopus californicus*. *Biochem. Genet.* 21, 239-251.

Cain, A. J., and Sheppard, P. M. (1950). Selection in the polymorphic land snail *Cepaea nemoralis*. *Heredity* 4, 275-294.

Cain, A. J., and Sheppard, P. M. (1954). Natural selection in *Cepaea*. *Genetics* 39, 89-116.

Capy, P., David, J. R., Carton, Y., and Pla, E. (1988). Grape breeding *Drosophila* communities in southern France: short range variation in ecological and genetic structure of natural populations. *Acta Oecol., Oecol. Gen.* 8, 435-440.

Carvalho, G. R., Shaw, P. W., Magurran, A. E., and Seghers, B. H. (1991). Marked genetic divergence revealed by allozymes among populations of the guppy *Poecilia reticulata* (Poeciliidae), in Trinidad. *Biol. J. Linn. Soc.* 42, 389-405.

Cavener, D. R., and Clegg, M. T. (1981). Evidence for biochemical and physiological differences between enzyme genotypes of *Drosophila melanogaster*. *Proc. Natl. Acad. Sci. USA* 78, 4444-4447.

Chambers, G. K. (1988). The *Drosophila* alcohol dehydrogenase gene-enzyme system. *In* "Advances in

Genetics" (E. W. Caspari and J. G. Scandalois, eds.), vol. 25, pp. 39-107. Academic Press, New York.

Chappell, M. A., and Snyder, L. R. G. (1984). Biochemical and physiological correlates of deer mouse a-chain hemoglobin polymorphisms. *Proc. Natl. Acad. Sci. USA* 81, 5484-5488.

Chappell, M. A., Hayes, J. P., and Snyder, L. R. G. (1988). Hemoglobin polymorphisms in deer mice (*Peromyscus maniculatus*): physiology of beta-globin variants and alpha-globin recombinants. *Evolution* 42, 681-688.

Charlesworth, B. (1990). Optimization models, quantitative genetics, and mutation. *Evolution* 44, 520-538.

Charlesworth, B. (1994). "Evolution in Age-Structured Populations." 2nd edition. Cambridge Univ. Press, Cambridge.

Charnov, E. L. (1989). Phenotypic evolution under Fisher's fundamental theorem of natural selection. *Heredity* 62, 113-116.

Clark, A. G., and Koehn, R. K. (1992). Enzymes and adaptation. *In* "Genes in Ecology" (R. J. Berry, T. J. Crawford, and G. M Hewitt, eds.), pp. 193-228. Blackwell, Oxford.

Clarke, C. A., and Sheppard, P. M. (1960a). The evolution of dominance under disruptive selection. *Heredity* 14, 73-87.

Clarke, C. A., and Sheppard, P. M. (1960b). Super-genes and mimicry. Heredity 14, 175-185.

Clarke, C. A., and Sheppard, P. M. (1960c). The evolution of mimicry in the butterfly *Papilio dardanus*. *Heredity* 14, 73-87.

Clutton-Brock, T. H. (1988). "Reproductive Success." University of Chicago Press, Chicago.

Cockerham, C. C. (1986). Modifications in estimating the number of genes for a quantitative character. *Genetics* 114, 659-664.

Coddington, J. A. (1988). Cladistic tests of adaptational hypotheses. *Cladistics* 4, 3-22.

Crespi, B. J., and Bookstein, F. L. (1989). A path-analytic model for the measurement of selection on morphology. *Evolution* 43, 18-28.

Crow, J. F. (1958). Some possibilities for measuring intensities in man. *Hum. Biol.* 30, 1-13.

Crow, J. F., and Nagylaki, T. (1976). The rate of change of a character correlated with fitness. *Amer. Nat.* 110, 207-213.

Darwin, C. (1859). "The Origin of Species." John Murray, London.

Darwin, C. (1868). "The Variation of Plants and Animals under Domestication." Orange Judd, New York.

David, J. R., Allemand, R., Van Herrewege, J., and Cohet, Y. (1983). Ecophysiology: abiotic factor. *In* "Genetics and Biology of *Drosophila*" (M. Ashburner, H. L. Carson, and J. N. Thompson, Jr., eds.), vol. 3, pp. 105-170. Academic Press, New York.

Dice, L. R. (1947). Effectiveness of selection by owls of deer mice (*Peromyscus maniculatus*) which contrast in colour with their background. Contrib. Lab. Vert. Biol., Univ. MI 34, 1-20.

Dobzhansky, T. (1937). "Genetics and the Origin of Species." Columbia University Press, New York.

Dobzhansky, T. (1948). Genetics of natural populations XVI: altitudinal and seasonal changes produced by natural selection in certain populations of *Drosophila pseudoobscura* and *Drosophila persimilis*. *Genetics* 33, 158-176.

Dunson, W. A., and Travis, J. (1991). The role of abiotic factors in community organization. *Amer. Nat.* 138, 1067-1091.

Dykhuizen, D. E., and Dean, A. M. (1990). Enzyme activity and fitness: evolution in solution. *Trends Ecol. Evol.* 5, 257-262.

Endler, J. A. (1978). A predator's view of animal color patterns. *Evol. Biol.* 11, 319-364.

Endler, J. A. (1986). "Natural Selection in the Wild." Princeton University Press, Princeton.

Fisher, R. A. (1930). "The Genetical Theory of Natural Selection." Clarendon Press, Oxford.

Fisher, R. A., and Ford, E. B. (1947). The spread of a gene in natural conditions in a colony of the moth *Panaxia dominula* L. *Heredity* 1, 143-174.

Ford, E. B. (1971). "Ecological Genetics." 3rd edition. Chapman and Hall, London.

Freriksen, A., De Ruiter, B. L. A., Groenenberg, H.-J., Scharloo, W., and Heinstra, P. W. H. (1994). A multilevel approach to the significance of genetic variation in alcohol dehydrogenase of *Drosophila*. *Evolution* 48, 781-790.

Fritz, R. S., and Simms, E. L., eds. (1992). "Plant Resistance to Herbivores and Pathogens." University of Chicago Press, Chicago.

Gadgil, M., and Bossert, P. W. (1970). Life historical consequences of natural selection. *Amer. Nat.* 104, 1-24.

Geer, B. W., Heinstra, P. W. H., and McKechnie, S. W. (1993). The biological basis of ethanol tolerance in *Drosophila*. *Comp. Biochem. Physiol.* 105B, 203-229.

Gibbs, H. L., and Grant, P. R. (1987). Oscillating selection on Darwin's finches. *Nature* 327, 511-513.

Gibson, J. B., May, T. W., and Wilks, A. V. (1981). Genetic variation at the alcohol dehydrogenase locus in *Drosophila melanogaster* in relation to environmental variation: ethanol levels in breeding sites and allozyme frequencies. *Oecologia* 51, 191-198.

Gould, S. J. (1977). "Ontogeny and Phylogeny." Harvard University Press, Cambridge, MA.

Gould, S. J., and Lewontin, R. C. (1979). The spandrels of San Marco and the Panglossian paradigm: a critique of the adaptationist programme. *Proc. Roy. Soc., London* B 205, 147-164.

Gould, S. J., and Vrba, E. S. (1982). Exaptation–a missing term in the science of form. *Paleobiology* 8(1), 4-15.

Grant, P. R. (1986). "Ecology and Evolution of Darwin's Finches." Princeton University Press, Princeton.

Grant B. R., and Grant, P. R. (1989). "Evolutionary Dynamics of a Natural Population of the Large Cactus Finch of the Galapagos." University of Chicago Press, Chicago.

Grant, P. R., and Grant, B. R. (1995). Predicting microevolutionary responses to directional selection on heritable variation. *Evolution* 49, 241-251.

Greene, H. W. (1986). Diet and arboreality in the emerald monitor, *Varanus prasinus*, with comments on the study of adaptation. *Field., Zool.* 31, 1-12.

Griffing, B. (1960). Theoretical consequences of truncation selection based on individual phenotype. *Aust. J. Biol. Sci.* 22, 43-53.

Haldane, J. B. S. (1932). "The Causes of Evolution." Longmans Green, London.

Haskins, C. P., Haskins, E. F., McLaughlin, J. J. A., and Hewitt, R. E. (1961). Polymorphism and population structure in *Lebistes reticulata*, a population study. *In* "Vertebrate Speciation" (W. F. Blair, ed.), pp. 320-395. University of Texas Press, Austin.

Henistra P. W. H, Thörig, G. E. W., Scharloo, W., Drenth, W., and Nolte, R. J. M. (1988). Kinetics and thermodynamics of ethanol oxidation catalyzed by genetic variants of the alcohol dehydrogenase from *Drosophila melanogaster* and *D. simulans*. *Biochim. Biophys. Acta* 967, 224-233.

Hoffman, A. A., and Parsons, A. A. (1984). Olfactory response and resource utilization in *Drosophila*: interspecific comparisons. *Biol. J. Linn. Soc.* 22, 43-53.

Hougoto, N., Lietaert, M. C., Libion-Mannaert, M., Feytmans, E., and Elens, A. (1982). Oviposition-site preference and *Adh* activity in *Drosophila melanogaster*. *Genetica* 58, 121-128.

Houle, D. (1991). Genetic covariance of fitness correlates: what genetic correlations are made of and why it matters. *Evolution* 45, 630-648.

James, F. C., Nesmith, C., and Laybourne, R. (1991). Geographic differentiation in the red-winged blackbird: a check on one of Lande's assumptions. Acta XX Congressus Internationalis Ornithologici Vol. IV, 2454-2461.

Jones, J. S., Leith, B. H., and Rawlings, P. (1977). Polymorphism in *Cepaea*: a problem with too many solutions? *Annu. Rev. Ecol. Syst.* 8, 109-143.

Kettlewell, H. B. D. (1955). Selection experiments on industrial melanism in the Lepidoptera. *Heredity* 9, 323-342.

Kettlewell, H. B. D. (1956). Further selection experiments on industrial melanism in the Lepidoptera. *Heredity* 10, 287-301.

Kettlewell, H. B. D. (1958). A survey of the frequencies of *Biston betularia* (L) (Lep.) and its melanic forms in Great Britain. *Heredity* 12, 51-72.

Kimura, M. (1965). A stochastic model concerning the maintenance of genetic variability in quantitative characters. *Proc. Natl. Acad. Sci. USA* 54, 731-736.

Koehn, R. K., and Hilbish, T. J. (1987). The adaptive importance of genetic variation. *Amer. Sci.* 75, 134-141.

Kozlowski, J., and Uchmansky, J. (1987). Optimal individual growth and reproduction in perennial species with indeterminate growth. *Evol. Ecol.* 1, 214-230.

Kreitman, M., and Hudson, R. R. (1991). Inferring the evolutionary theories of the *Adh* and *Adh-dup* loci in *Drosophila melanogaster* from patterns of polymorphism and divergence. *Genetics* 127, 565-582.

Kreitman, M., Sharrocks, B., and Dytham, C. (1992). Genes and ecology: two alternative perspectives using *Drosophila*. *In* "Genes in Ecology" (R. J. Berry, T. J. Crawford, and G. M Hewitt, eds.), pp. 281-312. Blackwell, Oxford.

Lande, R. (1976). Natural selection and random genetic drift in phenotypic evolution. *Evolution* 30, 314-334.

Lande, R. (1979). Quantitative genetic analysis of multivariate evolution, applied to brain: body size allometry. *Evolution* 33, 402-416.

Lande, R. (1988). Quantitative genetics and evolutionary theory. *In* "Proceedings of the Second International Conference on Quantitative Genetics" (B. S. Weir, E. J. Eisen, M. M. Goodman, and G. Namkoong, eds.), pp. 71-84. Sinauer, Sunderland, MA.

Lande, R., and Arnold, S. J. (1983). The measurement of selection on correlated characters. *Evolution* 37, 1210-1226.

Latter, B. D. H. (1960). Natural selection for an intermediate optimum. Aust. J. Biol. Sci. 13, 30-35.

Laurie-Ahlberg, C. C., Maroni, G., Bewley, G. C., Lucchesi, J. C., and Weir, B. S. (1980). Quantitative genetic variation of enzyme activities in natural populations of *Drosophila melanogaster*. *Proc. Natl. Acad. Sci.* USA 77, 1073-1077.

Law, R. (1979). Optimal life histories under age-specific predation. *Amer. Nat.* 114, 399-417.

Law, R., Bradshaw, A. F., and Putwain, P. D. (1977). Life history variation in *Poa annua Evolution* 31, 233-246.

LeBreton, J. D., Burnham, K. P., Clobert, J., and Anderson, D. R. (1992). Modeling survival and testing biological hypotheses using marked animals - a unified approach with case studies. *Ecol. Monogr.* 62, 67-118.

Lenski, R. E., and Service, P. M. (1982). The statistical analysis of population growth rates calculated from schedules of survivorship and fecundity. *Ecology* 63, 655-662.

Lewontin, R. C. (1974). "The Genetic Basis of Evolutionary Change." Columbia University Press, New York.

Liley, N. R., and Seghers, B. H. (1975). Factors affecting the morphology and behavior of guppies in Trinidad. *In* "Function and Evolution in Behavior" (G. P. Baerends, C. Beer, and A. Manning, eds.), pp. 92-118. Oxford University Press, Oxford.

Lofsvold, D. (1986). Quantitative genetics of morphological differentiation in *Peromyscus*. I. Test of the homogeneity of genetic covariance structure among species and subspecies. *Evolution* 40, 559-573.

Manly, B. F. J. (1985). "The Statistics of Natural Selection." Chapman and Hall, London.

Mattingly, H. T., and Butler, IV, M. J. (1994). Laboratory predation on the Trinidadian guppy: implications for the size-selective predation hypothesis and guppy life history evolution. *Oikos* 69, 54-64.

Maynard Smith, J. (1980). Selection for recombination in a polygenic model. *Genet. Res.* 35, 269-277.

Mayr, E. (1942). "Systematics and the Origin of Species." Columbia University Press, New York.

Mayr, E. (1988). "Toward a New Philosophy of Biology." Harvard University Press, Cambridge.

McDonald, J. F., Chambers, G. K., David, J., and Ayala, F. J. (1977). Adaptive response due to changes in gene regulation: a study with *Drosophila*. *Proc. Natl. Acad. Sci.* USA 74, 4562-4566.

McDonald, J. H., and Kreitman, M. (1991). Adaptive protein evolution at the *Adh* locus in *Drosophila*. *Nature* 351, 652-654.

McKechnie, S. W., and Morgan, P. (1982). Alcohol dehydrogenase polymorphism of *Drosophila melanogaster*: Aspects of alcohol and temperature variation in the larval environment. *Aust. J. Biol. Sci.* 35, 85-93.

McManus, M. (1994). Energy Allocation in the Sailfin Molly, *Poecilia latipinna*. Ph.D. dissertation. Florida State University, Tallahassee, FL.

McNeilly, T. (1968). Evolution in closely adjacent plant populations. III. *Agrostis tenius* on a small copper mine. *Heredity* 23, 99-108.

McNeilly, T., and Antonivics, J. (1968). Evolution in closely adjacent plant populations. IV. Barriers to gene flow. *Heredity* 23, 205-218.

Merçot, H., Defaye, D., Capy, P., Pla, E., and David, J. R. (1994). Alcohol tolerance, ADH activity, and ecological niche of *Drosophila* species. *Evolution* 48, 756-757.

Michod, R. E. (1979). Evolution of life histories in response to age-specific mortality factors. *Amer. Nat.* 113, 531-550.

Mitchell-Olds, T., and Shaw, R. G. (1987). Regression analysis of natural selection: statistical inference and biological interpretation. *Evolution* 41, 1149-1161.

Mousseau, T. A., and Dingle, H. (1991). Maternal effects in insect life histories. *Annu. Rev. Entomol.* 36, 511-534.

Oakeshott, J. G., May, T. W., Gibson, J. B., and Willcocks, D. A. (1982a). Resource partitioning in five domestic *Drosophila* species and its relationship to ethanol metabolism. *Aust. J. Zool.* 30, 547-556.

Oakeshott, J. G., Gibson, J. B., Anderson, P. R., Knibb, W. R., Anderson, D. G., and Chambers, G. K. (1982b). Alcohol dehydrogenase and glycerol-3-phosphate dehydrogenase clines in *Drosophila melanogaster* on different continents. *Evolution* 36, 86-96.

Orzack, S. H., and Tuljapurkar, S. (1989). Population dynamics in variable environments. VII. The demography and evolution of iteroparity. *Amer. Nat.* 133, 901-923.

Oudman, L., Van Delden, W., Kamping, A., and Bijlsma, R. (1991). Polymorphism in the *Adh* and *Gpdh* loci in *Drosophila melanogaster:* effects of rearing temperature on developmental rate, body weight, and some biochemical parameters. *Heredity* 67, 103-115.

Palmer, J. O., and Dingle, H. (1986). Direct and correlated responses to selection among life-history traits in milkweed bugs (*Oncopeltus fasciatus*). *Evolution* 40, 767-777.

Peters, R. H. (1976). Tautology in evolution and ecology. *Amer. Nat.* 110, 1-12.

Phillips, P. C., and Arnold, S. J. (1989). Visualizing multivariate selection. *Evolution* 43, 1209-1222.

Powers, D. A., Di Michele, L., and Place, A. R. (1983). The use of enzyme kinetics to predict differences in cellular metabolism, developmental rate, and swimming performance between LDH-B genotypes of the fish, *Fundulus heteroclitus. Isozymes: Current Topics in Biological and Medical Research* 10, 147-170.

Price, G. R. (1970). Selection and covariance. *Nature* 227, 520-521.

Price, G. R. (1972). Extension of covariance selection mathematics. *Annal. Hum. Gen.* 35, 485-490.

Provine, W. B. (1971). "The origins of theoretical population genetics." Univ. of Chicago Press, Chicago.

Ptacek, M. B., and Travis, J. (1996). Interpopulation variation in male mating behaviours in the sailfin mollie, *Poecilia latipinna. Anim. Behav.* in press.

Punnett, R. C. (1915). "Mimicry in Butterflies." Cambridge University Press, Cambridge.

Rausher, M. D., and Simms, E. L. (1989). The evolution of resistance to herbivory in *Ipomoea purpurea.* I. Attempts to detect selection. *Evolution* 43, 563-572.

Reeve, H. K., and Sherman, P. W. (1993). Adaptation and the goals of evolutionary research. *Quart. Rev. Biol.* 68(1), 1-32.

Reznick, D. N. (1981). "Grandfather effects": the genetics of interpopulation differences in offspring size in the mosquito fish *Gambusia affinis. Evolution* 35, 941-953.

Reznick, D. N. (1982a). Genetic determination of offspring size in the guppy (*Poecilia reticulata*). *Amer. Nat.* 120, 181-188.

Reznick, D. N. (1982b). The impact of predation on life history evolution in Trinidadian guppies: the genetic components of observed life history differences. *Evolution* 36, 1236-1250.

Reznick, D. N. (1989). Life history evolution in guppies. 2. Repeatability of field observations and the effects of season on life histories. *Evolution* 43, 1285-1297.

Reznick, D. N., and Bryga, H. (1987). Life history evolution in guppies. 1. Phenotypic and genetic changes in an introduction experiment. *Evolution* 41, 1370-1385.

Reznick, D. N., and Bryga, H. (1996). Life history evolution in guppies. 5. Genetic basis of parallelism in life histories. *Amer. Nat.* 147, 339-359.

Reznick, D. N., and Endler, J. A. (1982). The impact of predation on life history evolution in Trinidadian guppies (*Poecilia reticulata*). *Evolution* 36, 160-177.

Reznick, D. N., Rodd, F. H., and Cardenas, M. (1996a). Life history evolution in guppies (*Poecilia reticulata:* Poeciliidae). 4. Parallelism in life history phenotypes. *Amer. Nat.* 147, 319-338.

Reznick, D. N., Bryga, H., and Endler, J.A. (1990). Experimentally induced life-history evolution in a natural population. *Nature* 346, 357-359.

Reznick, D. N., Butler IV, M. V., and Rodd, H. (1996b). Differential mortality as a mechanism for natural selection in the guppy (*Poecilia reticulata*). *Evolution* in press.

Ridley, M. (1993). "Evolution." Blackwell Scientific Publications, Boston.

Roach, D. A., and Wulff, R. D. (1987). Maternal effects in plants. *Annu. Rev. Ecol. Syst.* 18, 209-236.

Robson, G. C., and Richards, O. W. (1936). "The Variations of Animals in Nature." Longmans Green, London.

Schluter, D. (1988). Estimating the form of natural selection on a quantitative trait. *Evolution* 42, 849-861.

Schmidt, K. P., and Levin, D. A. (1985). The comparative demography of reciprocally sown populations of *Phlox drummondii* Hook. I. Survivorships, fecundities, and finite rates of increase. *Evolution* 39, 396-404.

Seghers, B. H. (1973). An Analysis of Geographic Variation in the Antipredator Adaptations of the Guppy, *Poecilia reticulata.* Ph. D. thesis, University of British Colombia.

Service, P. M., and Lenski, R. E. (1982). Aphid genotypes, plant phenotypes, and genetic diversity: a demographic analysis of experimental data. *Evolution* 36, 1276-1282.

Sheppard, P. M. (1951). Fluctuations in the selective value of certain phenotypes in the polymorphic land snail *Cepaea nemoralis* (L.). *Heredity* 5, 125-134.

Simmons, G. M., Kreitman, M., Quattlebaum, W. F., and Miyashita, N. (1989). Molecular analysis of the alleles of alcohol dehydrogenase along a cline in *Drosophila melanogaster*. I. Maine, North Carolina, and Florida. *Evolution* 43, 393-409.

Simms, E. L., and Rausher, M. D. (1989). The evolution of resistance to herbivory in *Ipomoea purpurea*. II. Natural selection by insects and costs of resistance. *Evolution* 43, 573-585.

Snyder, L. R. G., Hayes, J. P., and Chappell, M. A. (1988). Alpha-chain hemoglobin polymorphisms are correlated with altitude in the deer mouse, *Peromyscus maniculatus*. *Evolution* 42, 689-697.

Sober, E. (1984). "The Nature of Selection." MIT Press, Cambridge.

Starmer, W. T., Barker, J. S. F., Phaff, H. J., and Fogleman, J. C. (1986). Adaptations of *Drosophila* and yeasts: their interactions with the volatile 2-propanol in the cactus-microorganism-*Drosophila* model system. *Aust. J. Biol. Sci.* 39, 69-77.

Travis, J. (1994a). The ecological genetics of life histories: variation and its evolutionary significance. *In* "Ecological Genetics" (L. A. Real, ed.), pp. 171-204. University of North Carolina Press, Chapel Hill.

Travis, J. (1994b). The interplay of life-history variation and sexual selection in sailfin mollies. *In* "Ecological Genetics" (L. A. Real, ed.), pp. 205-232. University of North Carolina Press, Chapel Hill.

Travis, J. (1994c). Size-dependent behavioral variation and its genetic control within and among populations. *In* "Quantitative Genetic Approaches to Animal Behavior" (C. M. Boake, ed.), pp. 165-187. University of Chicago Press, Chicago.

Travis, J., and Henrich, S. (1986). Some problems in estimating the intensity of selection through fertility differences in natural and experimental populations. *Evolution* 40, 786-790.

Travis, J., and Trexler, J. C. (1987). Regional variation in habitat requirements of the sailfin molly, with special reference to the Florida Keys. Florida Game and Fresh Water Fish Comm. Nongame Wildl. Program Tech. Rep. No. 3, Tallahassee, Florida.

Trexler, J. C. (1988). Hierarchical organization of genetic variation in the sailfin molly, *Poecilia latipinna* (Pisces, Poeciliidae). *Evolution* 42, 1006-1017.

Trexler, J. C., and Travis, J. (1990). Phenotypic plasticity in the sailfin molly, *Poecilia latipinna* (Pisces: Poeciliidae). I. Field experiment. *Evolution* 44, 143-156.

Trexler, J. C., Tempe, R. C., and Travis, J. (1994). Size-selective predation of sailfin mollies by two species of heron. *Oikos* 69, 250-258.

Trexler, J. C., Travis, J., and McManus, M. (1992). Effects of habitat and body size on mortality rates of *Poecilia latipinna*. *Ecology* 73, 2224-2236.

Trexler, J. C., Travis, J., and Trexler, M. (1990). Phenotypic plasticity in the sailfin molly, *Poecilia latipinna* (Pisces: Poeciliidae). II. Laboratory experiment. *Evolution* 44, 157-167.

Turelli, M. (1988). Population genetic models for polygenic variation and evolution. *In* "Proceedings of the Second International Conference on Quantitative Genetics" (B. S. Weir, E. J. Eisen, M. M. Goodman, and G. Namkoong, eds.), pp. 601-617. Sinauer, Sunderland, MA.

Turelli, M., and Barton, N. H. (1990). Dynamics of polygenic characters under selection. *Theor. Pop. Biol.* 38, 1-57.

Van Delden, W. (1984). The alcohol dehydrogenase polymorphism in *Drosophila melanogaster*, facts and problems. *In* "Population Biology and Evolution" (K. Wohrmann and V. Loescheke, eds.), pp. 127-142. Springer-Verlag, Berlin.

Van Delden, W., and Kamping, A. (1989). The association between the polymorphisms at the *Adh* and Gpdh loci and the In(2L)t inversion in *Drosophila melanogaster* in relation to temperature. *Evolution* 43, 775-793.

Van der Zel, A., Dadoo, R., Geer, B. W., and Heinstra, P. W. (1991). The involvement of catalase in alcohol metabolism in *Drosophila melanogaster* larvae. *Archiv. Biochem. Biophys.* 287, 121-127.

Van Noordwick, A. J., and De Jorg, G. (1986). Acquisition and allocation of resources: their influence on variation in life-history tactics. *Amer. Nat.* 128, 137-142.

Van Zandt-Brower, J. V. Z. (1958a). Experimental studies of mimicry in some North American butterflies. Part 1. The Monarch, *Danaus plexippus*, and Viceroy, *Limenitis archippus archippus*. *Evolution* 12, 32-47.

Van Zandt-Brower, J. V. Z. (1958b). Experimental studies of mimicry in some North American butterflies. Part 2. *Battus philenor* and *Papilio troilus*, *P. polyxenes* and *P. glaucus*. *Evolution* 12, 123-136.

Van Zandt-Brower, J. V. Z. (1958c). Experimental studies of mimicry in some North American butterflies. Part 3: *Danaus gilippus berenice* and *Limenitis archippus floridensis*. *Evolution* 12, 273-285.

Wade, M. J., and Arnold, S. J. (1980). The intensity of sexual selection in relation to male sexual behavior, female choice, and sperm precedence. *Anim. Behav.* 28, 446-461.

Wade, M. J., and Kalisz, S. (1990). The causes of natural selection. *Evolution* 44, 1947-1955.

Watt, W. (1977). Adaptation at specific loci. I. Natural selection on phosphoglucose isomerase of *Colias* butterflies: biochemical and population aspects. *Genetics* 87, 177-194.

Watt, W. (1983). Adaptation at specific loci. II. Demographic and biochemical elements in the maintenance of the *Colias* PGI polymorphism. *Genetics* 103, 691-724.

Watt, W., Cassin, R. C., and Swan, M. S. (1983). Adaptation as specific loci. III. Field behavior and survivorship differences among *Colias* PGI genotypes are predictable from in vitro biochemistry. *Genetics* 103, 725-739.

Weldon, W. F. R. (1893). On certain correlated variations in *Carcinus moenas. Proc. Roy. Soc., London* 54, 318-329.

Weldon, W. F. R. (1895). An attempt to measure the death-rate due to the selective destruction of *Carcinus moenas* with respect to a particular dimension. *Proc. Roy. Soc., London* 57, 360-379.

Weldon, W. F. R. (1899). Presidential address. British Association for the Advancement of Science, Report of meeting, Bristol, 1898. Transactions, Sec. D: pp. 887-902.

Whitlock, M. C., Phillips, P. C., Moore, F. B.-G., and Tonsor, S. J. (1995). Multiple fitness peaks and epistasis. *Annu. Rev. Ecol. Syst.* 26, 601-629.

Wickler, W. (1968). "Mimicry in Plants and Animals." McGraw-Hill, New York.

Wilkinson, G. S., Fowler, K., and Partridge, L. (1990). Resistance of genetic correlation structure to directional selection in *Drosophila melanogaster. Evolution* 44, 1990-2003.

Williams, G. C. (1957). Pleiotropy, natural selection and the evolution of senescence. *Evolution* 11, 398-411.

Williams, G. C. (1966). Natural selection, the costs of reproduction, and a refinement of Lack's principle. *Amer. Nat.* 100, 687-690.

Winn, A. A., and Evans, A. S. (1991). Variation among populations of *Prunella vulgaris* in plastic responses to light. *Funct. Ecol.* 5, 562-571.

CHAPTER 9

Molecular Population Genetics of Adaptation

RICHARD R. HUDSON

I. Introduction

Although a universally satisfactory definition of adaptation is probably impossible, a rather broad and simple definition is useful for addressing many issues: An adaptation is a character that helps its bearer to survive and reproduce. Usually, and in most cases of interest, this implies that an adaptation is a character that is maintained in the population by the direct action of natural selection. Thus, I take adaptation to be a result of an attractor that is a local maximum with respect to fitness in evolutionary space. For example, consider the case of purifying selection acting on a DNA sequence. The DNA sequence is considered adaptive if

i) Mutants will arise, at least occasionally, that result in modification of the sequence (or its elimination). This is equivalent to requiring that genetic variation will exist, at least from time to time.

ii) Such mutants must be selected against and hence typically eliminated from the population due to their effects on fitness.

If the evolutionary system is not at a maximum with respect to fitness, I will refer to changes of state that increase fitness as adaptive.

Thus, a reasonably complete understanding of an adaptive trait would require (a) information on the amount and nature of genetic variation in the trait, (b) knowledge of the fitness consequences of phenotypic variation in the trait. In other words, we need to know the connections among genotype, phenotype, and selection coefficients. With these things known (together with mutation, migration, and recombination rates as required), population genetics theory in principle should be able to predict gene frequency trajectories and equilibria. I say in principle because, with many interacting loci, variable environments, and finite population size, among other complications, the task may not be trivial. Obviously, the challenge is great and very few traits are thoroughly understood in all these respects (Lewontin, 1974; Wright, 1978).

In the following paragraphs, my focus will be studies of adaptation at the molecular level. The traits of interest are those associated with proteins, RNAs, and DNAs. In some cases, we will consider the sequences themselves to be the traits under study. In these cases the genetic bases for the trait is evident, although the effects on phenotypic characters may be completely obscure. I will briefly outline some of the methodology used to study molecular adaptation and some of the conclusions from such studies. As will be discussed below, there is little question that many of the nucleotide sequences which constitute the genomes of organisms were shaped by and are currently maintained by natural selection. However, there is controversy concerning the fraction of evolutionary changes at the molecular level that are in fact adaptive as opposed to being the result of mutation and random drift of selectively neutral variants (Kimura, 1983; Gillespie, 1991).

II. Functional Analysis of Protein Variation

A. Are the Protein Differences Between Different Species Adaptive?

To address this question, a number of investigators have purified proteins from species that live in different habitats and measured such enzymatic properties as V_{max}, K_m, k_{cat}, thermostability, or oxygen affinities at a variety of temperatures and pressures. (V_{max}, K_m, k_{cat} are parameters which are used to characterize the kinetic properties of an enzyme. K_m, the Michaelis constant, is a measure of the strength of binding of enzyme and substrate. k_{cat} is a measure of catalytic efficiency. The rate at which product is produced in an enzyme catalyzed reaction is often well approximated by a simple function of K_m, k_{cat} and the enzyme and substrate concentrations. V_{max} is the product of k_{cat} and the enzyme concentration, and gives the rate of reaction at high substrate concentrations.) By correlating the physico-chemical properties of the proteins with environmental parameters relevant to the species sampled, a plausible case for adaptive evolution of the proteins can be made in some cases. Gillespie (1991) provides a very readable and provocative summary of five such cases: (i) lactate dehydrogenase in fishes, (ii) phosphoglucose isomerase in bivalves, (iii) hemoglobin in a variety of organisms, (iv) insulin in guinea pigs and other mammals, and (v) lysozme c in ruminants and primates. These studies are not unlike many comparative studies of adaptation in morphological, physiological, or behavioral traits, and suffer from the same weaknesses and strengths discussed elsewhere in this book. In none of these studies was it feasible to actually manipulate traits and measure fitness or even any performance measure that might correlate with fitness. In most cases, the number of amino acid differences involved is not clear, so the results cannot really address the issue of what fraction of amino acid replacements might actually be adaptive.

Hemoglobin, which is perhaps the best studied protein from a functional and evolutionary standpoint, illustrates the difficulties very well. Many biochemical properties of the molecule (such as oxygen-binding properties, sensitivities to pH, and a variety of ligands) clearly have adaptive significance (Perutz, 1983). Many features that are easily disrupted by amino acid substitutions are strongly conserved or correlated with environmental conditions of different organisms and hence presumably maintained by natural selection and the long catalogue of human globin mutants that produce serious genetic diseases like thalassemia are unequivocal evidence for the adaptive nature of some sites in the protein (Vogel, 1969). Nevertheless, it remains unclear what fraction of amino acid substitutions, say between different mammal species, are adaptive in this molecule. The overall structure and shape of hemoglobin molecules appear to be conserved despite large numbers of amino acid substitutions that distinguish globins from different mammals (Perutz, 1983). This led Perutz (1983) to suggest that the majority of differences between the hemoglobins of mammals are the result of the fixation of neutral mutations by genetic drift, with only a small number of adaptive amino acid substitutions. Gillespie (1991), however, argues that the data are far from conclusive on this point and suggests the possibility that most amino acid substitutions have some adaptive significance. It is important to remember that very small selection coefficients can be extremely important in determining the fate of mutations, unless population sizes are quite small. [Recall the rule of thumb that if the product of effective population size and selection coefficient is much bigger than one, selection will dominate drift (Kimura, 1983). Hence selection coefficients as small as 10^{-3} or even 10^{-4} are likely to be evolutionarily significant.] Small physiological effects may have small but significant selective effects that then determine the evolutionary fate of sequences. The magnitude of the physiological and selective effects involved could be well below what could be detected directly in experiments in the lab or field. Evidence from studies of hemoglobin of closely related species or subspecies of mice and fish (reviewed by Gillespie, 1991) provide good evidence that subtle changes produced by single amino acid substitutions can be selectively important.

The case of lysozyme c is noteworthy as an example of evolutionary convergence in biochemical properties and sequence (reviewed also by Gillespie, 1991). In both ruminants and langurs (a leaf-eating primate), the enzyme has acquired a new function to aid in the digestion of bacteria that flow from the rumin into the stomach. Lysozyme has independently evolved in ruminants and langurs to function at low pH and to resist digestion in the stomach (Jollés et al., 1989). In the lineage leading to ruminants (since their most recent common ancestor with pigs), an unusually high number of amino acid substitutions occurred (Stewart and Wilson, 1987). (The number of substitutions was inferred by maximum parsimony methods.) This high number of substitutions is interpreted as being the result of adaptive evolution of the protein to function in its new role. Jollés et al. (1989) attribute functional significance to 21 of the 34 inferred substitutions on the ruminant lineage. Similarly, in the lineage leading to langurs since their divergence from baboon, there have occurred 10 amino acid substitutions while only 4 have occurred in the lineage leading to baboons. Remarkably, five of the substitutions in the langur lineage are to amino acids that are found at the homologous position in the cow lysozyme (Stewart and Wilson, 1987). Another example of convergence with a clear adaptive interpretation is the visual pigments in fish and humans. By comparing the DNA sequences of the human and fish visual pigment genes and knowing their phylogenetic relationships, Yokoyama and Yokoyama (1990) were able to show that the red pigments in humans and fish evolved from the green pigments independently by identical amino acid substitutions in a few key positions.

Even if we confine attention to Gillespie's small number of cases of adaptive protein evolution, we are left in the dark about whether these cases are representative of most protein evolution or not. Negative results concerning adaptation are presumably rarely published and in any case, negative results can always be explained away as due to the smallness of the effects involved. Thus, it appears to be impossible to generalize with confidence about the adaptive nature of most differences between species at the molecular level from studies seeking to find functional differences between proteins and then correlate them with environmental variables.

B. Are Protein Polymorphisms Selectively Maintained?

A number of protein polymorphisms within species have been investigated with approaches similar to those just described for addressing among species adaptation. Gillespie (1991) also summarizes several of these studies, including studies of lactate dehydrogenase in the fish *Fundulus heteroclitus*, phosphoglucose isomerase in *Colias* butterflies, hemoglobin in mice and birds, glutamate-pyruvate transaminase in a copepod, and aminopeptidase-1 in a mussel. [The last two mentioned enzymes are involved in osmoregulation.] In all of these cases, in addition to measuring biochemical properties of enzyme variants, some measure of performance, behavior, or fitness was estimated for different genotypes and interpreted in terms of the biochemical properties of the enzymes involved. Also, in several cases the polymorphisms exhibit geographic clines which are interpretable in terms of the biochemical properties of the relevant alleles. These clines by themselves suggest the action of natural selection, though historical factors can explain them also. These studies provide strong evidence that some amino acid substitutions have important phenotypic effects upon which natural selection can act.

Besides the cases reviewed by Gillespie to which I have just referred, I will briefly mention a few other cases where protein polymorphisms have been shown in a fairly direct way to be maintained by natural selection. Sickle cell anemia is the classic case of heterozygote superiority. In this case, biochemical properties, population genetic evidence (spatial distribution of allele frequencies and departure from Hardy-Weinberg equilibrium), and direct estimates of the fitness of sickle allele homozygotes all support the involvement of natural selection in the maintenance of this polymorphism.

Although the functional basis for the selection is typically not revealed, I will mention here that evolution in laboratory populations can be very useful in detecting the action of natural selection on protein polymorphism. Such studies are frequently difficult to interpret due to problems with inbreeding, linkage disequilibrium, and effects due to subjecting a population to a novel environment. A notable exception are the studies of Rose and his colleagues who have maintained large, outbred, laboratory-adapted populations of *Drosophila melanogaster*. They have examined protein polymorphism in their demographically selected lines and found that certain alleles of superoxide dismutase (SOD) are consistently shifted to lower frequency in replicate populations subject to selection for late reproduction compared to populations selected for early reproduction (Tyler et al., 1993). These observations clearly establish that selection operates on SOD variants, or variants tightly linked to the SOD locus.

Chemostat studies of five enzyme polymorphisms in *Escherichia coli* have been carried out by Hartl and Dykhuizen (1985). All the enzymes are involved in glycolysis fairly directly. Gillespie (1991) reviews these studies and concludes that no selective differences are detectable in the glucose limited chemostats. [Fitness differences as low as 0.4% would be detected in these experiments.] However, in chemostats where sugars other than glucose were supplied or when background mutations were introduced that might exaggerate the fitness effects of variation in the enzymes under study, selection was detected in 35% of the cases examined. The biochemical bases for these selective differences were not established.

Some recent work on HIV-1 demonstrates that two or three amino acid changes on a surface loop of the GP120 protein determines the ability to infect macrophage (Henkel et al., 1995). In addition, it has been shown that treatment with certain anti-viral drugs consistently selects for HIV variants with a particular base pair change at particular sites (Cleland et al., 1996). (See also Holmes et al., 1992).

All of these studies of protein polymorphism are intriguing and suggest the possibility that many amino acid substitutions result in kinetic differences in enzymes. In some of the cases investigated, these kinetic differences appear to have important physiological and, most importantly, fitness effects. Again, however, it is difficult to know what general conclusion to draw: are these loci rare exceptions or are they indicative of the general rule? As with the studies of between-species differences, negative results are probably rarely published and when they are, one is always left wondering if a bigger experiment would have changed the conclusion.

III. Indirect Inferences of Adaptation from Molecular Divergence and Polymorphism

The difficulty of directly measuring very small selection coefficients has led many investigators to examine the patterns of variation within and among species with the hope that patterns generated by evolution over long periods of time in large natural populations might be useful in distinguishing neutral from selective hypotheses concerning molecular evolution. Some of these studies will be described next. The following discussions will concentrate on protein and DNA sequence data. With minor exceptions, I will not discuss the statistical analyses of enzyme polymorphisms as developed in the pre-DNA-sequencing era. Although data on enzyme polymorphism have provided a great deal of information about the structure of populations and have provided a number of mysteries to explain, the analysis of this polymorphism has not by itself yielded much information about the nature of selection acting on that variation (Ewens, 1989; Gillespie, 1991).

A. Goldilocks Encounters the Sequence Data Base ("These proteins evolve too slow; these too fast; these are just right")

1. *Too slow:* I begin with a noncontroversial result. Histone H4 evolves at a very slow rate. Pea H4 and cow H4 differ in only two amino acid positions out of 100 positions (Isenberg, 1979). Since plants and animals diverged roughly 1 billion

years ago, it is safe to say that the histone sequence has been highly conserved. Over the billions of years of evolution of this molecule in eukaryotes, every amino acid position has experienced amino acid-changing mutations many many times and clearly almost all of these mutations have been deleterious and thus eliminated from populations by the action of natural selection. If any amino acid changes had been selectively advantageous, or even neutral, we would observe many more differences between histones from different species than we do. Thus, the histone protein sequence clearly satisfies the two criteria for one type of adaptation that are listed in the opening paragraph of this chapter: mutations occur in this sequence and they are rejected by natural selection. This provides strong evidence that the H4 sequence is in some sense optimal, although it should be noted that the optimality may only be local. By that I mean that single histone amino acid substitutions are no doubt almost always deleterious, but we cannot at present rule out the possibility that another sequence, say four mutational steps away from the extant sequence, might confer higher fitness on its bearer.

Histone is an extreme example of a very general phenomenon: amino acid substitutions occur at a relatively slow rate compared to the rates of evolution shown by silent sites and pseudogenes. The evidence is compelling that an important fraction of amino acid-changing mutations must be deleterious and hence that protein sequences are on the whole adaptive. This is not really surprising or controversial. What may be surprising to some is how weak the selection can be that preserves these sequences. Golding and Felsenstein (1990) have developed a maximum likelihood approach for estimating the strength of selection maintaining a particular base at a site. The basic idea is that given a phylogenetic relationship between taxa and a spontaneous mutation rate, a low level of divergence between taxa implies selective constraint. Golding (1994) has also used the method to estimate the strength of selection maintaining each hydrogen-bonded pair in the secondary structure of an rRNA molecule. The estimates are based on sequences from 51 taxa, including a range of eukaryotes, bacteria, and archaebacteria. In Fig. 1, Golding's estimates of selection coefficients are shown for each hydrogen bonded pair by a vertical line whose length is proportional to the strength of selection for that pair. What is actually estimated is the composite parameter, $4Ns$, where N is the effective population size and s is the deleterious effect of breaking a hydrogen bond. He found that $4Ns = 6.3$ was sufficiently large to account for complete conservation of hydrogen bonding at a site, i.e., for all 51 taxa to have a hydrogen-bonded pair at a site. Unless N is quite small, this means that $s = 10^{-4}$ is large enough to maintain a sequence over long evolutionary periods. On the other hand, with 10 taxa (out of 51) lacking hydrogen bonding at a site, the maximum likelihood estimate of $4Ns$ is still 1.88 and Golding finds that this degree of conservation is still highly unlikely under neutrality ($s=0$). Thus, $4Ns$ of 1.88 is small enough to allow some variation to appear among distantly related taxa, but is large enough to produce a significant conservation of sequence. Golding (1993) has also generalized the method to detect and estimate positive selection coefficients in cases were sequences evolve "too" fast.

Thus, while the conservation of protein sequences provides strong evidence for selection maintaining sequences, theory suggests that the selection need not be strong on average. For soluble enzyme loci, the observed frequencies of null alleles in natural populations of *Drosophila* suggests that selection against null

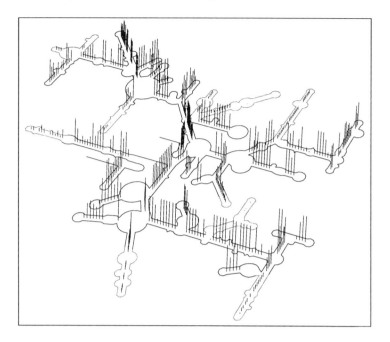

Figure I Estimated strength of selection on 16/18S rRNA molecules . The vertical lines indicate the strength of selection maintaining hydrogen bonding base pairs in the secondary structure of the rRNA molecule. Stronger selection is indicated by longer vertical lines. The estimates of the strength of selection are obtained with a maximum likelihood analysis of rRNA sequences from 51 species. Redrawn from Golding (1994).

alleles in heterozygotes is only about 10^{-3} (Langley et al., 1981). If null alleles are this weakly selected against in heterozygotes, then it is plausible that variants with minor effects on enzyme kinetics are even more weakly selected. In addition, though most of the 25 enzymes surveyed by Langley et al. were from intermediary metabolism, only 1 of the 13 loci for which null alleles were found was lethal as a homozygote. Thus, though protein sequences are clearly adaptive and conserved over evolutionary time, the selection coefficients involved are likely to be quite small in many cases.

2. *Just the right rate of evolution?:* In contrast to histones and protein coding sequences in general, pseudogenes are frequently put forward as sequences without adaptive significance that can freely accept mutations. (Pseudogenes are genes which are no longer expressed due to nonsense mutations and/or frameshift mutations. Pseudogenes are putatively the ultimate in junk DNA.) If it is true that pseudogenes can evolve without constraint, then the evolutionary divergence of pseudogenes in different species would give an accurate view of the spontaneous mutation process. This is because for a completely unconstrained sequence the substitution rate of the sequence is equal to the mutation rate (Kimura, 1983). One way to assess whether pseudogenes are in fact unconstrained would be to compare the pattern and rate of evolution of pseudogenes to directly measured patterns and rates of spontaneous mutation. Unfortunately, in those organisms with good pseudogenes for estimating the substitution process, there are at present no good estimates of the spontaneous mutation rate. Hence,

although pseudogenes are quite plausibly evolving free of constraint, the alternative that some constraint and even some adaptive evolution occurs in these sequences is difficult to rule out entirely at this time.

An intriguing observation is that silent sites are estimated to evolve at about the same rate as pseudogenes. We know that silent sites determine codon usage and can affect transcription rates and the secondary structure of mRNA, all of which suggest that natural selection may well play a significant role in substitutions at silent sites. In other words, there is likely to be some constraint on silent sites. In addition, the possibility of adaptive evolution at these sites must be considered. Perhaps the constraint is so slight that the evolutionary rate is indistinguishable from "unconstrained sequence" rates. Another possibility is that there is substantial constraint on silent sites, but adaptive evolution also occurs at some of these sites in such a way that the net rate of evolution is similar to pseudogene rates. Of course it is also possible that there are unknown constraints on pseudogene evolution which slow its rate of evolution to the same rate as the somewhat constrained silent sites.

The extent to which pseudogenes and silent sites are subject to natural selection is quite important. If it can be established that they are subject to minimal constraint then they can serve as a standard of comparison for all other classes of sites. The action of natural selection at other classes of sites could then be detected by a comparison of the patterns and rates of evolution at other classes of sites to the silent site and pseudogene patterns. The comparison of evolutionary patterns in different classes of sites is a powerful technique and will be discussed further below.

3. *Too fast:* There are a small number of protein coding loci for which protein altering changes occur more frequently than silent changes. Actually, in most cases, this pattern is characteristic of only a part of the protein involved. This pattern has been documented in the HLA loci of humans (Hughes and Nei, 1988), in immunoglobulin genes in rats (Frank et al., 1984), in the circumsporozoite protein of *Plasmodium falciparum* (Hughes, 1993), and in two acrosomal proteins of abalone (Lee and Vacquier, 1995; Swanson and Vacquier, 1995). In the first three cases, the rapidly evolving portions of the gene have been implicated in immune response or immune defense in host-pathogen interactions. In HLA the rapid protein evolution is concentrated in the antigen recognition region of the protein. The source of the selection pressure on acrosome proteins is less clear. In the case of HLA, there is not only rapid evolution of the proteins, but also very high levels of genetic variation within human populations. This pattern of variation will be revisited later.

The possibility of rapid molecular evolution as a consequence of positive selection has important implications for comparative molecular studies. Such comparative studies are frequently used to identify important functional sequences by searching for a strong conservation of sequences. Clearly such a methodology will occasionally miss functionally important sequences. It is a mistake to assume that a fast-evolving sequence is unconstrained.

4. *Too erratic:* An ardent neutralist view of protein evolution goes something like this: (1) most spontaneous mutations are strongly deleterious, and (2) of those mutants that are not deleterious, the vast majority are effectively neutral (in terms of fitness). Under this model, the rate of evolution of a locus will be $f\mu$, where μ is the spontaneous mutation rate and f is the fraction of spontaneous

mutants that are neutral for that locus. The parameter f is assumed to be variable from locus to locus, but essentially constant over evolutionary time for a given protein. It is the variation of f from locus to locus that accounts, under the neutral theory, for the variation in substitution rates between loci. Under this model, the expected number of substitutions that occur along an evolutionary lineage of length, t, is fμt. Because the process is stochastic, the actual number of substitutions is a random quantity and under the standard neutral model (with f fixed), the variance of the number of substitutions is equal to the mean, fμt. A number of authors have investigated the variance of the substitution process and have found that frequently the variance is too large to be compatible with the strict neutral model that I have just described. [See Gillespie (1991) for a review.] Gillespie has conducted perhaps the most robust analyses and finds that the amino acid substitution process is significantly too high in variance in 60% of the 20 protein loci that he examined. Gillespie suggests that this pattern is best accounted for by the action of natural selection producing bursts of adaptive changes in response to a changing environment. Neutral models with changing f have also been proposed to explain the high variance in the substitution process. If f values change drastically enough on the right time scale, clearly a high variance can be produced (Takahata, 1991).

My view is that changes in f are certainly plausible and very likely to have occurred; however, it is difficult to imagine them occurring without positive natural selection frequently being engendered at the same time. In particular, it is hard for me to see how large decreases in f can occur without large numbers of sites becoming subject to selection for adaptive changes. Recall that a large decrease in f means that many sites that had been free to evolve neutrally suddenly become constrained, such that most spontaneous mutations become deleterious. For all such mutations to be deleterious means that the extant sequence is locally optimum. In other words, a sequence that had been unconstrained and neutrally evolving, after some neutral change at another site (or a change in the environment) suddenly becomes optimally adapted to the new situation. The assumption is that in the new situation, very few of the previously unconstrained sites are in the deleterious state and hence are subject to favorable mutations that would substitute fairly quickly. It seems more plausible to me that the event of f decreasing substantially is instead likely to be accompanied by some adaptive evolution, subsequent to which the constraint might become high (and f small.)

The erratic nature of protein evolution continues to represent a serious difficulty for the neutral theory.

B. Analysis of Polymorphism Data

1. Geographic Patterns

In *D. melanogaster* a number of loci (and chromosomal arrangements) exhibit a correlation of allele or karyotype frequency with latitude (Oakeshott et al., 1982, 1984). Although clines can be generated in neutral characters by drift, in many of these cases the pattern is repeated on several continents in parallel ways that seem unlikely to be produced by chance founder effects. These parallel patterns provide strong evidence that the loci surveyed, or very closely linked loci, are subject to natural selection that correlates with latitude. How natural selection acts on the variation at these loci is unknown.

At the alcohol dehydrogenase (*Adh*) locus, a cline in frequency of two electrophoretic variants was tested for selective significance by Berry and Kreitman (1993). Along the eastern seaboard of the United States the frequency of Fast allele increases from about 10% in the south to more than 50% in the northern populations. If this differentiation of northern and southern populations is due to random drift of neutral alleles, through either founder effects or simply drift over time with limited migration between north and south, then one would expect other sites to have been similarly affected. In particular, one would expect Slow alleles in the north to have differentiated from slow alleles from the south. No such differentiation of Slow alleles is observed. From the similarity of frequencies of different slow haplotypes in the north and south, it appears that gene flow between the north and the south is high, sufficiently high to prevent differentiation of populations in the two regions. Berry and Kreitman concluded that the electrophoretic variation (or variation in linkage disequilibrium with it) is under natural selection that produces and maintains the cline in the face of high levels of gene flow.

Karl and Avise (1992) detected the action of natural selection using geographic information in a related way. They found strong differentiation of Atlantic and Gulf coast populations of the American oyster *Crassostrea virginica* when four nuclear DNA polymorphisms were surveyed. By contrast, all 14 electrophoretically detected protein polymorphisms showed relatively low allele frequency differences between the Atlantic and Gulf populations. Clearly both classes of polymorphisms cannot be evolving according to the neutral model. The DNA polymorphisms were restriction sites in essentially random pieces of DNA and are unlikely to be amino acid-changing polymorphisms. These polymorphisms are most likely in intergenic regions, introns, or are perhaps silent polymorphisms with a coding region. This led Karl and Avise (1992) to suggest that the DNA polymorphisms are indeed neutral markers and that the differentiation between the populations at these sites indicates that there is little gene flow between these two populations. The lack of differentiation at the enzyme loci was then attributed to some form of balancing selection which maintains similar allele frequencies in the two populations. McDonald (1994) provides a useful review of methods and problems in rigorously testing for the significance of heterogeneity in levels of differentiation.

2. Frequency Spectrum

With today's technology it is possible to sample a number of copies of a gene from a natural population and sequence the copies to reveal genetic variation at the DNA level. [Schaeffer and Miller (1993) sequenced around 100 copies of a 3500 base pair-long region including the *Adh* region of *D. pseudoobscura.*] Assuming that the sequences can be unambiguously aligned, we can examine each site of the gene and determine if the site is monomorphic or polymorphic in the sample. At polymorphic sites, we can characterize the variation by the number of different nucleotides present at the site in the sample (2, 3, or 4) and by the frequency in the sample of each nucleotide. The frequencies of the different nucleotides at a site will be referred to as the sample configuration at the site. Two hypothetical samples are illustrated in Fig. 2. For sites with just two nucleotides present, the sample configuration can be described by an unordered pair of numbers. For example, at site 1 of sample II, the configuration is denoted (1,6) to indicate that there are two alleles, one at frequency 1 (out of seven) and one allele at frequency 6 (out of seven.) The sample configurations can be useful in making inferences about selection acting on genetic variation. If all the observed polymorphism is due to slightly deleterious mutations, then at each polymorphic site there will be one nucleotide (the selectively favored nucleotide) at high frequency and one or more nucleotides (the deleterious mutants) present at low frequency. On the other hand, if some form of balancing selection is

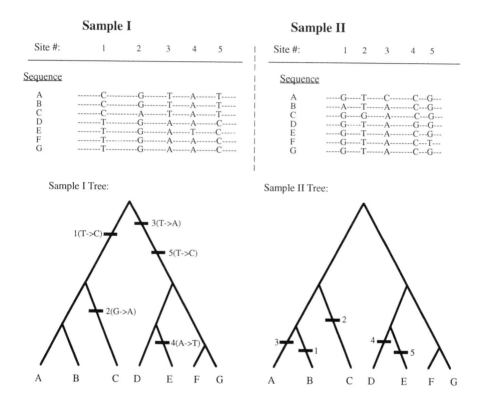

Figure 2 Two hypothetical samples of sequences with accompanying trees representing the ancestry of the sampled sequences. (The trees shown are not the only ones compatible with the data shown.) The bars on the trees indicate points at which changes occurred. The numbers next to the bars indicate which site changed, and on sample tree I, the actual nucleotide change is indicated. In both Sample I and Sample II there are seven sampled sequences and there are five sites that are polymorphic in the sample. In Sample I, sites 1, 3, and 5 exhibit the (3,4) sample configuration, and sites 2 and 4 exhibit the (1,6) configuration. In Sample II, all five polymorphic sites exhibit the (1,6) configuration.

occurring, two or more nucleotides at intermediate frequency might be observed at such polymorphic sites. Under a neutral model, at equilibrium the probabilities of various sample configurations can be calculated and an observed collection of sample configurations can be tested for compatibility with the equilibrium neutral model. The extent to which the polymorphisms tend to consist of one rare and one common allele, as opposed to two intermediate frequency alleles, can be quantified by a statistic, D, which can be used to test the equilibrium neutral model (Tajima, 1989) . If most polymorphisms consist of intermediate frequency variants, such as the (3,4) configurations of sample I, then D will be positive. If, on the other hand, most polymorphisms consist of one rare and one common variant, such as in sample II where all five polymorphic sites show a (1,6) configuration, then D will be negative. Under the equilibrium neutral model, D tends to be close to zero.

While Tajima's test has been quite useful for testing neutrality, it depends on an assumption that the population is at equilibrium under the neutral model. This is a serious drawback. Because equilibrium under neutrality can take very long to achieve in large populations, any rejection of the null model with this test can be plausibly explained by recent demographic events in the history of the population, instead of more interesting possibilities, including the action of natural selection. For example, if in using Tajima's test it is concluded that there are too many rare variants, this could be explained, as indicated above, by the presence of slightly deleterious mutations segregating in the population. However, an excess of rare variants is also consistent with a fairly recent founder event, in which all variation was lost, with the current variation then being due to recent mutations that have not yet reached the equilibrium distribution of frequencies. One way of testing an alternative hypothesis of recent demographic disturbances is to examine other loci or perhaps even other sites in the same locus. Since all sites in the genome would have been subject to the same demographic history, all neutral variation should exhibit the same pattern of variation. If, in our example, one finds that there is no excess of rare alleles at other loci, this suggest that the demographic explanation is not correct.

A particularly powerful test built on this idea of comparing the pattern of variation at different types of sites is the test of Sawyer et al. (1987). The basic idea of the test is to compare the sample configurations at two (or more) types of sites. If the two classes of sites exhibit significant differences in the frequency of different sample configurations, then it is reasonable to infer that there are substantial differences in the evolutionary forces acting on the two types of sites. For example, suppose that in Sample I of Fig. 2 that sites 2 and 4, which have sample configuration (1,6), were amino-changing polymorphisms and that sites 1, 3, and 5, which have the (3,4) configuration, were silent polymorphisms. These data would suggest (though the number of polymorphic sites are too small to be statistically significant) that silent polymorphisms and amino acid polymorphisms were subject to different selection pressures. If there is free recombination between the sites involved or if there is complete linkage between all the sites, a particularly robust test is possible using Fisher's exact test, or a simple Chi-square statistic. Sawyer et al. (1987) examined the sequences of the gnd locus in seven strains of *E. coli*. There were 12 amino acid polymorphisms, all of which showed the (1,6) configuration. There were 66 silent polymorphisms, with 34 of them having the (1,6) configuration. A simple 2x2 contingency table analysis is appropriate and shows that the amino-acid polymorphisms have a significantly different frequency spectrum than the silent-site polymorphism (p = 0.0026). Sawyer et al. (1987) suggest that the low frequency amino acid variants are slightly deleterious and show how to estimate the strength of selection against the variants under an assumption of equilibrium and free recombination between sites. It is worth reemphasizing that this test does not rely on equilibrium properties of the neutral model and clearly demonstrates that at least one type of site is not evolving according to the neutral model.

A conceptually similar test was used by Fitch et al. (1991), who examined influenza A virus evolution in about 20 isolates obtained at different times over the last 50 years. They reconstructed a "phylogenetic" tree and inferred the location on the tree of amino acid replacements in the hemagglutinin gene. They classified the amino acid replacements as antigenically significant or not. [This classification is conceptually analogous to the Sawyer et al. (1987) classification of polymorphisms as silent or not.] Fitch et al. (1991) then classified the replacements as occurring on the trunk of the tree (which leads to recently obtained isolates) or not. [This classification is analogous to the sample configuration classification used by Sawyer et al. (1987). This is because, as shown in Fig. 2, the location of a replacement on the tree determines the configuration of the polymorphism that results.] Fitch et al. (1991) chose this second classification on the basis that antigenic significant replacements are plausibly positively selected variants that played a role in the survival of the influenza virus that we observe today. A 2x2 contingency table analysis shows that amino acid replacements at antigenic sites indeed occurred significantly more often on the trunk than nonantigenic replacements. They concluded that positive selection had driven some of the antigenically significant replacements through the viral population.

McDonald and Kreitman (1991) also introduced a test that is similar conceptually to the Sawyer et al. (1987) test. The test of McDonald and Kreitman requires sequences from more than one species and more than one sequence from at least one species. With these sort of data there will usually be variation revealed both within species and among species. A slight reinterpretation of Sample I of Fig. 2 will illustrate this situation. In Sample I, suppose that sequences A, B, and C are from one species and that sequences D, E, F, and G are from another species. To carry out their test, McDonald and Kreitman classified variation as either amino-changing or silent, as did Sawyer et al. (1987). They also classified each variable site as producing polymorphism within species or producing divergence between species. In the newly interpreted Sample I data, the substitutions at sites 1, 3, and 5 are between-species changes and the substitutions at sites 2 and 4 produce within-species polymorphisms. Note, as with the other tests described, that this classification is essentially based on where the substitution occurs on the tree relating the sequences. Again a simple 2x2 contingency test is appropriate to decide if amino acid changing substitutions are distributed on the tree in the same way as silent substitutions. McDonald and Kreitman applied their test to sequences of the alcohol dehydrogenase locus (*Adh*) of several *Drosophila* species and rejected the neutral model. They found that there is too much amino acid divergence between species relative to the amount of amino acid polymorphism within species. The pattern is consistent with much of the amino acid sequence change between species having been driven by natural selection, i.e., adaptive protein evolution has been important in the divergence of *Adh* between closely related *Drosophila* species. Sawyer and Hartl (1992) have carried out an analysis of what patterns of polymorphism and divergence to expect under a variety of models and have provided methods for estimating the selection coefficients involved.

3. Peaks and Valleys of Variation

Unusually high or low levels of variation within populations can indicate natural selection. For example, it is known that balancing selection acting on a single nucleotide site that maintains a polymorphism for a long time at the site can result in greatly increased levels of neutral variation at linked sites (Hudson and Kaplan, 1988; Kaplan et al., 1988). This is observed at MHC loci in humans (Hughes and Nei, 1988), at plant self-incompatibility loci (Clark, 1993), and to some extent at the *Adh* locus of *D. melanogaster* (Hudson and Kaplan, 1988). Although few loci have been examined, it may be significant that other loci with protein polymorphisms have not shown dramatic evidence of such peaks of linked variation. This may be an indication that these polymorphisms are not maintained by balancing selection or that balancing selection tends to be short lived and hence that the observed polymorphisms are often recently arisen. Sickle cell variants in humans populations are a case in point. It is well established that sickle cell variants are maintained in some populations by balancing selection (where malaria is present)(Bodmer and Cavalli-Sforza, 1976), but sickle cell variants are all apparently recently arisen and have not resulted in large increases in variation at linked sites.

The rapid fixation of an advantageous mutant can sweep out variation at linked sites (Maynard Smith and Haigh, 1974; Kaplan et al., 1989) via a "hitchhiking" effect. This can result in a valley of variation, an unusually low level of variation in a region surrounding the site of the advantageous mutation. This hitchhiking effect of advantageous mutations has been invoked to explain the low levels of variation in regions of low recombination in D. melanogaster (Kaplan et al., 1989). The strength of the hitchhiking effect on linked levels of variation depends on the strength of selection on the advantageous mutations and the rate at which these mutations occur. Wiehe and Stephan (1993) have used the pattern of decreased variation to estimate these parameters.

It should be noted that these patterns of decreased variation in regions of low recombination in *Drosophila* may also be accounted for by so-called "background selection" on deleterious mutations occurring at linked sites (Charlesworth et al., 1993; Hudson and Kaplan, 1995). Under this hypothesis the continual production and selective elimination of deleterious mutations results in a lower effective population size which leads to lower levels of neutral variation at linked sites. This is an area of active investigation at the moment.

Though peaks and valleys in levels of variation may be an indication of current or recent selection, these patterns may also result from different levels of constraint at different loci. To distinguish the constraint hypothesis from other selective hypotheses, among-species comparisons of the sequences involved can be informative. Under the constraint hypothesis, levels of variation are high where constraint is low and hence the neutral mutation rate is high. At such less constrained loci, divergence between species is also expected to be high, because under the neutral model the divergence rate is equal to the neutral mutation rate. In other words, under the neutral model, the level of polymorphism and the amount of interspecific divergence should be strongly correlated. Demographic effects such as bottlenecks or population structure may perturb variation, but it remains true that loci with higher neutral mutation rates are expected to have higher levels of variation and higher rates of interspecific divergence than loci with low rates of neutral mutation.

This suggests that one might be able to test neutral models by comparing different loci in terms of their levels of polymorphism and amounts of interspecific divergence. Just such a test has been described by Hudson et al., (1987). Even under nonequilibrium situations, one expects under the neutral model that polymorphism and divergence will be correlated; however, their test statistic and the associated critical values were established by consideration of the equilibrium neutral model, and it is not clear how the test would be biased under various nonequilibrium neutral models. The test has been applied to the *Adh* region of *D. melanogaster*, where it shows that there is too much polymorphism in the *Adh* locus relative to other loci (Kreitman and Hudson, 1991). It has also been applied to compare polymorphism and divergence in loci in regions of high recombination versus loci in regions of low recombination. In this case, the results demonstrated that the low levels of variation in regions of low recombination cannot be due to simply a lower mutation rate or higher constraint (Begun and Aquadro, 1991).

4. Patterns of Codon Usage

There is redundancy in the genetic code, and hence most amino acids can be specified by more than one codon. The frequency with which various codons are used to specify each amino acid constitutes the codon usage pattern. Substantial evidence now shows that codon usage is in many cases strongly influenced by natural selection. This has in part been inferred from the correlation between codon usage and tRNA pool sizes (Ikemura, 1985), but also by comparison of codon usage in genes with high levels of expression compared to low levels of usage (Shields et al., 1988). [See Bulmer (1987) for some background theory.] In genes that are highly expressed and where one might expect selection to be strong for rapid and efficient translation, there is highly biased codon usage, different from the usage one would expect if different codons were functionally equivalent. By contrast, in less expressed genes, the pattern of codon usage is more consistent with the functional equivalence of the redundant codons.

Interestingly, in *Drosophila* it appears that codon usage is less biased in regions of low recombination than in regions of high recombination (Kliman and Hey, 1994). This suggests that the selection coefficients involved in codon usage are quite small. The argument for this is as follows. In order for selection to be strong enough to overcome mutation and drift, the product of effective population size and selection coefficients must be somewhat bigger than one (Kimura, 1983). If this product is less than one, drift will dominate selection. The codon usage bias in regions of high recombination suggests that the products of effective population size and selection coefficients are bigger than one for some substantial fraction of sites. Then how can we explain the low codon bias in regions of low recombination? It turns out that the effective population size may actually be smaller for sequences in regions of low recombination than in regions of low recombination. Recall that the observed levels of variation in regions of low recombination in *Drosophila*, are much lower than in other regions (by as much as a factor of ten). The lowered levels of variation are thought to be due to the effect of natural selection on variation in the regions, which arises because low levels of recombination affect large regions surrounding the actual site with fitness effects. This natural selection may be positive selection driving favorable mutations through the population (Wiehe and Stephan, 1993) or it may be selection against continually arising strongly deleterious mutations (Hudson and Kaplan, 1995). Both result in reduced effective population size. Other forms of natural selection, such as temporally varying selection, would plausibly have similar, although not identical effects. Thus, in regions of low recombination, the effective population size will be reduced, perhaps by a factor of ten. Since the codon usage is much less biased in these regions of low recombination, it is reasonable to infer that the product of effective population size (in regions of low recombination) and selection coefficients for optimal codons is often less than one. Since we have rough estimates of the effective population size we can make some inference about the strength of selection on codon usage as follows. Observed levels of variation in regions of high recombination suggest that the effective population size for these regions is roughly 10^6 (Kreitman, 1983). As indicated above, the effective population size in regions of low recombination is roughly ten times smaller than in regions of high recombination, and hence is about 10^5. It follows that the selection coefficients for codon choice are less than 10^{-4}.

IV. Conclusions

Parts of all proteins and many DNA sequences that regulate gene expression are clearly strongly conserved over the evolutionary time periods separating species. These sequences are clearly maintained in the face of mutation and drift, indicating the adaptive significance of the sequences. What is less clear is how to interpret changes that have occurred at those sites that have not been strongly conserved. The extent to which the differences between sequences are adaptive rather than the result of neutral mutation and drift is largely unresolved.

Functional studies of enzymes and proteins can certainly contribute to our understanding in cases where the functional effects are large. At present, we have a very small number of cases where enzyme or protein variation has clearly understood phenotypic and fitness effects. Generalizing from these studies is difficult. The alternative is to study patterns of sequence evolution. Studies of evolutionary patterns can be informative and can be applied to large numbers of loci. The irregular nature of the substitution process in proteins strongly suggests that proteins are not evolving according to a simple neutral model. Comparisons of patterns of polymorphism and divergence for different classes of sites point to the action of natural selection in a number of studies. The models needed to analyze sequence data are frequently complex and must make simplifying assumptions, but show promise of distinguishing between alternative hypotheses. These studies in themselves probably will not reveal the actual mechanism of natural selection, but they may well narrow the range of possibilities and suggest where to look experimentally . Several avenues of research will clearly continue to be fruitful in the future, including functional analysis of molecular variation within and among species and the statistical analysis of patterns of molecular variation at all levels.

References

Begun, D. J., and Aquadro, C. F. (1991). Molecular population genetics of the distal portion of the X chromosome in Drosophila: Evidence for genetic hitchhiking of the yellow-achaete region. *Genetics* 129, 1147-1158.

Berry, A., and Kreitman, M. (1993). Molecular analysis of an allozyme cline - alcohol dehydrogenase in *Drosophila melanogaster* on the east coast of North America. *Genetics* 134, 869-893.

Bodmer, W. F., and Cavalli-Sforza, L. L. (1976). "Genetics, Evolution and Man." W. H. Freeman, San Francisco.

Bulmer, M. (1987). Coevolution of codon usage and transfer RNA abundance. *Nature* 325, 728-730.

Charlesworth, C., Morgan, M. T., and Charlesworth, D. (1993). The effect of deleterious mutations on neutral molecular variation. *Genetics* 134, 1289-1303.

Clark, A. G. (1993). Evolutionary inferences from molecular characterization of self-incompatibility alleles *In* "Mechanisms of Molecular Evolution" (N. Takahata and A. G. Clark, eds.), pp. 79-108. Japan Scientific Societies Press, Tokyo.

Cleland, A., Watson, H. G., Robertson, P., Ludlam, C. A., and Leigh Brown, A. J. (1996). Evolution of zidovudine resistance-associated genotypes in human immunodeficiency virus type 1-infected patients. *J. Acq. Imm. Def. Synd.* (in press).

Ewens, W. J. (1989). Population genetics theory – the past and the future *In* "Mathematical and Statistical Problems of Evolutionary Theory" (S. Lessard, ed.), Kluwer Academic Publications, Dordrecht.

Fitch, W. M., Leiter, J. M., Li, X., and Palese, P. (1991). Positive Darwinian evolution in human influenza A viruses. *Proc. Natl. Acad. Sci. (USA)* 88, 4270-4274.

Frank, M. B., Besta, R. M., Baverstock, P. R., and Gutman, G. A. (1984). Kappa-chain constant-region gene sequences in genus *Rattus*: coding regions are diverging more rapidly than noncoding regions. *Mol. Biol. Evol.* 1, 489-501.

Gillespie, J. H. (1991). "The Causes of Molecular Evolution." Oxford University Press, New York.

Golding, B. (1993). Maximum-likelihood estimates of selection coefficients from DNA sequence data. *Evolution*, 47, 1420-1431.

Golding, B. (1994). Using maximum likelihood to infer selection from phylogenies *In* "Non-neutral Evolution" (B. Golding, ed.), pp. 126-139. Chapman & Hall, New York.

Golding, G. B., and Felsenstein, J. (1990). A maximum likelihood approach to the detection of selection from a phylogeny. *J. Mol. Evol.* 31, 511-523.

Hartl, D. L., and Dykhuizen, D. E. (1985). The neutral theory and the molecular basis of preadaptation, *In* "Population Genetics and Molecular Evolution" (T. Ohta and K. Aoki, eds.), pp. 107-124. Springer-Verlag, Berlin.

Henkel, T., Westervelt, P., and Ratner, L. (1995). HIV-1 V3 envelope sequences required for macrophage infection. *Aids* 9, 399-401.

Holmes E. C., Zhang, L. Q., Simmonds, P., Ludlam, C. A., and Brown, A. J. (1992). Convergent and divergent sequence evolution in the surface envelope glycoprotein of human immunodeficiency virus type 1 within a single infected patient. *Proc. Natl. Acad. Sci. (USA)* 89:4835-9.

Hudson, R. R., and Kaplan, N. L. (1988). The coalescent process in models with selection and recombination. *Genetics* 120, 831-840.

Hudson, R. R., and Kaplan, N. L. (1995). Deleterious background selection with recombination. *Genetics* 141, 1605-1617.

Hudson, R. R., Kreitman M., and Aguadé, M. (1987). A test of neutral molecular evolution based on nucleotide data. *Genetics* 116(1), 153-159.

Hughes, A. L. (1993). Coevolution of immunogenetic proteins of *Plasmodium falciparum* and the host's immune system *In* "Mechanisms of Molecular Evolution" (N. Takahata and A. G. Clark, eds.), pp. 109-128. Japan Scientific Societies Press, Tokyo.

Hughes, A. L., and Nei, M. (1988). Pattern of nucleotide substitution at major histocompatibility complex class I loci reveals overdominant selection. *Nature* 355, 167-170.

Ikemura, T. (1985). Codon usage and tRNA content in unicellular and multicellular organisms. *Mol.*

Biol. Evol. 2, 13-34.

Isenberg, I. (1979). Histones. *Annu. Rev. Biochem.* 48, 159-191.

Jollés, J., Jollés, P., Bowman, B. H., Prager, E. M., Stewart, C.-B.,and Wilson, A. C. (1989). Episodic evolution in the stomach lysozymes of ruminants. *Journal of Molecular Evolution* 28:528-535.

Kaplan, N. L., Darden, T., and Hudson, R. R. (1988). The coalescent process in models with selection. *Genetics* 120, 819-829.

Kaplan, N. L., Hudson, R. R., and Langley, C. H. (1989). The "hitchhiking effect" revisited. *Genetics* 123, 887-899.

Karl, S. A., and Avise, J. C. (1992). Balancing selection at allozyme loci in oysters: implications from nuclear RFLPs. Science 256(5053), 100-102.

Kimura, M. (1983). *The Neutral Theory of Molecular Evolution.* Cambridge University Press, Cambridge.

Kliman, R. M., and Hey, J. (1994). The effects of mutation and natural selection on codon bias in the genes of Drosophila. Genetics 137, 1049-1056.

Kreitman, M. (1983). Nucleotide polymorphism at the alcohol dehydrogenase locus of *Drosophila melanogaster. Nature* 304, 412-417.

Kreitman, M., and Hudson, R. R. (1991). Inferring the histories of the *Adh* and *Adh-dup* loci in *Drosophila melanogaster* from patterns of polymorphism and divergence. *Genetics* 127, 565-582.

Langley, C. H., Voelker, R. A., Leigh Brown, A. J., Ohnishi, S., Dickson, B., and Montgomery, E. (1981). Null allele frequencies at allozyme loci in natural populations of *Drosophila melanogaster. Genetics* 99, 151-156.

Lee, Y. H., and Vacquier, V. D. (1995). Positive selection is a general phenomenon in the evolution of abalone sperm lysin. *Mol. Biol. Evol.* 12, 231-8.

Lewontin, R. C.(1974). "The Genetic Basis of Evolutionary Change." Columbia University Press, New York.

Maynard Smith, J., and Haigh, J. (1974). The hitchhiking effect of a favorable gene. *Genet. Res.* 23, 23-35.

McDonald, J. H.(1994). Detecting natural selection by comparing geographic variation in protein and DNA polymorphisms *In* "Non-neutral Evolution" (B. Golding, ed.), pp. 88-100. Chapman & Hall, New York.

McDonald, J. H., and Kreitman, M. (1991). Adaptive protein evolution at the *Adh* locus in Drosophila. *Nature* 351, 652-654.

Oakeshott, J. B., McKechnie, S. W., and Chambers, G. K. (1984). Population genetics of the metabolically related *Adh, Gpdh* and *Tpi* polymorphisms in *Drosophila melanogaster.* I. Geographic variation in Gpdh and Tpi allele frequencies in different continents. *Genetica* 63, 21-29.

Oakeshott, J. G., Gibson, J. B., Anderson, P. R., Knibb, W. R., Anderson, D. G., and Chambers, G. K. (1982). Alcohol dehydrogenase and glycerol-3-phosphate dehydrogenase clines in *Drosophila melanogaster* on three continents. *Evolution* 36, 86-96.

Perutz, M. F. (1983). Species adaptation in a protein molecule. *Mol. Biol. Evol.* 1, 1-28.

Sawyer, S. A., Dykhuizen, D. E., and Hartl, D. L. (1987). Confidence interval for the number of selectively neutral amino acid polymorphisms. *Proc. Natl. Acad. Sci. USA* 84, 6225-6228.

Sawyer, S. A., and Hartl, D. L. (1992). Population genetics of polymorphism and divergence. *Genetics* 132, 1161-1176.

Schaeffer, S. W., and Miller, E. L. (1993). Estimates of linkage disequilibrium and the recombination parameter determined from segregating nucleotide sites in the alcohol dehydrogenase region of *Drosophila pseudoobscura. Genetics* 135, 541-552.

Shields, D. C., Sharp, P. M., Higgens, D. G., and Wright, F. (1988). Silent sites in Drosophila genes are not neutral: Evidence of selection among synonymous codons. *Mol. Biol. Evol.* 5, 704-716.

Stewart, C.-B., and Wilson, A. C. (1987). Sequence convergence and functional adaptation of stomach lysozymes from foregut fermenters. *Cold Spring Harbor Symp. Quant. Biol.* 52, 891-899.

Swanson, W. J., and Vacquier, V. D. (1995). Extraordinary divergence and positive Darwinian selection in a fusagenic protein coating the acrosomal process of abalone spermatozoa. *Proc. Natl. Acad. Sci. (USA)* 92, 4957-4961.

Tajima, F. (1989). Statistical method for testing the neutral mutation hypothesis by DNA polymorphism. *Genetics* 123, 585-595.

Takahata, N. (1991). Statistical models of the overdispersed molecular clock. *Theor. Popul. Biol.* 39, 329-344.

Tyler, R. H., Brar, H., Singh, M., Latorre, A., Graves, J. L., Mueller, L. D., Rose, M. R., and Ayala, F. J. (1993). The effect of superoxide dismutase alleles on aging in Drosophila. *Genetica* 91, 143-149.

Vogel, F. (1969). Point mutations and human hemoglobin variants. *Humangenetik* 8, 1-26.

Wiehe, T. H. E., and Stephan, W. (1993). Analysis of a genetic hitchhiking model, and its application to DNA polymorphism data from *Drosophila melanogaster*. *Mole. Biol. Evol.* 10, 842-854.

Wright, S. (1978)."Evolution and Genetics of Populations." Vol. 4. University of Chicago Press, Chicago.

Yokoyama, R., and Yokoyama, S. (1990). Convergent evolution of the red- and green-like visual pigment genes in fish, *Astyanax fasciatus*, and human. *Proc. Natl. Acad. Sci. (USA)* 87, 9315-9318.

CHAPTER 10

Paleontological Data and the
Study of Adaptation

MICHAEL J. NOVACEK

In the new hall of mammalian evolution at the American Museum of Natural History there is a moment of terror set some 15 million years ago on the Miocene savannah of North America. The gracile skeleton of *Ramoceros osborni*, a smaller and earlier relative of the living pronghorn antelope, is mounted as if the animal were in full gallop, nearly airborne. Close on its heels is a massive carnivore, the ancient bear-like dog *Amphicyon ingens*. No human, of course, ever witnessed this drama, but the completeness of the skeletons, their close resemblance to living beasts of some familiarity, and the naturalness of the match between movement and form tell us that the reality of this event can hardly be doubted (Fig. 1). Nor do these skeletons require embellishment, there are no accompanying fleshed-out sculptures or robots to convince us that these animals "were designed" to do exactly what is depicted. Thus, it is fitting that Darwin's (1859, p. 61) compelling words grace the glass panel in front of the skeletons:

"The structure of every organic being is related, in the most essential manner..., to all other organic beings with which it comes into competition..., from which it has to escape, or on which it preys."

This juxtaposition of words and objects is powerful and emblematic. But it also has some unfortunate capacity to obscure. True enough, we can learn much about fossils through use of analogy with living creatures. We can observe the feeding habits of a living organism, relate these behaviors to the architecture of jaws, teeth, and skeleton, and estimate lever arms and stress forces. We can then use these observations to reconstruct the various life-styles of organisms represented by fossils. Such an extrapolation, however, does not necessarily lead us to a greater understanding of biological adaptation. It does not establish whether our attempts to forge a testable set of theories on the evolution of adaptation are influenced by our knowledge of the fossil record. Is such an understanding demonstrable, or at least potentially retrievable, from

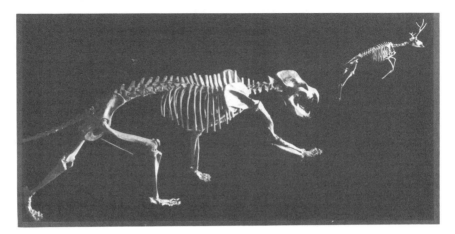

Figure 1 *Ramoceros osborni* flees *Amphicyon ingens*. Photograph courtesy of the American Museum of Natural History.

the study of fossils?

My attempt herein to address this question relies on scrutiny of the mission statement for the study of the adaptation, review of the limits and potentials of fossil evidence, and assessment of efforts to integrate basic paleontological inquiry with the study of adaptation. The goal here is not to defend the intrinsic merits of fossil data; from my own bias the value of fossils as evidence of biological history is self-evident (but see Nelson, 1978). Rather, I examine the possibility that fossil evidence uniquely contributes to the study of adaptation by providing insights that cannot be formulated from the study of living organisms alone. In some cases, this requires an assessment of the role of fossils in reconstructing phylogeny or change in functional design, areas that have been cited as a framework for the study of adaptation.

I. The Meanings of Adaptation

Evaluation of paleontological or any other kind of evidence must be framed by the defined purpose of the evidence. Unfortunately, the study of adaptation does not lend itself to such a straightforward definition. Reflecting on this obstacle as it relates to evolutionary morphology, D. Fisher (1985, p.120) remarked:

"The focus of most of the turmoil is the study of adaptation, and the nature of the disagreement is evidenced by charges that even our recognition of adaptation is based frequently on loose analogies among organisms and machines, anecdotal and unqualified reports of behavior and worst of all ad hoc rationalizations for any failure of predictions."

The amorphous nature of a mission statement for the study of adaptation has long frustrated biologists. Since the subject is not exclusive to this chapter, I will restrict discussion here to a few points, especially those that have preoccupied paleontologists. First of all, there is of course the problem of multifarious meanings for the word adaptation. It is used both for the process of becoming adapted and for the state of attaining adaptiveness. Mayr (1988, p. 135) offered a comprehensive definition which is either widely cited or independently echoed by others:

"Adaptedness is the morphological, physiological, and behavioral equipment of a species or of a member of a species that permits it to compete successfully with other members of its own species or with individuals of other species and that permits it to tolerate the extant physical environment. Adaptation is greater ecological-physiological efficiency than is achieved by other members of the population. Improved adaptedness may be due to a particular component of the phenotype, or to a single gene, or to the total genotype."

This popular characterization sidesteps some of the subtleties of the problem of definition. Has the distinction of *process* (of becoming adapted) and *state* (of attaining adaptiveness) been sufficiently clarified? Perhaps the need for this distinction has been overemphasized, and a structural analysis in evolutionary morphology *sensu* Lauder (1981) addresses both. Nonetheless, there are disadvantages in the use of the word adaptation to mean both a historical sequence of events leading to greater fitness and a state that enhances fitness regardless of historical origin. Hence, there are proposals for a refined terminology. Gould and Vrba (1982) suggested that adaptation be restricted to William's (1966) definition as any feature that enhances fitness and was selected for its current role. They substituted the term exaptation for "...characters, evolved for other usages (or no function at all), and later 'coopted' for their current role ...."

Whether or not one adopts this terminology, the distinction has heuristic value. We can think of many examples where a feature associated with some organism has a different function than the apparently same feature found in another organism. This contrast can be placed in historical context, with the input of phylogenetic data from fossil and/or extant organisms, to suggest that the role of a particular feature may have been modified during its evolutionary history.

But the problematic element of Gould and Vrba's terminology has nothing to do with this contrast. Their hierarchy of terms shows misplaced emphasis because exaptation lumps both features that either did or did not originate for a purpose. Should there be another term for a feature that *originates* through influences other than natural selection, regardless of whether or not it eventually comes to have a function? And, is not there more affinity between a feature whose original selected role is modified and one whose selected role remains the same than with a feature that had no originally selected role at all? [Actually Gould and Vrba applied the term nonaptation for the latter, but this category is paradoxically ranked as a subset of aptation (see their Table 1).] The most beneficial aspect of Gould and Vrba's argument seems to be the recognition of a history of changing function as opposed to a history of static function.

These refinements of definition do not eradicate several ambiguities. Evidence for extinction – one of the central (but not necessarily unique) contributions of paleontology – suggests, for example, that some individuals or species were "better adapted" than others, but is there any way other than retrodiction to determine what is "better adapted"? We may, as Fisher (1985, p.

125) pointed out, cite "... a particular behavior ('e.g., long legs are an adaptation for fast running')[but this] does not really clarify how an adaptation contributes to fitness (i.e., why slower individuals are at a reproductive disadvantage relative to a faster one)."

To confront this obstacle, one might measure the efficiency of systems in a given organism, match these to a set of environmental conditions, and compare different organisms or species along some scale of their effectiveness to fit these conditions. But these studies, denoted by Lauder (1981) as involving "equilibrium hypotheses," have limited predictive power because they assume stable or equilibrium conditions for the environment-organism interface. When we actually consider the process of adaptation against a dynamic environmental backdrop, such as the physical change documented in the fossil record, we run up against the converse to the problem frustrating "equilibrium analysis." The measurement of any index for efficiency (engineering, metabolism, reproductive biology etc.) is outside the scope of observation. Thus, we may be able to associate a behavior based on analogy with preserved structures in different organisms and document a pattern wherein some organisms (or species) endure and others do not, but fail to find the reason for this differential survival.

Another obstacle is the uncertainty over the nature of the "adaptive challenge" confronting organisms. If adaptation is widely regarded as a situation where the world poses problems that organisms need to solve (Lewontin, 1978), then explicit characterizations of these problems are in order. Yet when we fall back on the term *niche* to represent this problem set, we find paradoxical and insurmountable difficulties, for the niche cannot be extricated from the biological entity, the organism, itself. The difficulty is not breached by describing "empty" niches waiting for solution, even though this is possible by simply adding an arbitrary parameter. For example, one can envision the niche of a legless, burrowing, warm-blooded, hairy animal for which we have no known organism. This does not, however, address the question of adaptation as a process because: "Unless there is some preferred or natural way to subdivide the world into niches the concept loses all predictive and explanatory value"(Lewontin, 1978, p. 215).

One of the greatest problems presented to any investigation of adaptation - - and one certainly relevant to the paleontological perspective– is that many changes inferred to represent evolution need not be products of natural selection (Lewontin, 1978), despite alleged assumptions by "adaptationists" to the contrary (Lewontin, 1980). Whether or not this premise is accepted, the distinction of adapted from "non-adapted" things seems elusive to empirical work. For:

"There may be a question as to whether the fundamental postulates of synthetic theory are consistent with the notion of non-adapted parts. But this is not the same question as 'Are there non-adapted parts?' It cannot be answered by examining organisms, fossils or otherwise. It can only be examined by examining arguments" (Kitts, 1974, p.464).

Indeed, Lewontin (1978) and Gould and Lewontin (1979) have argued that functional morphology flirts with the untestable; an adaptive explanation that fails is often substituted with another explanation rather than rejecting the basic premise that the system under observation demands adaptive explanation.

II. Salvaging the "Adaptationist Program"

The forgoing objections, as well as many others, put to serious question the theoretical basis for the investigation of adaptedness of individuals and their characteristics, an approach labeled by Gould and Lewontin (1979) as the "adaptationist program." [A more extreme definition of the adaptationist program – one that includes the assumption that all observed traits of organisms are adaptive optimal solutions – has been offered by Lewontin (1979).] Despite the shortcomings these objections reveal, there is an intuitive sense that adaptation is a real element of evolution, too intrinsically fascinating and important to ignore. This situation leaves us looking for something to salvage for the purpose of scientific inquiry. A defense of the adaptationist program is built around several arguments. Lewontin (1978, p. 230), perhaps the most emphatic and influential critic, suggested redemption on rather weak grounds; namely, some of the assumptions of the adaptationist program "...can be tested in some cases" and because "adaptation is a real phenomenon," it makes no sense to abandon the notion entirely, opting for some other mechanism to explain observed specializations or historical change.

Mayr (1988) took the apologetic argument much further. While crediting Gould and Lewontin's (1979) criticisms of the adaptationist program, he vigorously objected to their more extreme characterization of the program as involving the atomization of traits in an organism and the explanation of these traits in terms of optimal design through natural selection. Mayr maintained instead that the basic principles of the adaptationist program are justified; only poor practices that borrow atomistic and deterministic thinking are the rightful targets for criticism. Mayr also emphasized that selection, as Dobzhansky (1956) long ago expressed, works on whole genotypes, not on individual traits. Nonetheless, he admitted (ibid, p.150) that discriminating the products of natural selection from those of stochastic processes presents a serious "epistemological dilemma." His solution is to explore all possibilities that the features in question would be favored by selection. Only failure in this quest allows one to consider "the unexplained residue tentatively as a product of chance."

Such a prescription naturally gives rise to doubts reflected in the above cited remarks by Kitts (1974), Lewontin (1978), and others. How do we know we have exhausted *all* possibilities in a search for the selective influence of features? For example, Fisher (1985) has also disputed the allegation that an adaptationist perspective necessitates the assumption "that all observed aspects of an organism are adaptive" (Lewontin, 1980, p. 242). Fisher's defense stresses that the motivation for exploration of a function does not assume that a function must exist. This defense fails, however, to suggest a means for empirically distinguishing the adaptive from the nonadaptive. To deal with this problem a fallback position is recommended: testing the general notion that a feature is adaptive is out of scope for the customary investigation of a particular hypothesis of adaptation (Maynard Smith, 1978). The latter only deals with the rejection of alternatives to a particular hypothesis. By analogy, the demonstration of greater evidence for one phylogenetic hypothesis over another is irrelevant to the possibility that neither of these alternatives are the product of descent with modification.

Another salvage operation on behalf of adaptationist research involves the allowance of pluralism: some cases exemplify the strong influence of adaptation, whereas other cases suggest alternative influences, such as developmental constraints (Gould and Lewontin, 1979). Mayr (1988) regarded this to be a false demarcation of causal factors, arguing that the putative alternatives to adaptive features described by Gould and Lewontin are ultimately influenced by natural selection. Likewise, Fisher (1985, p. 135) was only satisfied with "a stronger form of pluralism" than that accepted by Gould and Lewontin. Unfortunately, his expression of generosity does not seem to simplify the issue:

"...development and adaptation refer to sets of interactions located within broadly overlapping, even if not identical, portions of the causal plexus. The extended intergenerational plexus used to represent transformation is thus a complex interweaving of both development and adaptation. I do not deny that *some* features might owe their origin and/or maintenance entirely to either adaptation or development, simply passing through, or around, the filter of the other process, but many (or most) features will be influenced by both."

One lesson to draw from this account is that unweaving "the causal plexus" may be a remote prospect for empirical work. This might induce us to go easy in our struggles to define the adaptive/nonadaptive interface and turn our attention to other matters. One such alternative route is simply the refinement of hypotheses on the origin of particular adaptations. These involve specific cases, perhaps the transformation of a basic body plan or the development of "key innovations" (Liem, 1973) that might be identified to the origin and "adaptive radiation" of an important group of organisms. The applications here are parochial, but can be powerful – for example, they include questions on the diversification of body plans for "invertebrate" phyla, the multiple origins of flight, or hymenopteran-angiosperm coevolution. There has been a particular emphasis, of late, in setting such analyses within a phylogenetic framework (Lauder, 1981; Coddington, 1988, 1990; Baum and Larson, 1991; Wenzel and Carpenter, 1994), although this phylogenetic underpinning is explicitly rejected by others (Gutmann and Bonik, 1981). These and related matters are treated more extensively below. In order to consider the paleontological input in such studies or to identify areas where paleontology in a unique fashion defines the nature of the adaptationist program, I will first review some general qualities of fossil evidence.

III. The Meaning of Fossils

A fossil is defined here as any evidence for past life. As such, the fossil record can show extraordinary detail and texture. Fossils include bones, tests, plant parts, footprints, burrows, filmy traces of coelenterate membranes, impressions of skins and feathers, mineralized stomach contents, feces, insects preserved down to the odd vibrissa and embalmed in amber, and even DNA sequences from hardy genes. Despite this cornucopia, can we claim that fossils provide unique evidence for evolution and adaptation? This question deserves more than an intuitive answer. For example, one might argue that the reality of

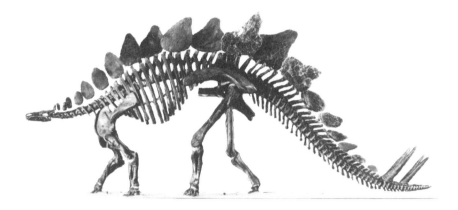

Figure 2 The enigma of the dorsal plates on *Stegosaurus*. Photograph courtesy of the American Museum of Natural History.

extinction, and its implications for competition, fitness, and adaptation, would be unknown without a fossil record. Extinction is, however, known from a record of extermination of species during human history. Also, the destruction of populations of species in the wake of expanding populations of human consumers, introduction of alien animals and plants, and catastrophic environmental changes and the like can, in themselves, lead us to conclusions concerning competition and fitness. What about the concept of preadaptation? Is not the fossil record necessary to validate such a concept? One might respond that the fossil record provides the exclusive evidence for ancestry that substantiates the *pre* in preadaptation. Yet it seems possible that a transformation in the function of a structure can be inferred from comparisons in living species that exhibit variation in function for the same part.

Thus, several key elements of historical hypotheses of adaptation are not, as is often alleged, the exclusive domain of paleontology. Our reliance on such data may be further diminished by the acknowledgment that many striking features of the fossil record have unclear adaptive implications. The ambiguity here is interesting and it brings to mind a famous example. Noting that there is still debate over whether the conspicuous dorsal plates of the dinosaur *Stegosaurus* (Fig. 2) were for defense, courtship, or thermoregulation, Mayr (1988, p. 138) stated, " However, when one views this energy-costly structure one cannot escape the conclusion that it could not have evolved except through natural selection because it provided some benefit to the possessors of this structure." Just the same, the function of the dorsal plates "in principle" cannot be determined with certainty (Lewontin, 1978, p. 218). Engineering experiments on possibly analogous structures in living creatures (e.g., the dewlap under the jaw of some lizards?) only tell us how the plates might have functioned. Moreover, such an argument for function tells us nothing directly about the action of natural selection and the origin of the plates as an adaptation. Lewontin *speculated* that if the plates had a thermoregulatory function then individuals with larger plates might have been able to gather food in the heat of the day. Hence, the larger-plated individuals might have been

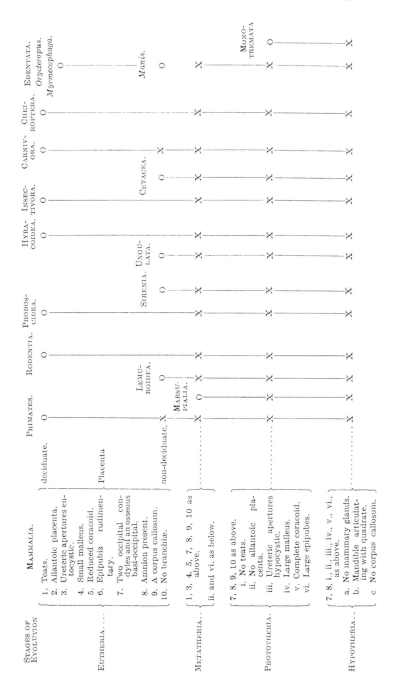

favored by selection. In turn, the larger plates may have been "preadapted" for defense. The evidence, unfortunately, is not sufficient to eliminate options to this scenario.

There is then some weakness attendant with the strength of fossil information. On the one hand, fossils provide us with wondrous structures and organisms unknown to the world of the present. On the other hand, a richer explanation of these items requires (but is not necessarily guaranteed by) a firm grasp of the present. As D. Kitts (1974, p. 458) remarked:

"Paleontologists often claim that fossils tell us something. But fossils in themselves tell us nothing; not even that they are fossils....When a paleontologist decides whether or not something found in the rock is the *remains* [italics in the original] of an organism he decides, in effect, whether or not it is necessary to invoke a biological event in the explanation of that thing."

The latter operation may seem cumbersome and arcane. Who would doubt that a *Ramoceros* skeleton is worthy of biological invocation? Yet the exercise is not gratuitous, as Kitts (ibid, p. 458) goes on to note "... Paleontologists encounter objects which do not resemble the parts of any living plant or animal and yet conclude that they are fossils." Such enigmas have checkered histories; they may eventually be assigned biological status for want of a better explanation. Conversely, some objects once thought to be fossils are eventually transferred to the nonbiological realm, as in Preston Cloud's (1960) argument that "jellyfish" fossils from the well-known ancient rock sequence of the Nankoweap group are actually gas bubble markings.

Interestingly, this uncertainty principle is at the core of molecular paleontology. A putative fossil DNA sequence might actually belong to the fossil from which it was isolated or it might represent a contaminant from some associated fossil material or, more insidiously, a contaminant from some extant specimen in the lab. The eventual ruling will come from elimination, or at least a downgrading, of some of these alternatives. Yet the value of fossil DNA data lies in this very ambiguity. Useful DNA of a 35-million-year-old termite cannot *perfectly* match a sequence from a living relative. In order to have phylogenetic content, the ancient DNA must have distinctive qualities, arrays of nucleotides that cannot be retrieved solely from living organisms (e.g., DeSalle et al., 1992).

These examples are instructive because enigmatic fossil evidence (whether morphological or molecular) may have some bearing on the potential of fossils as historical evidence. For if there is an element of the fossil record that cannot be explained in any straightforward manner based on our familiarity with the present, there is also some expectation that the fossil record can *uniquely* enhance our perspectives on evolutionary pattern and process. The question is then to what extent is this expectation demonstrably met? And, to what extent is this expectation demonstrably met for the purpose of testing historical theories of adaptation? After all, not all curiosities of the fossil record are necessarily informative in this context. The fact that the dorsal plates of *Stegosaurus* represent an enigma of the past does not ensure their value as distinctive evidence for the evolution of adaptation. To address these issues, I briefly

Figure 3 T. H. Huxley's theory for extreme parallelism in the multiple origins and evolution of mammalian lineages resulting from natural selection. "The sign 0 marks the places on the scheme which are occupied by known Mammals; while X indicates the groups of which nothing is known but the formed existence of which is deducible from the law of evolution" (Huxley, 1880, p. 658).

highlight the history of paleontological views on adaptation and then consider some areas of inquiry where the special role of paleontology has been implicated.

IV. Changing Paleontological Perspectives on Adaptation and Evolution

In the emergence of modern comparative biology, fossils were readily recruited to develop elaborate scenarios of adaptive change or to stress the role of natural selection in evolution. Thus, Huxley (1880) depicted natural selection as such a powerful force and adaptive change as such a dominant effect that it constrained descent along parallel pathways of evolution (Wyss, 1987). He accordingly drew "lawns" of evolutionary lineages for such groups as mammals (Fig. 3). These lines intersected adaptive stages of evolution, marked by conditions in many cases lacking empirical evidence, but proven to exist according to Huxley (1880) by "...the law of evolution." The resultant pattern is one of rampant parallelism but virtually no recognition of monophyly. One may fairly judge such models to be largely abandoned, although some of their essence persists in the adaptive grade concepts particularly popular in paleontology during the 1950s and 1960s (e.g., Simpson, 1959; Romer, 1966).

Stark alternatives to these applications of Darwinian evolution also emerged. For example, the influential and extraordinarily prolific Edward Drinker Cope disputed tenets of natural selection In this rejection, and in the assumption that acquired characteristics (those induced by environment or habit) could be inherited, Cope's (1896) perspective resembles Lamarck's alternative proposal for the evolution. The label applied to this view, "neo-Lamarckian theory," is, however, not entirely appropriate, as Cope dispensed with Lamarck's emphasis on a central "perfecting" trend in evolution and, at the same time, stressed the direct action of the environment on organismal structure, a view explicitly rejected by Lamarck (see Simpson, 1949). Neo-Lamarckianism has been credited with some contributions to later developments that embody the "New Synthesis." Simpson (1949, p. 270), for example, stated that "Its [Neo-Lamarckianism] emphasis on the relationship of structure-function-environment brought out a wealth of pertinent facts and made plain that the forces of evolution must effectively integrate the three."

Antiselectionist views also included theories that emphasized progressive improvement toward some ideal state. Notable here were orthogenetic theories readily adopted by paleontologists like H. F. Osborn. Orthogenesis was rooted in fundamental objections to Darwinian theory: if the initial step in evolution, namely variation, is random and the potentials for environments are unlimited, one could only expect a chaotic jumble of evolutionary phenomena. Since observations suggest more ordered patterns of evolution, the orthogenetic school opted for a situation where variation was not random, but was "pulled" along well-defined pathways of change. These pathways were mediated by a life-giving force, orthogenesis, or in German, Vervollkommungstrieb. Osborn (1929, 1934) documented change in extinct ungulates and other lineages as an expression of these orthogenetic pathways (Fig. 4).

Eventually, Osborn divorced major evolutionary patterns entirely from the influence of selection. He claimed (1934, p. 228), for example, in the case of proboscideans "... no species of elephant occupied the same geographic range as another species at any given period of geologic time; thus there was no competition between these species." Later revisions of proboscidean history (Maglio, 1973) demonstrate Osborn's observations to reflect an incomplete or distorted sampling of the fossil record. Nonetheless, the ordered occurrences of fossils, such as the large herbivorous mammals from the northern continents, were striking in their suggestion of some kind of directed evolutionary history (Fig. 4).

Preoccupation with the directionality of evolution was soon transformed into an extreme expression of progression, aristogenesis, a concept emphasizing that evolution involved trends toward "the best" of conditions. Osborn applied his knowledge of the fossil record to an explicit characterization of this concept. Reflecting on the observation that proboscideans and equids appear in the Eocene with rather similar four-coned molars, Osborn (1934, p. 219 and 220) stated, "But differentiating this visible or phenotypic similarity is the widely divergent aristogenic potentiality of the primitive proboscidean and primitive equine molar... [By the end of the Tertiary] the one ends in the marvelously complicated 27-plated molar of the wooly mammoth and the other is fated to evolve into the double-columned grinder of the horse." As Mayr (1988) noted, Osborn in making these statements was not advocating the predetermination of the fate of an evolutionary trend, but rather the potentiality of movement toward an optimum based on the emergence of the original trait. This acknowledgment of potentiality, however, did not include recognition of the variation in populations as a source for selection to act upon.

Orthogenesis was concurrent with other antiselectionist views of evolution, such as saltationism. In the latter, Darwin's corollary mandating gradual evolutionary change was rejected. There are ambiguities here because certain authors, like T. H. Huxley, championed Darwinian selection but opted for saltational evolution. Saltationism was ironically also challenged from the same emphasis on adaptation and selection that Huxley so fervently defended. Weismann (1882), for example, argued that abrupt transformation would produce many organisms incapable of existence, and that such fitful and rapid change is incompatible with the subtle differences among organisms with respect to function and adaptation. Although these points anticipated the pro-saltationists view of Bateson (1894) and DeVries (1901), the latter perspective did take some hold on the scientific community. Whereas gradualism - the assertion that all evolutionary innovation is the product of gradual transformations within populations - was largely adopted by "naturalists," saltationism was embraced by many embryologists, paleontologists, and early Mendelian geneticists (Mayr, 1988, p.412). Further articulation of the saltationist argument, notably represented by Goldschmidt (1940) and Schindewolf (1950), managed to survive the restructuring of Darwinian theory into Neo-Darwinism and the "New Synthesis" (Dobzhansky, 1937; Mayr, 1942; Simpson, 1944).

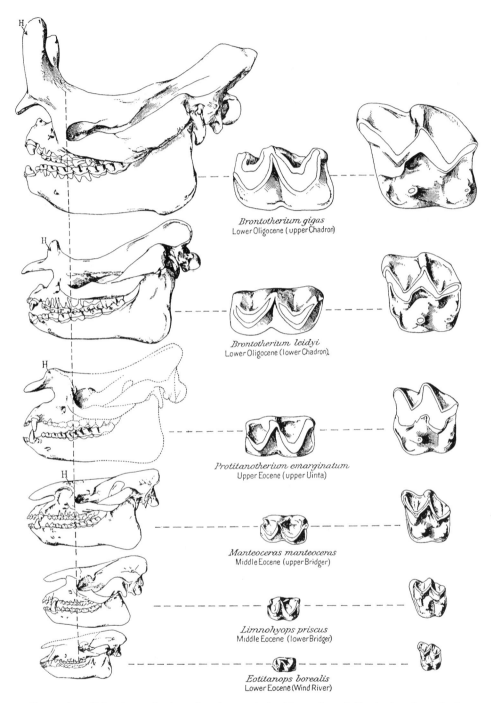

Brontotherium gigas
Lower Oligocene (upper Chadron)

Brontotherium leidyi
Lower Oligocene (lower Chadron)

Protitanotherium emarginatum
Upper Eocene (upper Uinta)

Manteoceras manteoceras
Middle Eocene (upper Bridger)

Limnohyops priscus
Middle Eocene (lower Bridger)

Eotitanops borealis
Lower Eocene (Wind River)

Figure 4   Osborn's application of orthogenesis in his studies of North American fossil taeniodonts. From Osborn (1929).

This was largely due to the saltationists' concern over empirical ambiguities, namely the lack of transitions among families, genera, orders, and higher taxa, and the preponderance of gaps in the fossil record. It is noteworthy that the later saltationist arguments did entail pluralism; Goldschmidt (1940), for example, ascribed to evolution involving gradual changes of an adaptive nature within species, but stressed the emergence of "hopeful monsters" through drastic mutation as a major cause for the origins of new taxa.

At this juncture, the views of George Gaylord Simpson, perhaps the most influential paleontologist for much of the 20th century, bear attention. For Simpson (1944, 1953), an ardent defender of Darwinian evolution through gradual population change, also at least temporarily held to an argument for quantum evolution. This in Simpson's view explained dramatic shifts to new "adaptive zones" and the differing evolutionary rates that he documented in various and sundry lineages. The apparent contradiction between gradualism and saltationism was erased by the assertion that a population-based mode of gradualism can also allow for varying rates of change, even successive phases of extremely fast change (tachytely) and of slow change (bradytely), or stasis. However, Simpson's (1944) claim for rapid change in large, widespread populations to account for new species was largely rejected, perhaps eventually even by the author himself (Mayr, 1988, p.409). As to the question of these evolutionary modes in relation to adaptation, Simpson recognized a basic paradox: the evolutionary shift from one adaptive zone to another required an interval of maladaptedness. Quantum evolution was at least a way of bridging this "no species land" as quickly as possible.

Simpson's impact is readily evident in the paleontological contributions of his contemporaries. A touchstone volume on the "new synthesis," "*Genetics, Paleontology, and Evolution*" (Jepsen et al., 1949), included a detailed case study by Patterson of a rather obscure group of fossil mammals, the taeniodonts. A cursory glance suggests a striking resemblance between Patterson's sketch of taeniodont history (Fig. 5) and Osborn's orthogenetic schemes for various ungulates (Fig. 4). But the resemblance here is merely pictorial; Patterson regarded taeniodonts as the apotheosis of Darwinian evolution in action, with an added boost from quantum evolution (Fig. 6). One of the taeniodont lineages, Patterson (1949) claimed, underwent a host of dramatic adaptive changes triggered by a single mutation that produced large, laterally compressed claws. Many of the other changes in the teeth, skull, and jaws evolved rapidly within this new adaptive zone. In reference to this lineage, the stylinodontine taeniodonts, Patterson (1949, p. 277) argued:

"Stylinodontine history is best interpreted in terms of the view that origin of new adaptive types involves abrupt shifts in adaptive direction and that such shifts are seldom preceded by inadaptive phases." And, (ibid , p. 277) "A preadaptive threshold is not involved when inception of the new adaptive type is the result of a single mutation. The prerequisite in this sub-mode is a genotype permitting ready integration of the new mutation."

On reflection, Patterson seems wildly audacious in extending his observations to arguments about the mode of quantum speciation and "preadaptive thresholds." Indeed, revisions of stylinodontine phylogeny are ill-matched with Patterson's reconstruction of ancestor-descendant succession (Schoch and Lucas, 1981). Patterson's study is of largely historical interest,

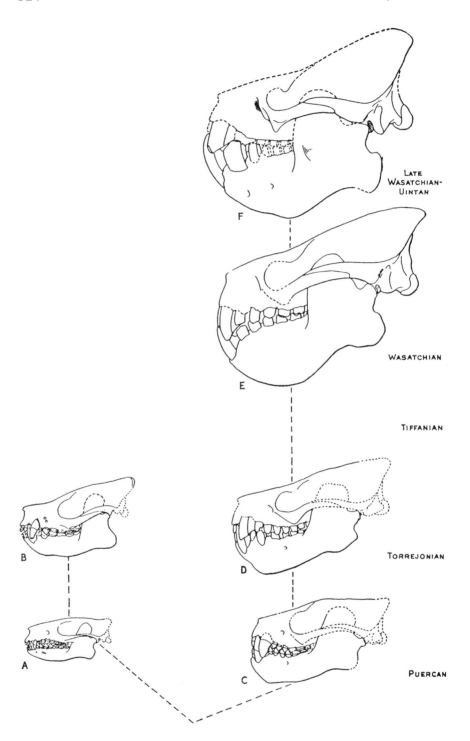

LATE
WASATCHIAN-
UINTAN

F

WASATCHIAN

E

TIFFANIAN

TORREJONIAN

D

B

A

C

PUERCAN

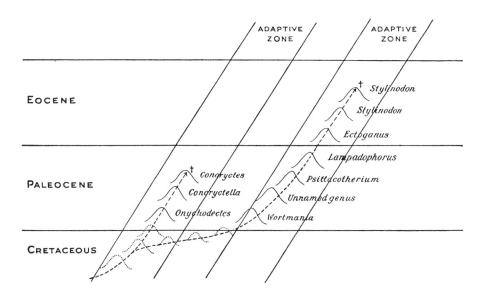

Figure 6 B. Patterson's theoretical interpretation, invoking quantum evolution, of taeniodont history. "The dashed line indicates phylogeny. The solid and dotted frequency curves represent, respectively, known and inferred stages" (Patterson, 1949, p. 268).

being recognized for the explicit way in which concepts of evolutionary rate, quantum change, and adaptation were addressed through paleontological analysis (see Simpson, 1953). But the success of such a blending in this case no longer seems so apparent.

One aspect of the taeniodont casebook, nonetheless, has some enduring influence. This has less to do with the opulent consideration of general evolutionary theory and concept than with the basic descriptive investment in systematics and morphological variation. By contrasting the inferred changes in two closely related groups, this study and others of its kind anticipated the current popular emphasis on comparisons of evolution and adaptation between sister taxa in a phylogenetic framework (e.g., Eldredge, 1979; Vrba, 1979, 1984).

The reality of quantum evolution and its adaptive implications reemerged with the highly influential proposals of Eldredge and Gould (1972) on punctuated equilibrium. Like Goldschmidt or Schindewolf, Eldredge and Gould argued that the notorious gaps in the fossil record were more than a sign of bad data; they may indeed have an evolutionary explanation. But Eldredge and Gould did not call upon the emergence of single mutations, "hopeful monsters," or the sudden origins of higher taxa to explain these gaps. They suggested that the apparent rapid speciation following prolonged periods of stasis in the fossil record (Fig. 7) was in full accordance with allopatric

Figure 5 Patterson's documentation of the paleontological history of the taeniodonts, conoryctines (A, B) and stylinodontines (C-F). From Patterson (1949).

founder effects that represented refinements in concepts of population biology and evolution (Mayr, 1988). Eldredge and Gould (1972) drew a distinction between their hierarchical approach to evolution and the conventional expressions of neo-Darwinism and the new synthesis. Predictably, the ensuing debate has involved several counterclaims, namely that hierarchical thinking is also not really new (Mayr, 1988, p. 417), that phyletic gradualism as the more traditional alternative to punctuated equilibrium has not been given its fair recognition (Gingerich, 1974), and that hierarchical views in the extreme sense, that is, the strict decoupling of speciation from what is normally labeled as microevolutionary (population-level) processes, is invalid (Hecht, 1974; Mayr, 1988, and others).

With regard to these objections, the "punctuationists," seem to be a moving target. In its pluralistic form, punctuated equilibrium does allow for phyletic gradualism with an uncertainty factor regarding the preponderance of one mode over another. Also, the interface between macroevolution and microevolution has been variously described (Stanley, 1979; Vrba, 1984). Punctuated equilibrium has, however, stimulated the desire to test two contrasting evolutionary patterns from an empirical analysis of the fossil record. There is some doubt if all, or overwhelmingly most, of the cases put forward as tests meet the minimum requirements for such evidence. Perhaps attainment can be found in at least some finer-scale studies of fossil groups (Theriot, 1992). As discussed below, these have implications for paleontological input on the origins of adaptations.

This brief reflection on theoretical developments in paleontology illustrates a few general tendencies. First, it is clear that many theories on the origin of adaptation are synonymous with theories on the mode of speciation or evolutionary rate. The two cannot be segregated; at least, an adaptive scenario is fully dependent on a construct provided by the favored historical theory, whether it be quantum speciation, orthogenesis, saltationism, gradualism, or punctuated equilibrium. Second, adaptive interpretation of some form is not the exclusive property of those who advocated Darwinian selection theory. Several notable figures rejecting this theory were taken with interpretations of adaptive trends and the relationship of structure, function, and environment (e.g., Cope, 1896). Third, many attempts to extrapolate from paleontological observation to general theories of evolution and adaptation have either wholly or appreciably failed. It is for example difficult to identify any current alliance with Patterson's (1949) illustration of quantum speciation, as writ in the taeniodont record. Likewise, Simpson at least tacitly abandoned his proposals for quantum speciation, and the existence of any paleontological support for it, in his later writings. Fourth, even retrospectively unfavorable judgment of certain ideas does not completely eradicate their continued influence. Thus we see a foreshadowing of comparisons between sister group lineages in Patterson's (1949) otherwise problematic effort. We will now consider the annealing of this legacy to more current discussions of paleontology and adaptation.

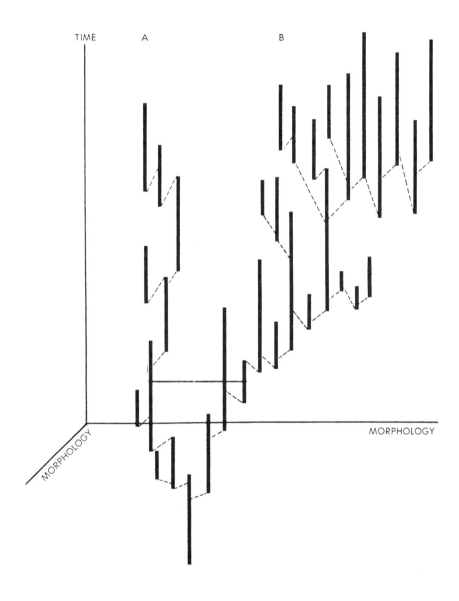

Figure 7 A three-dimensional scheme reflecting punctuated equilibrium in the fossil record. Contrasted here is a pattern of relative stability (A) with that of a trend (B). Note that the trend is not linear but is composed of rapid shifts in morphology (dashed lines) followed by prolonged periods of stasis (vertical bars). From Eldredge and Gould (1972).

V. Form, Function, Adaptation, and Change

Much that can be characterized as recent investigations of adaptation involves a refinement of functional analysis, overlapping with the enterprise labeled theoretical morphology. Extreme emphasis on the mechanical interactions of biological systems as a means of predicting adaptive transformation is recommended by the "constructivistic approach" promoted by Gutmann (1981), Gutmann and Bonik (1981), and others. Likely transformations are "examined" in "complete, enclosed structural systems" – models of organisms as functional wholes – and then compared to the diversity of relevant body plans in the real world. Thus, Gutmann (1981) argued that physical laws of hydrodynamics require that the notochord precedes the development of specialized dorsal longitudinal musculature, a dorsal hollow nerve cord, and gill slits, a sequence bearing on theories for the origin of chordates.

Among the problems with this approach is the suspicion that the hypothetical progression of systems cannot be so readily devised without a clear notion of diversity and relationships among organisms. The constructivistic approach assumes that theories of transformation can be developed purely from biomechanical principles. Yet the distribution of characters in related taxa could potentially refute this avowed transformation. Constructivists would deny this possibility with the argument that the distinction of primitive from derived states must be independently derived from such biomechanical laws *before* any phylogeny can be constructed. This puts severe limitations on our character assessments because there is a dearth of law-like properties associated with realistic levels of organismal complexity. Moreover, character transformation can be dealt with in a realistic fashion without such biomechanical laws. If we accept an equation between synapomorphy and homology (Patterson, 1982), then the hierarchical arrangement of characters provides a map for such transformations. It is a mistake to assume that this approach fails to provide an independent basis for assessment; the hierarchical map can be derived from a pattern of relationships that fits characters other than the system under study (see also Lauder, 1981; Wenzel and Carpenter, 1994; and comments below).

Primary focus on diversity and complexity underlies an alternative to constructivism, namely a comparative analysis of adaptation. One expression of this approach is to base hypotheses of adaptation on comparisons of organisms relying on various assumed random, equal probability, or related models (Felsenstein, 1985; Harvey and Pagel, 1991). Another approach draws on autoregressions or autocorrelations that are claimed to discriminate between adaptive and phylogenetic components (Cheverud et al., 1985; Gittleman and Kot, 1990). Here, expected correlations are related to taxonomic ranks in a Linnaean hierarchy, a practice that suffers from assuming some false notion of equivalence in comparisons of groups of similar Linnaean rank (Wenzel and Carpenter, 1994). A third comparative approach is more explicitly linked to a phylogenetic hypotheses; its qualities are examined later.

Another dimension to comparative studies of adaptation is provided by developmental data. Such information enhances character description, hypotheses of character transformation, and allows consideration of a greater (or more definably constrained) range of functional roles. As noted earlier, arguments that developmental constraints present alternatives to the influence of natural selection (Gould and Lewontin, 1979) have been disputed (Mayr, 1988). Nonetheless, a whole spectrum of developmental studies, such as those dealing with morphogenetic growth and organization, are often characterized as a "pre-Darwinian" approach clearly separate from studies of adaptive evolution (Shubin and Alberch, 1986).

Fossils have figured prominently in the construction of hypotheses involving historical change in the function and adaptive mode of structures. This largely reflects the common assumption that a temporal record may offer some distinction between ancestral and more advanced conditions. Early birds have toothed beaks, early horses have multiple toes, early whales have vestigial limbs, and so forth. Yet ambiguity invariably creeps into analysis at least on two counts. First, there is often a clear contradiction between the sequence of change suggested by fossils and the phylogenetic pattern that best fits the comparative character data. Second, many fossilized structures, like those notorious *Stegosaurus* plates, are not apt to reveal their functional roles. This does not only apply to bizarre structures for which analogy with living organisms is prohibitive. If a functional role for a fossilized structure cannot be narrowed to even one of two options, there is serious impediment to reconstructing a history of adaptation.

For example, consider a case which enjoys a rather widespread, if misguided, reputation as a robust theory of preadaptation ("exaptated aptations") and evolutionary origins. The fossil bird *Archaeopteryx* has been showcased as evidence for transition in function as well as structure (Fig. 8). The animal had feathers and wing structures, but it allegedly had only a rudimentary capacity for powered flight (Ostrom, 1974, 1979; Bakker, 1975). On the other hand, we can ascribe other purposes to feathers, notably the enhancement of insulation and thermoregulation critical to endothermic ("warm-blooded") animals. Hence, feathers could have originally evolved as structures for such purposes and later been "coopted" as an important element of the flight apparatus (Ostrom, 1979; Bakker, 1975; Gould and Vrba, 1982 and many other references).

This scenario, however, begs the question: why is the first appearance of feathers coincident, as far as we know, with some kind of wing structure? To this, Ostrom (1975) suggested that the wings in *Archaeopteryx*, and its as yet unknown precursors, were handy for catching insects and were implements for predation. This argument calls upon analogy with certain modern birds (black egrets, for example), where wings are used both in flying and in "mantling," an interesting behavior wherein the bird casts a shadow over the water with its wings in order to see its prey (McLachlan and Liversidge, 1978). The observation suggests a "characteristic behavior pattern with a genetic basis" that possibly relates to the flexibility in the use of wings for various purposes (Gould and Vrba, 1982, p. 8).

"Thus the basic design of feathers is an adaptation for thermoregulation and, later, an exaptation for catching insects. The development of large contour feathers and their arrangement on the arm arise as adaptations for insect catching and become exaptations for flight. Mantling behavior uses wings that arose as an adaptation for flight" (ibid, p.8).

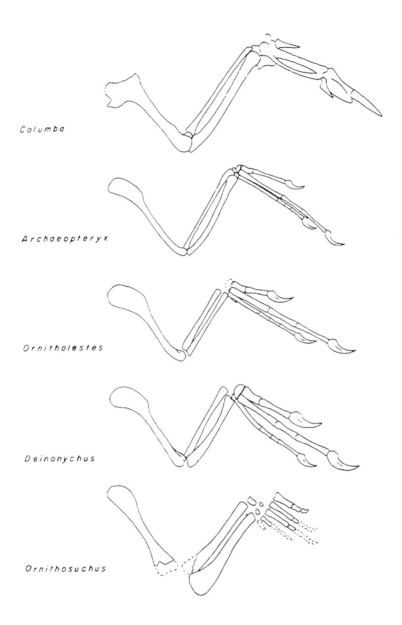

Columba

Archaeopteryx

Ornitholestes

Deinonychus

Ornithosuchus

Figure 8 A reconstruction of the forelimb of *Archaeopteryx* compared with that of a pigeon (*Columba*), two coelurosaurs (*Ornitholestes* and *Deinonychus*), and a "thecodont" (*Ornithosuchus*). From Ostrom (1994).

How does this scenario [which, in fairness, Gould and Vrba (1982) only cite for heuristic purposes] fulfill expectations for the use of fossils in describing the origin and history of adaptations? We can identify several basic problems. The first of these concerns age versus phylogenetic position. *Archaeopteryx*, one of the most compelling and complete fossils ever found, is also the earliest undisputed bird. Working from the argument of a close theropod - bird relationship (there are other, in my judgment, less reasonable alternatives, e.g., Feduccia and Wild, 1993), *Archaeopteryx* provides a striking mosaic of theropod and early avian characters (Ostrom, 1979, 1994). Yet the late Jurassic age of *Archaeopteryx* is apparently far too late for the divergence of its clade from the related theropod clade. Even though there is much doubt (see Ostrom, 1994) that the Triassic form *Protoavis* represents the earliest bird (Chatterjee, 1991), a good candidate would be expected to fill this "Triassic gap." Thus *Archaeopteryx* apparently represents the "next chapter in bird history" (Ostrom, 1994, p. 168).

This temporal gap is in itself not a serious liability; indeed to emphasize this bias would betray an overreliance on the age criterion (see further discussion below). *Archaeopteryx* could, after all, be part of an enduring lineage, preserving a basic combination of structures that might be expected in the earliest members of its clade. One finds, however, that *Archaeopteryx* does not satisfy such a conservative Bauplan, particularly with reference to the sequence of adaptive changes ascribed to the origin of birds. If feathers originally performed an insulative rather than a flight function, one would predict that some fossil, particularly a fossil filling the gap between *Archaeopteryx* and nonavian theropods, might have feathers but no modified wing-like structures. That fossils rarely preserve impressions of feathers greatly lowers our expectation for the discovery of this missing link. It does not allow us, on the other hand, to ignore the fact that such elusive evidence is necessary to test a transformation in function from thermoregulation to flight.

The second problem with the bird scenario resides in the ambiguity *in the original adaptation* for such evocative structures as feathers and wings. It is by no means evident that *Archaeopteryx* entirely lacked the ability for flight. Indeed, some analyses suggest (Ruben, 1991) that the flight limitations in *Archaeopteryx* may have been overstated, even though this animal was probably not capable of the extended flapping flight seen in modern birds (Speakman, 1993). In addition, the equation between feathers and thermoregulation relating to endothermy is not a given. Histological examination of bone in Cretaceous (and presumably feathered) birds reveals lines of arrested growth which possibly indicate ectothermy rather than the endothermy typical of modern birds (Chinsamy et al., 1994). These countervailing observations lead to a basic acknowledgment – the origin of feathers for the purpose of incipient flapping flight, rather than for endothermy, cannot be rejected (Randolph, 1994; Chinsamy et al., 1994). The paleontological evidence available (the combination of feathers and wing-like structures in a taxon that represents a transitional combination of theropod and bird features and also represents the earliest undisputed bird) is, in fact, more consistent with the emergence of feathers as a flight component.

As far as Ostrom's (1979) argument that the wing in *Archaeopteryx* was mainly employed in predation, we can only note that analogy with living organisms completely fails. There are apparently no living birds with well-developed wings that adopt these structures for predation and at the same time abandon flying behavior. Do these observations then allow us to reject the popular scenario of preadaptation for bird wings and feathers? Unfortunately no, at least not in a rigorous fashion. The existence of multiple purposes for feathers (Randolph, 1994) and the lack of critical theropod-bird intermediates suggests that the stepwise evolution in function cannot be entirely eliminated. All we can say is that the ambiguity frustrating this famous paleontological case is greater than widely acknowledged.

In citing this particular example, I do not wish to suggest that fossils fail to enhance our reconstruction of adaptive history. Although the fossil evidence may not be adequate to test multistep scenarios for the origin and changing function of wings and feathers, *Archaeopteryx* does carry critical information. The emergence of feathers and wings is identifiable in a form that shows a clear affinity with coelurosaurian theropod dinosaurs (Ostrom, 1975, 1994). This places our consideration of bird structure and function in a broader historical context. Accordingly, we may look to theropod diversity for the "Baupläne" that anticipated modifications of the forelimbs and eventually the design of the bird wing (Fig. 8). I shall argue below that fossils, if applied in an explicit phylogenetic context, do offer such insights and, in some cases, these insights cannot be recovered from information on living organisms. This argument first requires a review of the use of phylogenetic analysis in the study of adaptation.

VI. Phylogeny: A Map for the History of Adaptations?

Many of the comparative approaches discussed in the foregoing section either share a statistical emphasis or simply relate adaptive changes to salient features of the fossil record and broadbrush evolutionary patterns. These contrast with a third comparative approach that is more explicitly phylogenetic. Lauder's (1981) influential paper on the subject stressed that phylogenetic relationships demonstrated through cladistics have a critical bearing on hypotheses of adaptation. Such patterns, for example, allow us to distinguish whether a feature shared by two different organisms or taxa was independently acquired or inherited from a common ancestor:

"A phylogenetic hypothesis allows the reconstruction of the historical sequence of structural change through time and thus serves as a null hypothesis from which significant deviations may be detected and indicates the appropriate level of generality at which structure-environment correlations must be explained" (Lauder, 1981, pp. 431 and 432).

This statement has been interpreted to suggest that phylogeny reveals certain features, by virtue of their homoplasy, to be adaptations *and thus alternatives* to features inherited through common ancestry (Fisher, 1985). No such conclusion need be drawn from Lauder's point. Identifying an appropriate level of generality for a character has nothing to do with falsifying the notion that a character is adaptive.

An explicit phylogeny does not in itself address all subtleties of an adaptive history. The latter is enhanced through operations described above, namely experiments on function, descriptions of developmental pathways, and assessments of the limits of alternative design. These analyses may parallel in some ways the constructivist approach (Gutmann, 1981). But there is a radical departure here, as biomechanical or related studies are not isolated from the context provided by empirically tested phylogenies (e.g., Lauder, 1981). All these refinements have connections with paleontology, even though their primary application to fossil data may be limited.

Phylogenetic analysis increasingly has been promoted as an ordering framework for rigorous inspection of hypotheses on the origins of adaptations (Lauder, 1981; Coddington, 1988, 1990; Carpenter, 1989; Prum, 1990; Ross and Carpenter, 1990; Baum and Larson, 1991; Wenzel and Carpenter, 1994). Some authors (Felsenstein, 1985; Brooks and McLennan, 1991) have been emphatic about the salubrious nature of this trend. Indeed, the contributions of systematics in evolutionary biology are broadly recognized. As S. J. Gould stated (1991, p. 420):

"A major triumph of evolutionary studies in our generation has been, through the development of cladistics, the codification of a methodology for objective definition and determination of branching order."

It is noteworthy that, in the same breath, Gould (ibid, p. 420) warned evolutionists not to become "...intoxicated with the victory of cladistics..." and to avoid "...imperialistic..." tendencies in making "false extension of cladistic relationships to questions about the differential filling of morphospace." In a more explicit fashion, Frumhoff and Reeve (1994) also cautioned against what they view as the overextension of cladistics to analysis of adaptation.

These criticisms notwithstanding, phylogenetic analysis does seem to provide illumination for the inspection of adaptive histories. Some proposed liabilities of the phylogenetic approach have moot significance. For example, phylogenetics (Baum and Larson, 1991, p. 11) is thought to be vulnerable because it explicitly requires the assumption "...that the rate at which lineages move between selective regimes is low relative to the rate of lineage branching." Frumhoff and Reeve (ibid, pp. 175-176) argued that such an assumption is "...very unlikely to hold for many and perhaps most traits typically studied by evolutionary ecologists." But this point is debatable; in fact, de Queiroz and Wimberger (1993) found comparable consistency indices for behavioral and morphological traits in a large number of phylogenetic examples. Moreover, grossly unstable traits are not likely to be the evidence for the cladogram in the first place. Hair color and density of pelage may show high variations in certain mammal populations, and we might infer that such traits are highly labile in response to environmental change during the history of successive populations. But these varying states of hirsuteness are not customarily applied to mapping the relationships of the relevant mammalian species.

The very strength of a character in diagnosing a group suggests that it offers some simplification of adaptive history. The complex placenta and prolonged gestation that represents a synapomorphy of placental (eutherian) mammals has adaptive implications for all placental mammals regardless of variations in placental architecture, gestation time, and terrestrial and marine habitats (Luckett, 1977). At some level of generality, the match between "selective regime" and innovation in structure and function seems secure. Historical patterns based on cladograms deal with such levels of generality, not with highly labile characters and, by inference, highly unstable selective regimes. In this regard, I would suggest that the assumption expressed by Baum and Larson (1991, p 11) is not necessary for defending the phylogenetic analysis of adaptive history. A synapomorphy by virtue of its existence suggests the origin of an association among a structure, its function, and an inferred selective regime that is stable *relative* to many other changes and their inferred selective regimes.

VII. Paleontology Plus Phylogeny Equals History of Adaptation?

This defense of a core relationship between the phylogenetics and the inspection of hypotheses on the origins of adaptation leads us back to the original question regarding the role of paleontology. Do fossils provide uniquely important evidence for studies of phylogeny and adaptation? If, as argued earlier, the phylogenetic connection to studies of adaptation is valid, and if, in turn, fossils represent a unique contribution to developing and testing phylogenetic hypotheses, the importance of paleontology in this general context is secured. Addressing this question requires some review of relevant issues in systematics. A more analytical approach in establishing the value of fossil evidence has profited from an earlier phase of debate (Schaeffer et al., 1972; Nelson, 1978; see reviews in Eldredge and Cracraft, 1980, Eldredge and Novacek, 1985). This debate arose from negative reactions (Schaeffer et al., 1972; Nelson, 1978) to some traditional optimism concerning the capacity of the fossil record to faithfully represent phylogenetic history even at the level of direct ancestry and descent. Such an overly confident reading of the paleontological record fell easy prey to criticism; such traditional assumptions were accordingly abandoned in favor of a more conservative approach to the fossil evidence.

The skeptics then took their argument to the next level. Namely, fossils are simply additional data indistinguishable from Recent data, with the qualification that fossils were nearly always deficient relative to the latter because of their incomplete preservation (Patterson, 1981). This argument implies a secondary influence of fossil evidence in resolving phylogenies. Indeed it has been suggested that fossils be incorporated only after a branching pattern based on extant (often molecular) data has been established (Goodman, 1989). Thus, fossils could be relegated to the role of artifacts hung from the "Recent tree." This rather ignominious ranking implies a potential for fossil evidence that is always exceeded by data in extant organisms.

A rejoinder to this demotion of fossil data emerged from the same analytical approaches, namely cladistics, that inspired skepticism in the first place. Cladograms for higher amniotes, seed plants, mammals, and other groups (Gauthier et al., 1988; Donoghue et al. 1989; Novacek, 1992) demonstrated that fossils, their imperfections notwithstanding, can influence tree topologies in a primary way. This demonstration has been extended to molecular data in fossils (DeSalle et al., 1992).

Results of these studies have not brought all parties together. For example, Gauthier et al. (1988) corroborated a pattern (Gardiner, 1982) closely relating birds and mammals relative to other living amniotes when they restricted analysis to Recent data. Incorporation of fossil evidence, namely many extinct

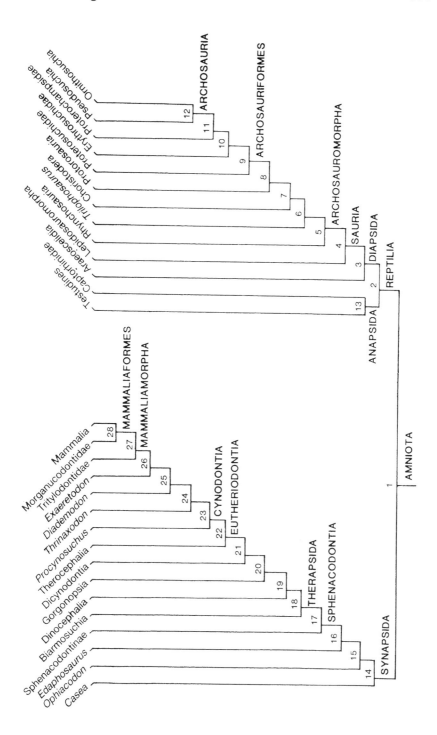

taxa that represent sister groups to living mammals, shifted mammals to a basal amniote branch (Fig. 9) excluded from a (((bird-crocodile) squamate) turtle) clade (Gauthier et al., 1988). In effect, it appeared that higher amniote phylogeny based on combined Recent and fossil data is particularly influenced by the latter. The latest iteration of the dialogue, nonetheless, is both a claim for the greater potential of Recent data and the close affinity, vis à vis other amniotes, of birds and mammals (Gardiner, 1993).

Whatever one's position on the amniote controversy (I am frankly disposed to the Gauthier et al. scheme), recent work recommends against relegation of fossil evidence to a rank below the expanding database on extant organisms. But do fossils, in themselves, provide any *unique* evidence (in the general sense) for phylogeny? At first pass, one is not persuaded that fossils represent such a distinct dimension. After all, any taxon, whether fossil or living, can carry with it traits that may affect the topology of its "family tree." Gauthier et al. (1988) were, however, emphatic about making a distinction between these categories, claiming that fossils are uniquely important because they, in effect, stopped evolving. They represent "points along the way" where character change has been truncated and "frozen" in the rock. Hence Gauthier et al. (1988) argued that fossils may recover information on ancient phylogenetic events not detectable through study of extant organisms. The latter are often deficient because of ambiguities (operationally equivalent to missing data) resulting from marked transformation (Gauthier et al., 1988; Donoghue et al., 1989).

This distinction for fossils requires an equation between ancientness and primitiveness that has long been embedded in the paleontological approach. Many early members of clades preserve characters radically altered in their more recent relatives. But the converse is also true; the paleontological equation is anything but law-like. Out of this uncertainty one might derive some probabilistic argument – fossils do seem to offer greater efficiency in finding characters bearing on remote cladistic events simply because they display a higher likelihood in preserving them. Norell and Novacek (1992a, b) showed that several case histories (Fig. 10) for different vertebrate groups varied markedly in the fit between stratigraphic occurrence (age of first appearance) and cladistic rank (*relative* time of splitting). But the fit was strikingly good for situations (e.g., large Holarctic mammalian herbivores) where the fossil record was judged to have merits for other reasons (more complete skeletal preservation, better stratigraphic control, more comprehensive geographic distribution of sites, etc.).

Fossil data thus warrant attention because of their high potential in recovering information on ancient phylogenetic events. This is not, however, a claim that fossils ordered stratigraphically might be used as a source of ultimate arbitration in phylogeny. Fisher (1994, p. 153), for example, cited Norell and Novacek (1992a) as documenting "generally high correlations between stratigraphic and cladistic order." The ratio of good fits to bad ones is actually more conservative; a factor that further erodes when hierarchy is accounted for – many of the good fits reside only within certain sectors (e.g., ungulates) of

Figure 9 Amniote phylogeny incorporating a large number of fossil clades. The remote position of Mammalia relative to birds and its sister taxa is anchored by the branches represented by fossil synapsids. From Gauthier et al. (1988).

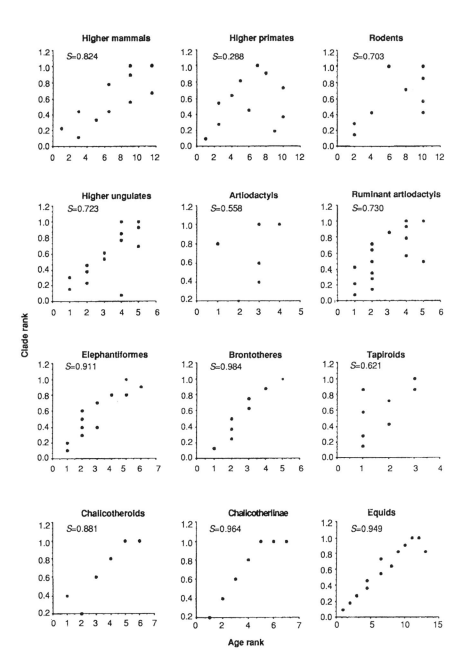

Figure 10 Rank correlations between the age of first occurrence in the fossil record and the relative cladistic rank (nodal position) based on cladograms strictly derived from morphological data (i.e., without reference to the known age of the fossil group). Correlations are based on a variety of vertebrate case studies. Note variance in correlation and the very tight fit between age and clade ranks in the case of equids (horses). From Norell and Novacek (1992a, b). Reprinted with permission. Copyright 1992 American Association for the Advancement of Science.

the vertebrate tree. Nonetheless, Fisher (1994, p. 157) further stated "if stratigraphic patterns are strong enough they may overturn cladograms (and unrooted networks) supported by morphology alone."

This assertion, integral to what Fisher dubbed "stratocladistics," represents an overextension of the results of comparisons of age and clade ranks (Norell and Novacek, 1992 a, b). Such studies do not validate a preference for cladistic or stratigraphic data (see also Rieppel and Grande, 1994). As for "overturning hypotheses," either domain of evidence has built-in properties for the purpose (e.g., homoplasy in cladistics; poor stratigraphic control in paleontology).

There is merit in identifying cases of high congruence between stratigraphic occurrence and cladistic rank (see also Smith and Littlewood, 1994). But this does not necessarily mean that such cases are, as Fisher suggested (1994), preferable to cases where an incompatibility among data sets is apparent. A basic compatibility between two data sets might emerge even if either or both data sets were regarded as intrinsically flawed. It seems that the categorization of data renders the matter of judging hypotheses against the scale of compatibility somewhat arbitrary [but see Smith and Littlewood (1994) for a different perspective]. Rather, the merits of cases of high compatibility lie in their heuristic potential. These cases offer a context to explore the implications of cladistic or stratigraphic pattern. A high correspondence in pattern instills an added element of confidence in this exploration. Horses are blessed with a good fossil record (Fig. 11), a robust cladistic hypothesis, and a high level of congruence between these data (Norell and Novacek, 1992 a, b). Horses, therefore, are a good choice for the examination of more subtle and elusive aspects of biological history, including the origin of adaptations (Fig. 12).

Returning to the original question, do fossils uniquely contribute to the development of phylogeny and thence to the theories on the origin and history of adaptation? The answer is a frustrating yes and no. In the strictest sense, fossils have *prima facie* no guarantee to contribute uniquely. The critical combination of character traits bearing on a particular hypothesis of monophyly and adaptive origin might reside just as well in a living organism. However, fossils demonstrably offer an increased efficiency in the search for evidence bearing on historical patterns. A preference for a bird-mammal affinity, as mandated by Gardiner's (1982) Recent data set, drastically affects our reconstruction of adaptive history for such motives as thermoregulation. But the identification of (1) a remote basal position of mammals, and (2) a connection between birds with coelurosaurian dinosaurs provides a constraint on adaptive change based on clades that are wholly extinct. Improving on the hypotheses for the origins and adaptation of feathers and wings in birds is now largely a matter of paleontological exploration. The notion that feathers may have an original use of thermoregulation but no relation to flight will not be established by securing information in extant creatures. (There are flightless, feathered birds today but phylogenetically these are readily identified as cases of secondary loss of flight.) A fossil could offer evidence of feathers, absence of wing structures, and a combination of other morphological structures that suggested not only an intermediate condition between that in *Archaeopteryx* and coelurosaurs, but also a sequence in the functional use of feathers. Under these conditions, fossils offer powerful contributions to the analysis of adaptive history.

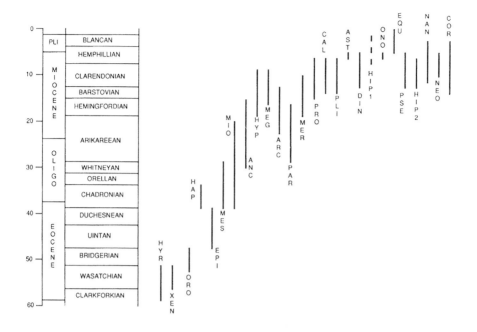

Figure 11 The sequence of fossil first occurrences for various clades of horses. Note the stepwise pattern of first occurrences. The pattern here is in nearly complete isometry with the independently derived cladograms for horses. From MacFadden (1992).

VIII. Broad-Scale Patterns of Evolution and Adaptation

Other areas of paleontological activity relate to studies of adaptation. One such area builds on the classic attempts by Simpson (1944) and others to measure the tempo and mode of evolution in different lineages. This has inspired the contemporary flourish of analyses (Fig. 13) on species turnover, origin, extinction, and survivorship over the Phanerozoic (Raup, 1976; Sepkoski, 1978; Stanley, 1979; and others). Much of the focus here concerns the matter of reliable statistical measurements of sampling, tests for random versus nonrandom patterns of change, estimates of time-scaled preservational bias (i.e., older fossil records tend to be less completely preserved and are therefore less reliable), and stratigraphic and global paleogeographic control. Aspects of this work have also raised questions and criticisms. If, for instance, the fossil record is inconsistent in preserving the actual time of divergence, are *minimum estimates* of origination indicated by fossil occurrence reliable indicators of taxon origin and taxon turnover (Novacek and Norell, 1982; Norell, 1992)?

Another concern is that many of the treatise-derived names and temporal ranges do not represent real, in the sense of monophyletic, groups (Patterson and Smith, 1987). If the groups estimated for evolutionary rate are not credible biological entities, then the meaning of broad-scale patterns for these arbitrary units is moot. Similar analyses on monophyletic units can show discrepancy with standard documentation of turnover and survivorship (Patterson and Smith, 1987). Other trends may seem robust enough to resist alteration in relation to modern systematic revision (Sepkoski, 1993), but that leaves open the question as to what we are measuring – it seems only logical to base such analyses on real, i.e., monophyletic, taxa. In the meantime, there are constructive methodologies for dealing with ghost lineages (see Fig. 14), that is, groups not directly represented but inferred from the branching patterns and stratigraphic distribution of related taxa (Norell, 1992). Plots of standing diversity that account for both known fossils and ghost lineages can differ significantly from widely accepted patterns of origination and extinction (Norell, 1992).

Moving beyond these caveats, we can examine the role such broad-scale studies play in historical analysis of adaptation. Such documented patterns are customarily wedded to adaptive scenarios. Often these refer to extrinsic factors and historical changes. The increase in diversity of certain marine groups is tied to changes in oceanic circulation that promoted diversification and spread of coral reefs; diversification (or extinction) of major terrestrial species is related to pulses of mountain building, changes in coastal configurations, glaciation, and so forth. Such casework is amply represented in an abundant literature, including many articles in journals like *Paleobiology*, wherein more synthetic comparisons of evolving marine and terrestrial communities have been fostered (for review, see DiMichele, 1994).

Unfortunately, attempts to go beyond these case histories and forge generalizations based on repeatable patterns in the fossil record show mixed results. Stanley (1979), for example, noted that certain taxa, such as graptolites, trilobites, and ammonites, in their rapid adaptive radiation and catastrophic extinction, contrast sharply with highly stable, enduring bivalves and gastropod mollusks, which took only moderate hits during the famed Permo-Triassic and Cretaceous-Tertiary extinction events. What accounts for the stability in the latter groups? In partial answer, Stanley (1979, p. 277) proposed:

"It is no accident that, in the basic pattern of macroevolution, the ammonites, the trilobites, and graptolites resemble terrestrial tetrapods of large body size. Here too, we may find a hint as to why many primitive taxa persist over long intervals of geologic time. Their low rate of speciation and extinction endow them with an inherent resistance to mass extinction."

To bolster this inference, Stanley (1979) cited differences between large and small mammals with respect to turnover during the later Cenozoic. Namely, smaller mammals seem to exhibit greater longevity over the past 3 million years, despite the fact that large mammals are better preserved. "This finding represents strong evidence that geologic longevity is inversely related to body size: Taxa of small animals normally exhibit low rates of extinction"(ibid., p. 277). In considering the selection parameters that might account for the correlation between body size and extinction pattern, Stanley (ibid., p. 278) offered the following:

"Among species of mammals in general as body size increases, average

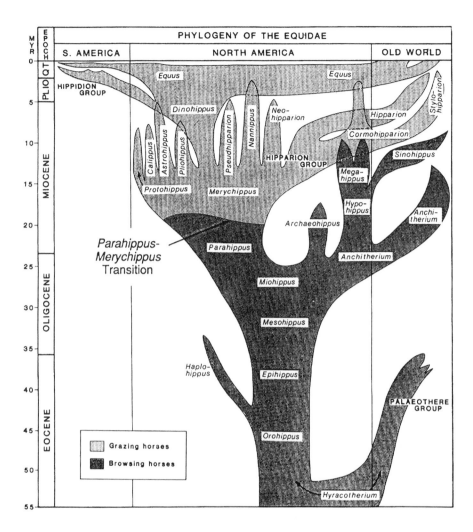

Figure 12 An evolutionary scenario for the history of horses. The pattern reflects modifications based on more recent paleontological and phylogenetic analyses, but some classic elements, such as a shift involving a reduction in toes and a development of high-crowned teeth appropriate for grazing, are consistent with both cladistic branching schemes and the fossil record. From MacFadden (1992).

population size decreases quite markedly... Given the obvious importance of body size to lineage survival, it is no surprise that small mammals exhibit lower average rates of extinction than large animals. On the other hand, we would expect the rate of speciation to be higher for small mammals because of weak dispersal... In fact, large mammals experience the environment in a fine-grained fashion, facing relatively few barriers to dispersal. Thus a typical species of large body size will contain relatively few individuals, but they will disperse

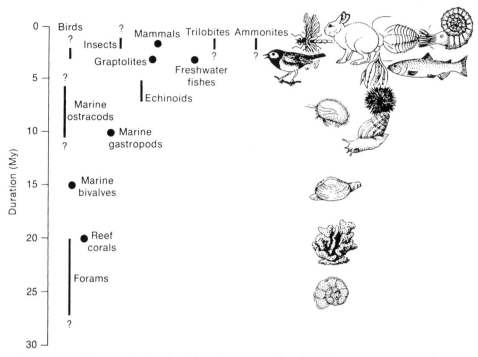

Figure 13   A proposal for estimated average species durations for some major taxa. From: Macroevolution by S. M. Stanley (1979). Copyright (c) 1979 by W. H. Freeman and Company. Used with permission.

rather readily over large areas. While its small numbers will render such species vulnerable to extinction, its ease of dispersal should suppress the rate of speciation. In contrast, while an average species of small body size should be relatively resistent to extinction by virtue of large total population size, its demes would occupy more restricted regions."

This series of posits constitutes a rather sweeping generalization about the evolutionary consequences of adaptive "motives" in organisms. Large body size putatively confers some adaptive advantage (presumably trophic opportunities, behavioral complexity, greater vagility, etc.), but has built-in liabilities (small effective breeding size, greater vulnerability to widespread environmental change). Thus, the selective pressures for the increase in body size allegedly consign the group affected to a rather bleak evolutionary fate, at least in comparison to those taxa that do not follow the route to greater body size. There are some echoes of earlier arguments here. Patterson (1949) related smaller population sizes in fossil mammals to more rapid evolution (in Patterson's terms, the extreme manifestation of high rate of evolution -- quantum speciation). However, in terms of generality, Stanley (1979) offers an explanation for an impressive slice of diversity, hundreds of millions of years of life history, and the major branches of metazoans.

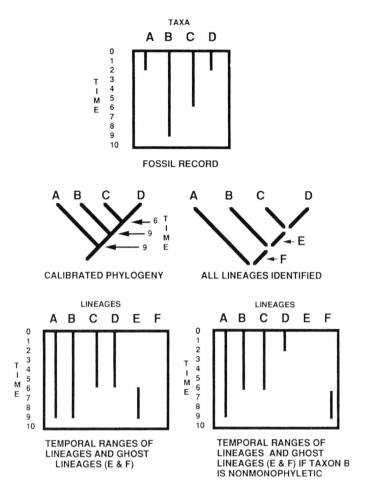

Figure 14 Fossil record and phylogeny calibrated to account for splitting sequences and ghost lineages. Note that the branching pattern (middle right) adjusts the duration of sister groups (e.g., C and D) given by the fossil record (top diagram). D must be at least as old as C (bottom left). Thus, both the fossil record and the phylogeny can be recalibrated (middle and bottom left). Ghost lineages E and F are inferred to exist from the branching sequence of A, B, C, and D (middle right). A less resolved branching pattern will yield a different calibration for the fossil record (bottom right). From Norell (1992). Used with permission.

Such a generalization would be dearly valued in biology, but there are questions concerning its actual utility. If large body size does predispose a taxon to a particular evolutionary fate, why is the pattern so difficult to summarize? There are, after all, taxa represented by fairly large species (crocodiles, conifers, and sharks) that persist in a healthy way through some of the major extinction episodes of the Phanerozoic. Even Stanley's focus on the mammalian record entails problems. The Paleocene and early Eocene were characterized by marked diversification and extinction of mammals, yet practically all the species during this interval were comparatively small in body size. One of the most dramatic cases of reduction and extinction involved the small multituberculates, a spectacularly diverse group of Mesozoic and Early Tertiary mammals whose demise coincided with the rise of their smallish, allegedly adaptive equivalents, the rodents (Krause, 1986). Furthermore, the contrast between large and small mammal survivorship during the last 3 million years of the Cenozoic could be explained by factors less overarching than the evolutionary potentials of population size. Possibly, many large mammal species were the victims of special and very effective predation by "largish" hominid mammals (Martin and Klein, 1984). Related to this is the fact that marked reduction in large mammal diversity is not a global phenomena; it characterizes certain islands and continents (e.g., Australia, New Zealand, North America) but not others (Africa).

Beyond this rather plaintive objection - namely, that the fossil record does not smoothly fit a theory relating body size to extinction rate - is additional doubt about the basic premise of the theory. Inconsistency in the fossil data notwithstanding, the theory proposes that somehow large body size *should* confer adaptive disadvantage with respect to evolutionary longevity. But why *should* this be so? The theory relates large body size to smaller population size, to greater dispersal capacity to broader environmental range, to greater probability of extinction. The last of these connections seems a *non sequitur*, even if one accepts the other correlates. There is no reason intuitively why a geographically widespread species should be more susceptible to extinction than a large population of small individuals confined to a particular region or environment. The vulnerability of endemics, regardless of their population size, in devastated habitats is, in fact, well documented (Wilson, 1992). Moreover, the capacity for dispersal in larger organisms could be scored as a counterbalance to small breeding population size; vagility ensures the possibility for interbreeding among fewer, more widely distributed individuals. There may be a correlation between marked speciation and diversification and the parameters associated with small body size (rapid turnover in generations, r-selected reproductive strategies, geographic heterogeneity, rates of genomic and morphological change, etc.) with *reference to particular taxonomic groups*, such as rodents. This is, however, a far cry from a general theory relating body size in diverse phyla to evolutionary potential. Unfortunately, such a theory lacks the built-in logic that would be helpful in envisioning the evidence necessary to test it.

This skepticism concerning a valiant attempt to correlate body size in taxa with broad-scale evolutionary patterns is not meant to exclude other generalizations drawn from the fossil record that might fare better. Instead of reviewing a litany of candidate theories, I offer two observations. First, such statements are predictably hard to come by; indeed hypotheses that aspire to embrace such a sweep of diversity and history are exceedingly rare. Second, where such generalizations do emerge, at least in recent years, questions related to phylogenetic patterns often frame the issues. Thus Eldredge (1979) contrasted the incidence of *stenotopy* (restricted "niche width," i.e., "ecological specialization") with *eurotopy* ("expanded niche width," "ecological generalization") in certain taxa and made predictions about the evolutionary consequences of these adaptive modes. Eurotopy is allegedly related to slow rates of speciation and taxic turnover (bradytely), and eurytopic groups exhibit "dampened" speciation rates and "retarded" morphological change (Eldredge,

1979, p. 14). Stenotopic taxa, on the other hand, can be connected with marked sympatry, high species diversity, and strong morphological divergence (Fryer and Iles, 1969; Eldredge, 1979).

The efficacy of this comparison in the broadest terms might be questioned, but Eldredge (1979) and others (Vrba, 1980, 1984) emphasize the importance of comparing stenotopy and eurotopy and their evolutionary correlates in *directly related groups* (i.e., sister taxa). This at least mitigates the hazards implicit in comparing apples to oranges (or, with reference to the above, comparing clams to ammonites, or large mammals to small mammals). Some of the onset variables can be neutralized by dealing with taxa that share a relatively close common ancestry. Certain comparisons of sister groups represented by fossil and living species (Vrba, 1984; Novacek, 1984) do support the above biological correlations with stenotopy and eurotopy (Fig. 15), even allowing a somewhat varying interpretation of these correlations (Vrba, 1984; Novacek, 1984).

IX. Adaptive Radiations: The Enigma of the Burgess Shale

Other adaptive/evolutionary studies focused on adaptive radiations certainly intertwine with phylogenetic issues. One of the most notable of this category is the analysis of the remarkable flourish of familiar, enigmatic, or simply bizarre animals of the early Cambrian radiation (Fig. 16), spectacularly preserved in the Burgess Shale(Walcott, 1908; Briggs, 1981; Whittington, 1985; Briggs and Fortey, 1989; Gould, 1989). The evolutionary significance of the Burgess Shale creatures is a matter of debate, whose earlier phase has been reviewed by McShea(1993). Gould (1989) proposed that the animals of the Burgess Shale are morphologically more disparate (i.e., show more variety in morphological features) than a comparable group of modern animals. The fossils, he argued, are variable in characters that today diagnose higher-level groups. This, Gould (1989, 1991) claimed, is possible evidence for a greater developmental flexibility in the early Cambrian animals, suggesting that "developmental entrenchment" may be a general evolutionary tendency.

This last audacious component suggests that indeed many of the disparate features that diagnose the early explosion of the arthropods are really a matter of some flux in developmental programs insulated from the action of selection on adaptation. McShea (1993, p. 399) characterized this argument as a major step in addressing the age-old issue of plasticity of morphology to natural selection vs. some internal dynamic, or the question of strong influence of "external forces" vs. the channeling of development, or simply "the nature-nurture problem." Nonetheless, several authors, notably Ridley (1990), took issue with Gould's (1989) perspective. The variety of objections can be boiled down to a few basic points: (1) there is no clear evidence for a significant reduction in morphological disparity between the early Cambrian arthropod groups and Recent arthropod taxa; (2) the impression of high disparity in the Cambrian forms is partly an artifact of dispatching lower level taxa to high ranks; (3) cladograms of the Cambrian groups do not reveal relatively high disparity in form; and (4) the argument for high disparity relies on a "retrospective fallacy" which overemphasizes variable traits that show constancy

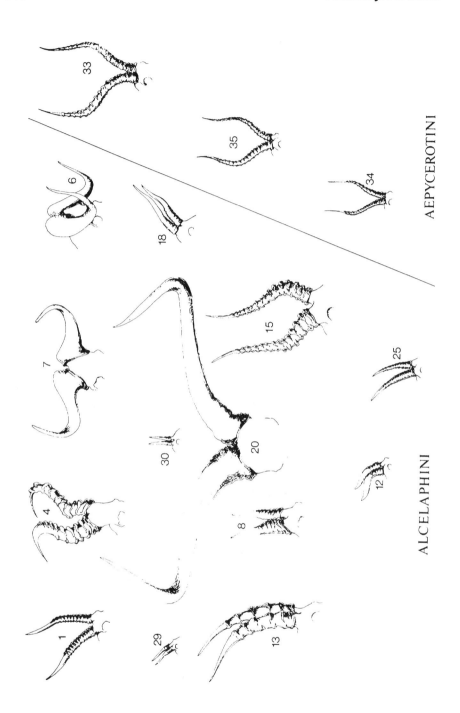

in later groups (Ridley, 1990; Briggs and Fortey, 1989).

These criticisms represent the "crux pitch" of Gould's assault on standard "adaptive" explanations of radiations. For he must convincingly show that high disparity in Cambrian arthropods is not simply a reflection of taxonomic bias and retrospective fallacy, that comparisons of disparity should derive from phenetics rather than cladistics, and that there is solvency to a hypothesis that suggests increasing developmental entrenchment in evolution. Unfortunately, his evidence for these points is largely anecdotal. In support of his reconstruction, Gould cited, for example, only Bard's (1990) learned but merely speculative view on possible mechanisms that initially permitted but later constrained phenotypic variation. On another front, Gould claimed that some of the Burgess Shale animals are more than merely odd, they lie outside the bounds of morphological range embodied by modern phyla. All would agree that the five-eyed "vacuum cleaner" *Opabinia* is weird, but this does not establish that the Burgess Shale creatures exceeded our coefficient for weirdness any more than a walking stick or a tardigrade. Indeed, recent revisions demonstrate that some of the most novel of the Burgess Shale forms are actually assignable to extant clades (see above-cited reviews). Gould rightfully points out that there is no reason to accept the standard view that the morphospace of life expands with "the cone of increasing diversity." The problem remains that the converse is not automatically justified; an evolutionary decrease of disparity is neither logically expected nor easy to demonstrate.

Central to Gould's (ibid, p. 418) position is the notion that a primary evolutionary signal should be tied to characters of "architectural depth," for whose identity we "share an intuition." Thus, he noted that we might regard segmentation and appendage number in arthropods as more fundamental than carapace color or the use of a gill branch in filtering or swimming. There is an essential problem here, one not very helpful to Gould's attempt to pose the question of morphological disparity and developmental retrenchment as a phenetic rather than cladistic argument. When I claimed above that features like hair color were not, in general, reliable in mapping higher mammalian relationships, this was not based on an *a priori* theory of the evolutionary valence of certain character systems. Nor was the claim based on "shared intuition." It instead stems from the difficulties in deriving stable phylogenetic patterns from these data *based on empirical work*. In some cases, such inspection may be tantamount to phenetics, for example, the observation of rampant variability for a trait within an OTU. But what if a trait, such as segment number, was highly stable for a given taxon? Would this mean that the characters had "architectural depth" and high reliability at a broader phylogenetic level? Not necessarily, because the demonstration of marked homoplasy in segment number could refute such an assertion.

The issue of homoplasy is related to Gould's (1991) point that the low consistency index (C. V.) of 0.384 in cladograms for Cambrian arthropods (Briggs and Fortey, 1989) is indeed evidence of greater disparity at the dawn of arthropod history. For this he is taken to task (Ridley, 1993) because a C. V. of

Figure 15 Comparisons of marked evolutionary diversification in the bovid group Alcelaphini with the more conservative pattern in its sister group Aepycerotini. From Vrba (1984).

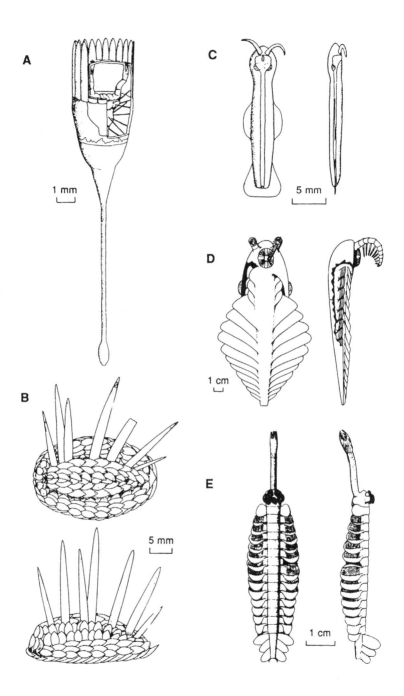

0.384 is expected given the 28 terminal taxa used in the Cambrian arthropod analysis (Sanderson and Donoghue, 1989). Gould (1993) accepted this correction, although he may be unduly acquiescent – many studies with taxon samples of this size show higher consistency values (Sanderson and Donoghue, 1989). Instead of further immersion in issues involving cladograms, Gould maintained that comparative phenetic (multivariate) analysis is the much needed pathway to resolution.

Even with the appearance of these recommended procedures, disparity of opinion over disparity of evolution persists. A principal components analysis of 134 characters revealed comparable levels of disparity between 21 modern and 25 Cambrian species of arthropods (Briggs et al., 1992). But Foote and Gould (1992) rejected this conclusion because it allegedly reflects a bias that exaggerates disparity in living arthropods. McShea (1993) weighed in on the side of early disparity because of the comparatively constrained range of morphology exhibited by certain arthropod subgroups despite 500 million years of evolution. Based on a protocol sensitive to taxonomic hierarchy, Foote (1993, 1994) proposed that a high initial morphological disparity and the subsequent decrease typified the evolutionary history of several groups, including trilobites and crinoids. Wills et al. (1994), however, in applying both phenetic and cladistic measures, maintained that Cambrian species of arthropods do not exhibit significantly higher levels of disparity than Recent species.

I have spent some time reviewing this issue, not simply because the Burgess Shale animals (Fig. 16) represent fascinating objects of paleontological discovery. The case draws us back to some of the essential points concerning the potentials and limitations of fossil evidence for studies of evolution and adaptation. The Burgess Shale fossils do represent an adaptive enigma that challenges our notion of the scope of morphological design based on our knowledge of living organisms. Does the example then challenge, at a fundamental level, certain evolutionary concepts? In a sense, Gould (1989, 1991) would say yes. He argued that the apparent high early disparity in Cambrian arthropods offers an alternative to the usual interpretation of adaptive radiations; namely the initial relaxation and subsequent channeling of development may have more to do with the history of radiations than the adaptive response of organisms to changing selection regimes. Gould's (1989, 1991) proposal is more provocative and more interesting, but unfortunately not necessarily more credible than the alternative. Evidence for an early phase of unsurpassed disparity in arthropod history is at best highly contestable. Thus, if one holds to the analysis of Wills et al. (1994, p. 121):

"Moreover, arguments for the operation of a special mechanism in the Cambrian have arisen from the premise that a much greater volume of morphospace was once occupied (with attendant implications for rates of evolution). In demonstrating these volumes to be equal, the proposal of a novel mechanism appears increasingly ad hoc."

Of course there is always some advantage to holding to the null hypothesis

Figure 16   Some examples of Early Cambrian organisms represented in the Burgess Shale: *Dinomischus* (A), *Wiwaxia* (B), *Amiskwia* (C), *Anomalocaris* (D), and *Opabinia* (E). From Conway Morris (1989). Reprinted with permission from Simon Conway Morris, University of Cambridge. Copyright 1989 American Association for the Advancement of Science

(no significant change in morphological disparity) because any indication of a real signal here might be susceptible to the numerous examples of bias associated with fossil evidence. But such bias is easy to invoke in most paleontological cases because there are usually obvious shortcomings with respect to preservation. The bias argument is not so automatic when the fossil record, as in the case of the Burgess Shale, provides us with an embarrassment of riches. Why then does ambiguity persist even in the face of a record so enriched with diversity and excellent preservation? The muddle here may have more to do with methodology than data. Both sides exhort the need for highly analytical tests of disparity. Despite the detailed analysis conducted (e.g., Foote, 1992, 1994; Briggs et al., 1992; Wills et al., 1994), "At present there is no agreed definition of disparity, much less any consensus on how to measure it" (Wills et al., 1994, p. 93).

A second problem with the theory of early disparity and developmental entrenchment is that the demonstration of the former does not necessarily substantiate the latter. Accordingly, Wills et al. (1994) stated that their tests of disparity, whatever the outcome, did not address the proposal for a special mechanism. The limitation applies even if morphological diversification outstripped taxonomic diversification in early arthropod history.

Thus, an anticipated demonstration that adaptive radiations show marked morphological disparity at their onset (Gould, 1993) begs, rather than identifies, an explanation of process. Both Ridley (1993) and Gould (1993) agreed that a productive avenue for seeking an explanation lies in the design of particular phenetic tests around a particular hypothesis, such as Gould's (1989) theory for developmental entrenchment. The nature of such tests, however, remains elusive despite the further refinement of various phenetic and cladistic analyses (Foote, 1992, 1994; Wills et al. 1994).

X. Fine-Scale Patterns of Evolution and Adaptation

While active debate continues over the meaning of broad-scale patterns in the fossil record, there seems to be a waning of paleontological attention to patterns of population change and speciation. I have admittedly not documented this downturn, but it is evident from the reduction in published papers alleging evidence for punctuated equilibrium or phyletic gradualism that seemed so abundant a few years ago. Perhaps this trend reflects the growing recognition of the enormous difficulty in documenting fine-scale evolutionary patterns. Perhaps the rising interest in phylogeny, merging molecular and morphological data sets, and the grander expressions of diversification in the fossil record have been distracting. Despite their rarity, fine-scale paleontological analyses are worthy of attention. If they really do meet expectations, these studies are the only hope for paleontological insight on the fine-tuning of organisms to changing environments through time. Ideal cases have potential bearing on attempts to describe and explain adaptive history.

Regrettably, ideal cases are hard to find. The impediments are the usual list of suspects: sequences of fossiliferous strata interrupted by gaps (unconformities, disconformities, etc.), ambiguous gaps between morphological centroids (they could represent punctuated patterns or merely deficiencies in sampling), the lack of geographic control (the interesting evolution always occurred somewhere else), or the lack of evidence for diagnosing or segregating different lineages under scrutiny (for review see Eldredge and Novacek, 1985).

Not too many years ago it was *en vogue* to call upon any example, imperfections notwithstanding, as part of a mounting case for either punctuated equilibrium or phyletic gradualism. This enthusiasm overran critics like Dingus and Sadler (1982) who reminded the chroniclers that the amount of critical sediment missing in a given rock sequence could approach, equal or exceed the amount available for sampling. Fortunately, critics (and reviewers) were eventually heard, and the standards for acceptabilty were raised, codified, and applied.

The approach here involves the review of the data in context of a theory which provides a novel interpretation of pattern and, at the same time, allows a wider variety of cases to be considered. A primary example is related to the punctuated equilibrium theory, as expressed by Eldredge and Gould (1972). The theory calls for rapid speciation followed by stasis in lineages. Unfortunately, the first component of the pattern, a sudden appearance of a new morph, can be ambiguous, as Eldredge and Novacek (1985) readily acknowledged. The new "species" could be the product of in situ remolding of the population, an invasion of a peripheral population, or simply an artifactual gap in the fossil sequence where any pattern for evolutionary change in the missing chunk of the lineage might have occurred.

The second part of the component is, however, a more straightforward matter of documentation. If a sequence of fossils through a reasonably continuous section shows no appreciable change, it is hard to think of any reason for the pattern other than the lack of significant evolutionary transformation in the lineage. In the words of the authors of the theory, "stasis is data" (Eldredege and Gould, 1972). Indeed, it is interesting how many fossil cases do show evidence of stasis over rock sequence representing hundreds of thousands or millions of years (Stanley, 1979). The strength of this signal has interesting implications for considerations of adaptation. It suggests that species do attain an adaptive equilibrium with various environmental and intrinsic factors that may persist beyond the "real time" of, for example, an ecological study.

A second opportunity for fine-scale analysis relies on a fastidious selection of examples. There are some situations where bias in preservation, continuity, and geography can be minimized, if not entirely eliminated. These offer data that go beyond a simple demonstration of stasis. In fact, they offer the only potential for confidently identifying the subtleties of morphological change (e.g., Bell et al., 1993) and the tempo and mode of population changes over a given temporal span. Marine benthic microorganisms often show global geographic distributions where shifts within a lineage can be discriminated from replacements by a lineage that evolved elsewhere. As we move toward the Recent we find, in general, that records are easier to monitor below the scale of millions of years, and thus some semblance of population change and subtle environmental modification is at least retrievable. This has been clearly demonstrated in the case of diatoms from freshwater lakes (Theriot, 1992), who extracted a striking pattern of gradual change between morphs in 12,000-

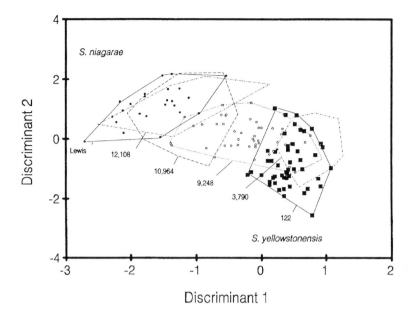

Figure 17 A plot of diatoms from a sedimentary sequence from Yellowstone Lake basin sediments. Numbers next to polygons are dates in years before present. Older morphs (open circles) resemble the "plesiomorphic species" *Stephanodiscus niagarae* (solid dots) inhabiting Lewis Lake, and later morphs (open circles) blend with the species *S. yellowstonensis* (solid squares) currently inhabiting the lake. An impressive suite of intermediate morphs can be identified for intervening years and there is no evidence of bimodality of form at a given time. From Theriot (1992).

year-old sediments of Yellowstone Lake. Older morphs resemble the "plesiomorphic species" *Stephanodiscus niagarae* while later morphs blend with the species *S. yellowstonensis* currently inhabiting the lake. An impressive suite of intermediate morphs can be identified for intervening years and there is no evidence of bimodality of form at a given time (Fig. 17). Theriot (1992) interpreted this result as in situ time transgressive change. The possibility of invasion from outside populations cannot be completely ruled out, but such seems highly unlikely because of the continuity of the record and the lack of intermediate morphs at any other lake in the region. In fact, other lakes have *S. niagarae* but lack *S. yellowstonensis*.

We see then that some aspects of the fossil record do approach the refinements required for discriminating patterns of change, perhaps even shedding some light on the nature of speciation. Diatoms in Yellowstone Lake, at any rate, would not be embraced as an example of punctuated equilibrium. What does such high resolution mean for studies of adaptation? Theriot (1992) addressed this question in a fundamental way. His analytical results, based on physiochemical and biological environmental data, support the notion (Mayr, 1982; Endler, 1989) that species show cohesiveness among their members and distinction from other species as a function of their "ecological uniqueness." The lakes with apomorphic endemics represent marginal habitats, whereas lakes representing both marginal and central habitats are occupied by *S. niagarae*. Theriot (1992, p. 150) concluded:

"Thus each of the three apomorphically diagnosed clusters empirically represents an ecological unit. However, the plesiomorphic or phenetic cluster called *S. niagarae* does not occupy a unique ecological range."

Of course this inference is drawn from a specific case and several issues are not addressed. Theriot pondered, for example, whether endemics were already marginal at the time of dispersal or were altered to become marginal and whether all modern habitats were historically occupied or some habitats experienced local extinction. He suggested optimistically, and I think justifiably, that "Answering these questions may reveal empirical detail relevant to the broader question of the role of environmental change in the evolution of apomorphy" (ibid, p.150).

The case of *Stephanodiscus* provides an unusual level of information, a level that far surpasses most alleged examples of fine-scale patterns. Nonetheless, we might expect this data base to improve. Freshwater lake, marine, and even fluvial sequences could offer such possibilities. It is clear, however, that these applications will have to meet some of the standards exemplified by the diatom study, standards that were perhaps largely ignored not so many years ago. The recognition of the groups must be integrated with phylogenetic analysis and referred to varying concepts of group recognition. The record must have identifiable attributes, such as extremely high resolution and stratigraphic completeness. Environmental parameters and their possibly subtle historical changes must be estimated through criteria independent of the groups in question. Finally, there must be some means of identifying and accounting for the influence of geographic factors (local extinction, immigration, etc.) on any temporal pattern.

XI. Summary and Conclusions

Consideration of adaptation and the paleontological role in understanding it intersects with questions regarding phylogeny, functional morphology, speciation models, developmental theories, estimates of evolutionary rates, population theory, ecology, and biostratigraphy. I have ranged freely among these topics, alighting on what I hope will be regarded as prominent examples drawn from an overwhelming scope of work that could not be fairly treated. From the onset, I intended to examine the limitations and potential of fossil data in uncovering the mysteries of adaptation. That, at any rate, was what I took to be the assignment. Obviously, the qualities of fossil evidence and analyses vary markedly, but several generalizations do emerge:

1) The concept of adaptation comes with some problematic terminological and methodological baggage, and the situation seems to cultivate misunderstanding or, worse, casual, untestable inferences. Not all is lost, however, if the effort is clearly demarcated. Historical theories of adaptation can be tested with evidence derived from both fossils and recent organisms under a framework that allows the examination of alternative pathways of change.

2) It is difficult to imagine our impoverished sense of the history of life

without a fossil record. Fossils do potentially promote a broader view of evolutionary patterns when they preserve problematic features or taxa that cannot be explained in a straightforward way simply by analogy with extant organisms. Such a distinction alone, however, does not ensure enlightenment. Many peculiar fossilized features resist explanation in terms of function or possible adaptation.

3) The history of paleontology illustrates a close integration of theories on the origin of adaptation with theories on the mode of speciation or evolutionary rate. But attempts to extrapolate from paleontological observation to general theories of evolution and adaptation have not, in general, fared well. What has endured is an increasingly analytical focus on paleontological pattern for the purposes of inferring evolutionary tempo and mode.

4) A basic element of research on adaptation concerns the determination of the function(s), through observation and experimentation, of a feature and the extrapolation of that information to the study of fossils. A stepwise theory for transformation of function (e.g., the origin of wings, flight, and feathers in birds) is then presented. Unfortunately, many of these studies are deficient because they fail to identify and eliminate alternative pathways for the origin and alteration of adaptations.

5) Evaluation of such alternative pathways depends on the historical framework explicitly provided by phylogenetic analysis. Despite some expressed misgivings (Frumhoff and Reeve, 1994), phylogenetic hypotheses have a salutary influence on developing historical arguments for adaptive origins and change. Combined with refined studies of function, behavior, or ecological roles, these historical investigations do have the potential to reveal patterns and perhaps processes basic to the history of adaptations.

6) If, as some claim, fossils uniquely contribute to the reconstruction of phylogeny and the latter is basic to understanding the history of adaptation, then the critical value of fossils to the study of adaptation is secured. The role of fossils in developing such theories is not, however, straightforward. In some cases, fossils are uniquely pivotal to identifying phylogenetic patterns. But fossils, by virtue of their appearance in the stratigraphic record, do not necessarily provide for a direct reading of the character change and, by inference, changing adaptation through time. Under optimal conditions, fossils are effective in placing structure and function in a broader historical context. Accordingly, we may look to extinct clades for the "baupläne" that anticipated modifications of design. Fossils, if applied in an explicit phylogenetic context, do offer such insights and, in some cases, these are insights that cannot be recovered from information on living organisms.

7) Alternatives to standard theories of adaptation, such as developmental retrenchment, have been inferred from broad-scale patterns of morphological disparity (e.g., Gould, 1989). Even if such disparity is demonstrable (the demonstration is in fact highly contentious), its bearing on an explanation for the pattern seems trivial. The theory of developmental retrenchment as an alternative to adaptation, if accessible to analysis at all, seems more in the domain of comparative development and genetics than paleontology. In this case, the fossil data merely inspire, rather than identify, an explanation of process.

8) At the other end of the spectrum, fine-scale patterns have been applied

to the study of lineage change and emerging adaptations. Case studies that qualify in this vein are exquisitely rare; they combine phylogenetic analysis, unambiguous group recognition, adequate samples for every stratum, a stratigraphic sequence of high resolution and completeness, environmental data, and some measure of biogeographic influences.

I have maintained throughout this essay that assessing the role of fossils in developing and testing theories of adaptation parallels such determinations in other areas of historical biology. The recent discussions over the weight of fossil evidence in reconstructing phylogeny are instructive in this regard, and several themes converge. In showing mosaics, sometimes combinations that at first appear enigmatic, fossils can offer data on a sequence of transitions with adaptive implications. It is not required that the fossils in question are perfectly ordered in the stratigraphic record; indeed sole reliance on such a criterion is simplistic and misleading. Fossils, however, offer high potential not only for informing a particular branching scheme, but for "filling in" the critical character changes that constrain inferences concerning adaptation.

This valuation may not satisfy paleontologists who have extrapolated directly from records of fossil occurrence to some rather profound notions of the evolutionary process. There seem to be limits, however, to the empirical justification of such theories – the adaptationist program, for instance, does not or cannot address the question of adaptive vs. nonadaptive change. As Darwin (1859) reflected with some sobriety, "...we have no right to expect to find, in our geological formations, an infinite number of those fine transitional forms which, on our theory, have connected all the past and present species of the same group into one long branching chain of life." Yet, a five-eyed Cambrian creature, a feathered *Archaeopteryx*, and a 12,000-year-old population of diatoms are triumphs of paleontological disclosure to be reckoned with.

Acknowledgments

I thank the editors for the invitation to contribute and for their useful criticisms and comments as well as the comments of Mark Norell, Louis Chiappe, Guillermo Rougier, and Gina Gould. Jim Carpenter provided a very detailed set of comments and editorial remarks that warrant special recognition. Aspects of this paper review research supported by NSF BSR 910868.

References

Bakker, R. T. (1975). Dinosaur renaissance. *Sci. Am.* 232, 58-78.

Bard, J.(1990). The fifth day of creation. *Bioessays* 12, 303-306.

Bateson, W. (1894). "Materials for the Study of Variation." Macmillan, London.

Baum, D. A., and Larson, A. (1991). Adaptation reviewed: a phylogenetic methodology for studying character macroevolutions. *Systematic Zoology* 40, 1-18.

Bell, M. A., Orti, G., Walker, J. A., and Koenigs, J. P. (1993). Evolution of pelvic reduction in threespine stickleback fish: a test of competing hypotheses. *Evolution* 47(3), 906-914.

Briggs, D. E. G. (1981). The arthropod *Odaraia alata* Walcott, Middle Cambrian, Burgess Shale, British Columbia. *Trans. Roy. Soc. London* B291, 541-585.

Briggs, D. E. G., and Fortey, R. A. (1989). The early radiation and relationships of the major arthropod groups. *Science* 246, 241-243.

Briggs, D. E. G., Fortey, R. A., and Wills, M. A. (1992). Morphological disparity in the Cambrian.

Science 256, 1670-1673.

Brooks, D. R., and McLennan, D. A. (1991). "Phylogeny, Ecology, and Behavior: A Research Program in Comparative Biology." University of Chicago Press, Chicago.

Carpenter, J. M. (1989). Testing scenarios: wasp social behavior. *Cladistics* 5, 131-144.

Chatterjee, S. (1991). Cranial anatomy and relationships of a new Triassic bird. *Phil. Trans. Roy. Soc. London* B332, 277-346.

Cheverud, J. M., Dow, M. M., and Leutenegger, W. (1985). The quantitative assessment of phylogenetic constraints in comparative analyses: sexual dimorphism in body weight among primates. *Evolution* 39, 1335-1351.

Chinsamy, A., Chiappe, L. M., and Dodson, P. (1994). Growth rings in Mesozoic birds. *Nature* 368, 196-197.

Cloud, P. (1960). Gas as a sedimentary and diagenetic agent. *Am. J. Sci.* 258A, 35-45.

Coddington, J. A. (1988). Cladistic tests of adaptationist hypotheses. *Cladistics* 4, 1-20.

Coddington, J. A. (1990). Bridges between evolutionary pattern and process. *Cladistics* 6, 379-386.

Conway Morris, S. (1989). Burgess Shale faunas and the Cambrian explosion. *Science* 246, 339-346.

Cope, E. D. (1896). "The Primary Factors of Organic Evolution." Open Court, Chicago.

Darwin, C. (1859). "On the Origin of Species by Means of Natural Selection; or, the Preservation of Favoured Races in the Struggle for Life." Murray, London.

de Queiroz, A., and Wimberger, P. (1993). The usefulness of behavior for phylogeny estimation: levels of homoplasy in behavioral and morphological characters. *Evolution* 47, 46-60.

DeSalle, R., Gatesy, J., Wheeler, W., and Grimaldi, D. (1992). DNA sequences from a fossil termite in Oligo-Miocene amber and their phylogenetic implications. *Science* 257, 1933-1936.

DeVries, H. (1901). Die Mutationstheories. Versuche und Beobachtungen uber die Entstehung der Arten im Planzenreich. Vol. 1, Die Entstehung der Arten durch Mutation. Leipzig; Veit. Eng. trans. J. B. Farmer and A. D. Darbishire. Open Court Publishing Company, 1909-1910, Chicago.

DiMichele, W. (1994). Ecological patterns in time and space. *Paleobiology* 20(2), 89-92.

Dingus, L., and Sadler, P.M. (1982). The effect of stratigraphic completeness on estimates of evolutionary rate. *Syst. Zool.* 31, 328-334.

Donoghue, M. J., Doyle, J., Gauthier, Kluge, A., and Rowe, T. (1989). The importance of fossils in phylogeny reconstruction. *Annu. Rev. Ecol. Syst.* 20, 431-460.

Dobzhansky, T. (1937). "Genetics and the Origin of Species." First Edition. Columbia University Press, New York.

Dobzhansky, T. (1956). What is an adaptive trait? *Amer. Nat.* 90, 337-347.

Eldredge, N. (1979). Alternative approaches to evolutionary theory *In* "Models and Methodologies in Evolutionary Theory" (J. H. Schwartz and H. B. Rollins, eds.), pp. 7-19. *Bull. Carnegie Mus. Nat. Hist.* 13, 1-105.

Eldredge, N. and Cracraft, J. (1980). "Phylogenetic Patterns and the Evolutionary Process." Columbia University Press, New York.

Eldredge, N., and Gould, S. J. (1972). Punctuated equilibria: an alternative to phyletic gradualism. *In* "Models in Paleobiology" (T. J. M. Schopf, ed.), pp. 82-115. Freeman, Cooper and Company, San Francisco.

Eldredge, N., and Novacek, M. J. (1985). Systematics and paleobiology. *Paleobiology* 11(1), 65-74.

Endler, J. A. (1989). Conceptual and other problems in speciation. *In* "Speciation and Its Consequences" (D. Otte and J. A. Endler, eds.) pp. 625-648. Sinauer, Sunderland, MA.

Feduccia, A., and Wild, R. (1993). Birdlike characters in the Triassic archosaur *Megalancosaurus*. *Naturwissenschaften* 80, 564-566.

Felsenstein, J. (1985). Phylogenies and the comparative method. *Amer. Nat.* 125, 1-15.

Fisher, D. (1985). Evolutionary morphology: beyond the analogous, the anectodal, and the ad hoc. *Paleobiology* 11(1), 120-138.

Fisher, D. C. (1994).Stratocladistics: morphological and temporal patterns and the relation to phylogenetic process. *In* "Interpreting the Hierarchy of Nature." (L. Grande and O. Rieppel, eds.), pp. 133-171.

Foote, M. (1992). Rarefaction analysis of morphological and taxonomic diversity. *Paleobiology* 18, 1-16.

Foote, M. (1993). Contributions of individual taxa to overall morphological disparity. *Paleobiology* 19(4), 403-419.

Foote, M. (1994). Morphological disparity in Ordovician-Devonian crinoids and the early saturation of morphological space. *Paleobiology* 20(3), 320-344.

Foote, M., and Gould, S. J. (1992). Cambrian and Recent morphological disparity. *Science* 258, 1816.

Frumhoff, P. C., and Reeve, H. K. (1994). Using phylogenies to test hypotheses of adaptation: a critique of some current proposals. *Evolution* 48(1), 172-180.

Fryer, G., and Iles, T. D. (1969). Alternative routes to evolutionary success as exhibited by African cichlid fishes of the genus *Tilapia* and the species flocks of the great lakes. *Evolution* 23, 359-369.

Gardiner, B. (1982). Tetrapod classification. *Zool. J. Linn. Soc. London* 74, 207-232.

Gardiner, B. (1993). Haematotheria: warm-blooded amniotes. *Cladistics* 9, 369-395.

Gauthier, J., Kluge, A. G., and Rowe, T. (1988). Amniote phylogeny and the importance of fossils. *Cladistics* 4, 105-209.

Gittleman, J. L., and Kot, M. (1990). Adaptation: statistics and a null model for estimating phylogenetic effects. *Sys. Zool.* 39, 227-241.

Gingerich, P. D. (1974). Stratigraphic record of Early Eocene *Hyopsodus* and the geometry of mammalian phylogeny. *Nature* 248, 107-109

Goldschmidt, R. (1940). "The Material Basis of Evolution." Yale University Press, New Haven.

Goodman, M. (1989). Emerging alliance of phylogenetic systematics and molecular biology: a new age of exploration. *In* "The Hierarchy of Life" (B. Fernholm, K. Bremer and H. Jornvall, eds.), p.43-61. Nobel Symposium 70, Elsevier Science Publications, Amsterdam.

Gould, S. J. (1989). "Wonderful Life." Norton, New York.

Gould, S. J. (1991). The disparity of the Burgess Shale arthropod fauna and the limits of cladistic analysis: why we must strive to quantify morphospace. *Paleobiology* 17(4), 411-423.

Gould, S. J. (1993). How to analyze the Burgess Shale disparity - a reply to Ridley. *Paleobiology* 19(4), 522-523.

Gould, S. J., and Lewontin, R. (1979). The spandrels of San Marco and the Panglossian paradigm: a critique of the adaptationist programme. *Proc. Roy. Soc. London* B205, 581-598.

Gould, S. J., and Vrba. E. S. (1982). Exaptation - a missing term in the science of form. *Paleobiology* 8(1), 4-15.

Gutmann, W.(1981). Relationships between invertebrate phlya based on functional-mechanical analysis of the hydrostatic skeleton. *Am. Zool.* 21, 63-81

Gutmann, W.F., and Bonik, K.(1981) Kritische Evolutionstheorie: Ein Beitrag zur Uberwindug altdarwinistischer Dogmen. Gerstenberg, Verlag, Hildesheim.

Harvey, P. H., and Pagel, M. D. (1991). "The Comparative Method in Evolutionary Biology." Oxford University Press, Oxford.

Hecht, M. (1974). Morphological transformation, the fossil record, and the mechanisms of evolution: A debate. Part II. The statement and the critique. *Evolutionary Biology* 7, 295-303.

Huxley, T. H. (1880). On the application of laws of evolution to the arrangement of the Vertebrata and more particularly of the Mammalia. *Proc. Zool. Soc. London* pp. 649-662.

Jepsen, G., Simpson, G. G. ,and Mayr, E. (eds.) (1949) "Genetics, Paleontology, and Evolution." Princeton University Press, Princeton, NJ.

Kitts, D. B. (1974). Paleontology and evolutionary theory. *Evolution* 28, 458-472.

Krause, D. W. (1986). Competitive exclusion and taxonomic displacement in the fossil record: the case of rodents and multituberculates in North America. *Contributions to Geology, University of Wyoming, Special Paper.* 3, 95-117.

Lauder, G. V. (1981). Form and function: structural analysis in evolutionary morphology. *Paleobiology* 7(4), 430-442.

Liem, K. F. (1973). Evolutionary strategies and morphological innovations: cichlid pharyngeal jaws. *Sys. Zool.* 22, 425-441.

Lewontin, R. (1978). Adaptation. *Sci. Am.* 239, 156-169.

Lewontin, R. (1979). Sociobiology as an adaptationist program. *Behav. Sci.* 24, 5-14.

Lewontin, R. (1980). Adaptation. The Encyclopedia Einaudi. Milan. *In* "Conceptual Issues in Evolutionary Biology" (E. Sober, ed.), pp. 234-251. MIT Press, Cambridge, MA.

Luckett, W. P. (1977). Ontogeny of amniote fetal membranes and their application to phylogeny. *In* "Major Patterns in Vertebrate Evolution" (M. K. Hecht, P. C. Goody, and B. M. Hecht, eds.), p. 439-516. Plenum Press, New York.

MacFadden, B. J. (1992). Interpreting extinctions from the fossil record; methods, assumptions, and case examples using horses (Family Equidae). *In* "Extinction and Phylogeny" (M. J. Novacek, and Q. D. Wheeler, eds.), pp. 17-45. Columbia University Press, New York.

McLachlan, G. R., and Liversidge,R. (1978). "Roberts' Birds of South Africa." 4th ed. John Voelcker Bird Book Fund. Cape Town.

McShea, D. W. (1993). Arguments, tests, and the Burgess Shale - a commentary on the debate. *Paleobiology* 19(4), 399-402.

Maglio, V. J. (1973). Origin and evolution of the Elephantidae. *Am. Phil. Soc. Trans.* 63, 1-149.

Martin, P., and Klein, R.(eds.) (1984). "Quaternary Extinctions: A Prehistoric Model." University of Arizona Press, Tucson.

Maynard Smith, J. (1978). Optimization theory in evolution. *Annu. Rev. Ecol. Syst.* 9, 31-56.

Mayr, E. (1942). "Systematics and the Origin of Species." Columbia University Press, New York.

Mayr, E. (1982). "The Growth of Biological Thought. Diversity, Evolution and Inheritance." Belknap Press, Cambridge, MA.

Mayr, E. (1988). "Toward a New Philosophy of Biology; Observations of an Evolutionist." Harvard University Press, Cambridge, MA.

Nelson, G. (1978). Ontogeny, phylogeny and biogenetic law. *Sys. Zool.* 27, 324-345.

Norell, M. A. (1992). Taxic origin and temporal diversity: the effect of phylogeny. *In* "Extinction and Phylogeny" (M. J. Novacek, and Q. D. Wheeler, eds.), pp. 89-118. Columbia University Press, New York

Norell, M. A., and Novacek, M. J. (1992a). The fossil record and evolution: comparing cladistic and paleontologic evidence for vertebrate history. *Science* 255, 1690-1693.

Norell, M. A., and Novacek, M. J. (1992b). Congruence between superpositional and phylogenetic patterns: Comparing cladistic patterns with fossil records. *Cladistics* 8, 319-337.

Novacek, M. J. (1984). Evolutionary stasis in the elephant-shrew, *Rhynchocyon* . *In* "Living Fossils" (N. Eldredge and S. Stanley, eds.), pp. 4-22.

Novacek, M. J. (1992). Fossils as critical data for phylogeny. *In* "Extinction and Phylogeny" (M. J. Novacek and Q. D. Wheeler, eds.), pp. 46-88. Columbia University Press, New York.

Novacek, M. J., and Norell, M. A. (1982). Fossils, phylogeny, and taxonomic rates of evolution. *Sys. Zool.* 31, 369-378.

Osborn, H. F. (1929). The titanotheres of ancient Wyoming, Dakota, and Nebraska. Volumes I and IT. Monograph 55. Department of Interior U. S. Geological Survey

Osborn, H. F. (1934). Aristogenesis, the creative principle in the origin of species. *Amer. Nat.* 68, 193-235.

Ostrom, J. H. (1974). *Archaeopteryx* and the origin of flight. *Quart. Rev. Biol.* 49, 27-47.

Ostrom, J. H. (1975). The origin of birds. *Annu. Rev. Earth and Planet. Sci.* 3, 55-77.

Ostrom, J. H. (1979). Bird flight: how did it begin? *Am. Sci.* 67, 46-56.

Ostrom, J. H. (1994). On the origin of birds and of avian flight. *In* "Major Features of Vertebrate Evolution" (D. R. Prothero, and R. M. Schoch, conveners, R. S. Spencer, ed.), Vol. 7, pp. 160-177.

Patterson, B. (1949). Rates of evolution in taeniodonts. *In* "Genetics, Paleontology and Evolution" (G. Jepsen, G. G. Simpson, and E. Mayr, eds.), pp. 243-278. Princeton University Press, Princeton N.J.

Patterson, C. (1981). Significance of fossils in determining evolutionary relationships. *Annu. Rev. Ecol. Syst.* 12, 195-223.

Patterson, C. (1982). Morphological characters and homology. *In* "Problems of Phylogenetic Reconstruction" (K. A. Joysey, and A. E. Friday, eds.), pp. 21-74. Academic Press, London.

Patterson, C., and Smith, A. B. (1987). Is periodicity of extinctions a taxonomic artifact? *Nature (London)* 330, 248-251.

Prum, R. O. (1990). Phylogenetic analysis of the evolution of display behavior in the neotropical manakins (Aves: Pipridae). *Ethology.* 84, 202-231.

Randolph, S. (1994). The relative timing of the origin of flight and endothermy: evidence from the comparative biology of birds and mammals. *Zool. J. Linn. Soc.* 112, 389-397.

Raup, D. M. (1976).Species diversity in the Phanerozoic: a tabulation. *Paleobiology* 2, 279-288.

Ridley, M. (1990). Dreadful beasts. *The London Review of Books* June 28, pp. 11-12.

Ridley, M. (1993). Analysis of the Burgess Shale. *Paleobiology* 19(4), 519-521.

Rieppel, O., and Grande, L. (1994). Summary and comments on systematic pattern and evolutionary process. *In* "Interpreting the Hierarchy of Nature" (L. Grande and O. Rieppel, eds.), pp. 227-255.

Romer, A. S. (1966). "Vertebrate Paleontology." 3rd Edition. University of Chicago Press, Chicago.

Ross, K. R., and Carpenter, J. M. (1990). Phylogenetic analysis and the evolution of queen number in eusocial Hymenoptera. *J. Evol. Biol.* 4, 117-130.

Ruben, J. (1991). Reptilian physiology and the flight capacity of *Archaeopteryx. Evolution* 45, 1-17.

Sanderson, M. J., and Donoghue, M. J. (1989). Patterns of variation in levels of homoplasy. *Evolution* 43,1 781-1795.

Schaeffer, B., Hecht, M., and Eldredge, N. (1972). Phylogeny and paleontology. *Evol. Biol.* 6, 31-46.

Schindewolf, O. H. (1950). "Grundfragen der Palaeontologie." Stuttgart: Schweizerbart.

Schoch, R., and Lucas, S. (1981). New conoryctines (Mammalia: Taeniodonta) from the Middle Paleocene (Torrejonian) of Western North America. *J. Mammal.* 62(4), 683-691.

Sepkoski, J. J. (1978). A kinetic model of Phanerozoic taxonomic diversity. I. Analysis of marine orders. *Paleobiology* 4, 223-251.

Sepkoski, J. J. (1993). Ten years in the library: new data confirm paleontological patterns. *Paleobiology* 19(1), 43-51.

Shubin, N., and Alberch, P. (1986). A morphological approach to the origin and basic organization of the tetrapod limb. *Evol. Biol.* 20, 319-387.

Simpson, G. G. (1944). "Tempo and Mode in Evolution." Columbia University Press, New York.

Simpson, G. G. (1949). "The Meaning of Evolution." Yale University Press, New Haven.

Simpson, G. G. (1953). "The Major Features of Evolution." Columbia University Press, New York.

Simpson, G. G. (1959). Mesozoic mammals and the polyphyletic origin of mammals. *Evolution* 13, 405-414.

Smith, A. B., and Littlewood, D. T. J. (1994). Paleontological data and molecular phylogenetic analysis. *Paleobiology* 20(3), 259-273.

Speakman, J. R. (1993). Flight Capabilities in *Archaeopteryx. Evolution* 47(1), 336-340.

Stanley, S. M. (1979). "Macroevolution, Pattern and Process." W. H. Freeman and Company, San Francisco.

Theriot, E. (1992). Clusters, species concepts and morphological evolution of diatoms. *Syst. Biol.* 41(2), 141-157.

Vrba, E. S. (1979). Phylogenetic analysis and classification of fossil and recent Alcelaphini (Family Bovidae, Mammalia) *J. Linn. Soc. (Zoology)* 11(3), 207-228.

Vrba, E. S. (1980). Evolution, species and fossils: how does life evolve? *S. African J. Sci.* 76, 61-84.

Vrba, E. S. (1984). Evolutionary pattern and process in the sister-group Alcelaphini - Aepycerotini (Mammalia: Bovidae). *In* "Living Fossils" (N. Eldredge and S. Stanley, eds.), pp. 62-79.

Walcott, C. D. (1908). Mount Stephen rocks and fossils. *Can. Alpine J.* 1, 232-248.

Weismann, A. (1882). "Uber die Berechtigung der Darwinschen Theorie." Engelmann, Leipzig.

Wenzel, J. W., and Carpenter, J. M. (1994). Comparing methods: adaptive traits and tests of adaptation. *In* "Phylogenetics and Ecology," pp. 79-101.

Whittington, H. B. (1985). "The Burgess Shale." Yale University Press, New Haven.

Wills, M. A., Briggs, D. E. G., and Fortey, R. A. (1994). Disparity as an evolutionary index: a comparison of Cambrian and Recent arthropods. *Paleobiology* 20(2), 93-130.

Williams, G. C. (1966). "Adaptation and Natural Selection." Princeton University Press, Princeton, NJ.

Wilson, E. O. (1992) "The Diversity of Life." The Belknap Press of Harvard University, Cambridge, MA.

Wyss, A. R. (1987). Notes on Prototheria, Insectivora and Thomas Huxley's contribution to mammalian systematics. *J. Mammal.* 68, 135-138.

PART III

Diversity of Adaptive Processes

Adaptation, in principle, is a perversely complex layer cake, with 300 base-pairs of selfish DNA at the bottom and vast clades of trilobites and dinosaurs at the top. We are used to thinking of adaptation in terms of the features of organisms that "work," perhaps because natural selection put them there. That is where Darwin started after all.

But even Darwin had to expand on this simple scheme. How else to make sense of sterile castes in social insect species? Darwin's solution was to invoke superordinate family selection. Today, we play theme and variations on that original intuition. Wade and Vermeij take us upward, from kin selection to interdemic selection to species selection. Vermeij takes the perspective of long eons of macroevolutionary time as frozen in paleontological data, while Wade offers a detailed intercalation of mathematical population genetics and laboratory insect experiments. Hurst, by contrast, gives us the microcosmic world of genomic parasites, in which the fundamental fitness interests of the organism may be entirely subverted. From either angle, it's hard to look at adaptation in the same way again.

Our concluding chapter provides a stimulating overview of the concept of adaptation. Frank shows us that, even as some evolutionary biologists have been losing faith in selection as an agent of transformation, neurobiologists, engineers, and computer scientists are discovering that it is one of the most powerful tools ever deployed by humanity.

CHAPTER 11

Adaptations of Clades:

Resistance and Response

GEERAT J. VERMEIJ

I. Introduction

What is the role of adaptation in evolution? In one form or another, this question has been central to the development of evolutionary thought during the last two centuries. It has taken on renewed interest with the realization that great regularity can arise as a cascade from chance events. Perhaps even more importantly, the emergence of macroevolution as a distinct subdiscipline of evolutionary biology has forced biologists to recognize that the comings and goings of apparently well-adapted biological entities are influenced by processes of sorting at each of several levels in the hierarchy of life, ranging from genes to individuals, species, more inclusive clades, and perhaps even ecosystems.

Gould (1985) eloquently expressed the doubts that many observers hold about the importance of adaptation when he asserted that processes of sorting acting at and above the level of the species are often antagonistic to, and in the long run are more important than, adaptation by natural selection operating among individual organisms. In other words, macroevolutionary processes that affect extinction and proliferation of species and more inclusive clades (evolutionary branches) are held to be primarily responsible for the pace and directions of evolution. A largely unanswered question is whether clade-level sorting can result in clade-level adaptations and, if so, whether such adaptation amplifies or negates adaptation at the level of individual organisms.

To my mind, it is pointless to argue about the evolutionary role of adaptation unless and until we specify precisely what is meant by adaptation and which criteria would be used for its verification or falsification. Accordingly, my first aim in this chapter is to present criteria for the recognition of adaptation. Building on this foundation, I shall ask whether the notion of adaptation applies to the level of species and beyond, and how processes of sorting at different levels of the

363

hierarchy affect evolution. Finally, I shall argue that the ability of individual organisms and of populations to respond to change increased among the more recently evolved major clades. This inference strengthens the view that the fates of clades and of individual organisms are intimately connected and are not decoupled, as some earlier observers have claimed.

II. Definition and Criteria

I define an adaptation as a heritable attribute of an entity that confers advantages in survival and reproduction of that entity in a given environment. This definition is fundamentally comparative; with the phrase "confers advantages in survival and reproduction" comes the implication that the attribute in question is better than some alternative. The definition also makes clear that the advantages are context dependent; under different circumstances, alternative attributes might be adaptive. I also use the word adaptation to mean the process by which an adaptation arises and is maintained.

This conception of adaptation is essentially identical to my previous views (Vermeij, 1987a) and to those of Reeve and Sherman (1993). The only difference is that whereas I used the word aptation for the attribute and adaptation for the process, I now prefer to use the same word adaptation for both. Gould and Vrba (1982) favored the term aptation to denote attributes because they wished to reserve the word adaptation to describe the historical improvement of a trait to perform a given function. They contrasted this concept of adaptation with exaptation, which refers to an attribute whose current function was taken on after the attribute evolved. Although this distinction is worth making, I believe the term adaptation should be used for both the beneficial attribute and for the process producing it.

Any entity that is capable of multiplication, variation, and heredity can become adapted and can possess adaptations. Maynard Smith (1991) refers to such entities as units of evolution. In the hierarchy of life, no genealogical entity is immune from adaptation; however, there is no *a priori* reason that entities of a given level in the hierarchy possess adaptations. The nature and extent of adaptation are matters of empirical inquiry.

Adaptive attributes must be heritable in some way, and a selective process (or nonrandom sorting) among heritable alternatives must cause some variants to have a reproductive or survival advantage. At the level of individual organisms, inheritance occurs by means of transmission by genes encoded in nucleic acids. Variations arise from mutations, which may affect a very small number of nucleotides, or from larger-scale reorganization of the genetic architecture. The selective process resulting in individual-level adaptation is called natural selection.

The concept of adaptation, however, does not require genes; it merely requires inheritance. Transmission can be effected by cultural means as well as by repeated infection, for example. The mechanism of inheritance cannot be known for organisms of the past. I therefore believe it is useful to distinguish clearly between functional benefit or adaptation on the one hand, and the mechanisms of transmission of the traits in question on the other.

The criterion of functional benefit requires that we establish a causal relationship among the purported adaptation, its bearer's reproductive performance or survivorship, and agencies that would reduce or enhance survival and reproduction. Merely showing that entities with attribute A survive or reproduce better than entities with an alternative attribute B is not enough, for such differential representation could arise either from chance or from a close but fortuitous genetic link with another trait. In other words, the causal connection between a trait and its purported benefit must include evidence that the adaptation enables the entity to cope with identifiable harmful agencies or to take advantage of identifiable favorable circumstances. Selection is caused by potentially harmful or beneficial agencies. The mechanical and physiological principles that determine how the adaptation confers benefits comprise the adaptation's function and are essential to the establishment of the causal link between the adaptation and the individual's performance (see also Lauder, 1995).

We do not need to understand the genetic architecture of traits to ascertain whether those traits are adaptations; however, there must be evidence that the traits are heritable and that their increased representation through the action of selection rests on the inferred or observed action of selective agencies. If, for example, we wish to argue that large body size is an adaptation of mammals to withstand the cold, evidence must be found showing that large size could or does provide a survival advantage under thermal conditions likely to be or to have been encountered, and there must be evidence from physics that large size functions either to shield the body from the cold or reduces the rate of heat loss from the body. It is important to emphasize that such causal links do not necessarily preclude other explanations, including additional adaptive benefits or a close genetic link with other adaptive traits. Without evidence linking representation with performance through the selective action of agencies that affect survival and reproduction, however, a claim of adaptation cannot be sustained. An understanding of the nature and tightness of linkage between traits is necessary in order to assess the likelihood and potential directions of adaptive change.

A hypothetical example will amplify this point. The secondary metabolites of a plant may be very effective against a particular herbivore, but they cannot be inferred to be adaptive until there is evidence that the herbivore is present in the plant's environment and that this herbivore is adversely affected by the chemicals. Secondary metabolites may have additional benefits and may be linked genetically to other traits affecting the plant's survival or reproduction. Polyphenols (or tannins), for example, are antiherbivorous compounds that often occur in high concentrations in dry-adapted plants. Without a causal link between tannins and desiccation, however, there is no reason to regard the presence of tannins as an adaptation to cope with desiccation.

Williams (1966, 1992) has advocated the view that adaptations be recognized by the criterion of conformity to design specifications. If attributes closely match expectations based on principles of mechanics or economics, they should be considered adaptive. The problem with this approach is that it confuses adaptation (a beneficial trait) with adaptedness (a measure of how effectively the trait functions). Many attributes are adaptive (that is, confer benefits) even if they fall far short of a "good" design. Adaptations, in other words, must be better than alternatives, but they need not be the best possible or even the best that could be cobbled together given the ecological constraints and the historical baggage of their bearers (see also Vermeij, 1987a; Dudley and Gans, 1991). In fact, as Williams (1992) and many others have fully appreciated, imperfection — falling short of the ideal — is very much the rule in the world of organisms. Sterrer's (1992) apt statement that "An organism is a hypothesis of its environment" succinctly captures the idea that adaptation can never be perfect or 100% effective. Organisms adapt to, and in turn influence, their environment. The extent to which an adaptation meets *a priori* specifications is an indication of adaptedness, but it cannot be used to exclude attributes from the category of adaptations.

Given that adaptation is a fundamentally comparative concept, we must be precise about how comparisons are to be made. There are essentially two options, one involving a comparison between co-occurring entities, the other involving a comparison between ancestor and descendant.

If adaptation refers to a current benefit, then the comparison must be done between entities that bear the purported adaptation and those that possess some alternative trait. Compared entities must co-occur (that is, be subjected to a similar environment) and be sufficiently similar that any difference in performance can reasonably be attributed to the presence or absence of the adaptation in question. For adaptations expressed in individual organisms, such contemporaneous comparisons are best done among individuals of the same species. Traits believed to influence the longevity or propagation of populations, species, or higher clades invite the kind of sister-group comparison advocated by Lauder (1981), Mitter et al. (1988), Farrell et al. (1991), Vermeij (1988), and many others. Another reasonable approach is to ascertain the fates over a specific interval of time of each species occurring in a fossil community, and testing hypotheses about whether the most persistent or the evolutionarily most prolific clades differ from clades that are either more prone to extinction or less likely to speciate. These methods have the virtue of comparing the fates among lineages living during a given fixed interval.

The alternative historical approach rests on the argument that an adaptation confers a benefit not enjoyed by the bearer's ancestors, which lack the trait in question. In this view, an adaptation is a derived (as opposed to an ancestral or conserved) trait (see Coddington, 1988).

I believe that this historical view of adaptation is too restrictive. Traits may confer benefits whether they are derived (unique to their bearers) or shared with other entities. Mutations that disrupt highly conserved traits are frequently deleterious, an indication that the traits in question are maintained at least in part by an adaptive mechanism and that they therefore confer contemporary benefits. Moreover, as Reeve and Sherman (1993) point out, it seems arbitrary to exclude traits from the category of adaptations simply because we do not know the history of those traits or their functions. In other words, the adaptational history of a trait should be kept separate from the functions of that trait.

This is not to say that the history of a trait is unimportant. Comparative biology, including phylogenetic analysis of sister-group and ancestor-descendant relationships, enables us to pinpoint when, where, and how adaptations arose; but it cannot identify the adaptations themselves.

In fact, history can occasionally be misleading. Consider, for example, the evolution of body size. Large size, insofar as it is heritable, is adaptive in many mammals because it confers competitive, defensive, and reproductive advantages (Van Valen, 1975). Elephants on small islands in the Mediterranean sea during the Pleistocene were smaller than their ancestral and sister species on adjacent continents. In other words, maximum adult size decreased in island populations.

The derived condition is small size. Does this mean that large size is no longer adaptive? It does not. During colonization of the islands, or when islands first became separated from the mainland, there may have been selection against the largest individuals; in other words, smaller size was adaptive. Eventually, however, a regime of selection in favor of large individuals was reestablished, but this time the distribution of body sizes was offset to a smaller mean. If a historical criterion had been applied in this instance, one might have concluded that large size was always disadvantageous among island elephants. If the criterion of contemporary function is used, large size might have been found to be adaptive. This example illustrates my point that adaptive benefit should not be conflated with the functional history of the trait and that history is an unreliable guide to contemporary function.

The establishment of criteria by which to accept or reject an interpretation of adaptation is essential to any treatment of the phenomenon at any level in the biological hierarchy. Gould and Lewontin's (1979) critique of adaptationism was successful because it exposed the lack of adequate criteria in much previous writing on the subject, but it failed on precisely the same grounds. Gould and Lewontin rejected adaptation-based explanations not because such explanations failed to meet carefully specified criteria, but because Gould and Lewontin wished to draw attention to the possibility that attributes can be conserved or even evolved without there being a functional benefit. This is an important point, whose value would be enhanced if criteria are employed to distinguish between functional and incidental or neutral traits.

III. Adaptation at the Species Level and Above

Most of the literature on adaptation and evolution is concerned with individual organisms. Adaptation at this level in the biological hierarchy explains in large measure why individuals appear to be well suited to their environment and accounts for the directions, if not the rate or timing of evolutionary change (Van Valen, 1983; Vermeij, 1987a, 1994).

As Gould (1982, 1985, 1990) has emphasized, however, much of the history of life is chronicled by the appearance and disappearance of clades. A clade consists of an ancestor and all of its descendants. It may possess traits that affect its evolutionary performance, which is measured as persistence (time interval between origin and extinction) and rate of proliferation of its components (the rate at which new daughter species are produced). The identification of such traits as adaptations of the clade requires that traits be transmitted in some heritable way. It also requires that there be a causal link among the trait, a clade's evolutionary performance, and agencies that potentially bring about or provide opportunities for extinction or speciation.

A. What Are Clade-level Traits?

Adaptations of clades may either be traits expressed in individual organisms or be characteristics of collections of individuals. Lloyd and Gould (1993) distinguish two categories of collective characteristics: emergent traits and aggregate traits. Emergent traits are population-level attributes of the clade as a whole. Examples include population size, population density, population subdivision, temporal variability in a number of individuals, number of species (for more inclusive clades), size of ecological and geographical range, number of breeding sites, and geographical location. Aggregate traits embody differences (heterogeneity) among components of the clade. An example is the variability in feeding habits or in the types of habitat occupied by individual organisms. In other words, the range of food or habits of the species as a whole is greater than that of any component individual. All these collective traits potentially affect a clade's evolutionary performance.

Hoffman (1989) believed that adaptations of clades should be looked for only among emergent or aggregate traits. He would exclude from consideration any traits of individual organisms, even if such traits profoundly affect or determine collective characteristics and the evolutionary fates of clades. Large body size, for example, would not be a clade-level adaptation in Hoffman's view, even though it typically is associated with such emergent properties as small population size, low population density, and low population turnover, all traits that could influence the probability of extinction and speciation of a clade. Hoffman argued that because nearly all collective traits are influenced and to some degree determined by the attributes of individuals, clade-level adaptations based on such aggregate or emergent manifestations are epiphenomena predictable from the properties of individuals. In the language of Vrba and Gould (1986), collective traits that affect a clade's evolutionary performance are sorted, not selected (see also Sober, 1984; Gould, 1990).

Although I agree with Hoffman that most collective characteristics arise from the attributes of individuals, I see no reason to enforce the requirement that clade-level adaptations be restricted to emergent and aggregate traits (see also Vermeij, 1989). If they were, important macroevolutionary phenomena would be missed for purely semantic reasons.

An example will make this point clear. Large body size in mammals and foraminifers (an organism-level trait) is very often favored in selection among individuals, but it is also associated with small populations and low fecundity. Populations of well-adapted large individuals are prone to extinction in part because their small numbers are greatly affected by stochastic changes in population size and because the population cannot respond rapidly to change. Therefore, although population size (an emergent property) is partly reducible to an individual level (body size), the evolutionary effects of the traits at these two levels of the genealogical hierarchy are opposite (Van Valen, 1975). The important question is not at which level a given trait is expressed, but whether and how it affects the evolutionary fate of clades (see also Van Valen, 1971, 1975; Williams, 1992).

B. Are Clade-level Traits Heritable?

A potentially significant problem with the interpretation of collective traits as adaptations is the heritability of those traits. Unlike many traits of individual organisms, for which well-known mechanisms of transmission exist, the collective traits of clades have no readily identifiable means of inheritance.

To my knowledge, only two studies have probed the heritability of collective traits. In both cases, the geographical range of species within clades was concluded to be heritable. Jablonski's (1987) study involved late Cretaceous gastropod and bivalve molluscs from the Atlantic and gulf coastal plains of the United States, whereas Ricklefs and Latham's (1992) analysis dealt with species in disjunctly distributed genera of herbs in eastern North America and eastern Asia. In order to establish heritability of geographical range, Jablonski compared the geographical ranges (measured in kilometers of coastline) of what he interpreted to be pairs of sister species. By calculating several nonparametric correlation coefficients, he showed that sister species tend to have geographical ranges more similar in magnitude than would randomly selected pairs of species. Moreover, for 15 out of 16 clades of 15 or more species, geographical range was tied to temporal persistence, with the more widely distributed species tending to have longer geological durations. According to Jablonski (1987), geographical range can be interpreted as a trait that satisfies all three conditions for adaptation — variation, heritability, and causal connection with fitness (species longevity in this case).

Unfortunately, Jablonski does not make clear whether the sister species, which he identified on the basis of statements in the descriptive taxonomic literature rather than from phylogenetic analysis, were allopatric (with nonoverlapping ranges) or at least partly sympatric (co-occurring in part of the range). If they were sympatric, the similarity in the magnitude of range could reflect either a conservative physiology (the same tolerances to temperature and salinity extremes, for example) or a similar response to local oceanographical conditions (comparable dispersibility and larval type, for example). Because such physiological and larval traits are expressed at the level of the individual mollusk, the collective property of the geographical range would be a species-level extension of heritable individual-level traits. If the sister species were allopatric, the similarity of sizes of geographical ranges would be less easily explained as the clade-level manifestation of an individual-level attribute.

Ricklefs and Latham's (1992) study uncovered heritability of geographical range among herbs, but not among comparable tree genera that have disjunct distributions in warm-temperate North America and Asia. Conservation of range in herbs may be enforced by unchanging physiological tolerances in the two regions. Ricklefs and Latham (1992) argued that whereas woody plants may evolve as climates vary over time, the ecologically more specialized herbs simply track the habitats to which they are already well adapted and therefore do not change physiological tolerances.

These studies may be criticized on several grounds. In neither case were geographical ranges studied in an explicitly phylogenetic context. The inference that species pairs were sister taxa may often have been correct, especially for the herbs, but no data were presented to make an independent evaluation of such relationships possible. In Jablonski's (1987) study, the geographical area covered was large, yet we know nothing about whether many of the species extended significantly beyond the geographical limits of the study, which were imposed by the absence of fossil-bearing sediments. This problem could easily be circumvented by repeating the analysis for the Recent fauna. Surprisingly, this has not been done. I suspect that such an analysis would reveal much lower, or perhaps no, heritability within many molluscan clades. In many gastropod families, such as the Turritellidae, Muricidae, Nassariidae, and Turridae, for example, widely distributed ancestors with planktonic larval dispersal stages have repeatedly given rise to narrowly distributed species in which the planktonic larval stage has been replaced by a benthic (or nonplanktonic) one (see, e.g., Ponder and Vokes, 1988; Bouchet, 1990; Lieberman et al., 1993; Gili and Martinell, 1994). As Russell and Lindberg (1988) point out, narrowly distributed species are easily overlooked in the fossil record because among Recent taxa there is a correlation between local abundance and the size of geographic range.

I conclude that the heritability of geographical range, insofar as the two studies pertaining to it have demonstrated it, arises largely as a species-level manifestation of traits that are inherited by individual organisms. This conclusion is likely to apply to other collective traits as well. A small population size, which increases a population's vulnerability to extinction and to some kinds of speciation, is a frequent population-level manifestation of such individual-level traits as large body size, high metabolic rate, a carnivorous diet, close resemblance to noxious or dangerous models in a system of mimicry, and pollination by highly mobile animal vectors. These traits are heritable, and if they are conserved during episodes of speciation, their population-level manifestations will also appear to be heritable.

C. Controls on Extinction and Speciation

Insight into the nature of adaptation of clades requires knowledge of the causes and conditions of extinction and speciation. Extinction occurs when all members of a clade die or when none is capable of leaving offspring. Something in the environment changes so rapidly or so radically that individuals and populations are unable to respond. This lack of response could have several causes. First, there may be insufficient genetic variation among individuals so that there is little raw material on which natural selection can act. This would be especially true for small populations because inbreeding among diploid individuals often forms lethal homozygotes of alleles that in a heterozygous condition could persist and contribute to a population's genetic response to a crisis (see Simberloff, 1986). Second, individual organisms may be unable to respond phenotypically by learning, by reducing the harmful effects of the change, or by altering their physiology or morphology in suitable directions. Finally, even if there were sufficient genetic variation, there might be insufficient time for recombination and natural selection to produce adapted individuals. If populations occur at low densities, many individuals will fail to find mates and therefore cannot contribute the heritable variants they possess to the next generation.

A species or population can persist during a crisis if one or more of the following conditions are met. (1) Some habitats to which members of the population were adapted are preserved within the original range of the population. In this case, no adaptive change is necessary. (2) There is sufficient heritable variation, or a mechanism to generate and maintain such variation, as well as enough time for variations to be recombined and selected. (3) Individuals are capable of responding by nongenetic means, including flexibility in physiology, morphology, and behavior. Persistence is most likely when populations are large (consisting of many reproducing individuals) and dense, geographically widespread, and ecologically diverse. The magnitude of change that would cause extinction would have to be very large in such cases. Small sparse populations of limited geographical and ecological range are most vulnerable to extinction, especially if individuals are phenotypically inflexible and if genetic variants are not exposed to selection.

It is important to distinguish between the characteristics that enable populations to resist extinction and the characteristics of populations that do not experience crises in the first place. Populations whose members are not affected by crises also persist, but persistence in such cases is by virtue of living in unchanged environments rather than successfully coping with change. I thus distinguish between resistance to extinction and refugial persistence. If adaptations against extinction exist, they must have the effect of providing resistance to crises.

Speciation occurs when part of a population becomes isolated from the parent population and then diverges. Isolation may be effected either by the extinction of populations linking the would-be isolate from the parent or by the dispersal of individuals to sites not previously occupied by the parental stock. Subsequent divergence involves a combination of the founder effect (the genetic constitution of the founding members of the isolate), genetic drift (stochastic changes in gene frequencies, especially characteristic of small populations), and selection in an environment different from that of the parent population. Speciation is thus most likely when the environment changes (leading to local extinction or to dispersal) and when individuals and populations are capable of response to change.

D. Ecological and Geographical Range

One of the earliest clade-level traits to be recognized as influencing evolutionary performance is ecological breadth. Although individuals may each occupy a relatively small number of habitats, the species as a whole often extends over a wide range of habitats and is then said to be eurytopic. Stenotopic species, by contrast, have a narrow ecological range. Empirical studies have associated eurytopy in molluscs and mammals with long persistence and with a low rate of speciation (Jackson, 1974; Hansen, 1978; Vrba, 1980). Stenotopic species have higher probabilities of extinction and may be more likely to speciate.

Related claims have been made about the effects of another collective trait, geographical range. Wide-ranging species are less susceptible to extinction than those with smaller geographical distributions (Jablonski, 1986; Vermeij and Petuch, 1986; Stanley, 1986a, b). These patterns are usually explained by asserting that species with broad ecological or geographical ranges generally have larger populations than those with smaller distributions. Regional or habitat-specific disasters could eliminate narrowly ranging populations, but would not bring down more broadly distributed ones.

Whether the aggregate trait of eurytopy can legitimately be considered adaptive must await analyses of its heritability. No tests of heritability of eurytopy have yet been undertaken. It is hard to see how collective ecological breadth can be reduced to traits of individual organisms, although one possibility is that it arises from phenotypically flexible responses by individuals to environmental change.

From this analysis it is clear that speciation and extinction are more likely when the population size is small and when conditions change. Circumstances that fragment populations therefore increase the likelihood of extinction as well as of founder speciation. This constitutes Stanley's (1979, 1986a, 1990) explanation for his observation that clades with high rates of extinction of component species also show high rates of species formation.

Susceptibility to speciation may not, however, always imply vulnerability to extinction. Some traits may, in fact, make it easier for populations to become isolated and to diverge as well as to provide resistance during crises.

Some evidence, for example, links eurytopy and a broad geographical range with high rates of species formation. Clades of tropical Pacific gastropods and of tropical American hummingbirds in which individuals are capable of wide dispersal are much richer in species than are clades whose members have sedentary habits and smaller geographical ranges (Vermeij, 1987b; Bleiweiss, 1990). Dispersibility provides many opportunities for the establishment of new daughter populations, which may then diverge rapidly enough to attain species status. At the same time, the ability to disperse widely is associated with a low probability of extinction.

At a higher level in the hierarchy of clades, eurytopy may also be associated with high rates of species formation. Clades of flowering plants with a high diversity of growth forms (trees, shrubs, herbs, and vines), modes of pollination, and methods of dispersal are richer in species than are clades whose species fall into only one group within each of these categories (Ricklefs and Renner, 1994). Again, it is unclear whether this kind of eurytopy can be considered adaptive because nothing is known about its heritability.

E. Variation and Responsiveness

Perhaps most important among traits providing resistance to extinction and susceptibility to species formation are those that affect the ability of individuals and populations to respond to change. Genetic responsiveness is enhanced when a larger amount of genetic variation is incorporated into an individual organism's genome and when more of the variation is exposed to natural selection. Buss (1987, 1988) noted that animal clades in which isolation (or sequestration) of the germ-line cells from other (somatic) cell types occurs late in ontogeny tend to be much richer in species than groups in which isolation occurs early. Late sequestration enables more heritable variation to be introduced through mutation into the germ-line as cells divide. Opportunity for variants to be incorporated is limited if few such cell divisions occur. Long asexual generations punctuated by sexual phases also allow the genome to pick up heritable variants.

Genetic variation is of little use unless it is subject to selection. If variation is selectively neutral or if births and deaths in a population occur randomly with respect to heritable variation, selection has little raw material on which to act. In an important and mostly overlooked paper, Mulcahy (1979) pointed out that there is much more intense selection on haploid microgametophytes in insect-pollinated flowering plants with closed carpels than in wind-pollinated or spore-bearing plants. In the latter plants, the time of arrival rather than intrinsic properties of the gametophyte is the most important factor determining whether pollination will occur. In insect-pollinated flowering plants, by contrast, it is the microgametophyte's growth rate of the pollen tube through the stylar tissues that is important. This growth rate is determined in part by the genes of the haploid gametophyte. Because many pollen grains land simultaneously on the stigma of insect-pollinated flowers, there is intense competition among pollen tubes. There is also evidence that growth rates of the gametophytes are correlated with the subsequent performance of fertilized diploid sporophytes. The important point is that the heritable variation existing among the gametophytes of insect-pollinated plants is subjected to intense selection, whereas any such variation existing in wind-pollinated or spore-bearing plants is largely neutral with respect to plant fitness.

I believe that a similar argument applies to fertilization in animals. In groups in which fertilization is external to the body, encounters between eggs and sperm are largely a matter of chance, or at least of proximity. Individuals can time releases of eggs and sperm so that these will coincide, but there is little opportunity for selection to operate on fitness-related variations in performance. In animals with internal fertilization or in those where release of gametes into the water occurs only in the presence of a particular mate, on the other hand, such opportunities are numerous. Not only must mates actively seek each other out, but there is an opportunity for them to assess each other's potential. Variations in mate attractiveness and performance in survival and reproduction are increasingly under the influence of selection and decreasingly subject to the vagaries of chance. Competition is extended to resources (mates) that essentially did not exist previously, and a whole new dimension along which selection takes place is created when fertilization becomes internal.

In fact, insect pollination in plants is a special case of internal fertilization. As in animals, it opens the possibility for selection related to mate choice. West-Eberhard (1983) has already drawn attention to the fact that groups in which sexual selection occurs are often highly diverse. Variation that in the absence of sexual selection might well be neutral takes on selective significance because it affects the recognition and attractiveness of mates. If cultural or genetic changes affecting mate choice occur in isolated populations, the latter are apt to diverge from parent populations. Such opportunities are far less likely to arise in the absence of sexual selection.

Another important consequence of internal fertilization is that populations can persist at low density. Raven (1977) and Regal (1977) pointed out that wind-pollinated plants must exist at high densities in order for plants to fertilize each other, whereas animal-pollinated species can exist at low densities because they rely on mobile vectors to shuttle between members of the same species. Similar arguments could apply to animals with internal fertilization as compared to those in which the union of gametes takes place by encounters outside the body. In both cases, dispersibility (by seeds or larvae) further adds to the capacity of populations to maintain viable populations at low densities (Regal, 1977). In various analyses of diversity patterns in flowering plants, the mode of dispersal has generally not been correlated with species numbers among families (Herrera, 1989; Ricklefs and Renner, 1994), but animal pollination is often associated with high species number (Eriksson and Bremer, 1992; Ricklefs and Renner, 1994). It may be that the effect of animal pollination on diversity is smaller than that of the capacity within families to take on different growth forms (Ricklefs and Renner, 1994), but the effect is still significant. The important point is that although low population density would normally place a species at a high risk of extinction, such a risk is substantially decreased for clades with internal fertilization because mating and the recombination of heritable variants are much more efficient and because more of the variants are exposed to selection. Such clades are more resistant to extinction and are also more likely to proliferate.

The expression of individuality is another manifestation of the ability to respond to change. At the level of individual organisms, this flexibility is achieved by phenotypic responses and by learning. Genetic individuality is made possible by recombination and sexual reproduction. As Sterrer (1992) points out, the fact that individuals differ from each other genetically as well as in their nongenetic (phenotypic) capacity not only increases the range of conditions under which individuals can thrive, but also makes the population or the species to which individuals belong less predictable for competitors. Individuality thus confers a collective advantage of diversity, which protects a target population from enemies that seek members as sources of energy (Sterrer, 1992). On longer time scales, diversity protects against events that occur too infrequently to be predictable by or adapted to by individual organisms.

At still higher levels in the genealogical hierarchy, diversity is manifested as flexibility in the rules governing the development of individuals. If the biological structure or behavior of an organism is so highly integrated that a change in one component has inevitable cascading consequences for many other components, only a small number of evolutionary pathways is available along which individuals and populations can change in response to changing circumstances. If, however, rules of construction or rules of behavior are flexible so that components can vary independently of one another, the range of potential responses is greatly increased. This so-called versatility (Vermeij, 1973) can be achieved when there are loose linkages among morphological traits or among the genes that encode them (see also Rosenzweig et al., 1987; Rosenzweig and McCord, 1991). Complex systems, such as those characterizing the organization of multicellular forms of life, have the capacity for both high integration and great versatility (Kauffman, 1993). Versatility is made possible when the ecological constraints that enforce stabilizing selection and that thus maintain the adaptive status quo are relaxed or removed and when the per capita availability of energy increases (Vermeij, 1987a). The macroevolutionary potential to respond thus requires the appropriate built-in flexibility inherent in a complex system as well as the right extrinsic conditions of reduced stabilizing selection and greater access to energy.

F. Clade-level Adaptation and the Historical Record

In the preceding section I have tried to show that clades can and do possess adaptations. Although most of these are expressed and inherited at the level of individual organisms, they have strong emergent advantages with respect to the persistence and proliferation of populations, species, and higher-level clades. In many cases, the status of clade-level adaptations requires formal comparative tests of the kind outlined in the first section of this chapter, but patterns uncovered thus far make adaptive interpretations highly plausible.

What, then, is the role of clade-level adaptations in the history of life? Are clades living today better adapted than clades of the past? Do clade-level adaptations conflict with traits that are beneficial only to individual organisms?

It has been known for some time that the rate of extinction of families of animals has generally declined through the Phanerozoic (Raup and Sepkoski, 1982; Van Valen, 1984, 1985; Benton, 1985; Valentine et al., 1991; Labandeira and Sepkoski, 1993). Mass extinctions have punctuated but not fundamentally altered this trend. On the other hand, vascular land plants are said to show the opposite pattern, in which more recently derived families are shorter lived (Niklas et al., 1983; Valentine et al., 1991).

Decreasing rates of extinction among families could be interpreted to mean that there have been improvements in the evolutionary performance of clades through time. Clades prone to extinction have, according to this interpretation, been replaced by extinction-resistant clades (Sepkoski, 1984; Flessa and Jablonski, 1985). This interpretation is consistent with the general increase in biological diversity through the Phanerozoic.

Such an increase (Signor, 1990) means that the number of extinctions has generally lagged behind the number of speciation events (see also Cracraft, 1985; Hoffman, 1989; Vermeij, 1989). The average clade may thus consist of a larger number of components (Valentine, 1969) and consequently be more resistant to extinction.

Such an analysis, however, ignores two important elements. The first is a time-dependent bias in preservation and taxonomic understanding. More recent biotas are both taxonomically better understood and more thoroughly sampled than older ones (Raup, 1979). Rare finds, which are more likely when sampling is more intense, extend the stratigraphic ranges of clades and thus influence estimates of the duration of clades. The second element is ecology. Many inclusive clades, such as Brachiopoda, Crinoidea, and Gymnospermae, have over time become restricted to habitats and regions that have generally been relatively unaffected by extinctions. Components of these clades inhabiting crisis-prone habitats have disappeared. In other words, the clades have changed ecologically (Vermeij, 1987a). If the biosphere has expanded through geological time, and if patterns of restriction to areas where extinction is always rare are widespread, then a greater resistance to extinction would be an inevitable consequence.

Unfortunately, studies of the fates of inclusive clades such as genera and families are uninformative about the fates of species. Moreover, most of the geological record is not good enough to estimate reliably the duration of species or the rate at which species are formed. Consequently, it is impossible to say whether there have been general improvements in evolutionary performance at the species level, even if such improvements did occur at higher levels in the hierarchy.

Even if the inference that more recently evolved clades have lower rates of extinction of their components is accepted, this fact alone cannot account for the observation that diversity has increased through time. Resistance to extinction is, after all, only one of two aspects of a clade's evolutionary performance. The other is proliferation. Ancient clades such as *Sphenodontia* (tuataras), *Limulida* (horseshoe crabs), and *Equisetales* (horsetails) offer compelling cases of persistence in the absence of significant diversification. It is persistence in concert with the capacity to form new lineages that sets more recently evolved clades apart from ancient ones.

The traits that are associated with the potential for diversification — sexual reproduction, internal fertilization, long interval between fertilization and germ-line sequestration, insect pollination, closed carpels, high metabolic rates, and body-plan versatility — are all derived conditions. So are traits that enable individuals to learn about, respond to, and predict the environment. There have, of course, been many reversals in these traits. Thus, obligate asexual reproduction (parthenogenesis) has evolved multiple times in sexually reproducing clades, but generally the groups characterized by this mode of reproduction are evolutionarily short lived or contain few species (Stanley, 1975b; Williams, 1975). Similarly, many gastropods in which the planktonic larval stage is secondarily lost have earlier sequestration of the germ-line than did their ancestors, and have as a result become less prone to founder speciation (Lieberman et al., 1993). These exceptions do not, however, invalidate the general trend for the more recently evolved inclusive clades to possess traits enabling these clades to form new lineages at high rates.

This evolutionary potential has been achieved in part because the usual risks of extinction associated with small populations have been compensated for by more efficient means of generating, conserving, and expressing variation as well as by responding individually to environmental change. Collectively and individually, and at many levels in the genealogical hierarchy, entities not only resist change, but more importantly they respond to it and, in many cases, even bring about changes themselves. Modern clades may be better hypotheses of their environments than were ancient ones. They have achieved this by virtue of adaptations, mainly expressed and inherited at the level of individual organisms.

If this interpretation is correct, we are faced with a curious apparent contradiction. The evolutionary performance of individual organisms may not have changed greatly over geological time, the environment in which they live may have become more rigorous or more unpredictable, and the adaptations that individuals have evolved to cope with this change may be more specialized and more energy-demanding than those of individuals in the past. But there is no compelling evidence that there has been, in this sense, evolutionary progress at the level of individual organisms (Vermeij, 1987a). Yet, at the level of clades (especially among the more inclusive ones), there is evidence of improvement in evolutionary performance, both in terms of resistance to extinction and potential for species formation. Such clade-level progress has been achieved through a synergistic combination of adaptation at the individual level and of collective adaptations related to the creation and expression of variation and, more generally, to the evolution of individuality (see also Buss, 1987; Sterrer, 1992).

This view of macroevolution is very different from that of Gould (1985) and many other observers. Gould believes that clade-level processes of sorting (that is, differential extinction and differential formation of species) are more important than, as well as generally antagonistic to, the process of natural selection that leads to organism-level adaptation. It is certainly true that, as the analysis of large body size in mammals exemplifies (Van Valen, 1975), a given trait may have benefits at one level in the hierarchy and disadvantages at another; but the characteristics that have the greatest macroevolutionary effects also provide benefits to individual organisms, and vice versa. This applies especially to traits that enable individuals and populations to respond to change. I suspect that Gould's views were influenced by traits whose main effect on clades is to confer resistance to extinction. Mine, on the other hand, emphasize the role of traits in the formation of new lineages. As I see it, the latter traits must on balance be more important than extinction-related ones because the diversity of life has risen generally, if unevenly, over the course of time.

The arguments marshaled in this chapter strongly imply that macroevolutionary processes of sorting are not, as some authors have claimed, decoupled (independent from) processes operating among individual organisms (see, e.g., Stanley, 1975a, 1979; Gould, 1982). Instead, there is a consistent and unifying theme of evolution discernible at all levels of the genealogical hierarchy, a theme of coping with and responding to environmental change. The adaptations of individual organisms are appropriate to commonplace events that affect the lives and deaths of organisms, whereas the adaptations of clades, though usually based on individual traits, are appropriate to the rare events that affect the formation and disappearance of species or higher-level clades. The basic requirements of coping with or responding to events remain the same. Some individuals must have the phenotypic and genetic resources to resist, and there must be some ability by individuals and populations to respond to change. The main, and perhaps only, difference between adaptation of organisms and adaptation of clades is the time scale. Adaptation with respect to crises or opportunities can occur only if such events take place at sufficient frequency to be encountered at least once during the average lifetime of the entity bearing the adaptation in question [see Hutchinson (1953) for a similar analysis of time scale in the adaptation of short-lived opportunistic individuals and competitively superior long-lived ones].

The history of evolutionary biology during much of the 20th century has been characterized by a wide gulf between those who study evolution and its mechanisms over very short time scales and those who approach it from the perspective of geological time. Perhaps these two widely differing conceptions can be reconciled if it is appreciated that evolution and adaptation operate on many different time scales and that the fundamental principles of selective survival and proliferation in a world of crises and opportunities are broadly applicable at all time scales.

References

Benton, M. J. (1985). Mass extinction among non-marine tetrapods. *Nature* (London) 316, 811-814.

Bleiweiss, R. (1990). Ecological causes of clade diversity in hummingbirds: a neontological perspective on the generation of diversity. *In* "Causes of Evolution: A Paleontological Perspective" (R. M. Ross and W. B. Allmon, eds.), pp. 354-380. University of Chicago Press, Chicago.

Bouchet, P. (1990). Turrid genera and mode of development: the use and abuse of protoconch morphology. *Malacologia* 32, 69-77.

Buss, L. W. (1987). "The Evolution of Individuality." Princeton University Press, Princeton.

Buss, L. W. (1988). Diversification and germ-line sequestration. *Paleobiology* 14, 313-321.

Coddington, J. A. (1988). Cladistic tests of adaptational hypotheses. *Cladistics* 4, 3-22.

Cracraft, J. (1985). Biological diversification and its causes. *Ann. Missouri Bot. Garden* 72, 794-822.

Dudley, R., and Gans, C. (1991). A critique of symmorphosis and optimality models in physiology. *Physiol. Zool.* 64, 627-637.

Eriksson, O., and Bremer, B. (1992). Pollination systems, dispersal modes, life forms, and diversification rates in angiosperm families. *Evolution* 46, 258-266.

Farrell, B. D., Dussourd, D. E., and Mitter, C. (1991). Escalation of plant defense: do latex and resin canals spur plant diversification? *Amer. Nat.* 138, 881-900.

Flessa, K. W., and Jablonski, D. (1985). Declining Phanerozoic background extinction rates: effect of taxonomic structure? *Nature* (London) 313, 216-218.

Gili, C., and Martinell, J. (1994). Relationship between species longevity and larval ecology in nassariid gastropods. *Lethaia* 27, 291-299.

Gould, S. J. (1982). Darwinism and the expansion of evolutionary theory. *Science* 216, 380-387.

Gould, S. J. (1985). The paradox of the first tier: an agenda for paleobiology. *Paleobiology* 11, 2-12.

Gould, S. J. (1990). Speciation and sorting as the source of evolutionary trends, or "things are seldom what they seem." *In* "Evolutionary Trends" (K. J. McNamara, ed.), pp. 3-27. University of Arizona Press, Tucson.

Gould, S. J., and Lewontin, R. C. (1979). The spandrels of San Marco and the Panglossian paradigm: a critique of the adaptationist programme. *Proc. Roy. Soc. Lond.* (B) 205, 581-598.

Gould, S. J., and Vrba, E. S. (1982). Exaptation - a missing term in the science of form. *Paleobiology* 8, 4-15.

Hansen, T. A. (1978). Larval dispersal and species longevity in Lower Tertiary gastropods. *Science* 199, 885-887.

Herrera, C. M. (1989). Seed dispersal by animals: a role in angiosperm diversification? *Amer. Nat.* 133, 309-322.

Hoffman, A. (1989). "Arguments on Evolution: A Paleontologists's Perspective." Oxford University Press, New York.

Hutchinson, G. E. (1953). The concept of pattern in ecology. *Proc. Acad. Nat. Sci. Philadelphia* 105, 1-12.

Jablonski, D. (1986). Larval ecology and macroevolution in marine invertebrates. *Bull. Mar. Sci.* 39, 565-587.

Jablonski, D. (1987). Heritability at the species level: analysis of geographical ranges of Cretaceous mollusks. *Science* 238, 360-363.

Jackson, J. B. C. (1974). Biogeographic consequences of eurytopy and stenotopy among marine bivalves and their evolutionary significance. *Amer. Nat.* 108, 541-560.

Kauffman, S. A. (1993). "The Origins of Order: Self-Organization and Selection in Evolution." Oxford University Press, New York.

Labandeira, C. C., and Sepkoski, J. J., Jr. (1993). Insect diversity in the fossil record. *Science* 261, 310-315.

Lauder, G. V. (1981). Form and function: structural analysis in evolutionary morphology. *Paleobiology* 7, 430-442.

Lauder, G. V. (1995). On the inference of function from structure. *In* "Functional Morphology in Vertebrate Paleontology" (J. Thomason, ed.), pp. 1-18. Cambridge University Press, Cambridge.

Lieberman, B. S., Allmon, W. B., and Eldredge, N. (1993). Levels of selection and macroevolutionary patterns in the turritellid gastropods. *Paleobiology* 19, 205-215.

Lloyd, E. A., and Gould, S. J. (1993). Species selection on variablility. *Proc. Natl. Acad. Sci. U. S. A.* 90, 595-599.

Maynard Smith, J. (1991). A Darwinian view of symbiosis. *In* "Symbiosis as a Source of Evolutionary Innovation: Speciation and Morphogenesis" (L. Margulis and R. Fester, eds.), pp. 26-39. MIT Press, Cambridge.

Mitter, C., Farrell, B. D., and Wiegmann, B. (1988). The phylogenetic study of adaptive zones: has phytophagy promoted insect diversification? *Amer. Nat.* 132, 107-128.

Mulcahy, D. L. (1979). The rise of the angiosperms: a genecological factor. *Science* 206, 20-23.

Niklas, K. J., Tiffney, B. H., and Knoll, A. H. (1983). Patterns in vascular land plant diversification. *Nature* (London) 303, 614-616.

Ponder, W. F., and Vokes, E. H. (1988). A revision of the Indo-West Pacific fossil and Recent species of *Murex* s. s. and *Haustellum* (Mollusca: Gastropoda: Muricidae). *Rec. Aust. Mus. Suppl.* 8, 1-160.

Raup, D. M. (1979). Biases in the fossil record of species and genera. *Bull. Carnegie Mus. Nat. Hist.* 13, 85-91.

Raup, D. M., and Sepkoski, J. J., Jr. (1982). Mass extinctions in the marine fossil record. *Science* 215, 1501-1503.

Raven, P. H. (1977). A suggestion concerning the Cretaceous rise to dominance of the angiosperms. *Evolution* 31, 451-452.

Reeve, H. K., and Sherman, P. W. (1993). Adaptation and the goals of evolutionary research. *Quart. Rev. Biol.* 68, 2-32.

Regal, P. J. (1977). Ecology and evolution of flowering plant dominance. *Science* 196, 622-629.

Ricklefs, R. E., and Latham, R. E. (1992). Intercontinental correlation of geographical ranges suggests stasis in ecological traits of relict genera of temperate perennial herbs. *Amer. Nat.* 139, 1305-1321.

Ricklefs, R. E., and Renner, S. S. (1994). Species richness within families of flowering plants. *Evolution* 48, 1619-1636.

Rosenzweig, M. L., and McCord, R. D. (1991). Incumbent replacement: evidence of long-term evolutionary progress. *Paleobiology* 17, 202-213.

Rosenzweig, M. L., Brown, J. S., and Vincent, T. L. (1987). Red Queens and ESS: the coevolution of evolutionary rates. *Evol. Ecol.* 1, 59-94.

Russell, M. P., and Lindberg, D. R. (1988). Real and random patterns associated with molluscan spatial and temporal distributions. *Paleobiology* 14, 322-330.

Sepkoski, J. J., Jr. (1984). A kinetic model of Phanerozoic taxonomic diversity. III. Post-Paleozoic families and mass extinctions. *Paleobiology* 10, 246-267.

Signor, P. W. (1990). The geologic history of diversity. *Annu. Rev. Ecol. Syst.* 21, 509-539.

Simberloff. D. S. (1986). The proximate causes of evolution. *In* "Patterns and Processes in the History of Life" (D. M. Raup and D. Jablonski, eds.), pp. 259-276. Springer, Berlin.

Sober, E. (1984). "The Nature of Selection: Evolutionary Theory in Philosophical Focus." MIT Press, Cambridge.

Stanley, S. M. (1975a). A theory of evolution above the species level. *Proc. Natl. Acad. Sci. U. S. A.* 72, 646-650.

Stanley, S. M. (1975b). Clades versus clones in evolution: why we have sex. *Science* 190, 382-383.

Stanley, S. M. (1979). "Macroevolution: Pattern and Process." W. H. Freeman, San Francisco.

Stanley, S. M. (1986a). Population size, extinction, and speciation: the fission effect in Neogene Bivalvia. *Paleobiology* 12, 89-110.

Stanley, S. M. (1986b). Anatomy of a regional mass extinction: Plio-Pleistocene decimation of the western Atlantic bivalve fauna. *Palaios* 1, 17-36.

Stanley, S. M. (1990). The general correlation between rate of speciation and rate of extinction: fortuitous causal linkages. *In* "Causes of Evolution: A Paleontological Perspective" (R. M. Ross and W. B. Allmon, eds.), pp. 103-127. University of Chicago Press, Chicago.

Sterrer, W. (1992). Prometheus and Proteus: the creative, unpredictable individual in evolution. *Evol. Cognit.* 1, 101-129.

Valentine, J. W. (1969). Patterns of taxonomic and ecological structure of the shelf benthos during Phanerozoic time. *Palaeontology* 12, 684-709.

Valentine, J. W., Tiffney, B. H., and Sepkoski, J. J., Jr. (1991). Evolutionary dynamics of plants and animals: a comparative approach. *Palaios* 6, 81-88.

Van Valen, L. (1971). Group selection and the evolution of dispersal. *Evolution* 25, 591-598.

Van Valen, L. (1975). Group selection, sex, and fossils. *Evolution* 29, 87-93.

Van Valen, L. (1983). How pervasive is coevolution? *In* "Coevolution" (M. H. Nitecki, ed.), pp. 1-19. University of Chicago Press, Chicago.

Van Valen, L. (1984). A resetting of Phanerozoic community evolution. *Nature* (London) 307, 50-52.

Van Valen, L. (1985). How constant is extinction? *Evol. Theory* 7, 93-106.

Vermeij, G. J. (1973). Adaptation, versatility, and evolution. *Syst. Zool.* 22, 466-477.

Vermeij, G. J. (1987a). "Evolution and Escalation: An Ecological History of Life." Princeton University Press, Princeton.

Vermeij, G. J. (1987b). The dispersal barrier in the tropical Pacific: implications for molluscan speciation and extinction. *Evolution* 41, 1046-1058.

Vermeij, G. J. (1988). The evolutionary success of passerines: question of semantics? *Syst. Zool.* 37: 69-71.

Vermeij, G. J. (1989). Evolution in the long run. *Paleobiology* 15, 199-203.

Vermeij, G. J. (1994). The evolutionary interaction among species: selection, escalation, and coevolution. *Annu. Rev. Ecol. Syst.* 25, 219-236.

Vermeij, G. J., and Petuch, E. J. (1986). Differential extinction in tropical American molluscs: endemism, architecture, and the Panama Land bridge. *Malacologia* 27, 29-41.

Vrba, E. S. (1980). Evolution, species and fossils: how does life evolve? *South Afr. J. Sci.* 76, 61-84.

Vrba, E. S., and Gould, S. J. (1986). The hierarchical expansion of sorting and selection: sorting and selection cannot be equated. *Paleobiology* 12, 217-228.

West-Eberhard, M. J. (1983). Sexual selection, social competition, and speciation. *Quart. Rev. Biol.* 581, 155-183.

Williams, G. C. (1966). "Adaptation and Natural Selection: A Critique of Some Current Evolutionary Thought." Princeton University Press, Princeton.

Williams, G. C. (1975). "Sex and Evolution." Princeton University Press, Princeton.

Williams, G. C. (1992). "Natural Selection: Domains, Levels, and Challenges." Oxford University Press, New York.

CHAPTER 12

Adaptation in Subdivided Populations:
Kin Selection and Interdemic Selection

MICHAEL J. WADE

I. Introduction

Natural populations of most species tend to be subdivided to varying degrees by geographic and physical barriers which constrain the movements of individuals. Limits to the dispersal of individuals and restricted gene flow between demes result in the clustering or clumping of individuals and genetic heterogeneity on a local spatial scale (Fig. 1). Moreover, habitat choice and conspecific queuing in the context of a patchy distribution of resources can result in the active aggregation of individuals. Whenever resources are patchily distributed in space, it is not at all unusual to find the organisms exploiting those resources also patchily distributed. Individuals within a patch of resource are more likely to interact with, compete with, and mate with one another than with individuals from other patches. In contrast, the membership of a nonsubdivided population is continuously and uniformly distributed like grass on a lawn or the equilibrium configuration of ideal gas molecules in a finite space. Most theoretical and experimental studies of adaptive processes have been conducted using the latter model of an idealized, nonsubdivided population, and the effects of population subdivision on adaptive processes have not been as extensively studied.

The spatial contagion in the activities and breeding habits of individuals can cause novel ecological and evolutionary processes to occur. For example, it is common in many species for females to aggregate at resources just prior to breeding. This tendency can be quantified using Lloyd's measure of microspatial aggregation, mean crowding (Lloyd, 1967). Whenever males attend these aggregations of females to mate, a process of sexual selection ensues where the intensity of sexual selection in males (Wade, 1979a; Wade and Arnold, 1980) is equivalent to the mean crowding of females (Wade, 1995). Furthermore, this sexual selection on males can lead to a runaway process in which females evolve to

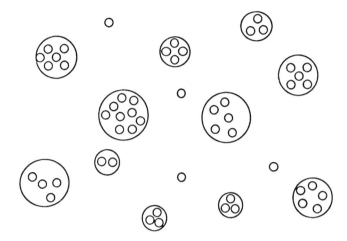

Figure 1 A schematic representation of a subdivided population. The smallest circles represent groups of related individuals such as families or matrilineages. The larger circles represent demes consisting of one to nine kin groups with random mating occurring across groups but within the deme.

overaggregate on resources beyond the ecological optimum (Wade, 1995). The overaggregation of females can itself open up a "mating niche," permitting the invasion of an alternative male mating strategy (Shuster and Wade, 1991; Wade and Dixon, in press). In this example, the simple tendency of individuals to move toward and aggregate at resources initiates a cascade of evolutionary events which ultimately influences the tendency to aggregate itself.

A second example of the novel effects of population subdivision involves the interaction of ecological processes and the genetic basis of adaptations. In spatially mosaic populations, local patches may fluctuate in numbers more or less independently of one another, depending upon rates of dispersal and dispersal mortality. The local population dynamics and long-term persistence of a species can be very sensitive to the balance among local extinction, dispersal, and colonization of new habitats (Harrison, 1991). In an ecological community, the balance among these processes may stabilize species interactions like predation and competition and permit the "regional coexistence" of species (Slatkin, 1974) that would otherwise exclude one another. At the same time, the colonization, extinction, and dispersal affect the population genetic structure by converting a fraction of the epistatic genetic variance into local additive genetic variance (Goodnight, 1988; Whitlock et al., 1993). This increased additive genetic variance within groups increases the potential for local adaptation. (This process will be discussed in more detail below.)

The genetic structure that emerges from spatial patchiness in mating and reproduction affects the kinds of adaptations that occur, the genetic architecture of those adaptations, the rate of adaptive evolution, and the origin of new species with novel phenotypes. Restricted migration (gene flow) among demes allows local adaptation to fine-scale environmental conditions to become an important force contributing to biodiversity. The greater the degree of genetic subdivision of a population, the more independent are the evolutionary trajectories of the local groups and the more diversifying are the effects of the various evolutionary forces, adaptive and nonadaptive alike. The following sections elaborate on each of these effects of population subdivision on the process of adaptation.

II. The Components of Population Structure: Breeding Groups and Interaction Groups

Population subdivision is often described in terms of hierarchical genetic patterns in the grouping of individuals, and the components of such a hierarchy are referred to as "levels" of population genetic structure. These genetic patterns are responsible, in large part, for the evolutionary consequences of population subdivision. It has proven useful to recognize two biologically important aspects of population structure: breeding groups and interaction groups (Wilson, 1975, 1980; Wade, 1980a). A *breeding group* or *deme* is a randomly mating group of individuals, which is more or less reproductively isolated from other such groups depending upon the amount and pattern of gene flow (Wade, 1982a). An array of such demes connected by migration is a *metapopulation* (Levins, 1970; Hastings and Harrison, 1994). An individual is more likely to mate with another member of its own deme than with a member of some other deme. Demic breeding structure limits the free exchange of genetic information across the metapopulation and, in combination with migration, determines the way in which genetic variation is distributed within and among demes.

A. Breeding Groups

The hierarchical pattern of genetic variation with respect to breeding groups can be quantified using Wright's F statistics (Wright, 1969). These measures describe the fraction of the total genetic variance that exists at a given level in the hierarchy of subdivision. For example, at a single locus with two alleles, the genetic variance among demes, σ_a^2, relative to the total genetic variance, $\sigma_T^2 = p(1-p) = \sigma_a^2 + \sigma_w^2$, is $F_{ST} = (\sigma_a^2)/(\sigma_T^2)$, where σ_w^2 is the average genetic variance segregating within demes. The parameter F has also been interpreted as the "fixation index" because $(1 - F)$ equals the local reduction in heterozygosity, H, due to the demic breeding structure. Lastly, there is a relationship between the genetic variance at one level and the genetic correlation at the lower level. Thus, the genetic variance among demes is equal to the genetic correlation among individuals within demes.

The value of F_{ST} is influenced by a number of factors (see below), primary among them are the effective numbers of breeding adults per deme, N_e, and the rate of gene flow among demes, m. In classic work, Wright (1969) showed that for random migration the equilibrium value of F_{ST} is approximately

$$F_{ST} = 1/(4N_e m + 1).\qquad [1]$$

This relationship between F_{ST} and $N_e m$ can be graphed (Fig. 2) and it is clear that F_{ST} declines rapidly with increasing numbers of migrants between demes. Because genetic variance among demes is a necessary precondition for interdemic

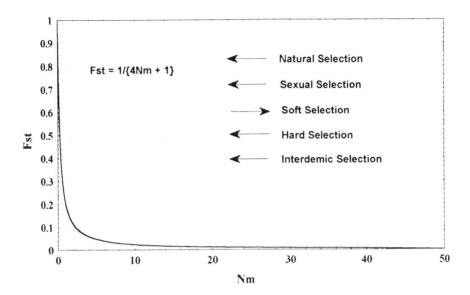

Figure 2   The relationship between F_{ST} and $N_e m$ and the adaptive processes that reduce or enhance the product, $N_e m_e$, and move a species up or down the curve of F_{ST}.

selection (Wade, 1978a; Wade and Goodnight, 1991), F_{ST} is also an important quantity. Figure 3, graphs the relationship between the genetic variance among deme means relative to the total genetic variance among individuals as a function of F_{ST} for an additively determined trait.

Given that its potential range is between 0 and 1, it is tempting to dismiss small values of F_{ST}, in the range of 0.03 to 0.15, as trivial levels of population subdivision. However, two issues are important to consider in evaluating the evolutionary significance of the genetic patterning measured by F_{ST}. The first stems from factors that serve to reduce N_e and m_e below the value of the census of individuals. These factors make N_e and m_e smaller, and hence F_{ST} greater, than expected based upon counting the local numbers of potentially breeding adults. The factors reducing N_e include most adaptive processes (Fig. 2; see below for detailed discussion). The second issue is that the biological importance of small values of F_{ST} is fundamentally an empirical question and not one that can be inferred from inspection of equation [1] or from its graph (Fig. 1).

Variation in N_e Due to Natural Selection: Wright emphasized that several biological factors tend to reduce N_e below the census of locally breeding individuals, N, including variation in offspring numbers, variation in breeding sex ratio, and fluctuations in the size of breeding groups (Wright, 1931, 1969). Individual variation in offspring numbers always accompanies natural selection and variation in family size is important to kin selection. Variation in the breeding sex ratio is almost synonymous with sexual selection (Wade, 1979a, 1995), and fluctuations in the size of local demes are the result of hard selection (Wade, 1985b) and interdemic selection (Wade and Goodnight, 1991). The magnitude of the reduction in N_e can be directly related to the strength of each of these

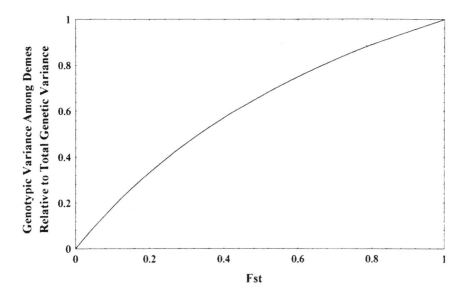

Figure 3 Change in the fraction of genotypic variance among demes in relation to F_{ST}.

adaptive processes.

In infinitely large, unsubdivided populations, linkage between selected loci can slow the rate of adaptation but not its ultimate outcome. In finite, subdivided populations the situation is fundamentally different, even when the loci act additively with respect to fitness (Hill and Robertson, 1966; Barton, 1995). The outcome of selection, expressed as the probability of fixation of other, selected loci, is reduced. This reduction is called the Hill-Robertson effect and is generally expressed as the inevitable reduction in N_e that accompanies natural selection.

The classic derivation of the sampling variance of gene frequencies assumes that the distribution of offspring numbers among parents is random with equal mean, O, and variance, σ_O^2. The general relationship between N_e and N is given by

$$N_e = (N - 1)/([\sigma_O^2/O^2] + [1/O]).\qquad [2]$$

For a stationary population without selection, $O = 2 = \sigma_O^2$, and N_e is approximately equal to N. If exactly two offspring are taken from each parent, so that $\sigma_O^2 = 0.00$, then N_e equals 2N, a fact often used in husbanding laboratory populations to mitigate the loss of genetic variation by random drift.

Whenever adaptation by natural selection is occurring, variance in fitness increases the variation in offspring numbers above random, by definition. The ratio, $[\sigma_O^2/O^2]$, equals the variance in relative fitness, σ_w^2, which determines the strength of natural selection. It was in this sense that Crow (1958, 1962) referred to this quantity, σ_w^2, as the intensity or opportunity for natural selection, I_{ns}. A different way to view this effect, based on ecological measures of crowding, is to say that natural selection reduces N_e below N because offspring numbers are patchily distributed in relation to the selected parents (Lloyd, 1967; Wade, 1995). Thus, equation [2] can be rewritten as

$$N_e = (N - 1)/([\ \sigma_w^2] + 0.5) \qquad\qquad [3a]$$

$$N_e = (N - 1)/([I_{ns}] + 0.5). \qquad\qquad [3b]$$

This effect of adaptation can also be described as a reduction in the probability of fixation of other loci by an amount equal to $(1/[1 + \{\sigma_w^2\}/\{2r^2\}])$, where r is the recombination rate between the two loci (Barton, 1995). The reduction in the probability of fixation at a locus due to the effects of selection on two flanking loci is multiplicative and greater than expected considering the effects separately (cf. Barton, 1995). Thus, selection at one locus increases the strength of random genetic drift at all other loci regardless of linkage. Linkage influences not only the rate but the outcome of adaptive evolution in subdivided populations, and natural selection at one locus "may substantially impede adaptation" at other, more weakly selected loci.

Variation in N_e due to Sexual Selection: Sex dimorphisms can evolve under sexual selection when the variance in male reproductive success exceeds that of females. When some males have multiple mates, the sex difference in the variance in offspring numbers (Wade, 1979a, 1995) is given implicitly by

$$\sigma_{males}^2 = \sigma_{females}^2 + (O^2)(\ \sigma_{mates}^2). \qquad\qquad [4]$$

Usually, selection on one sex is weaker than selection on both sexes because only one-half to one-third (lower values arising with sex linkage) of the genes in the parental generation are being screened by selection. However, the variance in mate number among males, σ_{mates}^2, or the intensity of sexual selection, I_{ss}, causes the sex difference in the strength or intensity of selection. Sexual selection can be much stronger than natural selection, i.e., $I_{ns} < I_{ss}$ (Arnold and Wade, 1984a, b). If we add $\sigma_{females}^2$ to both sides of equation [4] and divide by 2 to obtain the average variance in offspring numbers of parents of both sexes, we find that equations [3] change to

$$N_e = (N - 1)/([\ \sigma_w^2 + 0.5\sigma_{mates}^2] + 0.5) \qquad\qquad [5a]$$

$$N_e = (N - 1)/([I_{ns} + 0.5I_{ss}] + 0.5). \qquad\qquad [5b]$$

Thus, like natural selection, sexual selection reduces effective population size. (Note that the coefficient of I_{ss} is 0.50 because it is a sex-specific form of selection, affecting only the males.) The Hill-Robertson effect is stronger in this case because total selection is stronger.

Variation in N_e due to Soft Selection and Hard Selection: Temporal and spatial variation in the size of local breeding groups can reduce N_e. With temporal variation, N_e equals the harmonic mean local number of breeding adults. The harmonic mean (the reciprocal of the mean of reciprocals) is more heavily influenced by the generations with small numbers of breeding adults than it is by the generations with large ones. In Wright's famous example of this effect, a population that increases from 10 to 1,000,000, growing by a factor of 10 each of six generations, has an effective size equal to 54, less than 1/3000th of the arithmetic mean (185,185).

With soft selection, variation in offspring numbers due to variation in mean deme fitness is diminished by local density regulation (Wade, 1985b; Kelly, 1994a). We would expect that the total variance for any genetic trait, including fitness, in a genetically subdivided population would increase from σ_w^2 to $\{(1 - F_{ST})(\sigma_w^2) + (2 F_{ST})(\sigma_w^2)\}$ or, more simply, $(1 + F_{ST})(\sigma_w^2)$. If all demes begin with the same number of breeding adults, N, and variation among demes in fitness is completely suppressed by local density regulation, then

$$N_e = (N - 1)/([1 - F_{ST}][\sigma_w^2] + 1). \qquad [6]$$

The term $(2 F_{ST})(\sigma_w^2)$ is set equal to zero and N_e is *increased* above N by the factor $1/(1 - F_{ST})$ (Wright, 1943; Nei and Takahata, 1993). As shown earlier, reduction in the total variance in individual fitness, whether due to laboratory husbandry protocols or local density regulation by soft selection, increases N_e. (It is interesting to note in passing that this increase in N_e was viewed by theoreticians as a paradoxical effect of population subdivision. They did not appreciate that, in order to keep N constant after reproduction, despite individual variation, they were implicitly assuming a metapopulation with an ecology of soft selection.)

Under hard selection, although demes consist of equal numbers of breeding adults, N, at the start of each generation, they grow to different sizes due to local variations in mean fitness (Wade, 1985b). If the demes are genetically differentiated with respect to fitness to a degree, F_{ST}, then

$$N_e = (N - 1)/([1 - F_{ST}][\sigma_w^2] + [2 F_{ST}][\sigma_w^2] + 0.5). \qquad [7]$$

With hard selection, N_e is reduced by the factor $(1/[1 + F_{ST}])$ due to the added variance in individual fitness caused by the local genetic differentiation. This assumes that there is no correlation between the fitness of a deme in one generation and its fitness in the next, as would be expected with interdemic selection. If there is such a correlation, the reduction in N_e can be much larger.

Variation in N_e Due to Interdemic Selection: Like hard selection, interdemic selection in Wright's shifting balance theory (abbreviated SBT; Wright, 1931, 1969; Wade and Goodnight, 1991) requires local variation in the mean fitness of demes. Selection occurs by differential dispersion when migrants from demes of high local fitness disperse into demes of lower local fitness. There is a necessary relationship between demic mean fitness and the number of migrants originating in a deme. This relationship increases the individual variation in fitness because the members of high fitness demes have more offspring (as a result of outward migration) and the members of low fitness demes have relatively fewer offspring. Migration is *not random* as was assumed in deriving equation [1]. Furthermore, if the differential dispersal is based on genetic differences among the demes, then we would expect a correlation in demic fitness across generations: a high fitness deme should continue to send out migrants for many generations and a low fitness deme would continue to receive them (as observed experimentally by Wade and Goodnight, unpublished).

The effect of interdemic selection on F_{ST} can be viewed as an increase in the variance in the rate of migration above random and, thus, a lowering of the effective migration rate. The variance in the migration rate is proportional to the variance among demes in mean fitness or, in the terminology of quantitative genetics, to the interdemic selectional differential (Wade and Goodnight, unpublished). The persistence of adaptive peaks or demes with high fitness over generations means that the variance in migration will accumulate so that, from the perspective of neutral genes, F_{ST} will be further increased. It is very difficult to give general analytic expression to this effect of interdemic selection without assuming a specific relationship between migration and demic mean fitness. However, empirically, Wade and Goodnight (unpublished) observed that, on average, F_{ST} in a metapopulation with interdemic selection was 1.3 times greater in a metapopulation with the same average rate of random migration (Wade and Goodnight, 1991).

F_{ST} in Natural Populations: Loveless and Hamrick (1984) in a survey of the plant literature found that a typical value of F_{ST} for an outbreeding species of plants is approximately 0.118 while it averages 0.523 for a predominately selfing species. Eanes and Koehn (1978), Wright (1978), and McCauley (1987) surveyed F_{ST} in a number of species of insects and animals, including humans, and found median values in the neighborhood of 0.075. Singh and Rhomberg (1987a, b) surveyed 61 polymorphic loci in *Drosophila melanogaster* and found a modal value of F_{ST} between 0.08 and 0.10. The experimental metapopulations used by Wade and Goodnight (1991) to investigate the SBT exhibited comparable values of F_{ST}.

B. Interaction Groups

The second aspect of population subgrouping occurs within demes and is called an *interaction group*. Even with random mating, the membership of demes can be further substructured based on genetic contagion of interacting individuals, such as families, groups of families, dominance hierarchies, or other nonrandom clusterings. When individuals in these clusters interact in ways that affect their own fitness and those of other group members (e.g., Breden and Kelly, 1982; Wade, 1980b), D. S. Wilson (1980) has referred to these as *trait groups* to emphasize the effect that such a population structure has on the evolution of behaviors. This kind of subdemic structuring gives rise to kin selection as discussed in the next section.

III. Social Behaviors and Kin Selection

Whenever an individual exhibits a behavior that affects its own fitness as well as the fitnesses of conspecifics, the behavior can be considered a *social behavior* and its evolution will be affected by population subdivision (Wade, 1978a, b, 1979b, 1980a). Indeed, the degree of genetic subdivision alone can determine whether or not a behavior evolves in a species (Hamilton, 1964a, b; Wade, 1980a, 1985a; Wade and Breden, 1981, 1987).

This critical effect of population structure on social evolution becomes clearer when social behaviors are classified in terms of their fitness effects on the performer and the receiver(s) of the behavior (Fig. 4). A behavior that directly promotes the welfare of the performing individual and also benefits neighboring conspecifics is called *mutualistic*. And, somewhat more anthropomorphically, a

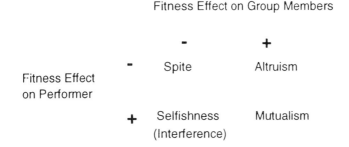

Figure 4 A classification of social behaviors based upon the fitness increments (+) and decrements (-) that accrue to the individuals performing the behavior and the group member(s) affected by the behavior.

behavior that diminishes the fitness of the performer and that of conspecifics is called *spiteful* The performer of an *altruistic* behavior lowers its own relative fitness but enhances the survivorship and/or reproduction of conspecifics. An extreme example of altruistic behavior is the existence of sterile castes in a colony of bees, ants, or wasps. These individuals completely sacrifice their own fitness in order to enhance the survivorship and reproductive success of other colony members. Darwin believed that the existence of such adaptations would be "fatal to my whole theory" of evolution by natural selection (Darwin, 1859, p. 236) because, by definition, an altruistic behavior is one that lowers the relative fitness of the performer but increases the relative fitness of the receiver. Natural selection should operate to eliminate such behaviors, yet they appear prevalent in some of the major taxonomic groups of insects and mammals.

Darwin's solution to this problem was to recognize that the adaptive function of the sterile castes in the social insects was to benefit the colony. He argued that *selection among colonies or families* permitted the evolution of these behaviors despite their apparent disadvantage to the individuals possessing them (Darwin, 1859, p. 237). The solution that he proposed explicitly postulated the existence of natural selection operating at the level of colonies and families in opposition to selection operating between individuals within groups.

Within-deme interaction groups may actively or passively involve genetic kinship between interacting individuals (Maynard Smith, 1964; West-Eberhard, 1975; Crozier, 1979; Wade and Breden, 1986; McCauley et al., 1988; Queller, 1989). Whenever individuals interact more often with kin than with unrelated individuals, the evolution of behaviors with beneficial fitness effects on conspecifics, like mutualism and altruism, is facilitated, as Darwin (1859) originally suggested. Conversely, the evolution of behaviors with negative fitness effects on conspecifics [*interference, selfishness,* or "social parasitism" in Wright's terms (1969)] is curtailed. When the groups of interacting individuals are genetic relatives, the evolutionary process that results is called *kin selection* (Maynard Smith, 1964; West-Eberhard, 1975; Crozier, 1979; Michod, 1982). This interesting interaction between population genetic structure and behavioral evolution has been experimentally demonstrated in laboratory populations of flour beetles (Wade, 1980a; see below), in field populations of willow leaf beetles (Breden and Wade,

1989; Wade, 1994), and studied comparatively in a large number of social insects (e.g., Ross and Matthews, 1989; Ross and Carpenter, 1991; Hughes et al., 1993; Ito, 1993).

The distinction between a breeding group and an interaction group may be easy to make for some species but difficult for others. In many instances, we expect the two to be correlated because the spatial geography and resource distribution often make it more likely that individuals interact with *and* mate with individuals from the same group. Whenever physical barriers to free movement constrain an individual throughout its lifetime, the genetic breeding structure and the genetic interaction structure can coincide. In other cases, the membership of a species may have a different kind or level of population subdivision at the time of reproduction than at other stages of development. Behavioral interactions may take place between individuals within patches, but matings may be more likely to take place between individuals of different patches as might happen with inbreeding avoidance mechanisms in plants (Charlesworth, 1995), dispersal phases in amphibians (Breden, 1987) and mammals (Hilborn, 1975; Shields, 1987), or incest taboos in humans (Breden and Wade, 1981; Wade and Breden, 1981). In some cases, the genetic variation among interaction groups may exceed that among breeding groups (cf. McCauley et al., 1988, for an empirical example).

The rate of behavioral evolution is more strongly affected by the population breeding structure than it is by the behavior interaction structure. The genetic interaction structure that accelerates the rate of adaptation 2-fold will accelerate it 10-fold if it affects the breeding structure (Wade and Breden, 1987). This differential effect of the components of population sructure was first shown in experimental studies of kin selection and the evolution of cannibalistic behavior in the flour beetle, *Tribolium confusum* (Wade, 1980a). In this study, three different interaction structures were created by varying the degree of genetic relatedness between larval cannibals and their potential egg victims from (1) full-sibs, to (2) half-sibs, to (3) no (random) relationship. These interaction structures were combined factorially with two different breeding structures, random mating and within-group mating. There was no artificial selection imposed in this study. Genetic and environmental variations in the larval cannibalistic behavior created selective forces and the different population structures determined the direction and rate of the evolutionary response. With within-group mating, cannibalism declined in proportion to the relationship between cannibals and victims, but no such differences among interaction treatments were observed with random mating. Although the strength of selection was similar in both breeding structures (Wade 1980a), within-group mating increased the genetic variance among groups (see discussion of "population heritability" below) and accelerated the response to selection.

Wade and Breden (1981), in a population genetic model of kin selection, contrasted the rate of evolution of an altruistic social behavior in a randomly mating population with that in populations with increasing levels of inbreeding by brother-sister mating (cf. Uyenoyama, 1984). They found that changing the breeding structure from random mating to complete inbreeding caused a 500-fold acceleration in the rate of behavioral evolution. Social evolution is similarly affected by variations in the breeding structure due to multiple mating, variations in paternity within broods, and *pleiometrosis*, the founding of colonies by multiple queens (Wade, 1985a). (In these theoretical and empirical studies, population density regulation occurred globally whereas local density effects can alter the conclusions. See remarks on "population viscosity" below.)

IV. Levels of Selection and Population Structure

In seminal work, Hamilton (1963, 1964a, b) described the necessary and sufficient conditions for an increase in the frequency of a behavior with negative effects on individual fitness but positive group fitness effects. Hamilton's rule states that altruistic behavior will evolve whenever the product of the fitness benefit, B, to the group members and the genetic relatedness within groups, r, exceeds the fitness cost, C, to the individual performing the behavior:

$$Br > C. \qquad\qquad [8]$$

Just as with breeding groups, the distribution of genetic variation within and among interaction groups is critical. The genetic variation among groups is the genetic relatedness or genetic covariation between individuals within groups. These are the same quantity viewed in different ways (Cockerham, 1954), similar to the relationship between the probability of identity by descent (covariance of genes in uniting gametes) and Wright's measure of genetic differentiation among populations, F_{ST} (Wright, 1969).

Kin selection has been shown formally to involve two (or more) different levels of selection between interaction groups (Hamilton, 1975; Michod, 1982; Wade, 1978a, 1979b, 1980a, b, 1982b, 1985b; Queller, 1992; Kelly 1992, 1994a, b). Selection occurs *within* interaction groups as a result of the differential effects of behaviors on the fitnesses of group members. Selection also occurs *among* interaction groups within the deme whenever interaction groups differ in the frequency or density of interacting individuals (Wade, 1978b, 1979b; Breden and Wade, 1989; Kelly, 1994a). When the behavioral phenotypes are determined by alternative alleles at a single locus (Breden and Wade, 1991; Wade, 1980b, 1982b, 1985b; Kelly, 1992, 1994a) or by many additively acting genes, each of small effect (Slatkin and Wade, 1978; Crow and Aoki, 1982), total selection can be partitioned formally into components, each corresponding to a different level of selection. In natural populations, the methods for analyzing phenotypic selection (Lande and Arnold, 1983; Arnold and Wade, 1984a, b; Wade, 1987; Wade and Kalisz, 1990; Goodnight et al., 1992) can be applied to structured populations. Breden and Wade (1989) used this approach to partition total phenotypic selection on cannibalism in willow leaf beetles into separate components of selection within and between kin groups. Similarly, Stevens et al. (1995) studied selection on several individual and group life history traits in jewelweed, *Impatiens capensis.*

Although the levels of selection are often opposing as suggested by Hamilton's rule, they need not be. Wade (1982b) provided a theoretical analysis of selection operating on an interference trait in a randomly mating population at *three* different and hierarchical levels of interaction groups: (1) selection between individuals within families; (2) selection between families within groups; and (3) selection between groups of families. Any of these three levels of selection can operate in concert with or in opposition to any other level depending upon the parameters describing the fitness effects of the behavior at each level. In this model, Hamilton's rule still governs the relative strengths of individual and family-level selection. However, the rule is no longer sufficient for determining whether or not the behavior will evolve because of the additional effects of the higher level of selection, namely, selection between groups of families (Wade, 1982b; Kelly, 1994a, b). This serves as a caution to those who would extrapolate from selection studies limited to a single pond of amphibian larval kin-groups, a single pride of lions, or a single troop of baboon matrilineages.

V. Conflict between Relative and Absolute Fitness

The population structure imposed by the interaction and breeding groups affect the ecology and evolution of social behaviors (Wade and Breden, 1987). Often, the local frequencies of randomly mating types within demes can be represented by simple functions of gene or genotype frequencies. However, if selection depends upon the interactions of conspecifics, then the relative fitnesses of individuals depend upon genotype frequencies (Wright, 1969, p. 120). The relative fitness with respect to local mean fitness governs selection between individuals within interaction groups, whereas mean absolute fitness determines how local population size changes (Wright, 1969, p. 478). If groups differ in mean fitness, then variance in local mean fitness provides an opportunity for intergroup selection whenever the numbers of migrants entering or leaving a group depends upon the local density. The greater the variation among demes in mean fitness, the greater the opportunity for interdemic selection via differential dispersal: phase three of Wright's shifting balance process (Wade and Goodnight, 1991; Wade, 1992). Because the maximum absolute fitness (global peak) does not correspond generally with the "selective peak," the maximum value of relative fitness within the group, evolution may not proceed to a gene frequency that maximizes the log of mean relative fitness. When the relative fitness and absolute fitness maxima do not coincide, then an evolutionary balance is reached that is different from that of either considered alone.

In many of the population genetic models of kin selection (see above), a simple kind of frequency dependence is assumed in which mean group fitness is a linear function of the frequency of a single allele affecting the social behavior of individuals. That is, group mean fitness increases with the numbers of individuals performing a social behavior. In these special cases, the group mean or inclusive fitness function is maximized (Wright, 1969, p. 121). For many other models of interactions (e.g., Lloyd, 1967), no such function exists and the general, uncritical use of the principle of maximization is not warranted.

In natural populations, ecological processes like competition and predation operate in addition to selection to change the numbers of individuals in local patches and to affect the dispersal of individuals among patches. Hamilton (1975) used the term *population viscosity* to describe the condition in which the dispersal of organisms between groups was limited without specifying the ecological consequences of limited dispersal for local competition. Limiting dispersal increases the genetic variation among groups and, without local density regulation, it enhances intergroup selection. In natural populations, however, it is reasonable to expect that the relationship between group mean fitness and the spatial variation in gene frequency will be complicated by the ecology of density regulation (Pollock, 1983; Kelly, 1994a; Queller, 1994). In many instances, density regulation is likely to operate most strongly in the largest groups, i.e., those with the highest mean absolute fitnesses (Lloyd, 1967). In the extreme case of strong local density regulation or soft selection (discussed above; see also Wallace, 1968; Christiansen, 1974; Wade, 1985b; Kelly, 1994a), the differences among interaction groups in mean absolute fitness caused by viability selection are completely canceled by local density regulation. Under soft selection, the initial relative sizes of all local groups, which are changed by selection, are reestablished after selection by competitive forces affecting all genotypes equally (Wade, 1985b; Kelly, 1994a). This kind of strong local density effect eliminates the component of selection between kin-groups (Wade, 1985b). Wade and Beeman (1994) showed how within-family density regulation can accelerate the evolution of maternal-effect selfish genes in flour beetles by removing the opposing force of between-family selection.

In contrast, with the absence of local density regulation giving hard selection (discussed above; see also Levene, 1953; Levins and MacArthur, 1966; Wade, 1985b), the mean absolute fitnesses of groups are not changed or opposed by local density regulation. Kelly (1994b) showed analytically how changes in the scale of density regulation change the effectiveness of kin selection. In natural populations, these complications of local density regulation can be addressed with the contextual analysis approach for studying intergroup selection (Goodnight et al., 1992; Stevens et al., 1995).

VI. The Ecology and Genetics of Metapopulations

Metapopulations experience novel *ecological processes* such as the extinction and colonization of local demes, deme fission and fusion, and the dispersal of migrants among demes (Fig. 5). The processes of local extinction, colonization, and deme fission are analogous to individual birth and death rates and they create an *age-structure* of demes within the metapopulation (Wade and McCauley, 1988; Whitlock and McCauley, 1990; Whitlock et al., 1993). Recently colonized demes are younger in age than long-established demes. Just as the net birth and death rates must be in balance for the persistence of a population, the net rates of colonization and deme fission must equal the rates of deme fusion and extinction in order for the metapopulation to persist (Hastings and Harrison, 1994).

More importantly, the rates and patterns of extinction and colonization determine the distribution of young and old demes in a metapopulation in a similar fashion to the effect of variable birth and death rates on the stable age distribution of individuals in a population. The consequences of demic age-structure in a metapopulation are similar to the demographic consequences of individual age-structure on population dynamics and life history evolution. When the likelihood of extinction is independent of deme age, a young or newly founded colony experiences the same risk of extinction as an older colony. In this case, the age-structure of the metapopulation is exponential and skewed toward demes of younger age. In other species, the older demes may exhaust local resources or accumulate predators and disease, and the risk of extinction may increase with deme age. These metapopulations have an age-structure more heavily skewed toward younger demes. On the other hand, in some species, most

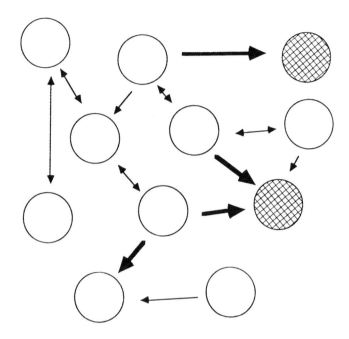

Figure 5 A schematic representation of a metapopulation and the dynamic processes of extinction, colonization, and dispersal that affect its member demes. Demes are represented by circles. The two hatched circles represent local extinctions and the shaded circle represents the opening of a new patch of habitat. The smaller arrows signify the dispersal of individuals between demes and the larger arrows are colonization events involving larger numbers of individuals than dispersal.

colonization attempts are unsuccessful but, once established, demes persist for long periods. When newly founded colonies experience much higher rates of extinction than established demes, the age-structure is skewed toward older demes.

   Migration and colonization represent biologically distinct kinds of dispersion among habitats or patches of resource and, in some species, may involve different life stages. *Migration* is generally considered to be the movement of individuals among extant demes (see also above). The number of migrants is often assumed to be small relative to the size of adult breeding groups. The migration rate is usually expressed as *m*, the fraction of N individuals breeding in a deme that originated from outside the deme. (Thus, *Nm* represents the numbers of migrants and 2*Nm* represents the numbers of migrating diploid autosomal genes or gene flow.) *Colonization*, on the other hand, is the establishment of a new deme or the recolonization of the site of an extinct deme. It can involve large or small numbers of individuals relative to average deme size and these may originate from one or more source demes. Two schematic *patterns* of dispersal are also useful to distinguish. The first was named *island model* migration by Sewall Wright (1978). With island model migration, a migrating individual is as likely to enter a deme near to its deme of birth as it is a deme much further away. Under this model, the aspects of the resource distribution and geography that are responsible for demic structure in no way constrain the movements of migrants. The second pattern of migration is called the *stepping stone* model. Here migrants are much more likely to enter neighboring demes than demes farther away and the rate of exchange of migrants between a pair of demes diminishes as the distance between the demes increases (Slatkin, 1977, 1985; Wade and McCauley, 1988; Whitlock and McCauley, 1990). Such isolation by distance is a common dispersal pattern.

The analogous extreme patterns of colonization have been designated the *migrant pool* and the *propagule pool*, respectively (Wade, 1975, 1978a; Slatkin, 1977; Whitlock, 1992). With migrant pool colonization, new demes are founded by groups of K colonists consisting of individuals from many different demes, each chosen at random from across the metapopulation (Wade and McCauley, 1988). In the propagule pool model, the K colonists founding a new deme are selected at random from within a single deme, which itself was randomly chosen from the metapopulation. Greater genetic differentiation of demes results with the propagule pool model than with the migrant pool model (Wade, 1978a; Wade and McCauley, 1988; McCauley, 1991, 1995). Whenever the number of founders, K, exceeds twice the number of migrants, $2Nm$, migrant pool colonization effectively enhances gene flow in the metapopulation and reduces genetic variation among demes (Slatkin, 1977, 1985; Wade and McCauley, 1988). In contrast, the random genetic drift associated with founding a new deme means that the genetic differentiation among demes is *always* enhanced by propagule pool colonization (Wade and McCauley, 1988; McCauley, 1991).

Clearly, these two modes of colonization represent extremes of a continuum as do island model and stepping stone migration. Whitlock and McCauley (1990) introduced a new parameter, the probability of common origin (pco, ϕ), which describes the essential genetic feature of any colonization process intermediate between these two extremes. The migrant pool model corresponds to a pco of 0.0 and the propagule pool to a value of 1.0. Intermediate values between 0.0 and 1.0 can result from variation in the numbers of source demes, variance in numbers of gametes per source deme, colonization by inseminated females, or from kin-structured colonization (Wade et al., 1994). The key condition for increased F_{ST} is given by

$$K < ([2N_e m]/[1 - \phi] + 0.5). \qquad [9]$$

"Kin founding," the founding of new demes by groups of relatives (Hedrick and Levin, 1984; Wade et al., 1994), enhances genetic differentiation to an even *greater* degree by reducing the effective number of genetic colonists. Kin founding is common when new breeding groups arise by the fissioning of older groups along matrilines (Ober et al., 1984; Wade and Breden, 1987) or when new habitats are colonized by multiple seeds from single fruits.

VII. A Novel Evolutionary Force in Metapopulations: Interdemic Selection and the Shifting Balance Process

Selection can occur at any level of biological organization, including demes, as long as the units in the hierarchy have three properties: (1) phenotypic variation among the units; (2) fitness variation among the units, i.e., the phenotypic variation must affect the likelihood that the units survive and reproduce; and (3) heritability, i.e., "offspring" units descended from a common parent(s) must resemble each other and the parent(s) with respect to survivorship and reproduction. Interdemic selection occurs in a metapopulation whenever ecological and genetic processes result in the differential extinction, colonization, or dispersal of local demes. The evolutionary response to interdemic selection depends upon the amount of genetic variation among demes and its relationship to the phenotypic differences among demes in extinction, colonization, and dispersal rates. The heritable fraction of the between-deme phenotypic variance, or *populational heritability*, determines the response to interdemic selection (Wade and McCauley, 1980, 1984; Slatkin, 1981; Wade, 1982a). Populational heritability is analogous to the concept of the "heritability of the family mean" (Falconer, 1981) but includes nonadditive components of genetic variation like epistasis and genotype interactions (e.g., Griffing, 1977, 1989; Goodnight, 1988, 1990).

In a series of experimental studies using laboratory populations of flour beetles, Wade and McCauley (1980, 1984, and references therein) explored how the populational heritability of mean fitness was affected by variations in the numbers of breeding adults and the amount and pattern of migration among demes. Measurable genetic differentiation among demes developed by random genetic drift within 10 to 15 generations of subdivision for breeding groups ranging from 6 to 96 adults and migration rates ranging from 0.0 to 0.20. They found that "even small amounts of genetic variation between demes can produce a high significant and biologically important degree of ecological differentiation" (Wade and McCauley, 1984, p. 1055). Even small values of F_{ST} in the range of 0.03 to 0.15 corresponded to estimates of population heritabilities in the range of 0.15 to 0.60. Heritability values in this range are considered moderate to high by quantitative geneticists (Falconer, 1981).

Once the genetic variation is partitioned among demes in the metapopulation, a single episode of interdemic selection by differential extinction can be sufficient to produce large changes in mean fitness (Wade and McCauley, 1980; Wade, 1982a). These genetic effects observed in laboratory metapopulations using outbred stocks also characterized the subdivision of recently collected wild populations of two species, *T. castaeum* and *T. confusum* (Wade, 1991). Furthermore, the observed changes in mean population growth rate persisted long after artificial interdemic selection was halted (Wade, 1984) and subdivision was shown to affect other ecologically important traits like adaptation to climate and competitive ability (Wade, 1990).

In Wright's shifting balance theory, genetic differences between demes give rise to differences between demes in mean absolute fitness as observed in the *Tribolium* experiments discussed above. However, the mechanism of interdemic selection is differential dispersal of migrants out from demes of high fitness and into demes of lower average fitness rather than differential extinction. Wade and Goodnight (1991) modeled Wright's process with three different experimental metapopulations of *Tribolium*. Each metapopulation had the same number of breeding adults per deme ($N = 20$) but differed from one another in the degree of population genetic structure because the island model migration imposed among demes varied in its frequency from every generation in one metapopulation, to every second generation, to once every third generation. Less frequent migration is expected to alter the balance between random genetic drift and gene flow and to enhance interdemic selection by increasing the genetic variation among demes. However, less frequent dispersion out of demes of high fitness into demes of lower fitness also reduces the interdemic selection differential and should weaken the overall response. We found surprisingly that the response to interdemic selection was nonlinear with respect to population genetic structure. When we regressed the standardized response to interdemic selection on the cumulative interdemic selection differential, we found that the largest "realized populational heritability" (Falconer, 1981; g^2 of Wade and McCauley, 1984) occurred with interdemic selection every two generations ($g^2 = 0.58$) instead of interdemic selection more or less often ($g^2 = 0.19$ in both cases). That is, the response to interdemic selection was not proportional to the cumulative interdemic selection differential or to the degree of genetic subdivision: an intermediately subdivided metapopulation gave the largest evolutionary response. These and other experiments on demes from these metapopulations (Wade and Griesemer, unpublished; Wade, unpublished) indicated that gene interactions were playing an important role in the genetic differentiation of these experimental populations.

VIII. Epistasis and Wright's Shifting Balance Theory

Most adaptations are not determined by single loci, but rather require the coordinated action of many genes to produce behavioral, morphological, or physiological phenotypes (Wright, 1969, 1978). Gene interaction or *epistasis* in biochemical genetics occurs when the "expression of one gene wipes out the phenotypic expression of another gene" (Lewin, 1990, p. 809). In evolutionary genetics, the *epistasic variance* (along with the dominance variance) is a populational quantity that measures the statistical effects of variations among individuals in gene combinations in relation to the total phenotypic variance. When multiple loci are involved in the determination of a trait, new features of selection within and among groups arise that are not captured by the single locus models (Wade, 1978a, 1992). These new features can enhance the evolutionary role of population subdivision and interdemic selection and diminish the role of selection within groups.

Wright (1931, 1978) proposed his shifting balance process as a means of directly selecting among different gene combinations. In the absence of population genetic structure, it is difficult to select directly among different gene combinations due to random mating and recombination. Parents tend to pass on genes rather than specific genotypic combinations to their offspring. When the variation between individuals is due to differences among them in gene combinations, the "heritability" of these individual differences is compromised by mating and recombination, diminishing the effectiveness of individual selection. For this reason, the additive genetic variance but not the epistasic variance is considered relevant to the response to artificial selection (Falconer, 1981): selected gains due to epistasis will be lost to recombination when selection is relaxed.

However, when demes are genetically isolated from one another, a favorable gene combination may achieve high frequency within a local deme by a combination of random genetic drift and adaptive selection. Wright referred to a locally fixed gene combination corresponding to a local maximum of relative fitness as an *adaptive peak*. When a gene combination reaches high frequency within a deme, or one gene in a co-adapted pair is fixed, recombination is weakened or mitigated as an opposing force to selection. By way of analogy, consider a deck of cards. A winning hand, consisting entirely of hearts, in one game is unlikely to be repeated in a subsequent game due to shuffling. However, if the deck itself should consist entirely of hearts, all hands from that deck will also consist entirely of hearts, despite shuffling.

Population subdivision always reduces the efficiency of recombination, even under conditions less extreme than the local fixation of gene combinations. If we view the association of alleles at two different loci as a genetic covariance (the disequilibrium or D), then it can be partitioned into within and among deme components in the same way that the genetic variance is partitioned. Population genetic structure impedes the approach to genetic equilibrium and the random combination of alleles across loci because the among-deme component of genetic covariance is not subject to recombination. Equivalently, the recombination distance between genes in a metapopulation is reduced by an amount approximately equal to the degree of population genetic subdivision, F_{ST}. This makes it easier for selection to favor specific gene combinations.

Selection between groups can be an effective way of spreading a favored gene combination through a species, at least under certain population structures (Wright, 1931, 1978; Wade and Goodnight, 1991; Wade, 1992). Local populations become genetically different from one another with respect to gene combinations corresponding to different fitness maxima. Interdemic selection permits a metapopulation to arrive at a global fitness maximum when favorable and fixed gene combinations are exported to demes of lower mean fitness by differential migration (Crow et al., 1990; Moore and Tonsor, 1994). Interestingly, Moore and Tonsor's (1994) simulation of the shifting balance process found that an intermediate level of migration was most favorable for causing the shift of a deme to a higher adaptive peak; a finding similar to the empirical observations of Wade and Goodnight (1991).

IX. The Additive Genetic Variance in Metapopulations

The components of genetic variance in a metapopulation are affected by the population genetic structure. The additive, dominance, and epistatic components of variance change with inbreeding, bottlenecks, founder events, or even strong directional selection on single genes (Cockerham and Tachida, 1988; Goodnight, 1988; Whitlock et al., 1993). Random genetic drift results in a decrease in the additive genetic variance within demes in the metapopulation relative to an outbred reference population. Wright (1951) showed that this decrease is proportional to F_{ST}, the probability that two alleles at a single locus are identical by descent in a randomly selected individual. This loss would appear to limit the opportunity for local adaptation to environmental differences among demes.

With dominance or within-locus allelic interactions, the additive variance within a deme can increase with F_{ST} (Robertson, 1952; Falconer, 1981; Willis and Orr, 1993) and the proportional increment in additive genetic variance is frequency dependent: rare recessive alleles exhibit a larger effect. This happens as a result of the increase in homozygosity associated with random genetic drift in finite breeding groups. Inbreeding with overdominance causes the same effect, although overdominance is not believed to be as common as dominance (Charlesworth and Charlesworth, 1987; Willis and Orr, 1993). Because rare recessives, especially those affecting fitness, are generally deleterious, increases in additive genetic variance due to dominance variance should be accompanied by inbreeding depression (Charlesworth and Charlesworth, 1987).

Epistatic variance can lead to a change in additive genetic variance within demes for less restrictive circumstances and without the inbreeding depression that attends the conversion of dominance variance. For F_{ST} values in the range most commonly observed in natural populations (0.01 to 0.40), the fraction of the epistatic variance, σ_{aa}^2, that is converted to local additive variance, σ_{a*}^2, in a metapopulation is approximately equal to $4(F_{ST})(1 - F_{ST})$ for recombination rates between 0.0 and 0.1 (Whitlock et al., 1993).

The average local additive genetic variance in the metapopulation is given by

$$\sigma_{a*}^2 = (1 - F_{ST})(\sigma_a^2) + 4(F_{ST})(1 - F_{ST})(\sigma_{aa}^2), \qquad [10]$$

which can be very different from that in a nonstructured population. Indeed, the epistatic components of fitness variance are not available for selection in a large homogeneously distributed population, but are converted into additive variance within local demes by random drift in the metapopulation. Thus, the potential for local adaptation is considerably greater in a metapopulation than in a homogeneously breeding population of the same size. At equilibrium, we can assume that the additive variance lost by subdivision is replaced by conversion of epistatic variance and we find that, at equilibrium,

$$\sigma_{a*}^2 = 4(1 - F_{ST})(\sigma_{aa}^2). \qquad [11]$$

If most mutations interact epistatically with genes in the background in which they first appear, then the conversion of epistatic to additive variance by drift could be an important source of genetic variation available for local adaptation.

The specific genes involved in adaptation are expected to vary from local deme to local deme because alleles segregating across the metapopulation will become fixed in some demes and lost in others. The same probability distribution that describes gene frequency variation among demes for a single locus also describes the distribution of the array of frequencies of segregating genes within a deme (Wright, 1931, 1968). Interpreting F_{ST} as a fixation index means that different genes will necessarily be fixed and contribute to the additive genetic variance differently in different local demes. With epistasis, the genetic outcome of local selection is inherently variable from one deme to another even when selection everywhere favors the same phenotype. Differently put, similar phenotypes from different localities could be very different in terms of the underlying genetic composition. Selection at the lowest level in the hierarchy of population subdivision thus becomes a genetically diversifying force rather than an homogenizing one (Wade, 1992).

X. Conclusion: Adaptation in Subdivided Populations

My focus in this chapter has been to describe how population genetic structure affects the microevolutionary forces that bring about adaptation. This is different from the perspective that a comparative biologist or systematist might have on adaptations based on interpreting fixed, functional, phenotypic differences between taxa. Given the constellation of microevolutionary forces, all operating simultaneously, it is very difficult to determine whether a difference observed between populations reflects an adaptive difference. Microevolutionary biologists most frequently equate an adaptation with a phenotype that, in the past or present, was the target of natural selection as opposed to other evolutionary forces. In the SBT, local natural selection works in concert with (not opposed to) random genetic drift and interdemic selection. Some of the "work" involved in fixing genes and exploring gene combinations is done by drift, some is done by interdemic migration, and some is done by natural selection within demes. Once you adopt the research posture that the central issue is not which evolutionary force is operating but rather to what extent is each one operating, it becomes harder to support the simple relation that adaptation is the phenotypic result of the process of natural selection. At the same time, the conspicuous "fit" of organisms to the environment is what interests me and the random forces (mutation and drift) cannot produce this fit.

There are four main points that I tried to make in this chapter in relation to local natural selection. First, in finite populations, natural selection at one locus interferes with natural selection at other loci in two ways. All other loci experience an increase in random genetic drift because selection, by definition, increases the variance in fitness and decreases N_e. (It is important to emphasize that this effect includes ALL sources of fitness variance, random, environmental, and genetic.) Linked loci, even those that act strictly additively on fitness, experience interference due to hitchhiking of their weaker alleles on the rise in frequency of alleles at the most strongly selected locus. The interference between selected loci manifests itself as a lowering of the probability of fixation at other loci of alleles good for fitness. Furthermore, the effect is multiplicative. This means that if adaptations consist of the coordinated action of many genes, then these groups of genes are built up one gene at a time by natural selection. This is different from the standard picture of all genes increasing in frequency all the time according to their net regression on total fitness (e.g., Williams, 1966, pp. 56-57). Instead, because of the interference, the alleles with the largest effects are selected first and many weaker, but favorable, alleles at other loci never get a turn at all.

Second, all adaptive processes, except soft selection, have this effect of reducing N_e (see Fig. 2 and Section II) and enhancing random genetic drift. For this reason, local natural selection will be a genetically diversifying force. The loci experiencing the strongest selection will vary from deme to deme and the remainder of the genome, weakly selected and neutral, will experience enhanced drift.

Third, random genetic drift converts epistatic genetic variance into local additive genetic variance. The effect may be more important in providing local additive genetic variance than the typical model of mutation-selection balance. Because the converted additive genetic variance will be based on different genes in different demes, local selection will be a genetically diversifying force, creating different gene combinations in different localities.

Fourth, interdemic selection is the only evolutionary force that is enhanced by the genetic diversification of local demes and that can operate directly on the epistatic components of genetic variance, if they manifest themselves as variations in mean deme fitness and differential migration.

As a final note, the human species appears to have undergone a number of range contractions and expansions due to repeated glaciations and has had a population structure characterized by repeated local extinctions and colonizations (Sokal et al., 1991). Eanes and Koehn (1978) report that the average value of F_{ST} among human populations on a worldwide scale is 0.15. Furthermore, many anthropological studies indicate that a variety of kin-group structures exist within different human populations due to differences in matrimonial systems, colony fissioning attributes, and local ecology (Klein, 1990). Inevitably, then, many of the social behaviors and adaptations that characterize humans have been influenced in significant ways by the population structure experienced over the course of what Klein (1990) has called the "Human Career." These data support Darwin's (1871, p. 160-167) supposition that selection among families and among tribes has been an important process in human evolution.

Acknowledgments

I thank John Kelly for his comments during the preparation of this manuscript. I also want to acknowledge the stimulating discussions of my graduate students (Felix Breden, Charles Goodnight, Lorraine Heisler, John Kelly, Susan Dudley, and Steve Tonsor), students of my colleagues (Patrick Phillips, Brad Shaffer, and John Willis), my postdocs (Chris Boake, Jim Griesemer, Dave McCauley, Steve Shuster, Lori Stevens, and Michael Whitlock), my faculty colleagues (Jeanne and Stuart Altmann, Stevan Arnold, Brian and Deborah Charlesworth, Russ Lande, Mathew Leibold, Monty Lloyd, Carole Ober, Bill Wimsatt, and Andy Abbott), and my advisors (Thomas Park and Montgomery Slatkin) throughout my laboratory, field, and mathematical explorations of evolution in subdivided populations.

References

Arnold, S. J., and Wade, M. J. (1984a). On the measurement of natural and sexual selection: theory. *Evolution* 38, 709-719.

Arnold, S. J., and Wade, M. J. (1984b). On the measurement of natural and sexual selection: applications. *Evolution* 38, 720-734.

Barton, N. H. (1995). Linkage and the limits to natural selection. *Genetics* 140, 821-841.

Breden, F. J. (1987). The effect of post-metamorphic dispersal on the population genetic structure of Fowler's toad, *Bufo woodhousei fowleri. Copeia* 2, 386-395.

Breden, F. J., and Kelly, C. H. (1982). The effect of conspecific interactions on metamorphosis in *Bufo americanus. Ecology* 63, 1682-1689.

Breden, F. J., and Wade, M. J. (1981). Inbreeding and evolution by kin selection. *Ethology and Sociobiology* 2, 3-16.

Breden, F. J., and Wade, M. J. (1989). Selection within and between kin groups in the imported willow leaf beetle. *Amer. Nat.* 134, 35-50.

Breden, F. J., and Wade, M. J. (1991). 'Run-away' social evolution: reinforcing selection for inbreeding and altruism. *J. Theor. Biol.* 153, 323-337.

Charlesworth, D. (1995). Multi-allelic self-incompatibility polymorphisms in plants. *BioEssays* 17, 31-38.

Charlesworth, D., and Charlesworth, B. C. (1987). Inbreeding depression and its evolutionary consequences. *Annu. Rev. Ecol. Syst.* 18, 237-268.

Christiansen, F. B. (1974). Sufficient conditions for protected polymorphism in a subdivided population. *Amer. Nat.* 108, 157-166.

Cockerham, C. C. (1954). An extension of the concept of partitioning hereditary variance for analysis of covariances among relatives when epistasis is present. *Genetics* 39, 859-82.

Cockerham, C. C., and Tachida, H. (1988). Permanency of response to selection for quantitative characters in finite populations. *Proc. Natl. Acad. Sci. USA* 85, 1563-1565.

Crow, J. F. (1958). Some possibilities for measuring the selection intensities in man. *Human Biology* 30, 1-13.

Crow, J. F. (1962). Population genetics: selection. *In* "Methodology in Human Genetics" (W. J. Burdette, ed.), pp. 53-75. Holden-Day, San Francisco, CA.

Crow, J. F., and Aoki, K. (1982). Group selection for a polygenic trait. *Proc. Natl. Acad. Sci. USA* 79, 2628-2631.

Crow, J. F., Engels, W. R., and Denniston, C. (1990). Phase three of Wright's shifting-balance theory. *Evolution* 44, 233-247.

Crozier, R. H. (1979). Genetics of sociality. *In* "Social Insects" (H. R. Hermann, ed.), Vol. I. Academic Press, New York.

Darwin, C. (1859). "On the Origin of Species by Means of Natural Selection, or, the Preservation of Favored Races in the Struggle for Life." John Murray, London.

Darwin, C. (1871). "The Descent of Man and Selection in Relation to Sex." Princeton University Press, Princeton.

Eanes, W. F., and Koehn, R. K. (1978). An analysis of genetic structure in the monarch butterfly, *Danaus plexippus* L. *Evolution* 32, 784-797.

Falconer, D. S., (1981). "Introduction to Quantitative Genetics." 2nd ed. Longman, London.

Goodnight, C. J. (1988). Epistasis and the effect of founder events on the additive genetic variance. *Evolution* 42, 441-454.

Goodnight, C. J. (1990). Experimental studies of community evolution. II. the ecological basis of the response to community selection. *Evolution* 44, 1625-1636.

Goodnight, C. J., Schwartz, J. M., and Stevens, L. (1992). Contextual analysis of models of group selection, soft selection, hard selection, and the evolution of altruism. *Amer. Nat.* 140, 743-761.

Griffing, B. (1977). Selection for populations of interacting genotypes. *In* "Proceedings of the International Congress on Quantitative Genetics, August 16-21, 1976" (E. Pollack, O. Kempthorne, and T. B. Bailey, eds.), pp. 413-434. Iowa State University Press, Ames, IA.

Griffing, B. (1989). Genetic analysis of plant mixtures. *Genetics* 122, 943-957.

Hamilton, W. D. (1963). The evolution of altruistic behavior. *Amer. Nat.* 97, 354-356.

Hamilton, W. D. (1964 a). The genetical evolution of social behavior. I. *J. Theor. Biology* 7, 1-16.

Hamilton, W. D. (1964b). The genetical evolution of social behavior. II. *J. Theor. Biology* 7, 17-52.

Hamilton, W. D. (1975). Innate social aptitudes of man: an approach from evolutionary genetics. *In* "Biosocial Anthropology" (R. Fox, ed.), pp. 133-155. John Wiley & Sons, New York.

Harrison, S. (1991). Local extinctions in a metapopulation context: an empirical evaluation. *Biol. J. Linn. Soc.* 42, 73-88.

Hastings, A., and Harrison, S. (1994). Metapopulation dynamics and genetics. *Annu. Rev. Ecol. Syst.* 25, 167-188.

Hedrick, P. W., and Levin, D. A. (1984). Kin-founding and the fixation of chromosomal variants. *Amer. Nat.* 124, 789-797.

Hilborn, R. (1975). Similarities in dispersal tendency among siblings in four species of voles (*Microtus*). *Ecology* 56, 1221-1225.

Hill, W. G., and Robertson, A. (1966). The effect of linkage on the limits to artificial selection. *Genet. Res.* 8, 269-294.

Hughes, C. R., Queller, D. C., and Strassmann, J. E. (1993). Relatedness and altruism in *Polistes* wasps. *Behav. Ecol.* 4, 128-137.

Ito, Y. (1993). "Behaviour and Social Evolution of Wasps." Oxford University Press, Oxford.

Kelly, J. K. (1992). Restricted migration and the evolution of altruism. *Evolution* 46, 1492-95.

Kelly, J. K. (1994a). The effect of scale dependent processes on kin selection: mating and density regulation. *Theor. Popul. Biol.* 46, 32-57.

Kelly, J. K. (1994b). A model for the evolution of communal foraging in hierarchically structured populations. *Behav. Ecol. Sociobiol.* 35, 205-212.

Klein, R. G. (1990). "The Human Career." University of Chicago Press, Chicago, IL.

Lande, R., and Arnold, S. J. (1983). The measurement of selection on correlated characters. *Evolution* 36, 1210-1226.

Levene, H. (1953). Genetic equilibrium when more than one niche is available. *Amer. Nat.* 87, 331-333.

Levins, R. (1970). Extinction. *In* "Some Mathematical Questions in Biology: Lectures on Mathematics in the Life Sciences." *Am. Mathematical Soc.* 2, 75-108.

Levins, R., and MacArthur, R. (1966). The maintenance of genetic polymorphism in a spatially heterogeneous environment: variations on a theme by Howard Levene. *Amer. Nat.* 100, 585-590.

Lewin, B. (1990). "Genes IV." Oxford University Press, New York.

Lloyd, M. (1967). Mean crowding. *J. Anim. Ecol.* 36:1-30.

Loveless, M. D., and Hamrick, J. L. (1984). Ecological determinants of genetic structure in plants. *Annu. Rev. Ecol. Syst.* 15, 65-95.

Maynard Smith, J. (1964). Group selection and kin selection. *Nature* (London) 201, 1145-1147.

McCauley, D. E. (1987). Population genetic consequences of local colonization: evidence from the milkweed beetle, *Tetraopes tetraophthalmus. Florida Entomologist* 70, 21-30.

McCauley, D. E. (1991). Genetic consequences of local population extinction and recolonization. *Trends in Ecol. Evol.* 6, 5-8.

McCauley, D. E. (1995). The effects of population dynamics on genetics in mosaic landscapes. *In* "Mosaic Landscapes and Ecological Processes" (L. Hansson, L. Fahrig, and G. Merriam, eds.), pp. 178-198. Chapman and Hall, London.

McCauley, D. E., Wade, M. J., Breden, F. J., and Wohltman, M. (1988). Spatial and temporal variation in group relatedness: evidence from the imported willow leaf beetle. *Evolution* 42, 184-192.

Michod, R. (1982). The theory of kin selection. *Annu. Rev. Ecol. Syst.* 13, 23-55.

Moore, F. B.-G., and Tonsor, S. J. (1994). A simulation of Wrights' shifting-balance process: migration and the three phases. *Evolution* 48, 69-80.

Nei, M., and Takahata, N. (1993). Effective population size, genetic diversity, and coalescence time in subdivided populations. *J. Mole. Biol.* 37, 240-244.

Ober, C., Olivier, T. J., Sade, D. S., Schneider, J. M., Cheverud, J., and Buettner-Janusch, J. (1984). Demographic components of gene frequency change in free-ranging macaques on Cayo Santiago. *Am. J. Phys. Anthrop.* 64, 223-232.

Pollock, G. (1983). Population viscosity and kin selection. *Amer. Nat.* 122, 817-829.

Queller, D. C. (1989). Inclusive fitness in a nutshell. *In* "Oxford Surveys in Evolutionary Biology" (P. H. Harvey and L. Partridge, eds.), Vol. 6. Oxford University Press, Oxford.

Queller, D. C. (1992). Quantitative genetics, inclusive fitness, and group selection. *Amer. Nat.* 139, 540-558.

Queller, D. C. (1994). Genetic relatedness in viscous populations. *Evol. Ecol.* 8, 70-73.

Robertson, A. (1952). The effect of inbreeding on the variation due to recessive genes. *Genetics* 37, 189-207.

Ross, K. G., and Matthews, R. W. (1989). Population genetic structure and social evolution in the sphecid wasp, *Microstigmus comes. Amer. Nat.* 134, 574-598.

Ross, K. G., and Carpenter, J. M. (1991). Population genetic structure, relatedness, and breeding systems. *In* "The Social Biology of Wasps" (K. G. Ross and R. W. Matthews, eds.), Cornell University Press, Ithaca, NY.

Shields, W. M. (1987). Dispersal and mating systems: investigating their causal connections. *In* "Mammalian Dispersal Patterns" (B. D. Chepko-Sade and Z. T. Halpin, eds.), pp. 3-24. University of Chicago Press, Chicago, IL.

Shuster, S. M., and Wade, M. J. (1991). Female copying and sexual selection in a marine isopod crustacean, *Paracerceis sculpta. Anim. Behav.* 42, 1071-1078.

Singh, R. S., and Rhomberg, L. R. (1987a). A comprehensive study of genic variation in natural populations of *Drosophila melanogaster.* I. Estimates of gene flow from rare alleles. *Genetics* 115, 313-322.

Singh, R. S., and Rhomberg, L. R. (1987b). A comprehensive study of genic variation in natural populations of *Drosophila melanogaster.* II. Estimates of heterozygosity and patterns of geographic differentiation. *Genetics* 117, 255-271.

Slatkin, M. (1974). Competition and regional coexistence. *Ecology* 55, 128-134.

Slatkin, M. (1977). Gene flow and genetic drift in a species subject to frequent local extinctions. *Theor. Popul. Biol.* 12, 253-262.

Slatkin, M. (1981). Populational heritability. *Evolution* 35, 859-871.

Slatkin, M. (1985). Gene flow in natural populations. *Annu. Rev. Ecol. Syst.* 16, 393-430.

Slatkin, M., and Wade, M. J. (1978). Group selection on a quantitative character. *Proc. Natl. Acad. Sci. USA* 75, 3531-3534.

Sokal, R. R., Oden, N. L., and Wilson, C. (1991). Genetic evidence for the spread of agriculture in Europe by demic diffusion. *Nature* (London) 351, 143-145.

Stevens, L., Goodnight, C. J., and Kalisz, S. (1995). Multi-level selection in natural populations of Impatiens capensis. Amer. Nat. 145, 513-526.

Uyenoyama, M. K. (1984). Inbreeding and the evolution of altruism under kin selection: effects on relatedness and group structure. *Evolution* 38, 778-795.

Wade, M. J. (1975). "An experimental study of group selection." Ph.D. thesis, University of Chicago.

Wade, M. J. (1978a). A critical review of the models of group selection. *Quart. Rev. Biol.* 53, 101-114.

Wade, M. J. (1978b). Kin selection: a classical approach and a general solution. *Proc. Natl. Acad. Sci. USA* 75, 6145-6158.

Wade, M. J. (1979a). Sexual selection and the variance in reproductive success. *Amer. Nat.* 114, 742-747.

Wade, M. J. (1979b). The evolution of social interactions by family selection. *Amer. Nat.* 113, 399-417.

Wade, M. J. (1980a). An experimental study of kin selection. *Evolution* 34, 844-855.

Wade, M. J. (1980b). Kin selection: its components. *Science* 210, 665-667.

Wade, M. J. (1982a). Group selection: migration and the differentiation of small populations. *Evolution* 36, 949-961.

Wade, M. J. (1982b). The evolution of interference competition by individual, family, and group selection. *Proc. Natl. Acad. Sci. USA* 79, 3575-3578.

Wade, M. J. (1984). The population biology of flour beetles, *Tribolium castaneum*, after interdemic selection for increased and decreased population growth rate. *Res. Popul. Ecol.* 26, 401-415.

Wade, M. J. (1985a). The influence of multiple inseminations and multiple foundresses on social evolution. *J. Theor. Biol.* 112, 109-121.

Wade, M. J. (1985b). Hard selection, soft selection, kin selection, and group selection. *Amer. Nat.* 125, 61-73.

Wade M. J. (1987). Measuring sexual selection. In "Sexual Selection: Testing the Alterantives." (Z. Halperin and D. Chepko-Sade, eds.), pp.197-207. University of Chicago Press. Chicago, IL.

Wade, M. J. (1990). Genotype-environment interaction for climate and competition in a natural population of flour beetles, *Tribolium castaneum. Evolution* 44, 2004-2011.

Wade M. J. (1991). Genetic variance for the rate of population increase in natural populations of flour beetles, *Tribolium* spp.*Evolution* 45, 1574-1584.

Wade, M. J. (1992). Sewall Wright: gene interaction and the Shifting Balance Theory. *In* "Oxford Series on Evolutionary Biology" (J. Antonovics and D. Futuyma, eds.), Vol. 8, pp. 35-62. Oxford University Press, Oxford.

Wade, M. J. (1994). The biology of the imported willow leaf beetle, *Plagiodera versicolora* (Laicharting). *In* "Novel Aspects of the Biology of the Chrysomelidae" (P. H. Jolivet, M. L. Cox, and E. Petitpierre, eds.), pp. 541-547. Kluwer Academic Publishers, The Netherlands.

Wade, M. J. (1995). The ecology of sexual selection: mean crowding of females and resource-defense polygyny. *Evolutionary Ecology* 9, 118-124.

Wade, M. J., and Arnold, S. J. (1980). The intensity of sexual selection in relation to male sexual behavior, female choice, and sperm precedence. *Anim. Behav.* 28, 446-461.

Wade, M. J., and Beeman, R. W. (1994). The population dynamics of maternal-effect selfish genes. *Genetics* 138, 1309-1314.

Wade, M. J., and Breden, F. J. (1981). Effect of inbreeding on the evolution of altruistic behaviors by kin selection. *Evolution* 35, 844-858.

Wade, M. J., and Breden, F. J. (1986). Life history of natural populations of the imported willow leaf beetle, *Plagiodera versicolora* (Coleoptera: Chrysomelidae). *Annal. Entomol. Soc. Am.* 79, 73-79.

Wade, M. J., and Breden, F. J. (1987). Kin selection in complex groups: mating structure, migration structure, and the evolution of social behaviors. *In* "Mammalian Dispersal Patterns" (B. D. Chepko-Sade and Z. T. Halpin, eds.), pp. 273-283. University of Chicago Press, Chicago, IL.

Wade, M. J., and Dixon, K. A. (1996). Invasibililty of a polygynous mating system by an alternative male mating strategy. J. Theor. Biol., in press.

Wade, M. J., and Goodnight, C. J. (1991). Wright's Shifting Balance Theory: an experimental study. *Science* 253, 1015-1018.

Wade, M. J., and Kalisz, S. (1990). The causes of natural selection. *Evolution* 44, 1947-1955.

Wade, M. J., and McCauley, D. E. (1980). Group selection: the phenotypic and genotypic differentiation of small populations. *Evolution* 34, 799-812.

Wade, M. J., and McCauley, D. E. (1984). Group selection: the interaction of local deme size and migration in the differentiation of small populations. *Evolution* 38, 1047-1058.

Wade, M. J., and McCauley, D. E. (1988). Extinction and recolonization: their effects on the genetic differentiation of local populations. *Evolution* 42, 995-1005.

Wade, M. J., McKnight, M. L., and Shaffer, H. B. (1994). The effects of kin-structured colonization on nuclear and cytoplasmic genetic diversity. *Evolution* 48, 1114-1120.

Wallace, B. (1968). Polymorphism, population size, and genetic load. *In* "Population Biology and Evolution" (R. C. Lewontin, ed.), pp. 87-108. Syracuse University Press, Syracuse, NY.

West-Eberhard, M. J. (1975). The evolution of social behavior by kin selection. *Quart. Rev. Biol.* 50, 1-33.

Whitlock, M. C. (1992). Nonequilibrium population structure in forked fungus beetles: extinction, colonization, and the genetic variance among populations. *Amer. Nat.* 139, 952-970.

Whitlock, M. C., and McCauley, D. E. (1990). Some population genetic consequences of colony formation and extinction: genetic correlations within founding groups. *Evolution* 44, 1717-1724.

Whitlock, M. C., Phillips, P. C., and Wade, M. J. (1993). Gene interaction affects the additive genetic variance in subdivided populations with migration and extinction. *Evolution* 47, 1758-1769.

Williams, G. C. (1966). "Adaptation and Natural Selection." Princeton University Press, Princeton, NJ.

Willis, J. H., and Orr, H. A. (1993). Increased heritable variation following bottlenecks: the role of dominance variance. *Evolution* 47, 949-956.

Wilson, D. S. (1975). A theory of group selection. *Proc. Natl. Acad. Sci. USA* 72, 143-146.

Wilson, D. S. (1980). "The Natural Selection of Populations and Communities." Benjamin/Cummings, Menlo Park, CA.

Wright, S. (1931). Evolution in Mendelian populations. *Genetics* 16, 97-159.

Wright, S. (1943). Isolation by distance. *Genetics* 28, 114-138.

Wright, S. (1951). The genetical structure of populations. *Annu. Rev. Eugenics* 15, 323-354.

Wright, S. (1969). "Evolution and the Genetics of Populations," Vol. 2. University of Chicago Press, Chicago, IL.

Wright, S. (1978). "Evolution and the Genetics of Populations," Vol. 4. University of Chicago Press, Chicago, IL.

CHAPTER 13

Adaptation and Selection of Genomic Parasites

LAURENCE D. HURST

I. Introduction

Stearns (1976) was not far from the truth when he defined fitness as that which everyone understands but no one can define. He might equally well have been discussing adaptation. Regardless of attempts to purify, co-opt, or monopolize its meaning, adaptation evades a unified definition. It is not, however, the purpose of this chapter to arbitrate between various meanings and to advocate one above another, nor to insist that genomic structures are, or are not, adaptations. Aside from showing that selection at the genomic level may be a commonplace and important phenomenon, one function of this chapter is a simple warning. At the genomic level, depending upon which usage one employs, there may be many clearly visible adaptations or there may be very few, possibly none. As a consequence, the usage of the language of adaptation should, at least in this context, be applied with caution.

II. What Is the Genomic Level?

One might imagine that adaptations at the genomic level must be commonplace. Mitochondria in insect flight muscle, for instance, have such tightly packed cisternae that they must clearly be adaptations to permit rapid respiration and energy conversion. This sort of adaptation, however, is not the principal subject of the chapter and not what I wish to consider as the genomic level. The adaptations that I wish to begin with are those of the "individuals within individuals."

A. Individuals within Individuals

Consider a tapeworm infecting an organism. The tapeworm is an individual adapted to parasitizing its host. The parasite has structures to attach to the intestinal lining, to take in predigested food, to release offspring, etc. Parasites may be able not only to "take" from their hosts but also to manipulate them to their own ends. The fluke *Dicrocoelium dendriticum*, for instance, burrows into the subesophageal ganglion of its ant host, so causing a change in the ant's behavior. Infected ants ascend to the top of grass stems when uninfected ants return underground. Once at the top of the stem, the ant clamps its jaws into the stem and remains there, immobile. In this position, grazing ungulates may then consume the ant and its fluke parasite. And so, in turn, the parasite enters its next host, thus allowing continuity of the life cycle.

Many similar examples could be given, some of which are still yet more elegant (see Dobson, 1988; Dawkins, 1990). Although these examples of host behavior need not be examples of adaptations on the part of the parasite [they may be side products or the response of the host (Minchella, 1985)], they could, at least in principle, be understood as the adaptations of the parasite. This chapter considers a particular class of parasites that manipulate their hosts and asks whether these manipulations may be considered as adaptations of these unusual parasites. The parasites in question are those that receive no horizontal transmission and may hence be considered as genomic parasites.

As an introduction to these genomic parasites, I shall first illustrate the logical continuity between manipulations forced by the ant's flukes and those of genomic parasites through discussion of the intermediary steps and boundary cases.

One tactic that many parasites use (though not typically those the size of tapeworms and flukes) is to infect the next generation of hosts by making their way into eggs and hence being maternally transmitted to offspring. Once a

TABLE OF ABBREVIATIONS

Abbreviation	Full term
CMS	Cytoplasmic male sterility
hCGb	beta subunit of human chorionic gonadotrophin
hLH	human luteinizing hormone
(h) PL	(human) Placental lactogen
Igf2	Insulin-like growth factor 2
Igf2r	receptor for Insulin-like growth factor 2
K_A/K_S	ratio of non-synonymous to synonymous substitutions
scat	Severe combined anaemia and thrombocytopaenia
SD	Segregation Distorter
SPE	Self-promoting element
SRY	Sex-region on the *Y*-chromosome
Ste	Stellate
Su(Ste)	Suppressor of Stellate

parasite has evolved such that it may exploit maternal transmission, then selection can also act to favor those parasites that do not harm females too much. The logic behind this is relatively simple. If a parasite can receive transmission through eggs, then it may be best off keeping the host alive (assuming the host is an egg producer) and acting such that the number of eggs that the female produces can be maximized. Thus the parasite's transmission rate can in turn be maximized. Put another way, the fitness of the host determines to a large degree the fitness of the parasite. This can be contrasted with those parasites that undergo horizontal transmission. Death of the host after (or as a consequence of) the escape of the parasite will have little, if any, bearing on the success of that parasite.

It need not, however, be the case that maternally transmitted parasites should not inflict some harm on their egg-producing host. If, for instance, reduced virulence is also associated with a reduced rate of maternal transmission, then selection need not favor the parasite that is entirely avirulent (if an increased transmission rate means a decreased maternal host fitness). Most particularly, this can be true if a parasite must compete with unrelated parasites for transmission. Under such conditions, selection could favor a parasite that is harmful to the host; if a benign parasite is outcompeted by more virulent ones, then it is often the case that the more virulent will spread in the population. The dynamics of this sort of process are not necessarily trivial and it is not always the case that the more virulent parasite will spread [if it is too virulent the host may, at the extreme, produce no eggs (or die), in which case selection can favor the relatively benign parasite] (Frank, 1992).

However when the parasite can only be transmitted via eggs, its fitness will strongly covary with that of the host, and under a broad range of conditions selection can favor an approach to avirulence. This is true for any parasite that to any degree requires the survival of its host for it to propagate and hence must be true for vertically transmitted parasites of any variety. However, as sperm are so small, parasites are typically not vertically transmitted when in a male (note that typically half the maternally transmitted parasites will end up in such a dead end). There are, however, a few exceptions (mostly viruses) (see, e.g., Fleuriet, 1988).

In a male that does not transmit the parasite directly to progeny (by sperm or other means), the parasite might hence be selectively favored if it can escape from the host to find a new host. This is indeed what a number of maternally transmitted parasites do (Hurst, 1993a). In mosquitoes, for instance, microsporidian parasites may receive maternal transmission. If the progeny that a microsporidian finds itself in is a male, the parasites kill the host and subsequently infect a secondary host (typically a copepod). In daughters, by contrast, the parasite may be avirulent and depend on maternal transmission to the offspring (see, e.g., Andreadis, 1988). Similarly, helminth parasites of numerous mammals are transmitted across the placenta and through the milk to the progeny. These helminths lie dormant in female offspring and only start to proliferate once they sense that the host is entering pregnancy (or a related cue). They then migrate to the positions in the female's body from which they may be transmitted to the progeny. By contrast, in males the helminths proliferate almost immediately after infection and, like the microsporidian parasites of mosquitoes, attempt to exit the host (unlike the mosquito example, male mortality does not usually follow) (Shoop, 1991).

In both of the above cases, harm to the male is not necessary and is most likely a side consequence of the mode of transmission (but note, in the mosquitoes example, there may also be an advantage to the microsporidians resident in females following from the death of males - see below). Once again, if a decrease in virulence is accompanied with a decrease in transmission rate, then selection need not favor parasites with reduced virulence.

B. Cytoplasmic Elements

It is now a short step to considering the best strategy of vertically transmitted parasites that, for whatever reason, do not have the option of escaping from males. All those heritable entities that receive exclusively maternal transmission will be referred to as cytoplasmic elements. These parasites are still at a dead end in males. All, however, is not lost for the parasite in this circumstance (see for review Hurst, 1993a). For example, isopod crustaceans, such as *Armadillidium vulgare*, bear maternally transmissible bacterial parasites that convert their male host into a female host (see, e.g., Juchault et al., 1992; Rigaud and Juchault, 1993). The frequency of females infected with the bacteria increases (relative to the frequency of uninfected individuals and individuals with nonfeminizing bacteria) because the bacterium finds itself in a transmitting host every generation, whereas the competing cytotype is transmitted only half the time. The inverse has also been described, i.e., an agent with paternal transmission that masculinizes females (Werren et al., 1987).

Many similar examples could be given. Inbred wasps, for instance, have cytoplasmic bacteria that convert normal sexual females (producing both males and females) into ones producing exclusively parthenogenetically derived female progeny (Stouthamer et al., 1990). These progeny in turn are capable of producing parthenogenetically derived female progeny, and so on. Other wasps have cytoplasmic elements that increase the proportion of eggs that are fertilized (being haplo-diploid, fertilized eggs are females) (Skinner, 1982).

Perhaps the least subtle strategy (and possibly also the most common) is that of the so-called cytoplasmic male killers (reviewed in Hurst and Majerus, 1993; Hurst, 1993a). These simply kill the host when they find themselves in a male. Unlike the microsporidians of mosquitoes, however, this appears not to permit horizontal transmission. In contrast to the above instances, the cytoplasmic element that kills a male does not enhance its own fitness directly - it too ends up in a dead host. So how can this cytoplasmic element spread in the population? The answer it seems is kin selection. Although the cytoplasmic element that kills males typically ends up in a dead male, the sisters of this dead male have clonal relatives of the element. Any fitness advantage that these females obtain from the death of males is effectively a fitness advantage to the cytoplasmic factor and hence allows the spread of this factor in the population. The possible fitness advantages are broadly of two types (assuming no horizontal transmission). First, the death of males may result in a decreased probability that the sisters will inbreed. If there is any inbreeding depression, the sister's daughters (who will also carry the cytoplasmic element) will be fitter than the average female in the population. In other cases, the death of males allows females to access resources which would otherwise have been denied them. In ladybirds, the death of males results in enhanced fitness of the sisters, both because females cannibalize their dead brothers and because, as their brothers are dead, the sisters have a higher probability that they will cannibalize rather than be cannibalized (Hurst et al., in preparation). It is hence noteworthy that many of the examples of male killers occur in insects that lay their eggs in clumps and have gregarious larvae (which hence are in competition with each other) and/or indulge in sib cannibalism.

A very closely related phenomenon in hermaphroditic plants is so-called cytoplasmic male sterility (CMS). Here male tissue is sterilized by cytoplasmic elements. The plants may then make the best of a bad job and redistribute resources to the ovules (again, reduced inbreeding levels may also enhance fitness of ovules) (Lewis, 1941; Frank, 1989; Saumitou-Laprade et al., 1994). As with male killing, the ovules contain relatives of the elements that sterilized the male tissue, and hence the trait spreads. Unlike the animal examples, which are usually caused by symbiotic bacteria, CMS is more often than not caused by a mitochondrial gene (Gouyon and Couvet, 1987; Saumitou-Laprade et al., 1994). So are these mitochondria parasites?

Whether we consider it desirable to refer to CMS-causing mitochondria as parasites is not really my concern. Rather, the point is to illustrate that components of the genome, such as the mitochondrial genome, may be able to manipulate their hosts and that in principle this manipulation is little different from that performed by more conventional parasites. It is equally legitimate therefore to at least think of an organelle as an individual that may be just as well adapted to its host as any more conventional parasite.

One defense for the comparison between mitochondria and more conventional parasites (such as bacterial pathogens) could be that it seems hard to know when a maternally transmitted bacterium stops being a bacterium and starts being an organelle (mitochondria and chloroplasts being almost certainly derived from prokaryotes). This defense is not necessary for the point to be addressed here. Regardless of the origin of organelles, the mathematical analysis of the spread of a mutant heritable factor causing some manipulation of the host (e.g., feminization) can be identical for any two replicators with the same transmission pattern. In sum, the adaptive manipulations that might be performed by mitochondrial genomes should, in many regards, be no different from the adaptations that a tapeworm or fluke might perform, were they similarly transmitted.

We need not restrict ourselves to discussion of the nonnuclear genetic components. Consider, for instance, the W chromosome in a butterfly. [N.B. the W chromosome of butterflies, like the Y chromosome of mammals and flies, is found only in one sex; for W chromosomes this is the female whereas for Y chromosomes this is the male.] Let us suppose that we found a species of butterfly in which, in a significant proportion of lines, all the male progeny died very young [several instances of this have in fact been described (Hurst, 1993a)]. Could we suspect a cytoplasmic male killer? Yes we could: a cytoplasmic element could spread were it to kill males just so long as some advantage goes to the dead sons' sisters. But what if the W chromosome could put a toxin into all eggs? And what if the same W also coded for the anti-toxin? For the heuristic point that is to be made, the reality of this scheme is irrelevant (it is in fact quite reasonable - a very similar autosomal factor has been described in both beetles and mice - see Section

VIII). If a strictly maternally transmitted cytoplasmic genome could spread due to the death of males, then so could the manipulative *W* chromosome. The population genetics of a maternally transmitted organelle or symbiotic genome is, as far is this issue is concerned, little different from that of the *W* chromosome. So, we must conclude, all else being equal, a *W* sex chromosome and a cytoplasmic genome may be just as adapted to transmission through eggs.

Some nuclear genes may hence be thought of as parasites capable of manipulating their hosts to their own advantage. Can the same hold for those genes that receive transmission through both sexes? There is from the above logic no reason not to think so. Consider for instance a gene on an autosome. In male meiosis, half of the sperm will contain one version of this gene and the other half will contain another allelic form (assuming the individual is heterozygous). These sperm are now in competition for an egg. If two individuals were in competition for any resource, would we not expect them to attempt to outrun each other or inhibit each other? Gametic competition between the meiotic progeny of one individual (meiotic drive) is well described in flies, mice, and ascomycete fungi and is suspected elsewhere (Lyttle, 1991; Lyttle et al., 1991).

Perhaps the best described example is segregation distorter (SD) [the other well understood example is the *t*-complex of mice (Silver, 1993)]. SD is a gene (in fact, two loci that segregate as one) close to the centromere on chromosome II in *D. melanogaster* (for reviews see Lyttle, 1991, 1993; Lyttle et al., 1991). If a male is heterozygous for this gene, then during meiosis SD produces a toxin that is placed in every presperm. SD-bearing sperm are immune to the toxin, but sperm carrying the wild-type allele are not (for problems with this oversimplified mechanistic model see Hiraizumi, 1990). As a consequence, the sperm containing the wild-type allele are killed and all (or nearly all) the eggs fertilized by the male contain SD. This "aggressive" gene thus wins the competition to fertilize the eggs, and can spread in a population despite it reducing male fertility. SD, like the cytoplasmic sex ratio distorters, is a parasite every bit as much as the tapeworm, it just happens that it is not horizontally transmitted.

III. Self-Promoting Elements

That elements of the genome may be parasitic was first clearly expounded by Östergren (1945) (but see also Lewis, 1941) in a discussion of what factors maintain B chromosomes. [A chromosomes are the regular components of the nuclear genome, the autosomes and sex chromosomes, while other not obviously necessary, often unpaired, chromosomes are classified as B chromosomes.] Since Östergren's paper, in which he explicitly refers to these extra chromosomes as parasites, the terminology has changed frequently. "Parasitic genetic elements" is a term sometimes employed, as is "genetic renegades", "outlaws," etc. (Alexander and Borgia, 1978; Dawkins, 1982). "Selfish genetic elements" is one of the most frequently applied (Werren et al., 1988). This latter term's sometime synonym, selfish gene, is presently of confused usage. In its original meaning (Dawkins, 1976) it was intended to apply to any gene that spread deterministically for whatever reason (parasitic or advantageous to the organism), but has since been converted to be equivalent to a selfish genetic element. Curiously in the Oxford English Dictionary the second meaning is given but is ascribed to Dawkins. The term ultra-selfish gene is also employed (Crow, 1988), presumably to contrast with the original definition (Dawkins, 1976) of selfish gene, but is equivalent to the later definition.

If a distinction exists between selfish genes (later usage) and selfish genetic elements it may be that some use selfish gene to include only those genes that are components of the genome. [I am grateful to Tim Prout for pointing this distinction out to me.] Selfish genetic elements may then either be reserved for extra-genomic components or as a term to encompass both the genomic and nongenomic parasitic elements. This usage is, however, further confused by the problem of knowing what the genome is, and is not. Are mitochondrial genomes part of the genome? Are B chromosomes part of the genome? And what of retroviruses?

I prefer to sidestep the ambiguities of these semantics, and shall discuss instead "self-promoting elements" (or "SPEs") (Hurst et al., 1996). A self-promoting element is any heritable element that deterministically spreads within a population but could, at the same time, be deleterious to the individual bearing it (such that the spread of the element creates the context for the spread of a suppressor) and that receives no significant degree of horizontal transmission (for further discussion see Hurst et al., 1996). Whether the element is embedded in the genome (e.g., SD) or of symbiotic origin (e.g., the feminizing cytoplasmic bacteria of isopods) is not relevant to the definition. Problems do, however, exist in defining just what a nonsignificant level of horizontal transmission means in this context. For the sake of argument I shall let it mean that the SPE experiences horizontal transmission at a rate that is so low that the SPE could not accumulate adaptations to this mode of transmission.

While a defining feature of SPEs is that they can spread, even if deleterious, it does not follow that all SPEs need be deleterious. Cultures of bacteria with transposable elements tend, for instance, to out-compete cultures without, at least under the artificial context of the chemostat (see, for discussion, Charlesworth, 1987). The same can also be shown in yeast (Wilkie and Adams, 1992). Similarly, B chromosomes need not necessarily be deleterious (though most probably are) (Shaw and Hewitt, 1990) and biased gene conversion may provide an advantage to sex (Bengtsson, 1985). The spread of these factors thus may have two components: selection on the "host" (i.e., what is usually considered as the individual) due to its bearing the self-promoting element and selection below the individual level. The two levels can be distinguished by whether a suppressor of the self-promoting activity, that acted within the same individual, would be able to spread. If the SPE had net individual-level fitness costs, then a suppressor could spread, but this need not be so if the SPE was beneficial to its host. For the instances that are to be discussed here, selection must be acting below the individual level.

IV. Genes and Interests

It would be desirable to have a shorthand for describing those actions that a gene might perform (feminizing the host, killing other sperm, being advantageous to the host, etc.) that would enable that gene, given its particular transmission pattern, to spread in the population. To some extent, this language already exists in human affairs: one can refer to the *interests* of genes (Dawkins, 1976, 1982). In this type of usage, to say that a cytoplasmic genome has "no interest" in a male is not to imply that the cytoplasmic genome has thought the issue through and decided that it would be bored. Rather it is a condensed way to express, for example, the notion that maternally transmitted cytoplasmic genes which turn males into females could spread. Similarly, when two genes do not share precisely the same set of interests, we might say that there is potentially a conflict of interest. This conflict can be precisely formulated in terms of the invasion of one gene (the self-promoting element) creating the context for the spread of a second gene (the suppressor) that acts within the same individual but opposes the action of the first.

What exactly can one mean by saying that a gene has interests? The gene in this context is not necessarily a segment of DNA coding for some protein (i.e., not the molecular biological definition of a gene). It is not that this usage is not important, just that it is not the logically precise usage that is necessary in this context. To approach what is meant by a gene in this evolutionary context it is useful to ask the question, how many different sets of interests might there be in an individual?

The formulation of the notion of a gene that is to be employed in this context can be understood by asking about where in the genome enhancers of a SPE could reside and spread and where suppressors of the SPE could reside and spread. If one chromosomal location could house a suppressor of an invading SPE and in this location the suppressor could spread (because of its suppressing activity), then we can conclude that these two locations must be in conflict and must have different interests.

As discussed above, the interests of a cytoplasmically-transmitted genome may be to distort the sex ratio such that females containing the manipulating gene are more common than they would be were the cytoplasmic factor not manipulating the host. However, an autosomal factor that performed the same manipulation would, under a broad range of conditions, not be able to spread (Lewis, 1941; Hamilton, 1967). Indeed, an autosomal suppressor of a cytoplasmic sex ratio distorter can invade under a broad range of conditions (Uyenoyama and Feldman, 1978; Werren, 1987). The equal sex ratio that Fisher (1930) derives is a stable sex ratio only if the interests of autosomes are considered. So in sexual species with nuclear and uniparentally transmitted cytoplasmic genomes, at least two different sets of interest must exist.

Let us also consider SD once again. It is in the interests of SD for SD to kill sperm not containing it. But what about the rest of the genome? The key question is whether other regions of the genome are likely to gain or lose by SD's activity. This will depend upon the extent to which they recombine with SD. Any gene unlinked to SD is as likely to find itself in the killed sperm as the surviving sperm. If, as seems likely, there is some cost to male fertility to the possession of SD, then these genes may be selected to suppress the action of SD. A suppressor in linkage equilibrium with SD could spread in the population because the suppressor would not find itself in a dead sperm as often as the allelic nonsuppressor.

So we may conclude that, at least in this context, that which has interests must be something akin to what classical genetics refers to as a linkage group, as these are domains in which linkage disequilibrium can be held for an adequate length of time to allow the accumulation of synergistic effects.

It is helpful in this context to illustrate again the notion that the potential for linkage disequilibrium is the central concern and not the absolute proximity of genes on chromosomes: a W chromosome and a cytoplasmic factor potentially have the same interests, as they can be in strong linkage disequilibrium despite the fact that they are not actually physically linked; it is cotransmission that guarantees linkage disequilibrium. This can be achieved by physical chromosomal proximity, but it is not an absolute requirement.

In sum then, for each linkage group there exists a possible set of common interests, while for unlinked genes there may exist an antagonistic set of interests. There must then be at least as many potentially independent interests as there are linkage groups. In general, we might say that in genetic systems the parties to any given initial manipulation/action are twofold: those genomic locations at which suppressors can spread and those locations at which enhancers can spread, typically those tightly linked to the original manipulator gene. For logical precision, however, it is necessary to discriminate two other parties. First, neutral ground. An infinitely small domain, defined by the recombination rate between the manipulator and the rest of the chromosome on which it resides, exists in which neither a suppressor nor an enhancer can spread. This is a mathematically definable point, but is probably irrelevant to most biological systems.

Aside from neutrals and the two conflicting parties, it is necessary to distinguish a fourth class of party, these being those linkage groups that are allelic to the SPE and that are hence in direct competition with the element (Cosmides and Tooby, 1981). The allelic competitors are different from the components of the genome that are in conflict with the SPE for a variety of reasons. First, an increase in frequency of the SPE must by definition cause a reduction in frequency of the allelic competitors. The same is not true of the parties in conflict with the element. This distinction is important in so much as the invasion conditions for suppressive modifiers of competing and conflicting alleles are very different. Further, I have defined the conditions that are to be discussed as those where the SPE and the suppressor act within the same individual. Unlike these conflicting parties, the competing alleles need not be in the same organism as the SPE. The competing party to a cytoplasmic sex ratio distorter is the set of nondistorting cytoplasms. If, as in the case of meiotic drive, the competing party is in the same individual, unlike the parties in conflict, the alleles of the self-promoting element have a possible tactic not open to unlinked suppressors, namely, cis acting insensitivity to distortion.

V. Fitness Co-variance and the Paradox of the Organism

If, as established above, an individual is composed of a set of linkage groups with potentially antagonistic interests, how can selection ever act at the individual level? Or to put it another way, in evolutionary biology organisms are often treated as though they are entities with, in a given situation, a unique interest. How can there be a unique interest if the genome is full of linkage groups with antagonistic interests, and how can an individual ever be considered as an entity with interests? The resolution of this, the paradox of the organism, follows logically from the discussion of the virulence of parasites presented briefly above (Dawkins, 1990).

Given that a potential parasite has, as a consequence of obligate vertical transmission, a high fitness covariance with its host, we may expect the parasite to evolve to be of reduced (but not necessarily zero) virulence. Cytoplasmic genes may be selected to harm males, but they are not normally selected to harm females (with the provisos noted above). The survival of a female host is necessary for the survival and transmission of cytoplasmic genes within that host. (Parenthetically, it is, at least for me, one of the abiding mysteries of evolution that mitochondria are ever "well behaved" in males.) Thus, we may conclude that a high fitness covariance exists between autosomal genes and cytoplasmic genes, at least when they are in a female. Similarly, autosomes in both males and females require the survival of the host to permit their transmission. All the nuclear linkage groups hence usually have an equal interest in the survival of the host. Dawkins (1990) put this point elegantly when noting that genes in organisms have "desiderata lists" (lists of their interests in given situations) and that most of the time in any given individual these lists overlap. If the notion that an individual has any interests is to have any currency, it is because of a consensus of interests. Within this domain of overlap, there is by definition one unique interest. That individuals might be treated as having interests is, however, only accurate to a first approximation and selection may favor the spread of SPEs that decrease individual fitness.

This relationship among cotransmission (or what more ecologically may be thought of as codispersion), fitness covariance, and the evolution of virulence (or avirulence) is at least implicit, and often explicit, in much of the early consideration of genetic conflicts and hierarchical selection (see, e.g., Lewis, 1941; Östergren, 1945; Hamilton, 1967; Price, 1970, 1972; Leigh, 1971b, 1983, 1991; Alexander and Borgia, 1978; Eberhard, 1980; Cosmides and Tooby, 1981; Frank, 1983; Wade, 1985). More recently the same understanding has made an impression on the understanding of more conventional parasites, i.e., those with horizontal transmission (reviewed in Ewald, 1988; Anderson and May, 1991).

VI. "Individual Level" Adaptations Are Mutualisms Between Linkage Groups

Not only might high fitness covariance select for avirulence, it might also select for mutualism (Yamamura, 1993). A parasite's best interests may be served by increasing the fitness of its host (perhaps, in the case of bacterial symbionts, by releasing metabolites otherwise unavailable to the host). At this point, the parasite would conventionally not be considered a parasite anymore, but rather a mutualist. This latter term is problematic, however, and it is useful here to clarify just what it can mean in this context.

Employing the logic that has been employed up to this point, a mutualism between two parties is an interaction in which a manipulation/action by one partner creates the context for the spread of *enhancers* of this effect coded for by the other party. A cytoplasmic element that increased the fitness of females will spread in the population and provide the conditions for nuclear genes that could boost this fitness advantage. This can be contrasted to parasitic interactions, in which the initial manipulation forced by the self-promoting element (e.g., a cytoplasmic agent that sterilizes male tissue) creates the conditions for the spread of suppressors (e.g., nuclear restorers in hermaphrodite plants). Note, however, that promoters of self-promoting elements can exist (usually if the two can be maintained in linkage disequilbrium) and the enhancer and SPE would be in a mutualistic interaction, while at the same time the SPE would be in a parasitic interaction with unlinked components.

The term mutualism (as formulated above) is specific to just one interaction between specific partners. The term mutualism is, however, more often than not applied as though it were the sum of all interactions between two partners (e.g., when we say that the algae and fungi in lichens are engaged in a mutualism). This summation loses many of the important details (notably the arenas of conflict). There may be a meaningful usage to it, however. One can ask whether a mutant in either party would be able to spread in the population, were it to become independent. If neither party would be better off were they independent, then the interaction as a whole may be considered a mutualism.

This question of whether one party is better off in or out of a relationship is, however, not precisely equivalent to the summation of all interactions. This is most particularly the case when the option of leaving is not open (i.e., if there is a constraint). It seems then slightly problematic to describe the mitochondrial/autosomal interrelationship as a whole as a mutualism. This is for the simple reason that typically neither party has an option anymore; they cannot simply opt out of the relationship. It seems useful to say that some components are mutualistic and others are parasitic, but it is unclear what can be gained by then referring to the relationship as a whole as a mutualism. Plant mitochondria and autosomes would then, for the most part, be in a mutualistic interrelationship in female tissue, while in male reproductive tissue the relationship has strong components (the cytoplasm's attempt to inhibit pollen production) that are conflictual and hence provide the conditions for the spread of autosomal genes that act to prevent the cytoplasmic factors from sterilizing male tissue.

In sum, a genome/individual may be thought of as an array of symbionts (linkage groups) with obligately vertical (though not necessarily maternal) transmission that may be engaged in mutualistic interactions, one with another. Thus, at the genomic level, adaptation may be all pervasive: all adaptations of individuals are at the genomic level in the sense that they can be framed as the strategies of linkage groups for their own transmission. Importantly, however, these actions do not create the context for the spread of suppressors in unlinked chromosomal domains. In this sense, the adaptations of flight muscle are presumptively evidence of adaptation at the genomic level (this logic is explicit in Williams, 1966, 1992; Dawkins, 1976). However, as the interests that are being served by such advantageous traits are those of the rest of the genome, which is more often thought of as individual level adaptation, the genomic nature of "the individual within the individual" is not evident and it seems inappropriate to include these as genomic adaptations.

As individual-level adaptation is treated elsewhere in this book, the rest of this chapter will only consider those instances in which the linkage group is being parasitic (i.e., when it is a self-promoting element). The notion of adaptation at the genomic level shall hence be restricted to asking about whether self-promoting elements are adapted to being parasitic in the same sense as tapeworms and flukes are adapted to manipulating their various hosts.

VII. Identifying Adaptations at the Genomic Level

Whether or not the manipulations performed by self-promoting elements are adaptations to a particular transmission route is dependent upon that which one considers to be an adaptation. Here I consider two different usages and contrast how adaptation may or may not be identified at the genomic level in each circumstance. The terminology (e.g. Brosius and Gould, 1992) that attempts to distinguish between original and co-opted function of segments of DNA (see also Graur, 1993) will not be considered.

The first usage may be useful in discussions concerning the relative importance of drift and selection. We may, for instance, wish to have a shorthand to describe those phenomena that are the product of the latter and not the former. We may hence say that adaptations are those traits that are the product of selection on that trait rather than the product of drift (it is usual also to exclude both hitchhiking and pleiotropy) (see Amundson, Chapter 1). The second definition that I would like to consider derives from the argument from design [as discussed by Williams (1966; 1992)]. Here the term is restricted exclusively to those characters that are so complex and bear such an obvious relationship between structure and function that they could not possibly be the result of chance evolution. This latter understanding could have numerous subtly different manifestations. I shall consider only one, that the structure's function should be deducible from the structure alone. For convenience, this second view will be referred to as the Williams argument.

Much of the time, the term adaptation is used in such a manner that both definitions would be in agreement and hence the distinction between the two is often unimportant. When it comes to phenomena at the genomic level, however, the distinction is very important. Selection at the genomic level can often be trivial to demonstrate, but only rarely can selection be inferred from amassed complexity and its relationship to function (to the best of my knowledge this argument was first made by Alan Grafen, but not published by him). I shall illustrate the point through a consideration of the standard means by which selection/adaptation may be demonstrated.

Under the second definition, Williams has argued that there is only one means to identify adaptations. The relationship between structure and function must be complex and evident. Under the first usage, however, selection may be inferred or demonstrated by at least four means. These being:

A. Direct Means

The observation of a change in gene frequency (or the dynamic balance of two precisely opposing changes) associated with a heritable character and demonstrable advantage. For example, selection is shown to cause an increase in frequency of a gene or to maintain an equilibrium.

B. Indirect Means

1. The Comparative Method

An argument that a character is an adaptation should predict the incidences and covariances with other characters or circumstances across various taxa.

2. Experimental Manipulation

If an entity is adapted to a particular set of circumstances, then one might be able to predict the response should the circumstances alter.

3. Adequate Complexity / Function

The final category is the same as that defining adaptations sensu Williams. These four approaches are discussed in turn.

VIII. Direct Observation of the Forces Affecting Gene Frequency

For most changes that are presumed to be the product of selection, a change in gene frequency associated with an allele of known phenotypic effect is not the way most amenable to demonstrating selection (but see Endler, 1986). In contrast, the defining characteristic of a SPE is what might generally be called over-representation. Selection below the individual level is hence both a defining feature and the most commonly observed feature of SPEs.

Consider, for example, cytoplasmic incompatibility (Rousset and Raymond, 1991). Within some populations of *Drosophila simulans,* there are two sorts of individuals: those harboring bacterial cytoplasmic symbionts of the genus *Wolbachia* and those which do not. These symbionts are vertically transmitted within eggs but not in sperm (or, at least, if they are transmitted through sperm, it is very infrequent). The symbionts seem to affect the response to sperm somehow such that if a male with the bacterium mates with a female without, the eggs all die. The pattern is consistent with a general toxin/anti-toxin scheme. Under this two locus configuration, the bacterial genome codes for a toxin that is put into the sperm. When in eggs it also provides an anti-toxin. Hence, female flies with the symbionts are compatible with males regardless of whether the male is infected. Similarly, an uninfected female is compatible with uninfected males, since the symbionts are not necessary for the development of the embryos. However, an uninfected female is incompatible with infected males (the female lays eggs which do not hatch). The frequency of females infected with the bacterium increases because of the decrease in the number of uninfected individuals. Often cytoplasmic incompatibility cytotypes go to high frequency but not to fixation. What balances the increase with a concomitant decrease to prevent the factor from going to fixation is not always clear. [Note that while the above two locus designation is heuristically accurate, it has yet to be proven and comparable one locus distorters are known (Wickner, 1991). The above mechanistic scenario should not hence be taken as fact.]

The effects of numerous cytoplasmic sex-ratio distorters are equally obvious. Feminizing bacteria in crustaceans find themselves in twice the number of females that a competing nonfeminizing cytotype would do under the same circumstance. The male-killing cytoplasmic elements in ladybirds have the very obvious effect of killing male eggs that the sisters then consume and hence develop faster. Meiotic drive genes such as SD and the *t-complex* in mice may be observed increasing in gene frequency every generation. Heterozygous fathers in both cases leave an excess of progeny carrying the distorter. In the case of SD, the death of half of a male's sperm can be observed cytologically. Many B chromosomes are similar in that the act of meiotic or premeiotic duplication can be seen in cell preparations. That the distorter does not go to fixation in the above two cases is because there is also a force that tends to decrease the SPE frequency: the sterility or inviability of homozygotes.

Self-promoting elements are thus unusual in that the role of selection (below the individual level) can often be relatively trivial to show. Find an overreplicating element and you have also (usually) found the reason for its spread. However, it does not follow that self-promoting elements are trivial to find. Experimental demonstration of the overrepresentation requires that a wild-type genomic state exists with which to compare genomes with the self-promoting element. What if the self-promoting factor is at fixation within a population? Two possibilities exist. First, one can alter individuals within the same population to regenerate the putative wild-type allele. Second, one can take closely related but unaffected populations and cross into these.

A difficulty with the latter procedure, however, is that the hybrid condition may not have been experienced by any extant self-promoting element, and hence interpretation of the phenotypic results may be difficult. This is particularly true for self-promoting elements that are expected to co-evolve with suppressors. The *t-complex* in mice for instance, though a known self-promoting element, is associated with male sterility in hybrids (Pilder et al., 1991, 1993). The case of sterility in populations of *Drosophila subobscura* is also informative (Hauschteck-Jungen, 1990). An X-linked driver is found in Tunisian populations of *D. subobscura* that occurs at a high and stable frequency (~30%) resulting in female-biased sex ratios. Crosses between Tunisian males with the driver and female laboratory stocks derived from wild European populations of the same species also produce female-biased sex ratios. However, when F1 hybrid females are back-crossed to the European stock, F2 male offspring carrying the drive chromosome are nearly always sterile (one would have expected them to show drive). Significantly, nondriving X chromosomes of Tunisian origin do not cause sterility. In this example, the gene causing sterility could be the driver itself (or a modifier), although the involvement of a tightly linked sterility gene cannot be discounted and preliminary analysis suggests that a linked sterility gene might be responsible. Why a sterility gene should be tightly linked to a drive gene is itself a problem worthy of further study.

This inability to predict the outcome of the placement of an SPE in a novel genetic environment need not beset all SPEs. Notably, those elements that are not expected to undergo co-evolution with the rest of the genome are more likely to behave "normally" in some novel circumstances. This, however, is not a hard and fast rule. For example, for SD in *Drosophila* to operate, the homologous chromosome must be sensitive to drive (in this instance a particular multicopy repeat, defined by an Xba1 restriction fragment site, is necessary in a particular range of copy numbers). An absence of this sensitivity on the homologous chromosome will ensure an absence of meiotic drive, despite the fact that the driver could be functioning as it normally does at the molelcular level.

Perhaps a lesser problem is that of obtaining false positives from hybrid crosses. Numerous angiosperm hybrids reveal cytoplasmic male sterility, although this is not witnessed within the hybridizing species (Kaul, 1988). Many times this is probably the release from suppression of a CMS gene (i.e., one that spread within the population because of the self-promoting effect) (see, e.g., Levy, 1991). But can we be sure it was not a *de novo* creation of a male sterile cytotype? Further investigation could control for this, but crossing into a different population may not be as simple a means as might be desirable.

These problems aside, the above two methods have provided evidence for two very similar self-promoting maternal effect lethals (and see also Taylor, 1994). The gene *scat*[+] causes severe combined anemia and thrombocytopenia (Scat) in mice (Hurst, 1993b; Peters and Barker, 1993), while *Medea* causes an analogous disease in beetles (Beeman et al., 1992). That they are considered to be self-promoting elements is mostly due to the interpretation of crossing experiments.

Scat was previously described as an autosomal recessive lethal condition (*scat*[-] being the recessive, *scat*[+] being the dominant). This view was overturned by the demonstration that *scat*[-]/*scat*[-] mice are viable just so long as the mother is also *scat*[-] homozygous (Peters and Barker, 1993). However, if the mother in which the *scat*[-]/*scat*[-] embryo develops has at least one copy of the dominant gene, then Scat will develop in the offspring. It is hypothesized that *scat*[+] (probably like *Medea*) is a two locus gene complex with one gene producing a toxin and another gene producing the anti-toxin. When both mother and fetus are *scat*[-] homozygous, no toxin is produced and hence no mortality is reported. However, if the mother is *scat*[+] she injects toxin into the fetus. Only if the fetus is *scat*[-]/*scat*[-] does it not code for the anti-toxin and hence only then does it die. This differential mortality ensures that *scat*[+] spreads at a cost, both to its competitor allele and to the unlinked parts of the genome; hence it is a self-promoting element. Whereas with Scat the putative wild-type allele can be maintained in inbred laboratory strains, the identification of *Medea* emerged through interpopulational crosses.

While following individual lines and comparing the transmission rates of genes at a particular locus provides good evidence that a gene may have spread/be maintained because of its self-promoting effects, one might object that the net fitness effects have not been assessed and that spread must actually be shown. This would require cage or field experiments. These experiments are rarely performed (but see, e.g., Fitz-Earle and Sakaguchi, 1986) and even if performed do not directly demonstrate the original reason for spread (Shaw and Hewitt, 1990).

IX. Indirect Means

Demonstrating that a gene is a self-promoting element by indirect means is not trivial and rarely convincing. Indirect evidence can, however, strengthen a case. I shall discuss the three indirect means in turn.

A. Comparative Methods

The comparative method attempts to identify independent evolutionary events and to correlate these with covariables (Harvey and Pagel, 1991). As regards the identification of self-promoting elements, the comparative approach is a valuable tool but can never be conclusive. I shall discuss three classes of evidence. First, if self-promoting elements are acting as postulated, their distribution might have particular phylogenetic patterns. Second, some classes of self-promoting elements have prototypical anatomical features. Analogy to these features provides suggestive data. Finally, I shall consider the prediction that some self-promoting elements should evolve relatively rapidly.

1. Comparative Predictions of the Occurrence of Self-promoting Elements

Self-promoting elements often require particular ecological, genetic and phenotypic conditions for their spread and maintenance. If a class of genes are self-promoting elements, then one might be able to predict their distribution as the function of required parameters. For many costly self-promoting elements to spread in the population there must, for example, be an adequate level of outbreeding (Hickey, 1982; Hickey and Rose, 1988). The cost of self-promotion would be suffered in full by the self-promoting agent (i.e., a high fitness covariance between the SPE and the rest of the genome). This logic has been applied to analyzing the issue of whether B chromosomes are, as Östergren (1945) first proposed, parasitic or whether they might, as Darlington presumed (Darlington and Thomas, 1941; Darlington and Upcott, 1941), be advantageous to the host.

Burt and Trivers (1993), in analyzing B chromosome number in angiosperms, report that inbred species have lower titers of B chromosomes than do outbred ones. From this one can conclude either that they are parasitic and require outbreeding to spread or that inbred individuals are less in need of their advantageous effects. Although the latter does not seem too appealing an explanation it cannot be ruled out from the comparative data alone. The generality of this conclusion can be questioned, however, as some B chromosomes are known to be advantageous (Miao et al., 1991) and at least one comparable intraspecific analysis has come to the opposite conclusion (Benito et al., 1992). For a related discussion of the relationship between inbreeding and selfish cytoplasmic factors, see Reboud and Zeyl (1994) and Hurst (1994a). For the possibility that introns may be self-promoting elements and at a lower titer in inbred organisms, see Hickey and Benkel (1986).

The same mode of analysis can be applied to the issue of the C-value paradox. The amount of DNA within the haploid genome, the C-value, varies enormously from organism to organism (Cavalier-Smith, 1985). Among eukaryotes it ranges from around 0.009 pg in the yeast *Saccharomyces cerevisiae* to over 700 pg in *Amoeba dubia*. This variation primarily reflects differences in the amount of noncoding DNA in the genome rather than differences in the number of genes. It has been suggested that such noncoding DNA possesses adaptive, nucleotypic properties, for instance it appears to determine nuclear volume. Variation in the C-value can thus be explained in terms of selection at the individual level, relying, for instance, on relationships among cell volume, nuclear volume, and hence genome size. However, the exclusivity of such explanations has been questioned.

Alternative theories rely on the ability of selfish DNA to copy and spread through both genomes and populations. Self-replicating elements will tend to increase the overall size of the genome, even, perhaps, above the level that is optimal for the host. C-values may therefore represent points of balance among opposing selective forces, those acting at the level of the gene, and those acting at the level of the individual organism (Doolittle and Sapienza, 1980; Orgel and Crick, 1980). If, as proposed, most of the variation in genome size is due to the accumulation of parasitic junk (selfish DNA), then, as before, genome size may vary as a function of inbreeding rate. At least within the angiosperms this appears not to be the case (Burt and Trivers, 1993; but see also Charlesworth et al., 1994). Alternative comparative analyses on different parameters have, however, reached the opposite conclusion (as regards salamander C-value see, e.g., Pagel and Johnstone, 1992). For further considerations of the evolutionary dynamics of repetitive DNA in eukaryotes and its relevance to the C-value paradox, see Charlesworth et al. (1994).

Not only might one predict variation within a group for the titer of self-promoting elements, one might also be able to predict the incidences of particular classes of self-promoting elements. Male-killing cytoplasmic elements, for instance, probably require a considerable transfer of resources from dead sons to sisters (and/or some level of inbreeding and inbreeding depression). Species with gregarious young are thus more likely to bear them than nongregarious ones, which indeed appears to be the case (Hurst and Majerus, 1993).

Consider also the hypothesis that paternally expressed imprinted genes are self-promoting growth factors (Haig and Westoby, 1989; Haig and Graham, 1991; Moore and Haig, 1991; Haig, 1992). If a mother has more than one mate, paternally derived genes in any given fetus will not necessarily be related to genes in fellow brood members or subsequent offspring of that same mother. In contrast, a maternally derived gene in any given fetus has a constant high probability ($p=0.5$) that its sibs will contain a clonal copy of it. As Haig (1992) notes, this tripartite asymmetry in relatedness among (1) paternally derived fetal genes (2) maternally derived fetal genes, and (3) genes in the mother (and hence in the fetus' sibs) creates a three-way conflict of interest over how much nutrition the fetus should demand from the mother. The mother would "prefer" to divide resources more or less equally and so maximize her net fitness, not the fitness of any given progeny. In contrast, if the paternally-derived genes in any given fetus are not going to be present in the sibs, then any decrement in maternal fitness resulting in reduced offspring production is, at the very least, irrelevant to these paternally-derived genes. Paternally-derived genes in a given fetus might thus prefer an allocation in excess of that preferred by the mother. The optimal amount of resources that the maternally-derived genes in the fetus should require will be intermediate between the optima for the paternally-derived genes (a large amount) and the amount the mother should be prepared to provide (a smaller amount).

This difference in optima between the maternally and paternally-derived genes in a fetus is proposed as an explanation for genomic imprinting. Although it is usually assumed that the parental derivation of a gene is irrelevant to the functioning of that gene, this is not always true. Some genes are only active if inherited from the mother (maternally imprinted) whereas others are the reverse. Imprinting of murine insulin-like growth factor 2 (*Igf2*) and the *Igf2* receptor (alias the mannose-6-phosphate receptor) [see Haig and Graham (1991) for interpretation and references] is the now classic example. Fetally expressed *Igf2* is one of the factors that are supposed to promote the acquisition of resources from the mother. Early in mammalian embryogenesis, as expected, the paternally-inherited copy of *Igf2* is expressed, while the maternally inherited copy remains silent. Instead, again as expected, the maternally-inherited copy of an antagonist to Igf2 (*Igf2r*) is expressed while the paternal copy of this remains silent. Alteration of expression of the above genes has the predicted consequences on fetal growth (Harvey and Kaye, 1992; Kiess et al., 1994).

According to this theory, for a system such as imprinting to operate in the sequestration of biological resources, (1) parental provisioning of offspring must be sex-biased, (2) resource allocation per offspring must be flexible, and (3) offspring must be able to manipulate the parental resource allocation. Imprinting of the above variety should thus be expected in those groups with intimate maternal/offspring connections. Eutherian mammals are one such group (for which all the best-known imprinted genes have been described). Most exchange of resources is done across the placenta and it is thus significant that imprinting affects many genes with placental expression. In marsupials, however, imprinting (non-random *X*-inactivation) predominantly affects muscle tissue; in this group the intimate contact is the mouth-nipple contact and hence suckling is most probably the means of manipulation of resource transfer. It would be useful to know about the presence or absence of imprinting in monotremes.

The other group known to have imprinting is the angiosperms in which, again, resources are not fixed and contact is maintained between mother and offspring (Haig and Westoby, 1989; Moore and Haig, 1991). Rather than gene-specific imprinting, other groups are known to have whole genome imprinting [e.g., heterochromatization of the paternal chromosome set of scale insects (Nur, 1989), differential methylation of the chloroplast genome of *Chlamydomonas* (Matagne, 1987)]. These are excluded from the present comparative analysis as they are regarded as having a different evolutionary explanation.

Although the preliminary comparative evidence supports the view that imprinting is the result of differential paternal and maternal interests, the case is far from proven, as the correlation is no demonstration of the process (for a review of alternatives see Hurst, 1996a). Alternative models predict the same pattern (see, e.g., Solter, 1988; Tycko, 1994; Thomas, 1995). The conditions necessary for imprinting are also those required for the maternal assessment of fetal condition. An advantage of having a placenta, according to an alternative hypothesis, is that it allows a mother to stop the provisioning of poor quality offspring and boost that of those of better quality. Imprinting might then be a means for the offspring to advertise its quality or to allow the early abortion of monosomics. This hypothesis predicts the same phylogenetic distribution as the previous hypothesis. But the second theory fails to explain why not all chromosomes are imprinted and would not predict that growth-promoting genes should be paternally expressed. Both of these theories can more easily explain why over- or under-expression of imprinted loci often results in embryonic mortality and why imprinted genes appear not to be rapidly evolving (see later).

2. Comparative Tests of Genetic Anatomy

Many classes of self-promoting elements may be expected to have particular chromosomal anatomies and molecular phenotypes. Many self-promoting elements have either been shown to comprise two loci or are consistent with the action of two loci, a toxin producer and the anti-toxin [e.g., meiotic drive genes (Lyttle, 1991), *scat⁺* (Hurst, 1993b), cytoplasmic incompatibility (Rousset and Raymond, 1991), bacterial killer systems (Gerdes et al., 1990), or cytoplasmic killers (Beale and Jurand, 1966; Puhalla, 1968; Somers and Bevan, 1969)]. For two locus systems that are in chromosomal domains in which recombination could be expected, very particular invasion conditions must be met. The toxin allele must initially appear in the population in linkage disequilibrium with the anti-toxin allele. If it does not, then it will provide the conditions for its own destruction and immediate elimination. However, the anti-toxin allele must not be too common initially. Were it common then spread would be impossible as most competing versions of the linkage group would be immune and the toxin producing allele would be selected against as it would simply suffer costs of toxin production while receiving none of the advantages. If, however, the anti-toxin allele is rare and the toxin allele must be in linkage disequilibrium with it, then the two must be closely associated and not be broken up by crossing over. Thus we may predict that two locus distorters may be more likely to accumulate near centromeres as these are the chromosomal domains with the lowest recombination rate.

scat⁺ is tightly linked to the centromere of murine chromosome 8. Similarly, all but one of the well described autosomal meiotic drive genes are linked to the centromere (Lyttle, 1991, 1993) (see Fig. 1). The one exception to the rule is a meiotic drive gene that acts in oogenesis and that is distant from the centromere (Agulnik et al., 1993) (but see also López-León et al., 1992). Like all the others, however, this driver comprises two loci held in an inversion which serves to reduce the recombination rate between the driver and the wild-type chromosome.

3. Comparative Rates of Evolution

A property of some, but by no means all, self-promoting elements is that they are rapidly evolving. This could be because they may be involved in an arms race with suppressors located elsewhere in the genome. Fast evolution thus may be suggestive of conflict of one variety or another. This need not, however, imply that the gene is a self-promoting element. The conflict may be between genomes, not within one. This is, for instance, the most likely explanation for why antigenic components of parasites and antigen recognition components of immune systems are so rapidly evolving (Hughes and Nei, 1988; Hughes et al., 1990; Hughes, 1991; Kuma et al., 1995) (Fig. 2). Intraspecific conflicts may also provoke rapid evolution. The rapid evolution of mammalian growth factors may be one example.

Haig (1993b) argues that many of the products secreted by mammalian fetuses into the maternal blood system are to manipulate the mother into providing greater fetal provisioning (see also above). He postulates, for instance, that human chorionic gonadotropin beta subunit (hCGb) is a fetal means to prevent maternal attempts to reject it. This involvement, he argues, is the reason for its fast evolution and divergence from its presumed ancestral protein, human luteinizing hormone (hLH). Both hLH and hCG are dimeric glycoproteins and share a common a-subunit but possess different b-subunits that are encoded by closely linked autosomal genes. There are six known hCGb genes and at least one of them is expressed in the placenta. The ancestral hCGb was probably derived from hLHb by two frame-shift mutations. The consequence of these mutations was the addition of a 24 amino acid tail to the C terminus. Haig notes that hCGb and hLHb share only about 80% amino acid identity and that their nucleotide sequences show a high proportion of nonsynonymous to synonymous changes (Talmadge et al., 1984). Intron sequence divergence appears to have gone at about the same rate as the silent substitutions.

Haig also proposes that, were fetal demands to go unopposed, the fetus would, for instance, remove more glucose from maternal blood than was in the maternal interest. He suggests that the mother's best interests are served by reducing her blood sugar to limit fetal uptake. Further, he argues that a mother and her fetus will compete after every meal for control of the blood sugar. Whereas the mother would prefer to take sugar out of the blood rapidly, so the fetus would prefer that the sugar remained in circulation and hence can be taken across the placenta. This conflict between mother and fetus has resulted, it is

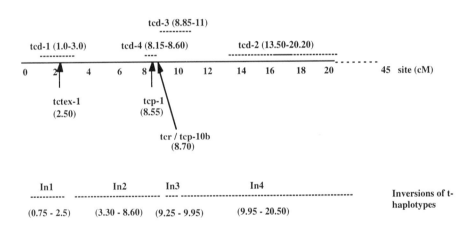

Figure 1 A map of mouse chromosome 17. The location of genes is shown in parentheses after the gene name. The chromosome is between 45 and 50 cM long (approximately 100 MB). The centromere is between *tcd-1* and site 0. Three well described distorters with additive effects are known and are indicated (tcd1-3). Tcd-4 may be a distorter but its status is disputed. There may be at least two other tcd loci. A putative candidate locus for tcd-1 is given, as is that for tcd-4 (tcp-1 is in fact a molecular chaperone). The responder locus (alias the insensitivity locus is tcr) which is thought to be equivalent to tcp-10b. Note that the *t*-complex is contained within four inversions and is effectively linked to the centromere. Inversions act as blocks to recombination and centromeric localities have very low recombination rates. This reduction in recombination rates is typical for two locus distorters as it is necessary to keep toxin and anti-toxin genes together.

proposed, in an evolutionary arms race in which a trait permitting a fetus to increases its output of anti-insulin hormone (keeping blood sugar high) will spread and, in response, a trait which sees the mother increase her production of insulin (reducing blood sugar levels) can also spread.

The product of the putative arms race is, it is proposed, a very high fetal production of an anti-insulin hormone and a correspondingly high maternal production of insulin. The large titers of these two hormones has practically no net effect on the flow of sugars into the fetus as the two effects cancel each other out.

In mechanistic terms, placental lactogen (PL) and placental growth hormone (PGH) are proposed to be the fetal anti-insulin hormones. If placental lactogens are the fetal means to manipulate sugar uptake, then we might expect rapid evolution of these sequences. Significantly then, Wallis (1993) has shown that placental lactogens of ruminants evolve at a "remarkably high rate" with a K_A/K_S value (ratio of nonsynonymous to synonymous substitutions) of 1.75. Similarly, a comparison between murine and rat placental lactogens I and II reveals K_A/K_S values of 0.42 and 0.41, respectively (discussed in Hurst, 1994c). While these values are not as high as those for ruminants, they are none the less significantly greater than the average of 0.14 (n=363) found in the mouse/rat comparison (Wolfe and Sharp, 1993). Protein identity for PLI and PLII across this comparison is 74.9 and 77.8%, respectively, which compares with an average of 93.9% (± 8.1;

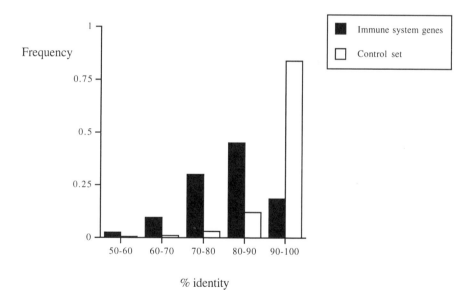

Figure 2 The frequency of immune (N= 46) and control genes (N=317) as a function of the percentage protein identity in the mouse-rat comparison. It is clear that as a class immune system genes are rapidly evolving, consistent with an antagonistic co-evolutionary arms race [analysis by the author (Hurst and Peck, 1996), data from Wolfe and Sharp (1993)]. In addition (data not shown), it can be shown that this high rate of evolution is not due to immune genes having a higher mutation rate than the control set.

n=363) (Wolfe and Sharp, 1993). A number of other genes that may well also be involved in maternal-fetal interactions have been shown to be relatively fast evolving, sometimes episodically. These include, growth hormone (Wallis, 1994), growth hormone releasing factor (Wolfe and Sharp, 1993), and prolactin (Wallis, 1981).

It should be noted that if the conflict is between fetus and mother, the gene for hPL is not strictly speaking a self-promoting element as the suppressor is not coded and expressed in the same individual as that coding for hPL. However, insulin is imprinted in mice [and probably in humans as well (Haig, 1994)] and hCGb is probably imprinted (Degroot et al., 1993). Both may hence be involved in intra-individual conflicts as well as between individual ones.

Maternal-fetal conflict has also been proposed as a means to explain the rapid evolution of the mammalian sex-determining gene SRY (for data on fast evolution see Tucker and Lundrigan, 1993; Whitfield et al., 1993). It is argued that the fitness of a mammalian zygote is affected by its probability of implantation and post-implantation maintenance, as well as the level of transplacental and transmammary uptake of resources. In a species in which females are not obligately monogamous, any cost-free Y-linked sequence that can positively alter any of the above parameters could spread in a population even if it harms the prospects of other embryos (Hurst, 1994b). It is argued that *Sry* could be one such gene. A suppressor of the selfish growth effects is expected to spread were it to have an appropriate parent-specific expression rule. As postulated (Hurst, 1994c), human *Sry* is known to have an X-linked suppressor (Bardoni et al., 1994) and this too is rapidly evolving (Dabovic et al., 1995). This finding is supportive of a role for *Sry* in some form of genetic conflict (as suggested by Whitfield et al., 1993), not necessarily maternal-fetal conflict.

One might also suppose that the conflict theory of imprinting would have predicted rapid sequence evolution of imprinted genes (the gene for a mutant Igf2 that can avoid binding to Igf2r should spread). In light of the above it is perhaps significant that no evidence for such an effect could be found when six imprinted genes were compared with a control set of over 350 genes in the mouse-rat comparison (Table 1). This may reasonably be taken as evidence against (but not falsification of) the conflict hypothesis for the evolution of imprinting.

B. Experimental Manipulation

If one includes under the umbrella of experimental manipulation all those experiments that reveal the self-promoting element as a linkage group with some means to gain overrepresentation, then manipulation is an important means for identifying self-promoting elements. Transferring *scat/scat* embryos into *scat⁺/scat⁺* mothers (Peters and Barker, 1993) would be one such manipulation. This class of verification is, however, no different from direct observation of selection and in this regard self-promoting elements are unusual (see Section VIII).

These manipulations may also provide more information about the self-promoting element. The male-killing spiroplasma in neotropical Drosophilids, for instance, has been confirmed as the agent responsible by the fulfilment of Koch's postulates for a disease agent (Hackett et al., 1986; Williamson et al., 1989). Knowing the causative element is interesting, but does not tell us much about the

TABLE I

Percentage Identity at the Protein Level of the Six Imprinted Genes
That Have Been Described in the Mouse-Rat Comparison.

Gene	% identity
Igf2	96.7‡
Insulin 1	93.5
Insulin 2	94.5
Mas-oncogene	97
Igf2r	>90†
SNRPN	100 §

For this comparison, the mean for an array of 360 genes is ~94% ± 8 (SD) (Wolfe and Sharp, 1993). Data compiled by the author. Although fast evolution (and low percentage identity) might have been expected, this is apparently not found. This either means that rapid evolution is not a defining feature of all systems in conflict or that imprinting is not the result of conflict.

‡ Note also, that the guinea pig sequence is identical to the human sequence (Levinovitz et al., 1992).

† About 400 of the approximately 2400 amino acids have yet to be sequenced in rat.

§ NB human sequence is also identical (see Schmauss et al., 1989).

spread of the trait.

Typically, within evolutionary studies manipulation is not thought of in the above terms. The manipulations discussed in Chapter 5 are more typical. This latter methodology has been successfully applied to animal behavior (e.g., optimal foraging analysis) and in the testing of sex-ratio theory. Manipulations of this subtlety are not a useful way to analyze self-promoting elements. Their behavior is just not flexible enough. In a few cases, however, where the spread of a supposed self-promoting element cannot be shown directly, the presumed self-promoting factor may have predictable qualitative effects and these can be experimentally investigated.

Consider again the hypothesis that paternally-expressed imprinted genes are self-promoting growth factors (Moore and Haig, 1991; Haig, 1992). Paternally-derived genes in a given fetus, it is postulated, should prefer an allocation in excess of that preferred by the mother, and maternally inherited genes might be selected to oppose this. This balance of demands is supported by a number of human imprinting disorders (e.g., Prader-Willi and Angelmanns syndrome) that have concomitant effects on growth consistent with the hypothesis. The hypothesis makes predictions about the size of a placenta if paternally or maternally-imprinted alleles are experimentally over- or underexpressed. The model predicts, for instance, that overexpression of the paternally-expressed (i.e., maternally imprinted) allele should result in a large fetus, while underexpression should have the opposite effect. This is supported by manipulations of mammalian fetuses in which an egg is given two sperm pro-nuclei (a large placenta develops) or two egg nuclei (a small placenta develops). Similarly, naturally produced paternal or maternal disomies result in changes in placental mass that typically conform to the prediction. There are, however, a few exceptions to this rule (Lyon, 1993; Thornhill and Burgoyne, 1993). Whether the exceptions are common enough to ensure that the claimed pattern is statistically significant is unknown and hence the above data should be treated with some caution.

Better data comes from manipulations (typically deletions) of individual genes. Eight genes have been investigated. *Igf2*, *Mas*, *Ins1*, and *Ins2* are all paternally expressed and enhance growth rates. *Igf2r*, *H19* and *p57KIP2* are all expressed off the maternally-derived genome and reduce growth rates. *Mash2* is expressed off the maternally derived genome and appears to enhance growth rates. The effect is however ambiguous, and excess *Mash2* has been conjectured to be responsible for the growth inhibition associated with a maternal duplication of the region within which it is held. Treating *Mash2* as a counter example, the correspondence between the direction of growth effects and direction of imprinting can be analyzed employing the G-test of independence with Williams' correction for a 2x2 table. This reveals Gadj=5.06, $0.05>p>0.01$. Taking this as the core analysis, it is found that changes to it sometimes alter the qualitative conclusion. First, if *Mash2* is treated as being a growth suppressor then the statistic becomes highly significant ($0.001>p>0.0001$). Second, if the two insulin genes are not counted as independent data points, the statistic remains significant. Finally, however, if *H19* is argued not to have any effects independent of *Igf2* (which may be the case) and is hence removed from the analysis, then the significance is removed. In sum then, it seems that there may be the predicted relationship among the direction of growth effects and the direction of imprinting but the case has yet to be proven.

However, many over- or underexpressed imprinted loci result in zygotic mortality rather than simply smaller or larger offspring [parthenogenetically-derived offspring for instance never survive, regardless of heterozygosity and this mortality is probably due to a deficiency in the products of imprinted genes (Walsh et al., 1994)]. The hypothesis would not obviously predict this. One might argue that such mortality could be due to co-evolution and that the mother tends to abort abnormally sized progeny. If so then, as with many conflict hypotheses, making unequivocal predictions may be very difficult.

C. Inference from a Clear Relationship Between Complex Structure and Function

Molecular machines now being made are clearly structures with functions and design. In principle then, there need be no problem considering small structures as being designed for functions. Can we, however, understand in similar terms small biological structures and can we derive function from structure? There are at least two reasons why the application of the argument from design at this level may be problematic.

First, we do not yet understand the operation of small biological structures and hence connecting structure to function is difficult. Or put another way, adaptation is very much in the eye of the beholder and, as our eye is a human eye, any specializations at the molecular level are out of our domain and hence we typically have less feel for them than we do of larger structures. This, in part, explains why our textbook example of adaptation is the eye rather than a protein import channel. Significantly, what are often considered "biochemical adaptations" (see, e.g., Hochachka and Somero, 1984) are not adaptations in this sense, as typically their designation involves *a posteriori* understanding of structure from an appreciation of a function rather than an *a priori* statement of function given the structure.

Even if we understood its structure-to-function relationship, however, could a protein import channel ever be a paradigm for adaptation or does small size impose limits that are problematic for the employment of the argument from design? The second problem with the application of the Williams argument is that even if function is derivable from structure, for small biological structures there need not be many steps to arrive at a structure and hence the probability that drift (chance) was responsible is hard to rule out simply from structural features. Small structures are not typically adequately complex.

To illustrate these two points, consider the hemoglobins of high-flying birds. These have convergently evolved towards very similar structures (often incorporating the same amino acids), almost certainly to enable efficient usage at low oxygen tensions (see Gillespie, 1991). Selection must have operated independently in the various lineages for a hemoglobin with the required oxygen-binding capacity. Could we have known that these hemoglobins were adapted to low oxygen tension without this data? We could have compared similar hemoglobins and, if we knew that one was from a high-flying bird, have proposed that the other might be from an organism adapted to low oxygen tension. This, however, is application of the comparative method, not an *a priori* statement about function from structure. More generally, methods currently employed to suggest the function of a polypeptide do not work out the function from the predicted secondary structure, rather they ascribe function by comparison with proteins of known function and structure. As with other small entities, the structure of the molecule does not contain (with our present understanding of protein functioning) enough unambiguous information about its function and the reason for this function. Furthermore, even if we did understand the function, it would be hard to claim that having a particular amino acid at a particular position must be the result of selection' drift could be just as good an explanation. The same is not true of the eye.

In general then, the application of the argument from design is problematic for small structures. To a large extent this must also be true of self-promoting elements. However, there is one feature that makes SPEs special in this regard. If one asks about why complexity is important in the argument from design, it is that to a large extent, a complex structure demands an explanation: it is too intricate to be considered a redundant characteristic (and is hence not consistent with drift). SPEs are unusual in that, although not necessarily complex, they are often clearly redundant when considered in terms of their utility for the individual (Section VIII). This is manifest not only in the apparent nonsense of the system (e.g., the death of half the sperm of a male or the death of the male progeny) that is accompanied by immediate changes in gene frequency, but also because we can often show a cost to this self-promotion. Associated with this redundancy, and to explain it away, we note the function of the self-promoting element (i.e., to self-promote). If a central concern of the argument from design is the necessity to infer redundancy (were a function not provided) and then explain away the redundancy by noting the function, costly self-promoting elements may hence fulfill the underlying conditions of the argument, but without the complexity and without a necessity for inference. That is to say, the problem of many steps can be

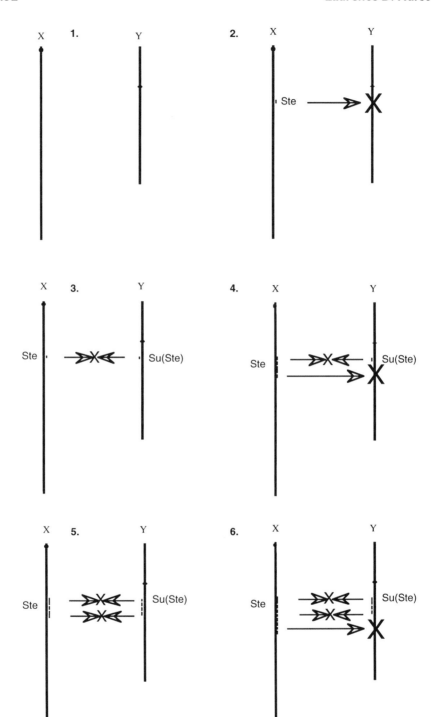

discounted as drift can often be discounted.

It is then worth noting a possibly significant exception to the generalization that the function of small structures cannot be found by analysis of their structure. *Stellate (Ste)* is a multicopy gene on the *X* chromosome of *D. melanogaster* whose transcription occurs only in the testes (Hardy et al., 1984). Copy number of the repeat varies among strains, up to about 200. In meiosis 1 *Stellate* expression is initiated, but expression of *Stellate* is suppressed in normal males by *Suppressor of Stellate (Su(Ste))* which is transcribed from the *Y* chromosome (Livak, 1984, 1990). That a *Stellate* sequence is not found in all Drosophilids is taken as an indication that the gene may not be necessary for normal spermatogenesis (Livak, 1990), but deletion analysis has yet to be performed. *XO* males of *D. melanogaster* are sterile and *Stellate* is implicated as one of the causes (the sperm of *XO* males have star-shaped crystals of the Stellate product, and it is from this shape that *Stellate* derives its name). The possibility that *Stellate* might be an *X* chromosome meiotic driver that has evolved in an arms race with its suppressor has been presented (Hurst, 1992) as a solution to the problem of how a gene that is possibly not required for spermatogenesis, and that renders the host sterile unless suppressed, could ever have evolved (Fig. 3).

The suggestion was that the original *Stellate* interfered with DNA packing in such a manner that the *Y* chromosome was more profoundly affected than the *X*. This might be simply due to the fact that the large *Y* had more DNA or heterochromatin requiring packing than the *X* or because the *Y* has more sites of interaction with the mutant protein. Whatever the mechanism, the consequence would be that when in low dose and with no suppression, an *X*-bearing *Stellate* would be present in more than 50% of the viable sperm. As long as host fertility was not reduced too dramatically, the gene would invade.

If the initial frequency of insensitivity is low (but note, this need not be the case for *X-Y* drive) then of all positions on the *X*, the centromere, being a region of low recombination, is the most likely location for a meiotic drive gene (see Section IX, A). The main body of *Stellate* copy repeats is not centromeric but it may be significant that two copies of *Stellate* are present at this position (Shevelyov, 1992).

As the driver invaded, so it would impose a cost, both reducing male fertility and biasing the sex ratio towards females. Thus a suppressor of this condition (*Su(Ste)*) on the *Y* chromosome could also invade and go to what, at least in the short term, would be a stable equilibrium. If this suppressor acted in a dose-dependent fashion, then a multiplication of the driver gene would produce a gene family that evaded suppression and once again produced drive. Selection would

Figure 3 A model for the evolution of *Stellate* (Hurst, 1992). (1) The sex chromosomes are "normal". (2) A copy of *Stellate* appears on the *X* (possibly next to the centromere - not as illustrated here). This acts to kill sperm from the same male containing the *Y* chromosome. The novel *Stellate* bearing *X* spreads as, although male fertility is reduced, the proportion of progeny with the *Stellate* bearing *X* goes up. The spread of this novel *X* creates (3) the context for the spread of a novel suppressor on the *Y*. If the suppression is dosage dependent then the spread of the suppressor provides (4) the context for the spread of an *X* with multiple copies of *Stellate*. This again provides the context for (5) a *Y* with even more copies of the suppressor, which in turn may give (6) yet more rounds of this arms race until a large body of repeats builds up that, if not suppressed, can cause sterility. When *Stellate* and its suppressor are balanced, loss of *Stellate* copies may be possible if their transcription is costly.

then act to increase suppressor copy number, and so on. If a balance between driver and suppressor was reached, selection would favor the deletion of driving genes, if the mere possession of drive genes has a cost. It can therefore be seen that the dynamics of driving genes are complex, with selection sometimes favoring an increase in copy number, at other times a decrease. Although the mechanism by which *Su(Ste)* inhibits the production of *Ste* product is not fully known, what is understood supports the assumption that *Su(Ste)* acts in a dose-dependent fashion.

This model has recently received very strong empirical support from analysis (Hurst, 1996b) of segregation data (Palumbo et al., 1994) from males with no *Su(Ste)* on the Y but that varied in copy number of *Stellate* repeats on the X. As predicted by the model, it was found that with a low copy number of *Stellate* (<35), males transmitted the X chromosome to significantly more than 50% of their progeny. Further, as also predicted (Hurst, 1992), but against previous empirically derived expectations (Hardy et al., 1984), it was found that the relative survival of X-bearing sperm, compared to that of Y-bearing sperm, goes up as the *Stellate* copy number goes up (Hurst, 1996b) (Fig. 4). It hence seems most reasonable to suppose that *Stellate* is, as predicted from the patterns of sterility and transcription, a meiotic drive gene. The case is not proven however and it remains to be seen whether variance in the number of copies of *Su(Ste)* has the predicted effect as well. In addition, it is curious that while there is considerable variation in the effect of a given *Stellate* copy number on the proportion of X-bearing sperm, there is very much less variation in the amount of nondysjunction (Robbins et al., 1996). The case may be proven if an X-chromosome lacking *Stellate* can be shown to have no distorting potential.

A similar arms race has been proposed for a multicopy repeat on the murine Y chromosome (Conway et al., 1994). Males with a deletion of the Y, removing some but not all copies of the repeat, produce a female-biased sex ratio. This is consistent with the action of an X versus Y meiotic drive gene. Whether the mouse X has a multicopy repeat responsible for the sex ratio effect has yet to be investigated. One clue, derived from analogy with *Stellate*, suggests that such a gene might, however, exist. Given that *Ste* and *Su(Ste)* are also homologous to some degree, the latter probably being derived from the former (Balakireva et al., 1992), it is then noteworthy that the multicopy repeat on the murine Y does have an X-linked homologue. It is yet more suggestive that this repeat is located very close to the X centromere (Laval and Boyd, 1993).

One might perhaps argue that the *Stellate* example, though not fully proven, shows that the argument from design can be applied at the genomic level [indeed there have been comparable attempts to identify conflicts in the absence of segregational data (Haig, 1993b)]. Thus we should remain optimistic about the application of selectionist thinking even at this level [note also that the genomic location of most two locus distorters was also predicted (Charlesworth and Hartl, 1978)].

X. Adaptations of Suppressors and the Evolution of Genetic Systems

Perhaps what is most convincing about conflict-based explanations in general (in the absence of segregation data) is that they convert nonsense into sense.

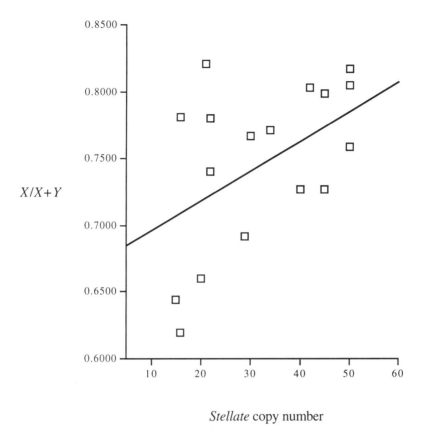

$X/X+Y$

Stellate copy number

Figure 4 The probability of survival of *X*-bearing sperm relative to that of *Y*-bearing sperm as a function of *Stellate* copy number (on the *X*) when the *Y* lacks the *Suppressor of Stellate*. A ratio of 0.5 is indicative of equal survival probabilities of *X* and *Y*-bearing sperm, whereas a figure greater than 0.5 is indicative of a greater probability of survival of *X*-bearing sperm. The slope of the line is significantly different from zero (P<0.05, n=16). If analysis is adjusted to take account of the number of data points contributing to each point used the correlation is more significant (Robbins et al., 1996). This hence shows that as *Stellate* copy number goes up so the proportional survival of *Y* bearing sperm, to that of the *X* goes down, as predicted by the hypothesis that *Stellate* is a meiotic drive gene (Hurst, 1996b). Note also that it is not in principle necessary that the best fit line should pass through *X/X+Y*=0.5 when Stellate copy number is zero. This is for the simple reason that there is expected to be a discontinuity between the presence and absence of a drive gene.

Unimportant facts are converted into sensible details (linkage to the centromere, placental activity, etc.). It is the elegance of this inversion that perhaps goes some way to convincing that the explanation might be correct. Elegance, however, need be no measure of truth. The conversion of what might have been put down to a baffling maladaptation (or even worse, assumed to be a constraint) into an exquisite adaptation has often been appealing to evolutionary biologists. Indeed, in much of evolutionary genetics it is the paradoxical aspects that attract the most attention. Sexual selection and the apparent redundancy of the peacock's tail is a particular case in point. But sexual selection is not just about males with large tails. It must also be about the evolution of the peahens' preferences. Similarly, as suggested for *Stellate* and numerous other instances, the putative conflict-initiating agent creates the context for the spread of a suppressor. So far this chapter has largely asked about the interests of the parasite within the genome. Linkage groups with alternative interests will be in conflict with these genes. Are these groups adapted to the task of preventing the spread of self-promoting elements and, if so, what are the consequences of such adaptation?

Suppressors, as with those of *Stellate* and well described for several meiotic drive genes, often simply prevent expression of the self-promoting element. The evolution of such suppressers is considered a major component of the factors maintaining Mendelian segregation (Crow, 1991). These suppressors will most obviously be shown by their increase in frequency compared to nonsuppressor allele and indirect means to demonstrate their appearance would be difficult.

An alternative form of response to the spread of a self-promoting element can, however, also be envisaged. Consider, for instance, the spread of a fast-replicating mitochondrial genome. If this genome is not too deleterious and the host has biparental inheritance of cytoplasmic genes, then the fast-replicating self-promoting element will spread in the population. Suppose that, as it spreads, a mutant nuclear gene enters the population. Let us further suppose that the nuclear gene kills the mitochondria provided by its mate. Will this mutant spread? If the mutant is initially associated with wild-type and not fast-replicating (and hence deleterious) mitochondria, the mutant can spread (Hoekstra, 1990; Hastings, 1992a). The model can also evoke mitochondrial meiotic drive analogues (Hurst and Hamilton, 1992). Once uniparental mitochondrial inheritance is established, any fast-replicating mitochondrial genome may spread within an individual, but not in the population. Is uniparental inheritance then an adaptation to prevent the spread of self-promoting cytoplasmic genes?

From the Williams argument, it again seems difficult to amass adequate complexity related to function. A variety of evidence suggests, however, that the above model is probably close to the truth. First, we find that, in isogamous organisms, nuclear genes do indeed control uniparental inheritance (Hurst and Hamilton, 1992; Hurst, 1995). Second, the model is competent to explain the apparent exceptions to the rule. For example, the model predicts that uniparental transmission should be the rule rather than simply uniparental inheritance. Under uniparental inheritance, the zygote contains cytoplasmic genomes from only one parent. Under uniparental transmission, a zygote may contain cytoplasmic genomes from both parents but transmit those from only one parent. The theory argues that it does not matter if a cytoplasmic genome over-replicates, if it is not to be transmitted. Hence it is a theory for uniparental transmission. In the best studied animal case of biparental inheritance (in the mussel *Mytilus*), this is indeed what is found, with two mitochondrial genomes being identified, one transmitted father to son, one transmitted mother to daughter (Skibinski et al., 1994; Zouros et al., 1994). Daughters then receive only maternally derived cytoplasmic genomes and transmit only these. Sons in contrast inherit both maternally and paternally-derived genomes but transmit only the latter (and only to sons). This is effectively two uniparental transmission lineages in which the sort of self-promoting cytoplasmic genome that was discussed above could not spread (but see, for alternative interpretation, Godelle and Reboud, 1995).

Third, several organisms do not fuse cytoplasms during conjugation and these appear to have mechanisms to prevent exchange of cytoplasmic genes. The best studied example is the ciliate *Paramecium*. In this group we find that the exchanged nuclear genes are tightly packed, the gap through which they are exchanged is very small, and the nuclear package must constrict to get through this gap (see Fig. 5). Cytoplasmic genes do not typically move from one partner to the other and the system is consistent with selection to prevent this movement (Hurst, 1990). Similarly, in basidiomycetes, hyphal fusion permits nuclear transfer but cytoplasmic transfer is strictly limited, suggesting selection opposing the transmission of cytoplasmic components (Hurst and Hamilton, 1992).

Regardless of the validity of the hypothesis, the model given above can be configured as a general one for the evolution of a means to control self-promoting elements. As noted above (Section IX, A), inbred lines are expected to be less vulnerable to deleterious self-promoting elements than comparable outbred ones. Asexuality is the extreme of the distribution (i.e., extreme inbreeding) and uniparentally inherited genomes are asexual. Read another way then, uniparental inheritance may have evolved as a means to enforce high fitness co-variance between cytoplasmic genes and the transmitting sex. Do genetic systems always evolve to maximize fitness co-variance and so hinder the spread of self-promoting elements? Clearly not. Sexuality greatly reduces fitness co-variance and provides the conditions for the spread of self-promoting elements [in this regard the spread of self promoting elements can be seen as an additional cost of sex (Hurst, 1990; Hastings, 1992a, b)]. But, given that sex is probably necessary for most larger eukaryotes, we may still ask whether genetic systems tend to evolve to minimize the costs enforced by self-promoting elements. Several authors have discussed situations in which this might be the case.

Maynard Smith and Szathmáry (1993), for instance, note that the evolution of chromosomes can be interpreted as a means to avoid the spread of fast-replicating deleterious components. They consider a population of cells some of which contain chromosomes, others allowing the genetic components of the chromosome to replicate independently. In each of the latter there is a probability that one of the components may start fast-replicating. The result after series of cell divisions (permitting random assortment of unlinked components) is that selection favors the cells with chromosomes.

A similar sort of analysis has been applied to the problem of the evolution of ciliate meiosis. Ciliates are unicellular protists that have sex by an unusual means (see Fig. 5). Rather than two gametes fusing, instead, as noted above, two cells pair up and exchange haploid nuclei. Each cell will have gone through meiosis prior to pairing to produce two such haploid nuclei, one of which remains in the cell and one of which is transferred. The manufacture of these nuclei is, however, strange. One might imagine that to make two haploid nuclei one might go through meiosis I, digest one nucleus, and resolve the other in meiosis II. Or perhaps one might go through meiosis to produce four haploid nuclei and digest two. Ciliates, however, produce four nuclei and digest three, then let the remaining one go through a division to produce two nuclei.

Why go to such lengths? It has been shown that just this sort of meiosis could

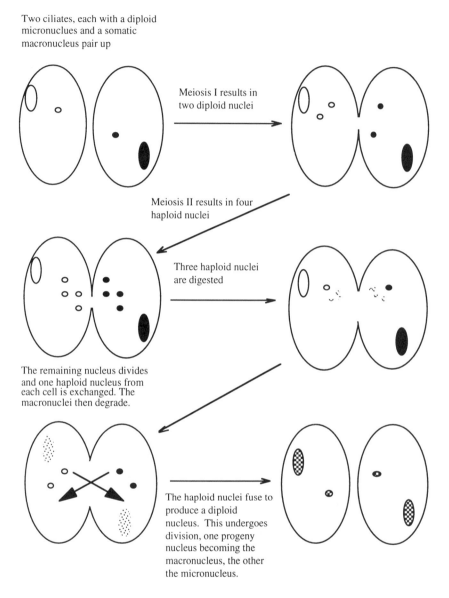

Figure 5 Diagrammatic representation of meiosis in ciliates. This meiosis is problematic in so much as the ciliates produce two haploid nuclei from meiosis, but do so in an apparently inefficient manner. They neither throw out one of the products of meiosis I, nor destroy only two products of meiosis II. Instead they destroy three products from meiosis II and duplicate the remaining one.

evolve, even if costly, due to the vulnerability of alternative types of meiosis to self-promoting elements (Reed and Hurst, 1996). Consider, for example, a nuclear linkage group that after meiosis produces both toxin and anti-toxin (i.e. a classic two-locus distorter). This toxin, it is assumed, is held within the cell in which it is manufactured. If the diploid cell is initially heterozygous for this gene, then with the simpler forms of meiosis the two nuclei produced may be different at this locus. If the toxin/anti-toxin nuclear gene is exchanged, then the cell that the nucleus leaves contains the toxin but possibly not the anti-toxin and may hence die [especially if the toxin is more stable than the anti-toxin, as is true in numerous other instances (Gerdes et al., 1990)]. The only cells that ever die are those that do not contain the toxin/anti-toxin complex. Hence, this mate-killing factor can invade a population under relatively broad conditions (if the death of the cell gives some advantage, the invasion conditions are particularly broad).

The spread of this element, however, creates the context for the spread of a nuclear modifier of the form of meiosis to that of current ciliate meiosis. It can be shown that, under various conditions (for example, close linkage to the wild-type allele), the modifier will invade and spread to fixation (Reed and Hurst, 1996). In effect, by doubling one product of meiosis, the system is geared such that the nucleus that remains in the cell is the identical twin of that which is passed into the partner cell. If then the nucleus being transferred has the killer trait, then so will that which remains (and hence it will also have the anti-toxin). If the nucleus that is transferred contains the wild-type allele, then so does the one that remains and neither toxin nor anti-toxin will be necessary.

Numerous similar examples exist in which selfish elements are hypothesized as the instigators of the conditions in which a suppressive response leads to a drastic change in the structure of the system concerned. For example, models exist for the initial evolution of cells, of mitosis, of meioses, of multicellularity and of germ lines (see Buss, 1987; Hurst, 1990; Haig and Grafen, 1991; Szathmáry, 1991; Haig, 1993a; Maynard Smith and Szathmáry, 1993; Frank, 1994). Not all of these, however, exist on solid logical ground (see, e.g., Buss, 1987) and more rigorous mathematical analysis is desirable in several instances.

A word of caution is necessary here. To understand the evolution of any given aspect of a genetic system, it is not adequate to note that this aspect may act as a defense against some selfish element. It is necessary to perform the necessary formal dynamical analysis (which is typically nontrivial) to work out whether the process could or could not occur.

While the above models establish plausible hypotheses for the evolution of genetic systems, the main problem facing these models is moving them out of the domain of plausibility alone and into a domain in which they can be tested. As with the identification of self-promoting elements, this can be done in one of four ways. An appeal to complexity, as with self-promoting elements, will probably not be convincing most of the time (although note that the idea that ciliate meiosis might need a special explanation is derived from the fact it is especially complex). Direct observations of major transitions in evolution are hardly practical. In the occasional instance, the natural experiment might have been performed. Consider for example the explanation for the peculiar form of meiosis in ciliates given above. This hypothesis has received some empirical support from experimental manipulations of ciliates with alternative meiotic forms (Reed and Hurst, 1996).

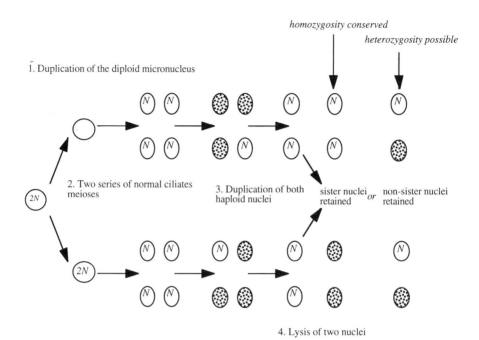

1. Duplication of the diploid micronucleus

2. Two series of normal ciliates meioses

3. Duplication of both haploid nuclei

sister nuclei *or* non-sister nuclei
retained   retained

4. Lysis of two nuclei

Figure 6 Meiosis in *Euplotes*. Note that after a parallel pair of ciliate meioses, rather than retaining two nuclei four nuclei are generated (step 3). By so doing *Euplotes* can control whether the nuclei to be inherited are either sister nuclei or non-sister nuclei. If the latter than heterozygosity can be maintained, if the former, homozygosity is guaranteed. It is found that in autogamous strains non-sister recovery is commonplace whereas in out breeders homozygosity is high. In at least one strain the preference changes dependent upon the form of reproduction. This pattern is consistent with the model for classical ciliate meiosis that envisages it to be a defense from nuclear localised mate killer elements (Reed and Hurst, 1996).

In species of the genus *Euplotes*, prior to meiosis there is an extra duplication of the DNA resulting in two diploid nuclei (Figure 6). Both of these nuclei then undergo the classical meiosis in parallel, producing a total of eight haploid nuclei. As per the classical ciliate meiosis, three nuclei from each meiosis are destroyed, leaving a total of two haploid products. The process does not, however, stop there. These two then undergo the postmeiotic duplication producing four products. Two of these again are destroyed, leaving just two pro-nuclei.

The incorporation of the final duplication and destruction ensures that this system could be equivalent to that of other ciliates as regards the relatedness between postexchange cells (and hence not greatly problematic to the theoretical outlook considered here). This would simply require that at the last stage the two destroyed nuclei were always sister nuclei (products of the same postmeiotic division). In contrast, however, were the digested pro-nuclei typically nonsister products, then *Euplotes* could be vulnerable to the form of selfish element that were envisaged. However, in autogamous *Euplotes* (ones in which the two haploid products of an individual fuse together and are not exchanged, effectively extreme selfing) this form of meiosis ensures that heterozygosity can be maintained (autogamy in *Euplotes* seems to be more common than in other ciliates) (see, e.g., Luporini and Dini, 1977). During selfing, self-promoting elements will not be a problem (see above).

An optimal solution might hence exist: in autogamous/selfing strains, nonsister nuclear recovery should be preferred (to maintain heterozygosity), but during outcrossing sister-nuclear recovery should be preferred (or at least should be found at a higher rate than in autogamous strains) so as to prevent nuclear mate-killers from invading. The available evidence provides support for the prediction (see Reed and Hurst, 1996). In some strains (Nobili and Luporini, 1967), meiosis is altered, in the expected direction, dependent upon whether the reproductive event is autogamous (preferential recovery of nonsister nuclei) or conjugative (higher rates of recovery of sister nuclei than in the autogamous case) (see also Luporini and Dini, 1977).

In several instances in which autogamy is a regular event (this can mean instances where autogamous and nonautogamous isolates are in sympatry and can intermate), heterozygosity is often maintained; hence the fusing pro-nuclei must more commonly be nonsisters. In contrast, the same pattern is not found in numerous isolates of *Euplotes*, in which autogamy appears not to be common. In these, homozygozity at a given locus is typically reported, indicating that sister pro-nuclear recovery goes on at a higher rate than in autogamous strains and often with a preference for sister-pro-nuclear recovery. This is all consistent with the hypothesis, but can hardly be considered proof.

This sort of natural experiment is not, however, available in most circumstances. Probably the best way forward in this circumstance is to apply the comparative method. This perhaps is somewhat unfortunate, as the comparative method can do no more than provide a correlation; it cannot demonstrate a mechanism. Furthermore, comparative analysis of systems in conflict is unusually problematic for, in the absence of information as to which interest will dominate, reliable prediction cannot be made.

Consider for example the problem of cytoplasmic genetics (Hurst, 1994a). It has been hypothesized that, starting with biparental inheritance, the spread of self-promoting cytoplasmic genomes (perhaps fast replicating ones or "killer" types) provides the conditions for the spread of nuclear genes that enforce uniparental inheritance of cytoplasmic genes (see above). However, not all cell fusions result in uniparental inheritance. Employing the same logic as the conflict hypothesis, it has been shown that, depending on the rate of in-/outbreeding and on which party is presumed to be in control, a variety of transmission patterns of cytoplasmic factors may be expected (Hurst, 1994a) and that all the possible outcomes can be found.

First, if nuclear genes are in control, then selection favors nuclear enforcement of uniparental inheritance. If this enforcement is costly, then it is not expected where, in a population with biparental inheritance, either selfish cytoplasmic genes cannot spread or the cost of those that do spread is always below the cost of nuclear suppression. These conditions are met when the rate of inbreeding is adequately high. Biparental inheritance may thus be expected in fusions between closely related cells. This expectation is consistent with data from sexual fusions in inbred fungi (yeast) and with somatic cell fusions in diverse taxa.

Second, if the cytoplasmic factors are in control, then, in an outbred species with enforced uniparental inheritance, a cytoplasmic factor that resists attempts to prevent its transmission may spread. By contrast, in the inbred condition the optimal strategy is for a cytoplasmic factor to allow its relative to be transmitted free of cost. If such resistance is possible, then in contrast to the above expectations biparental (or paternal) inheritance may be associated with outbreeding, whereas homothallic/selfing species could have relatively strict uniparental transmission. This expectation is consistent with data on cytoplasmic inheritance in higher plants and chlorophyte algae in which biparental inheritance (or significant leakage) is found in the outbred species but uniparental inheritance is typical of the inbred lineages.

In sum, all four possible states are found: both uni- and biparental inheritance are associated both with in and outbreeding. The conflict hypothesis is thus compatible with any outcome, and so no set of comparative data can directly falsify the model in the absence of extra information as regards who is in control. If we knew when to expect strong nuclear controls on cytoplasmic inheritance, then we might be able to understand why outbred animals typically have uniparental inheritance but why numerous outbred plants have biparental inheritance. Likewise we have a problem understanding why yeast has biparental inheritance, but other inbred fungi and algae have strict uniparental inheritance.

This problem aside, most of the time an assumption of "majority control", what Leigh calls "the parliament of the genes" (discussed in Leigh, 1971a, b, 1983, 1991), is probably valid, in which case prediction is possible. Likewise, Frank (1995) has shown that so long as groups of potential competitors are little-related, then mutual policing may evolve, in which case the system can evolve to look as though there is a majority interest that is being served. Most comparative predictions made in systems of conflict implicitly assume majority control (e.g., the optimal solution for outbred *Euplotes*, discussed above, is an optimal solution for the nondistorting components, not the distorting ones). Under this proviso, one might predict as a logical extension of Maynard Smith and Szathmáry's model for the evolution of chromosomes that organisms with numerous nuclei in one cell, all of which could potentially compete for the germ line, should evolve means to avoid the outbreak of fast-replicating nuclei. [This is indeed a prediction of any model arguing for selection on the majority to 'police' the minority.] The synchronous divisions of the nuclei in the syncytial cell of acellular slime molds may hence be postulated as an adaptation to ensure co-replication and hence high fitness co-variance. More phylogenetically independent data points would, however, be desirable.

In general, much suggestive data exist to allow one to suppose that genetic systems have been structured in response to selfish elements. It remains to be seen, however, whether these data are artifactual and misleading or a reliable indicator of what really shapes systems of inheritance.

XI. Summary

Parasites may be expected to adapt to their host as other organisms are expected to adapt to their environment. This adaptation may involve the parasite manipulating the host. Some parasites are never horizontally transmitted. The "interests" of these parasites are no different from any component of the genome with the same transmission genetics. The genome as a whole may hence be thought of as a grouping of potential parasites. By virtue of the absence of horizontal transmission, all the linkage groups in a genome have a high fitness covariance. This condition provides the context for selection to promote mutualism between these elements. All adaptation at the individual level is then the result of selection on these elements.

However, in a sexual organism there remain periods in which the fitness covariance is reduced. Maternally transmitted factors for instance are in an evolutionary dead end in males, but nuclear chromosomes are not. As a consequence, cytoplasmic factors that convert males into females have much broader invasion conditions than do nuclear alleles that perform the same manipulation. Similarly, a gene in one sperm may spread if it selectively inhibits those sperm containing alternative alleles at the same locus. Are these manipulations by self-promoting elements adaptations of the elements? The answer greatly depends on what one means by adaptation.

If it is intended to imply that the elements spread because of selection, then this is straightforward to demonstrate most of the time. [Note, however, the selection acts below the individual level.] Indeed, if a gene can be shown to be self-promoting, then the reason for its spread is typically obvious. Comparative analysis can give support to the view that a particular gene is a self-promoting element. This may be because it can be shown to be fast evolving, to have a titer dependent upon the level of inbreeding, to have a predictable phylogenetic distribution or may be compared, for instance as regards chromosomal location, to similar elements. If a self-promoting element is evidently self-promoting, then experimental manipulation need hardly be required. Manipulation may, however, be useful in instances in which the element cannot be observed spreading.

If one prefers that adaptations are structures of adequate complexity such that their fit to the environment can be extrapolated from this complexity, then adaptations of small things in general may be difficult to show. Implicit in the notion of complexity is the understanding that the structures under consideration would be redundant in the extreme were they not to have a function. They then demand explanation. Self-promoting elements then are unusual because they are associated with much more redundancy than most other genes. Indeed, one putative self-promoting element has been identified even in the absence of crossing data by virtue of the extreme apparent redundancy.

Self-promoting elements also provide the context for the spread of suppressors. It is equally valid to ask if these are genomic adaptations. This chapter concludes with a discussion of the possibility that one means of suppression is an alteration of genetic system. The evolution of uniparental inheritance of cytoplasmic genes and of ciliate meiosis are considered as case histories. As particular genetic systems may be the product of selection to counter the spread of self-promoting elements, they could be considered as adaptations. Testing such hypotheses by comparative analysis requires assumptions about which of the two conflicting parties is in control.

Acknowledgments

The author would like to acknowledge the help of Steve Frank, James Griesemer, Alan Grafen, Gilean McVean, John Barrett, Hywel Jones, Heather Carstens, and the editors. The author is supported by the Royal Society.

References

Agulnik, S. I., Sabantsev, I. D., Orlova, G. V. and Ruvinsky, A. O. (1993). Meiotic drive on aberrant chromosome 1 in the mouse is determined by a linked distorter. *Genet. Res.* 61, 91-96.

Alexander, R. D., and Borgia, G. (1978). Group selection, altruism, and the levels of the organisation of life. *Annu. Rev. Ecol. Syst.* 9, 449-474.

Anderson, R. M., and May, R. M. (1991) "Infectious Diseases of Humans." Oxford University Press, Oxford.

Andreadis, T. G. (1988). *Amblyospora connecticus* sp. nov. (Microsporidae: Amblyosporidae): Horizontal transmission studies in the mosquito *Aedes cantator* and formal description. *J. Invert. Pathol.* 52, 90-101.

Balakireva, M. D., Shevelyov, Y., Nurminsky, D. I., Livak, K. J., and Gvozdev, V. A. (1992). Structural organization and diversification of Y-linked sequences comprising *Su(Ste)* genes in *Drosophila melanogaster. Nucleic Acids Res.* 20, 3731-6.

Bardoni, B., Zanaria, E., Guioli, S., Floridia, G., Worley, K. C., Tonini, G., Ferrante, E., Chiumello, G., Mccabe, E., Fraccaro, M., Zuffardi, O., and Camerino, G. (1994). A dosage sensitive locus at chromosome Xp21 is involved in male to female sex reversal. *Nature Genet.* 7, 497-501.

Beale, G. H., and Jurand, A. (1966). Three different types of mate-killer (mu) particle in *Paramecium aurelia* (syngen 1). *J. Cell Sci.* 1, 31-34.

Beeman, R. W., Friesen, K. S., and Denell, R. E. (1992). Maternal-effect selfish genes in flour beetles. *Science* 256, 89-92.

Bengtsson, B. O. (1985). Biased gene conversion as the primary function of recombination. *Genet. Res.* 47, 77-80.

Benito, C., Romera, F., Diez, M., and Figueiras, A. M. (1992). Genic heterozygosity and fitness in rye populations with B chromosomes. *Heredity* 69, 406-411.

Brosius, J., and Gould, S. J. (1992). On genomenclature - a comprehensive (and respectful) taxonomy for pseudogenes and other junk DNA. *Proc. natl. Acad. Sci. USA* 89, 10706-10710.

Burt, A., and Trivers, R. (1993). Selfish DNA and the breeding system of flowering plants. *Abstracts of the Fourth Congress of the European Society for Evolutionary Biology* 62.

Buss, L. (1987) "The Evolution of Individuality." Princeton University Press, Princeton.

Cavalier-Smith, T. (1985) "The Evolution of Genome Size." John Wiley, Chichester.

Charlesworth, B. (1987). The population biology of transposable elements. *Trends Ecol. Evol.* 2, 21-23.

Charlesworth, B., and Hartl, D. L. (1978). Population dynamics of the segregation distorter polymorphism of *Drosophila melanogaster. Genetics* 89, 171-192.

Charlesworth, B., Sniegowski, P., and Stephan, W. (1994). The evolutionary dynamics of repetitive DNA in eukaryotes. *Nature* 371, 215-220.

Conway, S. J., Mahadevaiah, S. K., Darling, S. M., Capel, B., Rattigan, A. M., and Burgoyne, P. S. (1994). Y353/B: a candidate multiple copy spermiogenesis gene on the mouse Y chromosome. *Mammal. Genome* 5, 203-210.

Cosmides, L. M., and Tooby, J. (1981). Cytoplasmic inheritance and intragenomic conflict. *J. Theor. Biol.* 89, 83-129.

Crow, J. F. (1988). Anecdotal, historical and critical commentaries on genetics: the ultraselfish gene. *Genetics* 118, 389-391.

Crow, J. F. (1991). Why is Mendelian segregation so exact? *Bioessays* 13, 305-12.

Dabovic, B., Zanaria, E., Bardoni, B., Lisa, A., Bordignon, C., Russo, V., Matessi, C., Traversari, C., and Camerino, G. (1995). A family of rapidly evolving genes from the sex reversal critical region in xp21. *Mammal. Genome* 6, 571-580.

Darlington, C. D., and Thomas, P. T. (1941). Morbid mitosis and the activity of inert chromosomes in Sorgum. *Proc. R. Soc. London, B* 130, 127-150.

Darlington, C. D., and Upcott, M. B. (1941). The activity of inert chromosomes in *Zea Mays. J. Genet.* 41, 275-296.

Dawkins, R. (1976) "The Selfish Gene." Oxford University Press, Oxford.

Dawkins, R. (1982) "The Extended Phenotype." W. H. Freeman, Oxford.

Dawkins, R. (1990). Parasites, desiderata lists and the paradox of the organism. *Parasitology* 100, S63-S73.

Degroot, N., Goshen, R., Rachmilewitz, J., Gonik, B., Benhur, H., and Hochberg, A. (1993). Genomic imprinting and beta-chorionic gonadotropin. *Prenatal Diagnosis* 13, 1159-1160.

Dobson, A. P. (1988). The population biology of parasite induced changes in host Behavior. *Quart. Rev. Biol.* 63, 139-165.

Doolittle, W. F., and Sapienza, C. (1980). Selfish genes, the phenotype paradigm and genome evolution. *Nature* 284, 601-603.

Eberhard, W. G. (1980). Evolutionary consequences of intracellular organelle competition. *Quart. Rev. Biol.* 55, 231-249.

Endler, J. A. (1986) "Natural Selection in the Wild." Monographs in Population Biology, Princeton University Press, Princeton.

Ewald, P. W. (1988) Cultural vectors, virulence, and the emergence of evolutionary epidemiology. *In* "Oxford Surveys in Evolutionary Biology" (P. H. Harvey and L. Partridge, eds.), Vol. 5 pp. 215-246. Oxford University Press, Oxford.

Fisher, R. A. (1930) "The Genetical Theory of Natural Selection." Clarendon Press, Oxford.

Fitz-Earle, M., and Sakaguchi, B. (1986). Sex ratio distortion in populations and its possible role in insect suppression: Experimental studies with strains of *D. melanogaster* carrying cytoplasmically-inherited male kiling spiroplasmas. *Jap. J. Genet.* 61, 447-460.

Fleuriet, A. (1988) Maintenance of a hereditary virus. The sigma virus in populations of its host, *D. melanogaster. In* "Evolutionary Biology" (M. K. Hecht and Wallace, eds.), Vol. 23 pp. 1-30. Plenum, New York.

Frank, S. A. (1983). A hierarchical view of sex-ratio patterns. *Flor. Entomol.* 66, 42-75.

Frank, S. A. (1989). The evolutionary dynamics of cytoplasmic male sterility. *Am. Nat.* 133, 345-376.

Frank, S. A. (1992). A kin selection model for the evolution of virulence. *Proc. R. Soc. London B* 250, 195-197.

Frank, S. A. (1994). Kin selection and virulence in the evolution of protocells and parasites. *Proc. R. Soc. Lond. B.* 258, 153-161.

Frank, S. A. (1995). Mutual policing and repression of competition in the evolution of cooperative groups. *Nature* 377, 520-522.

Gerdes, K., Poulsen, L. K., Thisted, T., Nielsen, A. K., Martinussen, J., and Andreasen, P. H. (1990). The hok killer gene family in gram-negative bacteria. *New Biol.* 2, 946-956.

Gillespie, J. H. (1991) "The Causes of Molecular Evolution." Oxford Series in Ecology and Evolution, Oxford University Press, Oxford.

Godelle, B., and Reboud, X. (1995). Why are organelles uniparentally inherited? *Proc. R. Soc. Lond. B* 259, 27-33.

Gouyon, P. H., and Couvet, D. (1987) A conflict between two sexes, females and hermaphrodites. *In* "The Evolution of Sex and Its Consequences" (S. C. Stearns, ed.), Vol. pp. 245-261. Birkhauser-Verlag, Berlin.

Graur, D. (1993). Molecular deconstructivism. *Nature* 363, 490.

Hackett, K. J., Lynn, D. E., Williamson, D. L., Ginsberg, A. S., and Whitcomb, R. F. (1986). Cultivation of the Drosophila sex ratio spiroplasma. *Science* 232, 1253-1255.

Haig, D. (1992). Genomic imprinting and the theory of parent-offspring conflict. *Semin. Dev. Biol.* 3, 153-160.

Haig, D. (1993a). Alternatives to meiosis: the unusual genetics of red algae, Microsporidia, and others. *J. Theor. Biol.* 163, 15-31.

Haig, D. (1993b). Genetic conflicts in human-pregnancy. *Q. Rev. Biol.* 68, 495-532.

Haig, D. (1994). Is human insulin imprinted? *Nature Genet.* 7, 10.

Haig, D. and Grafen, A. (1991). Genetic scrambling as a defense against meiotic drive. *J. Theor. Biol.* 153, 531-558.

Haig, D., and Graham, C. (1991). Genomic imprinting and the strange case of the insulin-like growth factor II receptor. *Cell* 64, 1045-6.

Haig, D., and Westoby, M. (1989). Parent-specific gene expression and the triploid endosperm. *Am. Nat.* 134, 147-155.

Hamilton, W. D. (1967). Extraordinary sex ratios. *Science* 156, 477-488.

Hardy, R. W., Lindsley, D. L., Livak, K. J., Lewis, B., Silversten, A. L., Joslyn, G. L., Edwards, J., and Bonaccorsi, S. (1984). Cytogenetic analysis of a segment of the Y chromosome of *Drosophila melanogaster*. *Genetics* 107, 591-610.

Harvey, M. B., and Kaye, P. L. (1992). Igf-2 stimulates growth and metabolism of early mouse embryos. *Mech. Dev.* 38, 169-174.

Harvey, P. H. and Pagel, M. D. (1991). "The Comparative Method in Evolutionary Biology." Oxford Series in Ecology and Evolution, Oxford University Press, Oxford.

Hastings, I. M. (1992a). Population genetic aspects of deleterious cytoplasmic genomes and their effect on the evolution of sexual reproduction. *Genet. Res.* 59, 215-225.

Hastings, I. M. (1992b). Why is sex so frequent? *Trends Ecol. Evol.* 7, 278-279.

Hauschteck-Jungen, E. (1990). Postmating reproductive isolation and modification of the "sex rati" trait in *Drosophila subobscura* induced by the sex chromosome gene arrangement $A_{2+3+5+7}$. *Genetica* 83, 31-44.

Hickey, D. A., and Benkel, B. (1986). Introns as relict retrotransposons: implications for the evolutionary origin of eukaryotic mRNA splicing mechanisms. *J. Theor. Biol.* 121, 283-291.

Hickey, D. H. (1982). Selfish DNA: a sexually transmitted nuclear parasite. *Genetics* 101, 519-531.

Hickey, D. H., and Rose, M. R. (1988) The role of gene transfer in the evolution of eukaryotic sex. *In* "The Evolution of Sex: An Examination of Current Ideas" (R. E. Michod and B. R. Levin, ed.), Vol. pp. 161-175. Sinauer, Sunderland.

Hiraizumi, Y. (1990). Negative segregation distortion in the SD system of *Drosophila melanogaster*: a challenge to the concept of differential sensitivity of Rsp alleles. *Genetics* 125, 515-525.

Hochachka, P. W., and Somero, G. N. (1984). "Biochemical Adaptation." Princeton University Press, Princeton, NJ.

Hoekstra, R. F. (1990). Evolution of uniparental inheritance of cytoplasmic DNA. *In* "Organisational Constraints on the Dynamics of Evolution" (J. Maynard Smith and G. Vida, ed.), Vol. pp. 269-278. Manchester University Press, Manchester.

Hughes, A. L. (1991). Circumsporozoite protein genes of malaria parasites (Plasmodium spp.): evidence for positive selection on immunogenic regions. *Genetics* 127, 345-353.

Hughes, A. L., and Nei, M. (1988). Pattern of nucleotide substitution at major histocompatibility complex class I loci: evidence for overdominant selection. *Nature* 335, 167-170.

Hughes, A. L., Ota, T., and Nei, M. (1990). Positive Darwinian selection promotes charge profile diversity in the antigen binding cleft of class I MHC molecules. *Mol Biol Evol* 7, 515-524.

Hurst, G. D. D., and Majerus, M. E. N. (1993). Why do maternally inherited microorganisms kill males? *Heredity* 71, 81-95.

Hurst, L. D. (1990). Parasite diversity and the evolution of diploidy, multicellularity and anisogamy. *J. Theor. Biol.* 144, 429-443.

Hurst, L. D. (1992). Is *Stellate* a relict meiotic driver? *Genetics* 130, 229-30.

Hurst, L. D. (1993a). The incidences, mechanisms and evolution of cytoplasmic sex ratio distorters in animals. *Biol. Rev.* 68, 121-193.

Hurst, L. D. (1993b). *scat⁺* is a selfish gene analogous to *Medea* of *Tribolium castaneum*. *Cell* 75, 407-408.

Hurst, L. D. (1994a). Cytoplasmic genetics under inbreeding and outbreeding. *Proc. R. Soc. London B* 258, 287-298.

Hurst, L. D. (1994b). Embryonic growth and the evolution of the mammalian Y chromosome .I. The Y as an attractor for selfish growth factors. *Heredity* 73, 223-232.

Hurst, L. D. (1994c). Embryonic growth and the evolution of the mammalian Y chromosome .II. Suppression of selfish Y-linked growth factors may explain escape from X-inactivation and for rapid evolution of *Sry*. *Heredity* 73, 233-243.

Hurst, L. D. (1995). Selfish genetic elements and their role in evolution: the evolution of sex and some of what that entails. *Phil. Trans. R. Soc. B* 349, 321-332.

Hurst, L. D. (1996a) Evolutionary theories of genomic imprinting *In* "Genomic Imprinting" (W. Reik and A. Surani, ed.), Vol. pp. in press. Oxford University Press, Oxford.

Hurst, L. D. (1996b). Further evidence consistent with *Stellate*'s involvement in meiotic drive. *Genetics* 142, 641-643.

Hurst, L. D., Atlan, A., and Bengtsson, B. O. (1996). Genetic conflicts. *Quart. Rev. Biol.* in press.

Hurst, L. D., and Hamilton, W. D. (1992). Cytoplasmic fusion and the nature of sexes. *Proc. R. Soc. Lond. B* 247, 189-194.

Hurst, L. D., and Peck, J. R. (1996). Recent advances in understanding the evolution and maintenance of sex. *Trends Ecol. Evol.* 11, 46-52.

Juchault, P., Rigaud, T., and Mocquard, J.-P. (1992). Evolution of sex-determining mechanisms in a wild population of *Armadillidium vulgare* Latr. (Crustacea, Isopoda): competition between two feminizing parasitic factors. *Heredity* 69, 382-390.

Kaul, M. L. H. (1988) " Male-Sterility in Higher Plants." Monographs in theoretical and applied genetics, Springer-Verlag, Berlin.

Kiess, W., Yang, Y., Kessler, U. and Hoeflich, A. (1994). Insulin-like growth-factor-ii (igf-ii) and the igf-ii mannose-6-phosphate receptor - the myth continues. *Horm. Res.* 41, 66-73.

Kuma, K., Iwabe, N., and Miyata, T. (1995). Functional constraints against variations on molecules from the tissue-level - slowly evolving brain-specific genes demonstrated by protein-kinase and immunoglobulin supergene families. *Mol. Biol. Evol.* 12, 123-130.

Laval, S. H., and Boyd, Y. (1993). Novel sequences conserved on the human and mouse X chromosomes. *Genomics* 15, 483-491.

Leigh, E. G. J. (1971a) "Adaptation and Diversity." Freeman, Cooper and co., San Francisco.

Leigh, E. G. J. (1971b). How does selection reconcile individual advantage with the good of the group? *Proc. Natl. Acad. Sci. USA* 74, 4542-4546.

Leigh, E. G. J. (1983). When does the good of the group override the advantage of the individual? *Proc. Natl. Acad. Sci. USA* 80, 2985-2989.

Leigh, E. G. J. (1991). Genes, bees and ecosystems: the evolution of a common interest among individuals. *Trends Ecol. Evol.* 6, 257-262.

Levinovitz, A., Norstedt, G., van den Berg, S., Robinson, I. C., and Ekstrom, T. J. (1992). Isolation of an insulin-like growth factor II cDNA from guinea pig liver: expression and developmental regulation. *Mol. Cell. Endocrinol.* 89, 105-110.

Levy, F. (1991). A genetic analysis of reproductive barriers in *Phacelia dubia*. *Heredity* 67, 331-345.

Lewis, D. (1941). Male sterility in natural populations of hermaphroditic plants. *New Phytol.* 40, 158-160.

Livak, K. J. (1984). Organisation and mapping of a sequence on the *Drosophila melanogaster* X and Y chromosomes that is transcribed during spermatogenesis. *Genetics* 107, 611-634.

Livak, K. J. (1990). Detailed structure of the *Drosophila melanogaster Stellate* genes and their transcript. *Genetics* 124, 303-316.

López-León, M. D., Cabrero, J., and Camacho, J. P. (1992). Male and female segregation distortion for heterochromatic supernumerary segments on the S8 chromosome of the grasshopper *Chorthippus jacobsi*. *Chromosoma* 101, 511-516.

Luporini, P., and Dini, F. (1977). The breeding system and the genetic relationship between autogamous and non-autogamous sympatric populations of *Euplotes crassus* (Dujardin) (Ciliate Hypotrichida). *Monitore Zool Ital (NS)* 11, 119-154.

Lyon, M. F. (1993). Epigenetic inheritance in mammals. *Trends Genet.* 9, 123-128.

Lyttle, T. W. (1991). Segregation distorters. *Annu. Rev. Genet.* 25, 511-557.

Lyttle, T. W. (1993). Cheaters sometimes prosper: distortion of mendelian segregation by meiotic drive. *Trends Genet.* 9, 205-10.

Lyttle, T. W., Sandler, L. M., Prout, T., and Perkins, D. D. (1991). The genetics and evolutionary biology of meiotic drive. *Am. Nat.* 137, 1-456.

Matagne, R. F. (1987). Chloroplast gene transmission in *Chlamydomonas reinhardtii*: a model for its control by the mating-type locus. *Curr. Genet.* 12, 251-256.

Maynard Smith, J., and Szathmáry, E. (1993). The origin of chromosomes I. Selection for linkage. *J. Theor. Biol.* 164, 437-446.

Miao, V. P. W., Matthews, D. E., and Vanetten, H. D. (1991). Identification and chromosomal locations of a family of cytochrome P-450 genes for pisatin detoxification in the fungus *Nectria haematococca*. *Mol. Gen. Genet.* 226, 214-223.

Minchella, D. J. (1985). Host life-history variation in response to parasitism. *Parasitology* 90, 205-216.

Moore, T., and Haig, D. (1991). Genomic imprinting in mammalian development: a parental tug-of-war. *Trends Genet.* 7, 45-9.

Nobili, R., and Luporini, P. (1967). Maintenance of heterozygosity at the mt locus after autogamy in *Euplotes minuta* (Ciliata, Hypotrichida). *Genet. Res.* 10, 35-43.

Nur, U. (1989) Chromosomes, sex ratios, and sex determination *In* "Armored Scale Insects, Their Biology, Natural Enemies and Control" (D. Rosen, ed.), Vol. A pp. 179-190. Elsevier Science Publishers, Amsterdam.

Orgel, L. E., and Crick, F. H. C. (1980). Selfish DNA: the ultimate parasite. *Nature* 284, 604-607.

Östergren, G. (1945). Parasitic nature of extra fragment chromosomes. *Bot. Not.* 2, 157-163.

Pagel, M. D., and Johnstone, R. A. (1992). Variation across species in the size of the nuclear genome supports the junk-DNA explanation for the C-value paradox. *Proc. R. Soc. London B* 249, 119-124.

Palumbo, G., Bonaccorsi, S., Robbins, L. G., and Pimpinelli, S. (1994). Genetic-analysis of *Stellate* elements of *Drosophila melanogaster*. *Genetics* 138, 1181-1197.

Peters, L. L., and Barker, J. E. (1993). Novel inheritance of the murine severe combined anemia and thrombocytopenia (Scat) phenotype. *Cell* 74, 135-142.

Pilder, S. H., Hammer, M. F., and Silver, L. M. (1991). A novel mouse chromosome 17 hybrid sterility locus: implications for the origin of t haplotypes. *Genetics* 129, 237-246.

Pilder, S. H., Olds, C. P., Phillips, D. M., and Silver, L. M. (1993). Hybrid sterility-6: a mouse t complex locus controlling sperm flagellar assembly and movement. *Dev. Biol.* 159, 631-42.

Price, G. R. (1970). Selection and covariance. *Nature* 227, 520-521.

Price, G. R. (1972). Extension of covariance selection mathematics. *Annu. Hum. Genet.* 35, 485-490.

Puhalla, J. E. (1968). Compatibility reactions on solid medium and interstrain inhibition in *Ustilago maydis*. *Genetics* 60, 461-477.

Reboud, X., and Zeyl, C. (1994). Organelle inheritance in plants. *Heredity* 72, 132-140.

Reed, J. N., and Hurst, L. D. (1996). Dynamic analysis of the evolution of a novel genetic system: the evolution of ciliate meiosis. *J. Theor. Biol.* 178, 355-368.

Rigaud, T., and Juchault, P. (1993). Conflict between feminizing sex ratio distorters and an autosomal masculinizing gene in the terrestrial isopod *Armadillidium vulgare* Latr. *Genetics* 133, 247-52.

Robbins, L. G., Palumbo, G., Bonaccorsi, S., and Pimpinelli, S. (1996). Measuring meiotic drive. *Genetics* 142, 645-647.

Rousset, F. and Raymond, M. (1991). Cytoplasmic incompatibility in insects: why sterilize females? *Trends Ecol. Evol.* 6, 54-57.

Saumitou-Laprade, P., Cuguen, J., and Vernet, P. (1994). Cytoplasmic male sterility in plants: molecular evidence and the nucleocytoplasmic conflict. *Trends Ecol. Evol.* 9, 431-435.

Schmauss, C., Ohosone, Y., Hardin, J. A., and Lerner, M. R. (1989). A comparison of sn-RNP-associated Sm-autoantigens: human N, rat N and human B/B'. *Nucleic Acids Res* 17, 1733-1743.

Shaw, M. W., and Hewitt, G. M. (1990). B chromosomes, selfish DNA and theoretical models: where next? *Oxf. Revs. Evol. Biol.* 7, 197-223.

Shevelyov, Y. Y. (1992). Copies of a *Stellate* gene variant are located in the X heterochromatin of *Drosophila melanogaster* and are probably expressed. *Genetics* 132, 1033-7.

Shoop, W. L. (1991). Vertical transmission of helminths: Hypobiosis and amphiparatenesis. *Parasitol. Today* 7, 51-54.

Silver, L. M. (1993). The peculiar journey of a selfish chromosome: mouse t haplotypes and meiotic drive. *Trends Genet.* 9, 250-4.

Skibinski, D. O. F., Gallagher, C., and Beynon, C. M. (1994). Sex-limited mitochondrial DNA transmission in the marine mussel *Mytilus edulis*. *Genetics* 138, 801-809.

Skinner, S. W. (1982). Maternally inherited sex ratio in the parasitoid wasp *Nasonia vitripennis*. *Science* 215, 1133-1134.

Solter, D. (1988). Differential imprinting and expression of maternal and paternal genomes. *Annu. Rev. Genet.* 22, 127-146.

Somers, J. M., and Bevan, E. A. (1969). The inheritance of the killer character in yeast. *Genet. Res.* 13, 71-83.

Stearns, S. C. (1976). Life history tactics: a review of the ideas. *Quart. Rev. Biol.* 51, 3-47.

Stouthamer, R., Luck, R. F., and Hamilton, W. D. (1990). Antibiotics cause parthenogenetic *Trichogramma* (Hymenoptera/Trichogrammatidae) to revert to sex. *Proc. Natl. Acad. Sci. USA* 87, 2424-2427.

Szathmáry, E. (1991). Common interest and novel evolutionary units. *Trends Ecol. Evol.* 6, 407-408.

Talmadge, K., Vamvakopoulos, N. C., and Fiddes, J. C. (1984). Evolution of the genes for the b subunits of human chorionic gonadotropin and luteinizing hormone. *Nature* 307, 37-40.

Taylor, D. R. (1994). Sex ratio in hybrids between *Silene alba* and *Silene dioica*: evidence for Y-linked restorers. *Heredity* 74, 518-526.

Thomas, J. H. (1995). Genomic imprinting proposed as a surveillance mechanism for chromosome loss. *Proc. Natl. Acad. Sci. USA* 92, 480-482.

Thornhill, A. R., and Burgoyne, P. S. (1993). A paternally imprinted X chromosome retards the development of the early mouse embryo. *Development* 118, 171-174.

Tucker, P. K., and Lundrigan, B. L. (1993). Rapid evolution of the sex determining locus in Old World mice and rats. *Nature* 364, 715-717.

Tycko, B. (1994). Genomic imprinting - mechanism and role in human pathology. *Am. J. Pathol.* 144, 431-443.

Uyenoyama, M. K., and Feldman, M. W. (1978). The genetics of sex ratio distortion by cytoplasmic infection under maternal and contagious transmission: An epidemiological study. *Theor. Popul. Biol.* 14, 471-497.

Wade, M. J. (1985). Soft selection, hard selection, kin selection, and group selection. *Am. Nat.* 125, 61-73.

Wallis, M. (1981). The molecular evolution of pituitary growth hormone prolactin and placental lactogen: a protein family showing variable rates of evolution. *J. Mol. Evol.* 17, 10-18.

Wallis, M. (1993). Remarkably high rate of molecular evolution of ruminant placental lactogens. *J. Mol. Evol.* 37, 86-88.

Wallis, M. (1994). Variable evolutionary rates in the molecular evolution of mammalian growth hormones. *J. Mol. Evol.* 38, 619-627.

Walsh, C., Glaser, A., Fundele, R., Ferguson-Smith, A., Barton, S., Surani, M. A., and Ohlsson, R. (1994). The non-viability of uniparental mouse conceptuses correlates with the loss of the products of imprinted genes. *Mech. Dev.* 46, 55-62.

Werren, J. H. (1987). The coevolution of autosomal and cytoplasmic sex ratio factors. *J. Theor. Biol.* 124, 317-334.

Werren, J. H., Nur, U., and Eickbush, D. (1987). An extrachromosomal factor causing loss of paternal chromosomes. *Nature* 327, 75-76.

Werren, J. H., Nur, U., and Wu, C.-I. (1988). Selfish genetic elements. *Trends Ecol. Evol.* 3, 297-302.

Whitfield, L. S., Lovell-Badge, R., and Goodfellow, P. N. (1993). Rapid sequence evolution of the mammalian sex-determining gene SRY. *Nature* 364, 713-5.

Wickner, R. B. (1991) Yeast RNA virology: the killer systems *In* "The Molecular and Cellular Biology of the Yeast Saccharomyces: Genome Dynamics, Protein Synthesis, and Energetics" Vol. 1 pp. 263-296. Cold Spring Harbor Laboratory Press.

Wilkie, C. M., and Adams, J. (1992). Fitness effects of Ty Transposition in *Saccharomyces cerevisiae*. *Genetics* 131, 31-42.

Williams, G. C. (1966). "Adaptation and Natural Selection, a Critique of Some Current Evolutionary Thought." University Press, Princeton, NJ.

Williams, G. C. (1992) "Natural Selection: Domains, Levels and Challenges." Oxford Series in Ecology and Evolution, Oxford University Press, New York.

Williamson, D. L., Hackett, K. J., Wagner, A. G., and Cohen, A. J. (1989). Pathogenicity of cultivated *Drosophila willistoni* spiroplasmas. *Curr. Microbiol.* 19, 53-56.

Wolfe, K. H., and Sharp, P. M. (1993). Mammalian gene evolution: nucleotide sequence divergence between mouse and rat. *J. Mol. Evol.* 37, 441-456.

Yamamura, N. (1993). Vertical transmission and evolution of mutualism from parasitism. *Theor. Popul. Biol.* 44, 95-109.

Zouros, E., Ball, A. O., Saavedra, C., and Freeman, K. R. (1994). An unusual type of mitochondrial inheritance in the blue mussel *Mytilus*. *Proc. Natl. Acad. Sci. USA* 91, 7463-7467.

CHAPTER 14

The Design of Natural and Artificial Adaptive Systems

STEVEN A. FRANK

I. Introduction

The design of adaptive systems will be among the key research problems of the 21st century. This new field is emerging from several distinct lines of work.

- Modern immunology is based on the theory of clonal selection and adaptive immunity. The remarkable recognition abilities of the vertebrate immune system depend on the programmed mechanisms of antibody variation and selection that occur within each individual.
- The design of intelligent computer systems and robots depends on a balance between adaptive improvement by exploration and efficient exploitation of known solutions. Many of the current computer implementations use evolutionary algorithms to achieve adaptation to novel or changing environments.
- The adaptive response of genetic systems to environmental challenge depends strongly on the tempo and mode of sex and recombination. Sexual systems vary widely in nature. Which processes have shaped this variation is a major puzzle in evolutionary biology.
- Wiring a brain during development and using that brain to learn are great problems of information management. Recent studies in neuroscience suggest that programmed mechanisms of stochastic variation and controlled selection guide neural development and learning. If true, then nature has solved these informational problems by using somatic adaptive systems that are programmed to work in the same way as natural selection.

What do these different fields have in common? Will there be a new science of adaptation shared by biology and engineering? Can a unified theory guide the study of so many different phenomena? What will be the central tenets of such a theory?

These are difficult questions. To make a start, I survey the range of adaptive systems as they are currently understood: adaptive immunity, learning, development, culture and symbiosis, the origin and evolution of genetic systems, and artificial adaptive systems in engineering. The facts that I present in my survey, fascinating in their own right, provide the database from which more general insights must be built.

II. Challenges to Adaptive Systems

Before starting on the survey, it is useful to have a conceptual framework. I begin with a rough definition. An "adaptive system" is a population of entities that satisfy the three conditions of natural selection: the entities vary, they have continuity (heritability), and they differ in their success. The entities in the population can be genes, competing computer programs, or anything else that satisfies the three conditions.

A. *Types of challenge*

In this section I propose a classification of the challenges that have shaped adaptive systems and the ways in which adaptive systems have responded to these challenges. A surprisingly small number of challenges and responses cover the main features of adaptive systems ranging from genetics to robotics. I illustrate the concepts with brief examples that will be discussed in more detail during the survey.

Information decay is one kind of challenge. For example, genetic systems suffer information decay when random mutations occur. If mutations accumulate too rapidly then adaptive improvement by natural selection is impossible. The population suffers an "error catastrophe" (Eigen, 1992) or "mutational meltdown" (Lynch et al., 1993).

Predictable complexity is another type of challenge. For example, the information required to specify the point-to-point neural connections of a human brain greatly exceeds the amount of information encoded by the genome. Thus the genetic system must cope with the problem of creating a complex pattern during development from a relatively limited set of instructions.

Unpredictable challenges are the third type of problem faced by adaptive systems. For example, parasites vary unpredictably over space and time. To give an engineering example, a robot engaged in war cannot have prewired responses for all possible attack strategies that the enemies may use. A successful robot must adjust to unpredictable events.

I will argue during my survey that this small list—information decay, complexity, and unpredictability—describes the main challenges faced by adaptive systems (Table 1). The next step is to consider how adaptive systems respond to these challenges (Table 2).

TABLE I

Challenges to Adaptive Systems

Challenge	Comments
Information Decay	A ubiquitous "tax" on information storage and transmission.
Predictable complexity	Challenge is to learn predictable pattern or achieve predictable form, but complexity of pattern greatly exceeds the available information storage.
Unpredictable challenge	(a) Environmental—abiotic challenges and biotic interactions without feedback. (b) Coevolutionary—biotic interactions with feedback between systems.

B. Responses of adaptive systems

Enhancing *transmission fidelity* is one way to overcome the problem of information decay. For example, Bernstein et al. (1988) suggest that sex is the genetic system's way of enhancing transmission fidelity in response to the information decay imposed by mutation. In their theory sex brings together two different copies of the genetic material, which allows a damaged copy to be corrected by the undamaged copy.

The problem of balancing *exploration versus exploitation* recurs in all adaptive systems (Holland, 1975). Exploration of new ways to solve problems often carries a cost because competitors may devote more energy to the efficient exploitation of known solutions. For example, sex increases genetic variability among offspring compared with asexual reproduction. Greater variability improves the chances that some of the offspring will have genotypes that match an unpredictable environment. Thus sexual systems may be a form of exploration, but this exploration is costly because asexuality is usually a more efficient mode of reproduction. There is much controversy among evolutionary biologists about whether sexual systems have evolved as a method of exploration in response to unpredictable challenge or as a method to enhance transmission fidelity in response to the to challenge of information decay.

TABLE 2

Responses to Challenge

Response	Comments
Transmission fidelity	Mechanisms to reduce errors in the storage and transmission of encoded solutions.
Exploration versus exploitation	The balance between costly exploration for improved efficiency and the cheap exploitation of known solutions.
Generative rules	Simple rules to generate complex phenotypes. Genotypes do not specify explicit blueprints for structure.
Instructional subsystem	Mechanism to store information obtained directly from the environment.
Adaptive subsystem	A system of variation and selection "spawned" by an evolving system to solve a particular problem.
Symbiosis	Cooperation between separate evolving entities to achieve greater group efficiency. Conflicts among group members often arise.

The transmissible information (genotype) of an adaptive system often contains *generative rules* for the design of phenotypic structure (Thompson, 1961; Lindenmayer, 1971). In organisms each detail of morphology and behavior is not coded by an explicit DNA sequence; there is no blueprint for design. Fingerprints are generated by the biochemical rules of morphogenesis contained in the genome. Those rules may be fairly simple, but the outcome is complex and partly influenced by chance.

Simple environmental patterns may directly influence the internal information store through an *instructional subsystem.* For example, repeated stimulation of some neurons causes an increase in the stimulus required to evoke a response. The instructional subsystem takes a direct measure of environmental pattern.

Adaptive systems may spawn *adaptive subsystems* to handle difficult challenges (Gell-Mann, 1994). For example, the immune system of vertebrates has a specialized set of mechanisms to generate variability among recognition molecules and a second set of mechanisms to select and amplify recognition molecules that react with invading parasites. These controls of the adaptive immune system are specified by the underlying genetic system, or, put another way, the genetic system has spawned an adaptive subsystem to handle the unpredictable challenges of parasitic attack. In later sections I will discuss certain aspects of development and learning as adaptive subsystems spawned by the genetic system.

Symbiosis is the living together of two or more dissimilar organisms. An interesting theory about the origin of life illustrates the importance of symbiosis (Eigen, 1992). Information decay was a severe problem for the first replicating molecules because of high mutation rates. The mutation rate sets an "error threshold" that determines the upper limit on the size of informational molecules and thus the storage capacity of genetic systems. The early replicators were limited to very small genome sizes because of the error threshold. This creates a paradox: small genomes do not have sufficient information to code for an error-correcting replication machinery; without error correction larger genomes cannot evolve.

Symbiosis appears to be the solution. A set of small replicators, each below the error threshold, may have cooperated to produce error-correcting enzymes. This symbiotic group, with a reduced rate of transmission errors, could then increase in size and complexity.

Cooperation among early replicators was the first successful symbiosis. The most recent example of symbiosis in adaptive systems comes from research on robot design. Teamwork among robots boosts efficiency for tasks that require division of labor and specialization, such as automated manufacturing, search and rescue, or surveillance (Parker, 1993). Both biological symbiosis and robot teamwork must resolve the tension between the autonomy of components and the control of the symbiotic group. I will discuss this problem for both genomes and robots in later sections.

C. Outline of survey

I turn next to my survey of adaptive systems. I start with the transformation of genotype into phenotype. The first of these   sections describes vertebrate immunity, an adaptive subsystem of variation and selection that occurs within each individual's body. The following section considers the problems of neural development and learning. I raise the possibility that adaptive subsystems play a role in these complex informational processes. The final section of this group focuses on morphology. I contrast simple generative rules for development with more complex processes of developmental variation and selection.

After the genotype-phenotype transformations, I turn to the evolution of genetic systems. There is a natural tendency to view a genetic system as a stable, well-defined core of hereditary information. But each apparent system is actually a complex symbiosis of partly conflicting and partly cooperating hereditary systems. Each has its own pattern of continuity (transmission) and its own generative rules for the production of phenotype. Sex and recombination define one widespread pattern of hereditary mixture and symbiosis. I consider how sex fits into the recurring challenges and responses of adaptive systems outlined in Tables 1 and 2.

The final section places some new aspects of human engineering in the framework of adaptive systems. At one level these new methods are simply the use of variation and selection as an engineering tool for problems such as robotics. The effective use of selection follows in many ways the design of natural adaptive systems. At another level the new forms of "artificial life," with their new symbioses and their higher-order adaptive subsystems spawned by humans, are simply the next historical stage in the evolution of adaptive systems.

III. Adaptive Immunity in Vertebrates

How does a host recognize foreign molecules that signal a parasitic invasion? How does an individual distinguish self from nonself to avoid attacking its own tissues? A vertebrate host solves these problems with an adaptive system that causes evolutionary change among the populations of cells within its body. These evolutionary changes within the body—somatic evolution—are controlled by a complex set of mechanisms that are encoded within the genome. In my language of environmental challenge and adaptive response, the genetic system spawned an adaptive subsystem to handle the unpredictable challenges of parasitic invasion.

In this section I summarize many details about vertebrate immunity. The details are fascinating and provide the basis on which theories of adaptive systems must be built. But in a general survey of adaptive systems the details can also be overwhelming. So in this introduction I provide a link between the abstract discussion about challenges and responses in the previous section and the details of adaptive immunity that follow.

The response of an adaptive subsystem depends on two levels of evolutionary change. At the somatic level of vertebrate immunity, cellular clones undergo programmed genetic recombination and enhanced mutation during particular periods. These mechanisms create variation in the ability to recognize and bind foreign molecules. Variants that bind invaders are amplified by programmed controls that enhance the replication rate of some cellular clones while reducing the replication of other clones. The mechanisms that control somatic recombination, mutation, and selection (amplification) are coded at the genetic level. Thus evolutionary modifications at the genetic level ultimately control the responses of the somatic system.

This two-level system provides special opportunities to study the forces that shape adaptive systems. The challenge—parasitic invasion—is clearly defined. The response requires recognition of invaders. Adaptive immunity uses a number of techniques to adjust exploration for better recognition of invaders versus exploitation of existing recognition tools. This dynamic balancing between exploration and exploitation occurs on short time scales. Although studying immunology is not easy, we will see that other adaptive systems rarely provide such clear challenge-response couples.

The vertebrate immune system is actually a complex mixture of adaptive subsystems and traits that are encoded directly by the genome. For example, several important aspects of recognition depend on subsystems of random variation and selective amplification, but at least one key aspect of recognition is controlled directly by a genetically encoded set of alleles (MHC). This mix of somatic exploration and the exploitation of fixed recognition presents several interesting and unsolved problems in this two-level adaptive system. I will discuss these problems later in the section, but first some biological background is needed to set the stage.

A. Positive Selection and Clonal Expansion

I will describe a measles infection to introduce some of the details of adaptive immunity. Measles viruses invade the upper respiratory tract. Toward the end of the 10- to 12-day incubation period, the first symptoms of headache, fever, and sore throat appear. At this time the viral population within the host is large and rising rapidly. Viral particles enter the blood and spread, forming secondary infections in the skin that lead to the characteristic measles rash (Davis et al., 1990).

The body maintains a vast array of nonself detectors—the antibodies. Each antibody recognizes a particular molecular pattern. When a measles virus invades the body, only a few antibodies can recognize the surface molecules of the viral coat. Recognition stimulates division of the B cells that produce matching antibody. This process, called clonal expansion, generates a large population of antibody-producing cells that are specific for the measles virus. (I present a simplified description of the immune system. Good introductions are given by Mims, 1987; Golub and Green, 1991; Mims et al., 1993.)

Antibody can bind and neutralize free virus particles. However, the host has few antibodies that can react with the measles virus on first encounter, thus the virus enters cells and begins rapid multiplication. Meanwhile, the antibodies that react with the virus stimulate clonal expansion of B cells. After several days the antibody titer is high. At this stage antibodies alone cannot clear the infection, perhaps because many infected cells harbor the virus internally.

A second defense, the killer T cells, destroys host cells that harbor viruses. T cells have dynamics similar to the B cells. The large population of T cells can recognize many different kinds of foreign molecules, but only very few T cells recognize a measles virus on first infection. Those T cells that recognize the virus stimulate clonal expansion. Members of this expanded clone, specific for measles, can clear the infection.

Upon reinfection with measles, the host can mount a rapid antibody and T cell response that clears the virus. This immunological memory lasts throughout life. There are several controversial theories about how immunological memory is maintained but, at present, there is not enough evidence to end the debate.

Evolution by natural selection occurs when there is variation, selection, and transmission. The clonal expansion of B and T cells, leading to infection clearance and immunological memory, satisfies the requirements for adaptive evolution caused by natural selection. Cell lines (clones) vary in their recognition properties; reproductive rate varies according to these recognition properties; and offspring cells resemble their parents. This adaptive system causes evolutionary change among the populations of cells within the body. The idea that adaptive immunity is based on natural selection was first proposed by Jerne (1955; see also Burnet, 1959).

B. Diversity: Somatic Recombination and Mutation

Clonal expansion of specific B and T cells in response to challenge by foreign molecules is easy enough to imagine. But how does the body generate sufficient variation so that each new invader can be recognized?

The remarkable mechanisms that generate clonal diversity of B and T cells have been worked out over the past two decades (Golub and Green, 1991; Janeway, 1993; Nossal, 1993). The process differs slightly for B and T cells. I describe the generation of B cell (antibody) diversity (see Fig. 1).

Each antibody molecule has two kinds of amino acid chains, the heavy chains and the light chains. A heavy chain has three regions that affect recognition: variable (V), diversity (D), and joining (J). A light chain has only the V and J regions. There are 100 different V genes, 12 D genes, and 4 J genes.

Each progenitor of a B cell clone undergoes a special type of DNA recombination that brings together a V–D–J combination to form a heavy chain coding region. There are 100 x 12 x 4 = 4800 V–D–J combinations. A separate recombination event creates a V–J combination for the light chain, of which there are 100 x 4=400 combinations. The independent formation of heavy and light chains creates the potential for 4800 x 400=1,920,000 different antibodies. In addition, randomly chosen DNA bases are added between the segments that are brought together by recombination, greatly increasing the total number of antibody types.

The mechanism for generating the diversity on which selection acts switches from recombination to mutation during the course of an infection (Fig. 2). Recombination creates a large number of very different antibodies. Initially, each of these antibodies is rare. Upon infection one of these rare types may match, stimulating selective amplification of the B cell clone. The matching B cells increase their mutation rate, creating many slightly different antibodies that vary in their affinity to the invader. Those mutant cells that bind more tightly are stimulated to divide more rapidly. This evolutionary fine-tuning of the B cell population is called affinity maturation (Golub and Green, 1991).

C. Negative Selection and Self versus Nonself Discrimination

If recognition sequences are generated randomly, then how does the host discriminate self from nonself? Another selective mechanism solves this problem.

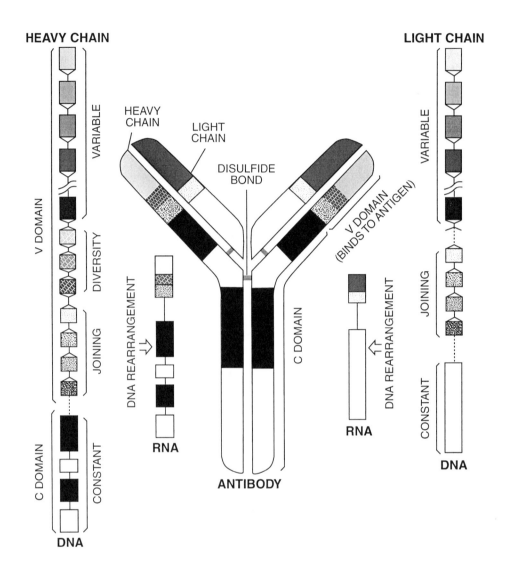

Figure 1 The coding and assembly of antibody molecules. Randomly chosen alternatives are used from different DNA modules to construct a "recombined" RNA transcript, which is then translated into a protein chain. Two heavy and two light chains are assembled into an antibody molecule. Redrawn from Janeway (1993).

The T cells mature in the (T)hymus. The randomly generated recognition type of each maturing T cell is tested against the molecules of the body before the cell is released. The cell dies if it recognizes self molecules. Thus random generation of variation followed by selective death creates circulating T cells that react only with nonself molecules. The processes by which B cells are prevented from reacting with self molecules are not fully understood. It may be that the absence of self-

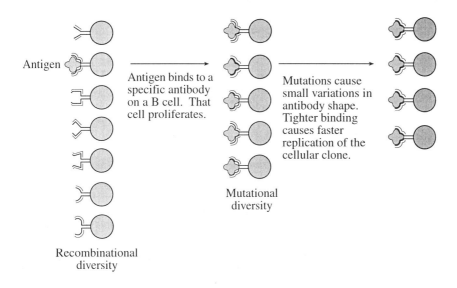

Figure 2 Clonal selection of B cells to produce antibody that matches an invading antigen. Recombinational mechanisms produce a wide variety of different antibody molecules (Fig. 1). All B cells of a particular clone are derived from a single ancestral cell that underwent antibody recombination. Members of a clone express only a single antibody type. Cells are stimulated to divide rapidly when an antigen matches the antibody receptor. This creates a large population of B cells that can bind the antigen. These cells undergo increased mutation in their antibody gene during cell division, producing a set of antibodies that vary slightly in their binding properties. Tighter binding causes more rapid cellular reproduction. Thus this second stage of selection (affinity maturation) enhances the antibody-antigen fit. Modified from Golub and Green (1991).

reactive T cells is sufficient to prevent clonal expansion of self-reactive B cells.

The use of a selection mechanism to create the self versus nonself distinction is costly because many of the newly formed T cells react with self and never mature. However, it is difficult to imagine how the vast array of recognition sequences could be generated without random combinatorial mechanisms. Janeway (1993) notes that the total antibody diversity is particularly impressive because a human has approximately 100,000 genes, but, at any one time, the 10 trillion B cells in an individual can make more than 100 million distinct antibody proteins. Thus random variation and selection appear to be a very good solution to the problems of information storage and response to unpredictable challenges.

Thus far I have described adaptive immunity caused by selection of cell clones within an individual. The adaptive immune system is itself a product of genetical evolution by natural selection. The evolutionary analysis of adaptive immunity therefore requires attention to two levels of selective processes, genetical and cellular.

D. Genetical Evolution of Adaptive Immunity

The adaptive immune system has a complex set of control mechanisms that generate variation, destroy T cells that react with self (negative selection), amplify cellular clones that react with invaders (positive selection), and maintain the ability to react quickly to reinfection by past invaders (memory). These controls of adaptive immunity are inherited (innate) traits produced by genetical evolution.

The immune system has, in addition to the controls of adaptive immunity, many other traits that are innate. Perhaps the best understood is the major histocompatibility complex (MHC), which I now describe.

T cells destroy an infected host cell if they can recognize the infection. In order to signal the T cells, host cells continually cut up intracellular proteins and present these fragments on the cell surface. The circulating T cells distinguish between presented fragments that are self or nonself and respond accordingly.

The molecules that bind intracellular protein fragments and bring them to the surface are coded by genes that reside within the MHC region. Each antigen-presenting molecule from the MHC has a groove that accommodates a peptide of 9 amino acids. Each particular MHC molecule can recognize and present on the cell surface only a subset of protein fragments (peptides). An individual has several different MHC types that, taken together, determine the set of peptides that can be recognized and carried to the cell surface for presentation. (Nine amino acids may seem, at first glance, to be too few for a discriminating recognition system. But there are 20 amino acids and $20^9=512,000,000,000$ different peptides with 9 amino acids.)

There are both costs and benefits to having a large number of MHC types (Fig. 3; Nowak et al., 1992; Mitchison, 1993). The MHC molecules, which are found on cell surfaces typically bound to self peptides, define tissues as self. As T cells mature they are tested against the innate repertoire of MHC-self peptide complexes. A developing T cell dies if it would destroy a cell with self MHC. Thus the greater the MHC repertoire, the larger the number of T cells that are destroyed during development. If the MHC repertoire is too broad, then too few T cells would be able to develop.

On the other hand, if too few MHC types were present, then the host would not be able to recognize and present the protein fragments from many pathogens. The optimal number of MHC types must strike a balance between the costs of too broad a definition of self, causing a narrow T cell repertoire, and the benefits of recognizing a wide array of invaders. This type of optimality argument can help to define the forces that have influenced the genetical evolution of innate components in the immune system. There are, of course, many other factors that may have influenced the evolution of the MHC loci, such as the processes of gene duplication by which these loci have multiplied from a single ancestral locus.

The MHC loci are highly polymorphic, with between 10 and 80 different alleles known for each locus. Two lines of evidence suggest that resistance to particular diseases can strongly affect the frequency of MHC alleles. First, most of the variation among alleles occurs in the groove that binds protein fragments—the specific recognition area. Second, a few cases are known in which there is a strong spatial correlation between endemic diseases and MHC alleles that are associated with resistance to those diseases. For example, the allele HLA–B53 is associated with resistance to a severe strain of malaria that occurs in children in The Gambia. HLA–B53 occurs at a frequency of 25% in this west African nation; by contrast, the frequency of this allele in Europe is 1% (McMichael, 1993). Other MHC alleles are implicated in resistance to HIV, the cause of AIDS, and to Epstein–Barr virus, the cause of various cancers. Disease correlations with MHC alleles suggest that selective pressures continue to influence the genetical evolution of the immune system (Thomson, 1991; Mitchison, 1993).

I close this section by summarizing four cases in which genetical evolution influences adaptive immunity.

(1) Genetical evolution of MHC loci affects the control of adaptive immunity. The number of MHC loci and the level of polymorphism at each locus determine the balance between negative selection of T cell clones and the ability to present foreign protein fragments on the cell surface.

(2) The immune system uses adaptive mechanisms for some types of recognition (B and T cells) and direct (genetically encoded MHC) recognition for the presentation of protein fragments. This raises some interesting questions. From the point of view of an optimally designed immune system, is this particular mix of adaptive and innate recognition ideal? Or, would genetical evolution favor a shift toward adaptive recognition of protein fragments if suitable genetic variation existed?

(3) Regulation of the immune response is another form of innate control over adaptive immunity. Deployment of defenses is often costly in terms of energy spent on the production of new cells and toxic substances. In addition, the battle against invaders may lead to inflammation or local swelling because the methods used to clear infection can also damage the host tissues. Thus regulation of the components of the immune system and the setpoint for triggering a response are under strong selective pressures. These regulatory aspects of the vertebrate immune system are not well understood at present.

Induction of defense is a complex subject. It may be useful to look at much simpler forms of inducible defense to understand the problems involved. Harvell (1990a, b) has written excellent reviews of the inducible defenses that occur in a wide variety of organisms. The selective pressures on the setpoint for induction have been studied in two recent papers (Clark and Harvell, 1992; Frank, 1993). A phylogenetic perspective of the evolution of immune responses is presented by Klein (1986).

(4) The process of affinity maturation, discussed above, is another example of innate control. Affinity maturation occurs when the mutation rate of a B cell clone increases in response to a match between the clone's antibody and a foreign molecule. Those mutant cells that bind more tightly are stimulated to divide more rapidly. A quantitative trait such as enhanced mutation rate is very likely to be influenced by the processes of genetic variation and natural selection—put another way, the quantitative controls of affinity maturation are the product of adaptive processes at the genetic (innate) level.

Affinity maturation dynamically balances exploration versus exploitation in adaptive immunity. Initially the system explores widely by recombination to meet unpredictable challenges. After a close match is found the system exploits the match by reducing the information decay that further recombination would cause while simultaneously exploring for small improvements by mutation.

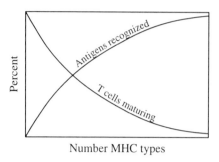

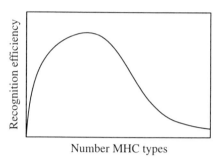

Number MHC types   Number MHC types

Figure 3 Optimal number of MHC types to recognize foreign antigen. The left panel shows two opposing forces. On the one hand, increasing numbers of MHC types enhance the probability that a foreign antigen will be recognized by the array of MHC specificities. On the other hand, more MHC types reduce the number of T cells that mature because self-reactive T cells are destroyed. The right panel shows that, by combining these two effects, there is an intermediate optimum that maximizes the recognition efficiency per T cell produced. There are not enough data to estimate accurately the magnitude of the two processes and the actual number of MHC types expressed (Nowak et al., 1992). The best estimate from current data is that 1-7% of T cells are removed by negative selection per MHC type. Each MHC type can bind a set of nine amino acid peptides with varying affinities. Thus each MHC type probably binds a fraction between 10^{-8} and 10^{-10} of the 5.12×10^{11} possible antigens. The number of MHC types expressed is estimated between 8-40 for humans. The actual number expressed in many different cell types is probably closer to the low end of the range.

The questions I have discussed about the immune system can be phrased more generally to apply to any two-level adaptive system. What is the optimal mix of innate (closed) and adaptive (open) mechanisms for particular problems in pattern recognition? What is an efficient mode of selection in order to achieve rapid learning and accurate memory? How is this selective mode achieved by the genetical evolution of innate controls of the adaptive process? What forces influence the setpoint that triggers a response to a recognized pattern? How are the problems of information storage and transmission solved?

I will show in the following sections that these questions, introduced with the immune system, apply to a variety of other adaptive systems.

IV. Learning

A newborn organism, faced with the world for the first time, begins to receive signals about the environment. Simple organisms often detect light, temperature, and chemical gradients. A baby mammal, endowed with a rich array of feature detectors, receives a tremendous amount of information. At first, sensory input has only the limited meaning encoded in the genome that guides the preliminary wiring of the nervous system—the primary repertoire. As the organism interacts with its environment during early development the neural connections undergo rapid changes, leading to an altered neural wiring pattern—the secondary repertoire.

All aspects of perception, meaning, and even consciousness must derive from the organism's physiological and neural interactions with the environment. Thus the problem is to understand how genes shape the primary repertoire, how interactions between the primary repertoire and the environment lead to the secondary repertoire, and how learning is embodied in neural-environment interactions.

Consider an analogy with the immune system. The problem in immunity is to recognize all molecules within the body and categorize those molecules as self or nonself. The nonself molecules must further be categorized according to the appropriate type of defensive response that should be induced. Finally, the immune system remembers its categorizations over long time periods.

The number of molecules to be recognized and placed into different categories greatly exceeds the informational capacity of the genome. Thus there are only two ways for the body to obtain the information to recognize the diversity of molecules encountered: instruction or selection.

The instructional idea, popular in immunology in the 1930s and 1940s, suggests that new antibody (recognition) is achieved by shaping the antibody to the foreign template (see Golub and Green, 1991, pp. 8-12). Instruction directly from the environment is possible in principle, but is not known to occur. Under direct instruction, the body must be able to build its informational molecules to match a template—this requires a sufficiently malleable informational structure that can handle whatever external template is posed. In addition, the external information, now encoded in a similar internal molecule, must be transmitted successfully within the body. This seems to require that internal communication encodes information in a way that matches unpredictable forms of external information. As discussed in the previous section, the immune system uses selection rather than instruction to recognize and categorize the world.

Recognition and categorization, central to immunity, are also the fundamental problems of perception. The external world contains too much information and poses too many problems for all information and solutions to be coded directly into the genome. Once again, the alternatives for acquiring information are instruction or selection.

A. Instruction versus Selection

Consider an instructional model of learning (Fig. 4). The genotype codes for generative rules that specify how to build a neural network. The network has four inputs that react differently to features of the environment. Environment A stimulates (1, 2, 4). Environment C stimulates (1, 3, 4). Learning rules change synaptic connections between nodes (neurons) according to correlations in activation. Eventually the network learns to categorize features (1, 2, 4) as A and (1, 3, 4) as C by stimulating an internal node representing the environmental state. These internal nodes stimulate, in turn, other nodes that trigger appropriate action for each environment.

The network has learned by acquiring information about correlated features of the environment. However, the network architecture, the learning rules, and the interpretation of A and C must be strictly specified by the genotype for this learned information to be useful. Simple types of associative learning may be achieved in this way, with a specific module built to obtain information (instruction) directly from the environment.

The internal nodes A and C may be part of a larger network. Environmental categorization could then occur at a deeper level, without the need to specify directly the interpretation and action at such a fine scale. But this simply pushes the problem back without solving it. At some internal level, meaning and interpretation with respect to fitness consequences must be encoded by the genotype. There is no way for a network to self-organize toward an unspecified goal such as reproduction.

The instructional model in Fig. 4 seems to require too much of the genotype in terms of specifying the architecture, learning rules and meanings for each instructional module of a complex brain. Selection may be the only way to build a complex and meaningful information system from simple rules. The vertebrate immune system is an excellent example.

How could there be a selective, adaptive subsystem in the brain built by simple rules encoded in the genotype? The answer differs in only a few details from the instructional model of Fig. 4. Figure 5 shows a typical neural network fantasy that has the key elements of a selective system: random variation, continuity, and differential success.

The top row of nodes (1, 2, 3) are stimulated by three detectors of environmental state. For example, if the environment has states (1, 3), then the (1, 3) nodes will be stimulated in each of the three neuronal groups. In this case the group on the left will pass on the strongest stimulus to the next layer, which in turn stimulates the coordinating center A. Note that A is simply a richly connected region where neuronal groups tend to converge; there is no intrinsic meaning to the A region.

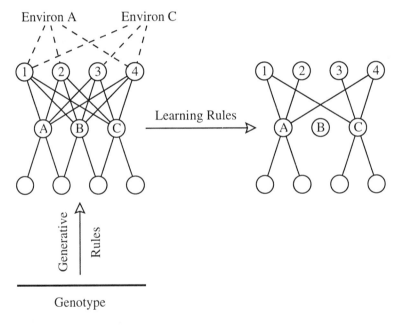

Figure 4 An instructional model of learning. This is a simple neural network example. See Hertz et al. (1991) for an introduction to the literature on neural processing of information and simple learning models.

When A is stimulated then connections are activated to two complex centers, which are treated here as "black boxes." The Move Forward center triggers simple motions that, in this case, result in detection of water. The Thirst center is a genetically coded region that returns positive stimulus when water is detected or acquired.

We can now trace the events that follow detection of environmental state (1, 3) along the bold connections. Starting at the top, the neuronal group at the left responds most strongly to the stimulus and passes a relatively stronger signal to the connection center at A. This switching point has connections to both the motion and thirst centers. In this case behaviors are triggered that cause water detection, sending a positive signal to the thirst center. The thirst center initiates a return pathway of stimulus, following the lines of the most strongly active connections. Internal rules of synaptic change cause the bold pathway to be strengthened—a form of credit allocation or fitness assignment for relative success with respect to the internal goal of satisfying thirst.

This is a simple neural network with basic learning rules. But it differs from the instructional model in three important ways. First, the system begins with a population of neuronal groups that respond differently to the same input. Second, the initial structure of each neuronal group is uncorrelated (random) with respect to environmental challenge and "meaning." Third, categorization of environments arises spontaneously as a result of the differential success of neuronal groups. The categorizations and success are subordinate to innate (genetically encoded) goals.

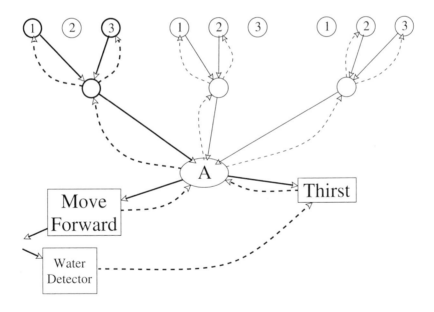

Figure 5 A selection model of learning. Note at the top that the system begins with a population of neuronal groups, where each group responds in a different way to the environmental features labeled (1, 2, 3).

B. Neural Darwinism

I now turn to a summary of the selective theory of brain development and learning recently proposed by Edelman (1987, 1988, 1989, 1992; see Edelman, 1987, pp. 14-22 for a review of earlier, related ideas by J. Z. Young, J.-P. Changeux and others). This theory of "neural darwinism" shares many features with neural network models (Rumelhart et al., 1986; McClelland et al., 1986; Hertz et al., 1991) and ideas from other areas of neurosciences. But Edelman's work differs in its explicit emphasis on selective ideas and a strong effort to tie every detail of the theory directly to biological observations of neural development and structure.

The basic components of neural darwinism are:

1. *Genes plus stochastic variation plus developmental selection create the primary repertoire.* The morphology (wiring) of the vertebrate nervous system at birth is an immensely complex structure. The information in the wiring pattern far exceeds the amount of information contained in the genome. Yet even the most helpless organism at birth has considerable sensory and motor abilities. Edelman (1992, p. 64) suggests that the wiring at birth—the primary repertoire—forms by a process of developmental selection:

> Imagine now this epigenetic drama in which sheets of nerve cells in the developing brain form a neighborhood. Neighbors in that neighborhood exchange signals as they are linked... They send processes out in a profuse fashion, sometimes bunched together in bundles called fascicles. When they reach other neighborhoods and sheets they stimulate target cells. These in turn release diffusible substances or signals which, if the ingrowing processes have correlated signals, allow them to branch and make attachments. Those that do not either pass on or retract. Indeed, if they do not meet their targets, their parent cells may die. Finally, as growth and selection operate, a mapped neural structure with a function may form. The number of cells being made, dying, and becoming incorporated is huge. The entire situation is a dynamic one, depending on signals, genes, proteins, cell movement, division, and death, all interacting at many levels.

Chance plays a large role in the actual neural connections formed. Even identical twins are predicted to have different neural maps at a fine scale. Indeed, this variability is inevitable given the fact that the information in the primary repertoire greatly exceeds the information in the genome.

The variability in the primary repertoire provides the basis for neuronal group selection to form the secondary repertoire.

2. *Selection of neuronal groups leads to perceptual categorization, memory, and learning.* A neuronal group is, roughly speaking, a set of neurons activated together in response to a particular stimulus. The selective processes act on populations of neuronal groups in the manner illustrated previously in Fig. 5.

For this selective process of learning to work effectively, the primary repertoire must contain many partly coupled groups whose connection strengths can subsequently be altered by rules of synaptic change. The initial synaptic diversity and subsequent change correspond to variation and selection in

Edelman's theory.

Two points about neuronal group selection must be stressed. First, the theory focuses on neuronal groups as the appropriate unit of analysis. This makes sense because individual neurons or synapses have too little information to provide a basis for selective differentiation of their performance. Second, selection of neuronal groups leads to altered synaptic strengths of connection rather than differential reproduction. "Group selection proceeds through synaptic modifications induced by the correlated activation of cells within a group" (Edelman, 1987, p. 169; see also Merzenich et al., 1988).

Group competition plays a central role in the theory. Roughly speaking, groups compete for members and for feedback connections from other groups: those groups poised to receive frequent correlated stimulus among all members will be constantly strengthened relative to other groups that less frequently receive correlated stimulus. Groups do not have clear boundaries because they are defined simply by correlated firing of local neurons; thus individual neurons may be in several groups and their associations necessarily change as a result of neuronal group selection.

3. *All learning and memory are embodied in the interaction between the environment and neural physiology as modulated by the innate value systems.* Categorization occurs by correlations of firing among neuronal groups in response to external stimuli. The types of category formed by this process depend on the sensitivity of the feature detectors and any biases in the primary repertoire. Thus even the simplest process of category formation has a strong innate (genetic) component that is subject to evolutionary modification by natural selection. However, this simple form of categorization is not sufficient to explain the types of goal-directed learning and behavior observed.

Edelman suggests that there are internal value systems or hedonic centers that are themselves embodied in neuronal groups, for example, thirst in Fig. 5. Other hedonic centers include hunger, sex drive, and curiosity. These centers are linked extensively throughout the brain to other neuronal groups by feedback connections. These control centers establish the basis for assigning "fitness" in competition among neuronal groups. These controls are directly influenced by the genotype and are subject to evolutionary modification.

Selective ideas about the brain are exciting, not because they are clearly true—it is too early to say yet—but because they could form a relatively complete theory of the nervous system and of behavior. Edelman has made a serious attempt to account for observed details of development and plasticity in neural maps, the causation of behavior, and the evolutionary modification of internal value systems (Plotkin, 1987). The idea that classification and behavior are controlled by synaptic modification of neural maps is not new. The novelty is an emphasis on selective systems and a consistent vision that explains many aspects of neurobiology and behavior—a global brain theory.

The special aspects of selective systems for neurobiology are sometimes difficult to keep clear. Therefore I close this section by emphasizing the distinction between selective and instructive theories. These theories are not strict alternatives about brain development and function. Any complex nervous system is likely to use both systems. The empirical issue is the relative importance of the two systems for the traits we wish to understand.

A selective (adaptive) system acquires information by selecting from *variation* in a *population* of individual architectures generated *independently* of the environmental challenge. An instructional systems acquires information by *varying* a *single* architecture in a manner *correlated* with environmental challenge. The distinction leads to different predictions about neural development and learning. For example, selective systems require stochastic variation during development to provide the necessary variation for selection. Instructional systems will generally perform more poorly as stochastic variation increases in the development of the initial architecture.

C. Genetical Evolution of Learning

There are many ways in which genetic information interacts with the environment to affect behavior and learning (e.g., Drickamer, 1992; Alcock, 1993). For example, the external sensors (vision, olfaction, etc.) are mainly innate, imposing particular channels of communication between the organism and its environment. There are closed behavioral programs that follow a fixed pattern once invoked by environmental cues—these closed programs are also mainly innate. There are open behavioral programs subject to modification and learning through interaction with the environment. Learning is inevitably guided by internal value systems (McFarland and Bösser, 1993).

Interesting questions about the role of genetical evolution in learning include: What sorts of environments favor a closed program in which all information is stored in the genome? How does the genetic system discover closed programs? What environments favor an open program that causes behaviors, with initial biases fixed by the genome, to change with experience? What forces shape the innate biases and genetic controls of learning?

These evolutionary questions about the control of learning parallel those questions that I raised about the controls of immunity. Some controls of the immune response are innate, such as the specific set of MHC types (closed program). By contrast, other parts of the immune system are built by the genetic specification of controls on the adaptive subsystem, such as the negative selection against self recognition and the positive selection of clones with antibody that matches foreign molecules (open program).

I summarize two key issues for an evolutionary analysis of learning.

1. Learning Accelerates Genetical Evolution—the Baldwin Effect

How can a complex behavioral sequence be favored by natural selection if each isolated part of the sequence is of little value? A genetic mutation that caused the whole sequence is unlikely to occur all at once, and each genetic variant for part of the sequence will not be favored in isolation.

Baldwin (1896) suggested that learning can help to overcome natural selection's limited ability to discover complex behavioral traits, thereby accelerating the rate of evolutionary change. The idea that phenotypic modifications such as learning can feed back to inherited changes suggests a lamarckian mechanism of evolution that has essentially been disproved. The association between Baldwin's ideas and the discredited lamarckian mechanism confined acceptance of these ideas to a minority of evolutionary biologists.

However, recent work has shown that learning can indeed greatly accelerate evolutionary change without appeal to lamarckian inheritance (Hinton and Nowlan, 1987; Maynard Smith, 1987; Ackley and Littman, 1992; Fontanari and Meir, 1990; French and Messinger, 1994; Anderson, 1995).

Learning provides information about how close the genotype is to a good solution (Fig. 6). Imagine that a complex behavioral sequence (phenotype) has a high fitness but that slightly altered sequences are no better than random behaviors. If there is no learning then a genotype has to encode exactly the right sequence to gain any fitness advantage; nearly correct genotypes are no better than random. The chance of the favored genotype arising from a background of random behaviors is vanishingly small.

Now suppose that some learning occurs. Learning can be thought of as an exploration of behavioral sequences similar to the genetically encoded sequence, where sequences with improved performance are adopted by the animal. An animal's chance of finding the correct behavioral sequence depends on how near it is to the correct sequence initially. Fitness therefore drops off gradually from a peak at which the genotype encodes the optimal behavioral sequence, the height dropping with the number of behavioral changes that must be discovered to find the optimum. Natural selection is very good at pushing genotypic composition

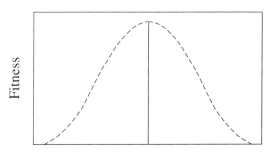

Initial, genotypically encoded behavior

Figure 6 The Baldwin effect. Without learning, each organism expresses its genetically encoded behavior. Only a particular, complex behavioral sequence has high fitness, all other behavioral sequences have equally low fitness. This is shown by the solid line for a narrow fitness peak among possible genotypes. An exceptionally rare mutation is required to improve fitness. All other variations are equally bad with respect to the optimum. Thus natural selection will not move the population closer to the optimal behavioral sequence. Now suppose that learning occurs, with initial behavior determined by genotype. Those genotypes near the peak have a high probability of finding and learning the optimal sequence in a reasonable period of time. Genotypes more distant from the peak have a lower probability of learning the optimal sequence. Thus fitness increases smoothly with decreasing genetic distance from the optimum, allowing natural selection to cause steady improvement over time (dashed curve).

steadily up a slope of improving fitness. Thus learning, by providing clues about the distance to the favored behavioral sequence, greatly accelerates the rate of evolutionary change.

2. What environmental challenges favor learning?

Learning provides a genotype with a method of phenotypic exploration. The Baldwin effect shows that exploration can ultimately cause the transfer of learned behaviors into an innate genetic program. Innate (closed) genetic solutions have an advantage over learned (open) solutions in a fixed environment because no energy is wasted on failed explorations. If the Baldwin effect were the only force operating, all unchanging problems would be solved by closed programs that exploit known solutions rather than open programs that explore opportunities for improvement. Learning would disappear.

What types of environmental challenge favor learning? There is no coherent body of theory on this important question (see preliminary efforts in Holland, 1975; Boyd and Richerson, 1985, chapter 4; Todd and Wilson, 1993).

Questions about learning can be stated in a more general way when learning is viewed as an emergent adaptive system that has evolved by natural selection of genetic variants: What types of challenge favor an adaptive system (genetics) to spawn a subsystem of variation and selection (learning)? In what ways will adaptive subsystems such as learning differ when they have been shaped by different kinds of challenge?

These questions can be addressed only when cognitive mechanisms are viewed both as controls of adaptive learning systems and as evolved adaptations of the underlying genetic system. This dual view of learning has a long history (Richards, 1987), but has only recently gained attention in evolutionary biology (Real, 1992, 1993), psychology (Barkow et al., 1992) and computer science (Meyer et al., 1993).

Computer models of learning are perhaps the greatest spur to conceptual work. Computers can be used to test which learning programs and types of cognition perform best under different kinds of environmental challenge.

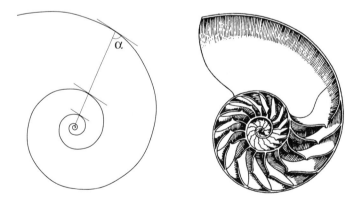

Figure 7 The logarithmic spiral. The *Nautilus* shell on right is copied from Thompson (1961, p. 173).

Advances in robotics and engineering control also depend on a clear understanding of how different types of learning and cognition influence performance. Although many recent papers have focused on these topics (Meyer et al., 1993; Brooks and Maes, 1994; Cliff et al., 1994), there are no general conclusions yet.

The problem once again focuses attention on the specific kinds of challenge that shape adaptive systems. Referring to Table 1, each of the three main challenges can favor learning. Information decay in the genome can be corrected phenotypically by an adaptive subsystem that learns. Predictable complexity may be an important challenge when the motor controls and input-output connections for a behavioral sequence contain too much information to store within the limited genome. Young animals sometimes use trial-and-error learning periods to develop behaviors that eventually converge to a fairly routine, species-typical sequence. This type of "developmental" learning may be a response to the challenge of information storage for complex behaviors. Finally, unpredictable challenges can be met with an adaptive subsystem that learns by using processes of variation and selection.

V. Development

Parasitic attack or unpredictable abiotic environments require flexibility. Many animals produce phenotypic solutions with adaptive subsystems, such as immunity and learning. Immunity and learning are two aspects of development—the creation of a phenotype from the information encoded in the genotype. Morphology is another, more traditional, domain of developmental study. The problem is how the one-dimensional information in the genotype is transformed into the three-dimensional structure.

A. Morphology

Mollusc shells develop according to simple generative rules (Thompson, 1961; Meinhardt, 1995). The left panel of Fig. 7 shows a logarithmic spiral, which is a nearly perfect match for the coiling pattern of the *Nautilus* shell shown on the right. A logarithmic spiral is produced by drawing a line from the center to the current tip of the spiral, and then adding to the spiral such that the angle between the radial line and direction of growth is constant. Shells can grow only by adding new material to the leading edge. New growth typically follows a constant angle relative to the radius, causing a logarithmic coiling pattern.

Variation in the angle of growth explains variation in coiling patterns in the radial direction, as shown in the top row of Fig. 8. Shells also vary in the tightness of coiling with respect to height, which can be explained by the relative rate of growth down from the radial direction (bottom row of Fig. 8).

B. Developmental Selection

The full range of physicochemical processes involved in shell growth are complex. Yet there is a certain determinism in the rules of growth when compared with

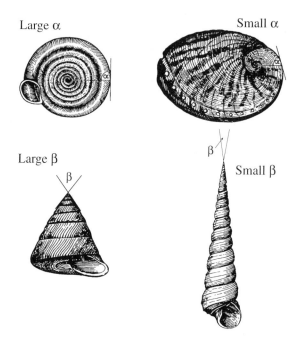

Figure 8 Logarithmic coiling patterns of shells. The angle α controls coiling in the radial direction, and the angle β controls coiling in the vertical direction. From Thompson (1961, p. 192).

immunity and learning. Figure 9 shows the contrast between these direct rules of morphology and the indirect route by which immunity and learning affect the phenotype.

The shell example suggests that morphological pattern formation follows direct generative rules specified by the genotype. However, there is considerable controversy about the details of those rules (Hall, 1992; Goodwin et al., 1993). Several authors have proposed, as an alternative to direct rules of growth, that pattern formation is best viewed as the outcome of developmental selection. This theory emphasizes stochastic variation among a population of growth trajectories coupled with selection of particular trajectories that meet critical design criteria. (Recent suggestions of this theory include Edelman, 1988; Sachs, 1988; Wagner and Misof, 1993.)

Developmental selection determines the innate wiring pattern of brains in Edelman's theory of neural darwinism. The process depends on stochastic fluctuations in the movement and proliferation of cells to generate variability in structure. Two selective systems act on the variability generated by stochastic fluctuations. Positive selective controls stimulate cell division and the formation of particular neural connections. Negative selective controls cause rapid cell death

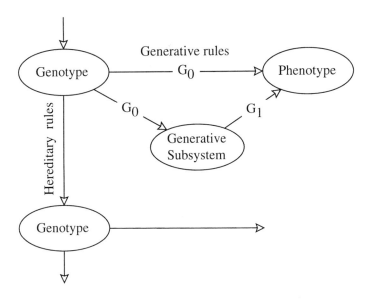

Figure 9 Generative rules for the development of the phenotype. The genotype is passed from generation to generation according to the hereditary rules for transmission, as in traditional models of biological populations. The success of each genotype depends on its phenotype. The genotype specifies properties of the phenotype through generative rules for development. The genotype directly encodes the generative rules G_0. These rules may lead directly to the phenotype, such as specific genes for metabolic enzymes or laws of growth for morphological development. Alternatively, G_0 may produce a generative subsystem, which in turn specifies a second set of generative rules, G_1. The vertebrate immune system is an adaptive subsystem that specifies generative rules for variation and selection. Simple learning rules are generative rules that change the phenotype in response to the environment; adaptive learning rules produce phenotypic change by variation and selection. A language module may be a generative subsystem that influences the basic grammatical rules used to generate sentences. The grammatical rules, and all aspects of phenotype, are also influenced by the environment (not shown).

among those neurons that fail to make selectively favored movements and connections. The primary repertoire created by developmental selection is highly variable at the level of particular cell-cell connections. This variability forms the substrate on which neuronal group selection shapes the secondary repertoire of the learning organism.

Developmental selection is a reasonable hypothesis for neuronal development. But the nervous system is so complex that it is difficult to compare selective theories with other ideas. Two simpler developmental problems provide a clearer view of the conceptual issues.

Distribution of stomata on leaf surfaces.—Plants exchange gases with the air through small openings (stomata) on the leaf surface (Fig. 10). The distribution of stomata on the leaf surface is a classic example of spatial patterning in development (Sachs, 1991). Pairs of stomata rarely develop adjacent to each other, and the distances between neighboring stomata are larger than if locations were determined randomly. But the developmental program is not a deterministic

Figure 10 Stomata on the surface of a *Begonia* leaf. The stomata are opened and closed by the changing shape of the small guard cells that surround the pore. The spatial distribution is regulated developmentally to prevent neighboring cells from differentiating into stomata. But the particular cell lineages that differentiate into stomata and the distances between neighboring stomata appear to be strongly influenced by chance events. From Sachs (1991, p 7), copyright © 1991. Reprinted with the permission of Cambridge University Press.

unfolding of pattern—the fate of particular cellular lineages and the ultimate location of stomata depend on many chance events.

Kagan et al. (1992) have proposed a model of developmental selection to explain stomatal patterns. In the model each cellular lineage is an *individual* growth trajectory. Pattern is determined by variation and selection among the *population* of cellular lineages that contribute to the leaf surface.

Each stoma typically develops from differentiation of a single cellular lineage. Differentiation begins with an unequal cellular division. The smaller product may divide a second time, more or less unequally. The process continues until an equal division forms the two guard cells of the stoma (Fig. 10).

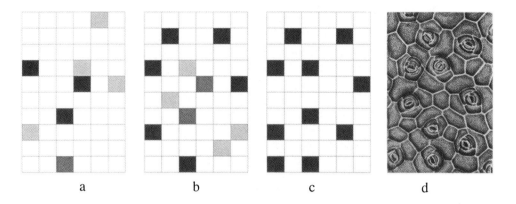

a   b   c   d

Figure 11 A model of variation and selection that develops a spatial pattern similar to the stomata on a *Begonia* leaf. Each square represents a cellular lineage. All lineages begin as normal epidermal cells (white). In each time step a normal lineage switches to a growth trajectory leading to stomatal development with probability 0.05. The early stage is coded by the lightest gray and assigned a value of one. A second stage of differentiation is dark gray, with a value of two. A late stage is black, with a value of three. In each time step, a sum of values is calculated for the eight neighbors of each cell, discounting the four corners by one-half. (The edges are connected in a torrus so that all cells have eight neighbors.) A sum of less than one allows an intermediate stage to progress to the next stage. A sum between one and three causes reversion to the previous stage. A sum greater than three resets the lineage to the normal state. (Reversion may often be caused by death of a lineage and filling in by neighbors.) The panels a, b, and c show the state of a sample run after 5, 15, and 30 time steps. At the final step in panel c all intermediate stages revert to normal cells, leaving only normal and final, stomatal stages. This type of model is known as a cellular automata, popularized in the *Game of Life* and found on many computer screen savers.

The development of a single stoma depends on programs internal to the cell lineage. But observations suggest that variation and selection among lineages play an important role in final pattern (Kagan et al., 1992). The initial unequal divisions appear to arise spontaneously among lineages without any particular pattern. When neighboring lineages begin differentiation, one or both revert to a typical developmental program that leads to normal epidermal cells. Figure 11 shows how a simple model of variation and selection can produce a spatial pattern typical of stomata.

A model with random variation and local selection seems so simple that it hardly needs justification. Indeed, the classic reaction-diffusion model for spatial pattern depends on similar principles (Murray, 1989). In each location chemical reactions cause the increase or decrease of a particular substance. Diffusion of the reactants causes spatial interactions between neighboring sites. Simple models produce patterns that match zebra stripes, leopard spots, and coloring on sea shells (Murray, 1989; Meinhardt, 1995).

Reaction-diffusion models differ from population models of variation and selection in several ways. Reaction-diffusion is a continuous chemical process that leads to a highly ordered final state. Population models have discrete births and deaths of cellular lineages. The final pattern is ordered but the growth trajectories and details of final state are unpredictable. In population models, form depends on the range of variant growth trajectories that are generated, and the control processes that select among alternative trajectories.

Competition among shoots for root resources.—A particular shoot on a plant typically obtains resources from a corresponding subset of roots. The association between roots and shoots can change over time by modification of vascular contacts among neighboring shoots.

Sachs et al. (1993) suggested that a plant contains a population of competing shoots. Those shoots in the best condition outcompete their neighbors and obtain additional root support by modification of vascular connections.

The individual shoot "modules" compete for resources based on variable growth rates. The developmental program, which evolved by genetical selection, controls the flow of resources among members of the shoot population. Thus simple, local generative rules can develop an efficient, large-scale phenotype by imposing selection on uncontrollable aspects of cellular and modular variation. Indeed, developmental selection works only if there is significant variation among alternative, competing developmental trajectories.

C. Summary

This section concludes my summary of adaptive subsystems and generative rules. These processes transform hereditary information (genotype) into mechanisms that can interact with the environment (phenotype).

I have assumed that the genotype varies in its coding for particular generative rules, such as the generation of antibody variants or the number of MHC types. But I have not considered the processes that influence the kinds of hereditary information that are bound together to form a genotype. How is such a complex unit of information formed by evolutionary processes? Why is information regularly mixed between units by sex in order to produce subsequent generations? Why do different evolving populations—adaptive systems—often mix to form higher-order groups that cooperate and compete? In the next two section I turn to these questions on the evolution of hereditary information.

VI. Symbiosis

Adaptive systems are highly social. That may seem a strange statement. But the fact is that adaptive systems often compete with one another and often join forces in cooperative communities. This social structure is perhaps the most important and difficult problem in understanding the natural history of adaptive systems.

In later sections I will discuss some of the consequences of interactions among adaptive systems. In this section I briefly describe the natural history of symbiosis. My goal here is to discuss the main concepts and to provide an introduction to the literature. I begin with an analogy between culture and bacterial gut symbionts. This surprising analogy emphasizes that the boundaries between separate evolving systems blur in real life. This is a fact that must be faced squarely by historical descriptions of adaptive systems and by any theory that attempts to explain the main properties of such systems.

A. Culture

Culture is the ideas, facts, attitudes, and beliefs that are *transmitted* from one member of the society to another. Dawkins (1976a) coined the term "memes" to refer to individual units of culturally transmitted information. A word of a language is an example of a meme. Darwin and many others have noted that languages evolve by differential success of words and rules of composition.

Memes are analogous to genes because both have (1) temporal continuity transcending their containers (bodies or brains), (2) particular patterns of transmission, (3) imperfect transmission that generates variability (mutation), and (4) differential rates of transmission (reproduction). Thus cultural units—memes—form an evolutionary population that has its component frequencies determined mainly by a system of variation and selection (Dennett, 1995).

Cultural selection is different from the somatic selective systems of immunity and learning discussed earlier. The somatic selective systems are simply phenotypic mechanisms by which an organism meets environmental challenge, just as regulation of body temperature is a phenotypic mechanism that can enhance survival and reproduction. Because the somatic systems are governed by innate (genetic) controls, the apparent goal-directed nature of these somatic systems is wholly subordinate to the goal-directed nature of the underlying genetic system (Plotkin and Odling-Smee, 1981).

Memes, by contrast, have a continuity that transcends a single body and a transmission system that differs from genes. In short, culture has a life of its own.

Genetic systems are familiar, but memes may seem a bit strange at first. Memes live in bodies but can be passed from parent to offspring, from teacher to student, among friends, among enemies, or from child to grandparent. Memes are transmitted like the symbiotic flora and fauna that live in digestive tracts. Gut bacteria, and memes, face two opposing selective pressures. Bacteria or memes that enhance host survival also enhance their own survival—the cooperative side of the relationship. By contrast, a trait that enhances transmission of a bacteria or meme from host to host, but also harms the host, could increase in frequency.

The potentially harmful effects of gut symbionts are illustrated by gut bacteria that cause diarrhea. Natural selection favors an increase in the virulence of the bacterial flora when diarrhea can increase host-to-host bacterial transmission at a rate sufficient to offset the reduced survival of the bacteria caused by harm to the host. Natural selection favors a decline in virulence when diarrhea does not increase transmission sufficiently to offset reduced survival (Anderson and May, 1982; Ewald, 1994; Frank, 1996).

A simple meme example is less easy to find. Dawkins (1976a, p. 198) discusses the meme for human celibacy as a case in which a meme maintains itself in spite of reducing the genetic success of the host. A truly celibate priest does not reproduce, yet his celibacy meme has managed to transmit itself to enough young men to maintain its numbers over many hundreds of years. Although celibacy may have alternative explanations, a meme that did increase its host-to-host transmission at the expense of host survival is a virulent meme in the same way that diarrhea is caused by virulent symbionts increasing their host-to-host transmission. Both memes and bacteria can be helpful symbionts or harmful parasites.

Memes, like bacteria, create their own apparent goal-directedness because they form a selective system with replicators whose permanence transcends individual bodies. But a meme's ability to be transmitted—literally, to infect a mind—depends on the structure of minds. Minds, in turn, have some innate (genetic) controls, so the genes of the host and the symbiotic memes cannot be wholly independent. People disagree about the extent that genes, by shaping the structure of minds, can constrain the types of memes (cultures) that can succeed (Boyd and Richerson, 1985; Barkow et al., 1992). I briefly mention some of the issues.

The tension between the direction of evolution favored by genes versus memes has two components (Cavalli-Sforza and Feldman, 1981; Boyd and Richerson, 1985). The first concerns the extent to which these directions will differ. The second concerns how tensions are resolved when selection of genes versus memes favors different traits. I examine each of these components in turn.

If a meme is transmitted only from parent to offspring, then the inheritance patterns of genes and memes are symmetric (Boyd and Richerson, 1985). In this case any trait that enhances meme transmission also enhances gene transmission and vice versa. All members of the symbiotic community favor the same direction of evolutionary change. One can therefore analyze which traits are favored by selection from either a purely genetic or a purely memetic point of view. The conclusion from either analysis is that selection favors traits that increase relative reproductive success. (See Feldman and Zhivotovsky, 1992, for a more sophisticated theory of symmetric transmission.)

Memes frequently have a pattern of transmission that differs from genetic transmission. Genes and memes may favor different directions of evolutionary change with asymmetric inheritance. The diarrhea and celibacy examples show that a tension arises when symbionts (memes or bacteria) enhance their host-to-host transmission at the expense of host survival and reproduction.

Who wins when there is a conflict between host and symbiont? One line of thought suggests that the host has the upper hand. The idea can once again be described by a parallel with gut symbionts.

Ruminants have an additional niche for symbionts in their second stomach chamber. The additional chamber was probably favored by natural selection of genetic variants because the symbionts enhance genetic transmission of the host, or at least they did at some time in the past. Put another way, the structure and physiology of the additional stomach probably evolved to use symbionts in a way that enhances host fitness.

Minds, to the extent that they contain and transmit memes, are a niche for memetic symbionts. The structure of minds evolved, and continues to evolve, by selection of genetic variants that favor enhanced genetic transmission. Thus genes, by controlling the structure of the mind, may be able to constrain the types of memes that can succeed. If so, then joint selection of gene-culture interactions should favor traits that enhance the relative reproductive success of genes.

On the other hand, some people argue that once a niche for cultural transmission evolves in minds, cultural evolution can proceed unconstrained. Thus culture can be understood without reference to the historical reasons for the evolution of the cultural niche. In terms of the ruminant, once the second stomach exists, bacterial evolution is unconstrained by reproductive consequences for the host.

The truth is undoubtedly between the extremes of wholly unconstrained cultural evolution and cultural change purely subordinate to genetical evolution. The point I wish to emphasize here is that the symbiosis between bodies and culture is influenced by the same evolutionary processes as the symbiosis between bodies and gut flora. A body contains a large community of many different evolving systems. These systems share common interests only to the extent that their transmission patterns overlap. Memes are simply one type of inhabitant in this complex community.

B. Symbiosis in the Genome

The genome itself is also a community that contains both common and conflicting interests. The factors that influence conflict and cooperation among parts of the genome have been discussed extensively (Hurst, this volume). I present one brief example to illustrate the problem.

Most organisms inherit mitochondrial DNA from their mother, with no input from their father. By contrast, most other genetic material is obtained equally from the mother and father. For most traits these different modes of transmission, matrilineal versus biparental, have no consequences for the direction of evolutionary change favored by selection. For example, efficient respiration increases both matrilineal and biparental transmission.

The allocation of resources to sons and daughters affects matrilineal and biparental transmission differently. Traits that enhance the production of daughters at the expense of sons always increase the transmission of matrilineally inherited genes. For example, in some hermaphroditic plants the mitochondrial genes may inhibit pollen development and simultaneously enhance the production of seeds (Edwardson, 1970; Hanson, 1991). Selection of genetic variants in the mitochondria would favor complete loss of pollen production in exchange for a small increase in seed production because the mitochondrial genes are transmitted only through seeds (Lewis, 1941). Such reallocation of reproduction would greatly reduce the transmission of biparental genes because biparental transmission depends on the sum of the success through seeds and pollen. Thus there is a conflict of interest between the mitochondrial and nuclear genes over the allocation of resources to male (pollen) and female (ovules) reproduction (Gouyon and Couvet, 1985; Frank, 1989).

This conflict within the genome is similar in structure to the tension between genes and memes. Conflict occurs whenever a trait can cause differences in the rate of transmission of subgroups within the organism (Dawkins, 1982). Different traits partition the organism in different ways. In the example given here, respiration unifies the whole community, whereas resource allocation to sons and daughters splits the community among subgroups that are inherited matrilineally, patrilineally and biparentally (Fig. 12).

There are many similar types of conflict that occur within genomes (Hurst, this volume). Leigh (1977) has referred to this aggregate of common and competing interests as the "parliament of the genes." I discussed the mitochondrial example because it also illustrates my theme of symbiosis. At one time in evolutionary history cells existed but none had mitochondria. Several hundred million years ago a cell formed a successful symbiosis with an

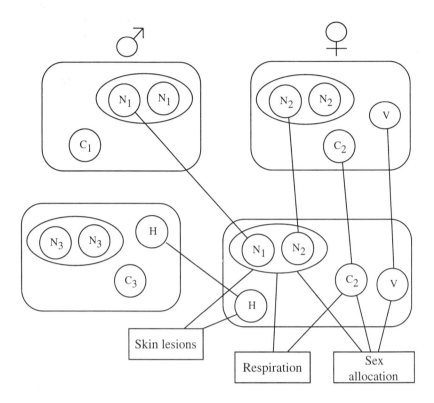

Figure 12 The mosaic nature of hereditary transmission, symbiosis, and the generation of phenotype. The top male and female individuals, which are really communities of symbionts, produce the offspring on the lower right by sexual reproduction. One-half of the nuclear (N) genes come from each parent. The cytoplasmic (C) elements, including the mitochondria, come only from the mother. Vertically (V) transmitted symbionts, which are typically viruses and bacteria, also follow the matriline. The individual on the lower left carries a horizontally (H) transmitted parasite, which infects the new offspring. This offspring expresses several traits, three of which are shown. Each trait is influenced by different hereditary subsets of the community. For example, the horizontal parasite produces skin lesions that enhance its rate of transmission, and the nuclear genes produce traits that attempt to clear the infection and heal the lesions. All components of the community, (N, C, V), are in conflict with the horizontal parasite (H) over the lesions because only the parasite is transmitted through lesions. The C and N components directly affect respiration. All components favor efficient respiration because all gain longer life and greater transmission from this trait. The matrilineally transmitted C and V components favor production of daughters. The biparentally transmitted N components favor equal production of sons and daughters. The transmission of H is unaffected by sex allocation.

intracellular bacteria. The bacteria probably had a metabolic pathway absent in its host (Margulis, 1981; see Khakhina, 1992, and Cavalier-Smith, 1993, for history of this research).

C. The Organism as a Community

An organism is a complex community. Conflicts occur within the genome; tissues house a vast array of bacteria, viruses, and other symbionts; and in some organisms the brain provides a home to memes. If the genome is a parliament, then the "organism" is a loose federation of states. Each particular challenge defines how interests overlap or conflict.

Although an organism is a complex symbiosis, the simpler, traditional view of organisms as unitary is adequate for many questions. For example, a trait such as efficient respiration unites the community with a common goal. In addition, most genes are transmitted in the usual biparental way. These genes often dominate the community because they can influence the substrate (tissues and brain) for all other symbionts.

It is a simple fact that adaptive systems tend to form symbioses. In the next section I examine the role symbiosis plays in genetic systems.

VII. The Origin and Evolution of Genetic Systems

Oparin (1924) and Haldane (1929) asked: How did life begin? Of course they were not the first to ask that question, but their ideas formed the first plausible theories for how replicating molecules arose from the prebiotic environment. The tradition established by Oparin and Haldane has led to many interesting studies on the chemistry of early life (Deamer and Fleischaker, 1994). Three questions dominate this work. Which organic molecules arose spontaneously in the prebiotic environment? What are the properties of these molecules in terms of stability, capacity for information storage, and a tendency to replicate? Which are the precursors of life?

The chemical research assumed that natural selection sorted among early competitors and refined the replication process. However, chemical problems dominated the research and few thought explicitly about the origin of life as the evolution of an adaptive system. Eigen (1971) initiated a new era by examining the balance between information decay in replication (mutation) and improvement by natural selection. Although this tension between mutation and selection had been widely studied in population genetics, application to the origin of life identified special problems about early evolution that stimulated a new wave of research.

Meanwhile, books by Williams (1975) and Maynard Smith (1978) on the evolution of sex focused attention on the other end of the problem. This work starts at the present, with the great diversity of genetic systems that exists in nature, and asks about the evolutionary forces in the past that have shaped this diversity. The well-developed theories from this work have recently been applied to problems near the origin of life, leading to many new insights about the evolution of genetic systems.

A. The Error Threshold and the Origin of Life

The first replicating molecules were copied without the aid of error-reducing replication enzymes. This undoubtedly led to high mutation rates. Evolutionary improvements are difficult with high mutation rates because replication errors erode any gains in reproductive efficiency caused by natural selection. Roughly, if more than one mutation occurred in a molecular sequence in each round of replication, then the population would eventually consist of random sequences. If the mutation rate were less than one per molecule, sequences with a reproductive advantage would spread in the population (Eigen, 1971; Eigen and Schuster, 1977; see reviews in Maynard Smith, 1979; Eigen, 1992).

The mutation rate sets a limit on the size of evolving molecules. If the error rate per site in a molecular sequence is μ, then molecules of length greater than $1/\mu$ will have more than one error per round of replication. Only molecules smaller than this "error threshold" can be improved by natural selection.

Eigen and Schuster (1977, 1978a,b) estimate that an RNA molecule of between 10 and 100 base pairs is above the error threshold if there are no replication enzymes to enhance accuracy. Information to produce a replicase probably requires between 1000 and 10,000 base pairs. This size range is above the error threshold, thus a replicase cannot evolve by natural selection. The error threshold presents a barrier to the evolution of large genomes with efficient replication. As Maynard Smith (1983) said, "No large genomes without enzymes, and no enzymes without a large genome."

Eigen and Schuster (1977, 1978a,b; Eigen, 1992) suggested the hypercycle model to explain how replication enzymes evolved in spite of the constraint imposed by the error threshold. Maynard Smith (1979), in his elegant example of the hypercycle, describes replication of the message GOD SAVE THE QUEEN. Suppose that each letter is coded by five bits in a sequence of 0's and 1's. (A string of five 0's and 1's is needed to code for the 26 letters of the alphabet because $2^4=16$ and $2^5=32$.) The 15 letters require 75 bits. If the mutation rate is $1/50$, then the processes of mutation, replication, and selection of the strings closest to the target message would lead to random strings because the error rate overpowers the rate of selective improvement. The message is too long.

Now suppose that each word is encoded on a separate molecule. The largest word needs only 25 bits, which is less than the error threshold of 50. Mutational decay is no longer a problem. But separation of words causes a different problem: the replication rates of the words are likely to differ. Because the molecules for each word compete for substrates, the population would ultimately consist of only the fastest replicating word and the mutants derived from that word. The whole message cannot succeed by independent evolution of the individual words.

The hypercycle solves the problem of coordinating a symbiotic group that is composed of competing subunits. Suppose an increase in the number of GODs increases the rate of replication of the SAVEs, an increase in the SAVEs aids the replication of the THEs, the THEs enhance the QUEENs, and the QUEENs complete the cycle by enhancing the replication rate of the GODs, yielding: GOD → SAVE → THE → QUEEN → GOD →...

The hypercycle is stabilized by the coupling of replication rates among words. A member of a cycle can outcompete any isolated word, and an efficiently coupled cycle can outcompete less efficient cycles. The cycles act as individuals that compete against other cycles. Each word in a cycle is like a gene in an individual.

The components of a cycle could cooperate in producing a replication enzyme that decreased the error rate in copying molecules. The replicase would then reduce the mutation rate, making large genomes possible. Compartmentalization into protocells is the next evolutionary step. Compartmental cycles with increasing genome size lead to the cellular (or viral) forms of life that exist today and that appear in the earliest fossils.

This view emphasizes the role of symbiosis in the evolution of adaptive systems. The very strength of the hypercycle theory, the power of symbiosis among small replicators to produce complex function, also turns out to be its greatest weakness.

B. Symbiosis and Early Evolution

Conflict frequently destroys symbiotic relationships even when there is great potential for mutual benefit and an overall increase in the efficiency of the system. The major evolutionary increases in complexity have occurred on those few occasions when the conflicting interests of symbionts were partly subjugated to the overall benefit of the association (Maynard Smith, 1988; Maynard Smith and Szathmáry, 1995). Examples include populations of replicating molecules that cooperate in protocells, symbiosis among prokaryotic cells to form the modern eukaryotic cell, cooperation of cells to form multicellular organisms, cooperation of individuals to form social groups, and gene-culture symbiosis.

Hypercycles provide an excellent introduction to the conceptual problems of symbiosis. Maynard Smith (1979) showed that the basic hypercycle can be invaded by parasitic components that destroy the overall efficiency of the system (see also Bresch et al., 1980).

In the GOD SAVE THE QUEEN example, each word has two important functions. The enzymatic replicase enhances the reproductive rate of the next word. The target function affects a molecule's ability to use the replicase of the previous word.

Mutations that enhance target efficiency spread because they increase self-replication. Mutations that increase replicase efficiency are neutral in a randomly mixing population. For example, suppose a mutant GOB produces a better replicase for SAVE than does GOD. The better replicase enhances the reproduction of SAVE, which enhances THE, which enhances QUEEN. More QUEEN means more replicase for the GOD/GOB species. But the GOD and GOB subspecies benefit equally from the additional replicase, so the more efficient producer, GOB, does not increase relative to GOD. A similar argument shows that GOB will also be neutral if it produces a poorer replicase than GOD. Thus hypercycles cannot develop in a mixed population because the replicase is a neutral trait.

An established hypercycle also has a problem. Suppose that GOB produces no replicase for SAVE and has an enhanced target affinity for the replicase from QUEEN. The lack of replicase is a neutral trait, but the greater target affinity will cause GOB to outcompete GOD. The cycle will collapse because of parasitism. The basic hypercycle fails.

What can explain the origin and maintenance of symbiotic replicators in the first protocells? Perhaps the order of events must be switched. Eigen's hypercycle theory suggests that successful symbioses (hypercycles) are followed by compartmentalization into protocells. But compartments of replicating molecules may have come first, followed by cooperation among replicators (Maynard Smith, 1979; Bresch et al., 1980).

If the replicators of a developing cycle share a compartment, then the success of each replicator depends on two levels of selection. A parasite can spread within its compartment, but that parasite's success may be low because its compartment will be outcompeted in the population. For example, if a parasite takes over its own compartment it will have increased in frequency locally. But the compartment's rate of division may drop to zero because the parasite disrupts the orderly functioning of the protocell. The parasite, by damaging its container, dooms itself to extinction.

The higher, compartment level of selection can potentially screen off the lower level of competition within the compartment (Brandon, 1984). This is a form of group selection. The effective formation of an evolutionary unit at the compartment level requires that compartments differ significantly in their rate of division. Roughly speaking, the rate at which selection increases the frequency of parasite-free compartments must be greater than the rate at which parasites can take over their own compartmental lineage (Szathmáry and Demeter, 1987; Szathmáry, 1989a,b).

Maynard Smith and Szathmáry (1993) extend these ideas to show that the evolutionary origin of chromosomes depends on a similar sort of group selection and formation of a new evolutionary level. A chromosome is a set of physically linked replicators (genes). The problem is how genes that were initially separate became linked.

In Maynard Smith and Szathmáry's model, linked pairs of genes suffer a disadvantage within cells because large chromosomes replicate more slowly than single genes. Thus the frequency of chromosomes declines within a single lineage. This disadvantage for linkage may be offset by the positive synergistic effect of pairs of genes. If a cell lacking one of the two genes functioned poorly, then the chromosomes would have the advantage that they never end up in cells lacking one of the genes. Whether chromosomes succeed depends on the rate at which unlinked genes can take over their own compartmental lineage compared with the frequency and reproductive disadvantage of cells that lack one of the synergistic pair of genes.

These models for the evolution of cooperation within genomes assume that the transmission of the symbionts is purely vertical, confined entirely within a lineage of dividing compartments. However, compartments are bound by simple membranes and the symbionts may be transmitted horizontally between lineages. For example, different compartments may occasionally fuse, mixing the symbionts from two groups, or individual replicators may occasionally be freed from compartments and picked up by another compartment.

Horizontal transmission of symbionts between compartments changes the evolutionary dynamics. A parasite can succeed if its rate of horizontal transmission is large enough to offset the reduced efficiency that it imposes on its host compartment. This is the problem of the evolution of virulence that I discussed previously in the context of culture and gut flora.

It seems inevitable that horizontal transmission and parasitism were key features in the origin and evolution of genetic systems (Bremermann, 1983; Frank, 1996). For example, a replicator might contribute nothing to the functioning and reproduction of the cell, but instead use all of its coding information for two parasitic functions. The first is rapid replication within cells and the ability to outcompete the other replicators for limited substrates. The second is enhanced horizontal transmission by either release into the environment and absorption into other cells or by increasing the rate at which the host cell fuses with other cells. Cellular fusion causes mixing of genomes and a primitive form of sex (Hickey and Rose, 1988).

Ideas about hypercycles, chromosomes, and parasites raise many interesting questions. What role did the mutation rate and the error threshold play at various stages in the origin and evolution of genetic systems? How did the tensions between levels of selection shape genetics? Were genomic parasites and horizontal transmission common? Was defense against parasitic invasion an important challenge? Unfortunately there is no way to study directly the early evolution of genetic systems.

C. Artificial Life

Ray (1992) suggested that artificial life in computer models may provide clues about the evolution of genetic systems. Ray's creatures live in the memory of a computer. The location in memory can be thought of as a compartmentalized cell. Each creature is a set of instructions that influences survival, reproduction, and interaction with other creatures. Replication produces a daughter copy next to the parent. Mutations may occur during replication.

These artificial creatures evolve as replicating algorithms that compete for CPU time and memory space. The algorithms are coded in the Tierran language, which has only 32 different instructions. This is approximately the size of the alphabet used to build proteins: there are 64 DNA triplets that are translated into 20 amino acids. The language is composed mostly of typical machine instructions for a computer: flipping bits, copying bit strings, tracking locations in a sequence, and so on.

The mode of addressing is a special feature of the language. A computer system associates a numeric address with a physical location. Tierran addressing is based on a biological analogy. Molecules diffuse and interactions occur when two molecules have complementary physical structures. Thus Tierran finds addresses by template matching; an instruction to jump to an address causes a search for a template match among physically close creatures. This allows for simple types of recognition.

Ray's model is not designed to study the origin of life but rather early evolution once replicating molecules exist. Thus he had to seed his system with a self-replicating program. In most runs he used a seed (ancestor) that is 80 instructions long and has only the minimal capacity of self-replication. No specific evolutionary potential was designed into this ancestor, it simply replicates itself indefinitely when there is no mutation.

The system proceeds by the following cycle. Each individual (algorithm) is allowed in turn to execute some of its instructions. Lifespan is determined by a queue. Newborns enter the bottom, and death is imposed at the top to keep empty a specified fraction of the environment (memory). Individuals move up the queue as additional births are added at the bottom. Errors in executing code can accelerate movement toward the reaper at the top. Mutation occurs by a low rate of bit-flipping in all organisms, and an additional error rate during replication. Size mutations also occur during the replication process. Thus genome length can evolve.

A run starts with the ancestor sequence and follows the life cycle. The system quickly diversifies and forms complex communities. These communities can be difficult to analyze in detail because the algorithms, composed of bit flips and memory jumps, are not easy to read. Ray, in his preliminary work, has identified several ecological types.

Parasites with short genomes cannot self-replicate but use the code of other creatures to specify how to reproduce. Hyper-parasites attack parasites. A hyper-parasite gets its own address into the copy pointer of the parasite, so a parasite replicates the hyper-parasite's genome rather than its own.

Social hyper-parasites can only replicate in aggregations. Each individual needs the code of a genetically similar neighbor to reproduce. The fact that an offspring is placed close to its parent may cause spatial aggregation of closely related creatures that aids the evolution of cooperation. Spatial aggregation has the same effect as compartmentalization and can lead to higher-level evolutionary units as discussed for the hypercycle and related models. Those earlier models suggested that higher evolutionary units are prone to internal parasites. As expected, Ray found cheaters that he calls hyper-hyper-parasites. These cheaters position themselves between social hyper-parasites and gain the benefits of neighbor-aided replication without reciprocating.

Ray briefly mentioned host immunity and parasite countermeasures to avoid detection. Recognition among cooperating and competing symbionts probably plays an important role in the coevolutionary dynamics of the system.

Ray made only a brief comment on the role of mutation (information decay). In a few runs the community became dominated by creatures with 700 to 1400 instructions per genome. These communities died because creatures in this size range exceed the error threshold that sets an upper limit on genome size (see Maynard Smith, 1992). It would be interesting to test whether larger and more complex symbiotic genomes could evolve in Ray's system if some form of expanded compartmentalization were introduced. This type of analysis would allow one to study jointly information decay (mutation) and symbiosis, the two main forces that influenced the early evolution of genetic systems.

D. The Evolution of Sex

I have described how symbiosis and information decay influenced the early evolution of genetic systems. In this section I add a new theme to the discussion, the role of exploration versus exploitation.

The biological problem is sex. In eukaryotes (nonbacteria), sex typically causes the orderly mixing of genes from two parents to form an offspring. How did complex systems of genetic mixing arise? What kinds of challenge to adaptive systems maintain sex relative to nonmixing, asexual systems? There are many theories about the origin and maintenance of sex (Maynard Smith, 1978; Michod and Levin, 1988; Kondrashov, 1993). I will briefly summarize the most prominent theories as they relate to my themes about the evolution of adaptive systems.

Theories for the origin of sex focus on prokaryotes (bacteria). The prokaryotes have simple forms of genetic mixing that, presumably, are similar to the types of mixing that occurred during early evolution.

One theory focuses on the challenge of information decay (Bernstein et al., 1988; Michod, 1993). In this theory genetic mixing brings together two copies of homologous DNA by fusion of haploid cells. The paired DNA allows one strand to correct damage to the other strand, greatly reducing the rate of deleterious mutation.

A competing theory for the origin of sex focuses on symbiosis. As discussed above, genomic parasites can spread if their rate of horizontal transfer overcomes their reduced vertical transmission within the host's lineage. Hickey and Rose (1988) have suggested that horizontal transfer by parasites led to the mixing of whole genomes, the first step in the sexual cycle. This idea cannot be tested directly because the origin of mixis occurred in the past. In support of the theory, mixis in modern prokaryotes is caused by horizontally transmitted subgenomic plasmids (Hurst, 1991).

Bell (1993) has extended the parasite theory for the origin of sex. Eukaryotes have two distinct phases in their life cycle: a vegetative phase of growth and reproduction and a sexual phase of genetic mixing followed by genetic segregation. Bell argued that the characteristic features of the eukaryote genome arose from the entrainment of parasitic genetic elements into the life cycle. In Bell's theory, mixis originated by the Hickey-Rose model of parasitic transmission. Bell also argues that mating type genes and centromeres, part of the machinery of orderly segregation and meiosis, had a parasitic origin. See Hurst (this volume) for more on genomic parasites and the evolution of genetic systems.

The maintenance of sex poses a different kind of problem. Asexual reproduction is a more efficient mode of reproduction than sex, so why are most systems sexual? Sex requires the time-consuming processes of mating, mixing of genetic material in diploid offspring, and the orderly reduction of chromosomes to form haploid gametes. Sex also breaks up coadapted gene complexes.

The most spectacular puzzle concerns the "twofold cost" of sex (Williams, 1975; Maynard Smith, 1978). Multicellular species typically have large gametes (females) and small gametes (males). The small gametes contribute only genes to the offspring but no resources—in effect, small gametes are parasitic on the reproductive effort of the large gametes. Because sexual females invest all of the resources but only one-half of the genes in offspring, their rate of genetic propagation is one-half that of an asexually reproducing individual that transmits all of its genes to offspring.

TABLE 3

Classification of Theories to Explain the Maintenance of Sex

	Decay	Explore vs exploit
Species	Muller's ratchet	Adaptive radiation
Gene	Mutation clearance	Variable environment

Many theories attempt to explain why sex is maintained in spite of a twofold disadvantage (Maynard Smith, 1978; Kondrashov, 1993). I summarize the four leading theories. These theories can be classified in two ways. One division splits the models by the challenge to the adaptive system, either exploration versus exploitation or information decay (mutation). The second division splits according to the level of selection, either long-term effects and species selection or short-term effects and genic selection. Table 3 shows this classification.

Muller's ratchet causes deleterious mutations to accumulate in small, asexual populations (Muller, 1964; Maynard Smith, 1978). The effects on population fitness can be understood by following the rise in the number of deleterious mutations carried by the best chromosomes in the population. Label each chromosome in the population by the number of deleterious mutations that it carries. Suppose initially that some chromosomes have zero mutations. Occasionally, by chance, all of the surviving replicates of a zero chromosome will have one or more mutations, transferring this chromosome lineage to the class with one mutation. In a small population the rate at which chromosome lineages increase the number of mutations carried outpaces the rate at which selection favors chromosomes with fewer mutations. Eventually all chromosomes with zero mutations will be lost and the best class will have one mutation—the ratchet has turned. The process continues over time, with population fitness steadily declining.

Sex and genetic recombination can prevent the ratchet. Two chromosomes, each with a different mutation, can recombine to form progeny with zero mutations. The rate of chromosomal improvement by recombination is usually sufficient to prevent the decline of population fitness. Thus sexual populations can outcompete asexual populations over long periods of time. The problem with this theory is that an asexual individual within a sexual population has a twofold reproductive advantage because it avoids the cost of sex.

Muller's ratchet is a sufficient explanation for sex only if the rate at which asexual populations suffer "mutational meltdown" (Lynch et al., 1993) is sufficient to overcome asexuality's short-term advantage within populations. This is, once again, the problem of two competing levels of selection, similar to the origin of compartmentalized protocells struggling against the lower level of genomic parasites.

The adaptive radiation theory is another species-level model. In this case sexual species have an advantage because they generate a wider diversity of genotypes and can adapt more quickly to new habitats than asexual species (Fisher, 1958). Thus asexual genotypes, which have a short-term advantage, lose when new environmental challenges arise. In this model sexual species gain because they are better at exploring and discovering new solutions to new problems. Asexual species gain because they are better at exploiting a fixed environment.

The two gene-level theories assume that species-level advantages for sex are not sufficient. They explain how sex can have a short-term advantage over asexuality in spite of the twofold cost of sex. The mutation clearance model focuses on information decay; the variable environment model examines exploration versus exploitation in rapidly changing environments. Both models can be explained by an elegant theory that is hidden in the appendix of Haldane's (1932) classic book *The Causes of Evolution*. Haldane's model examines intense

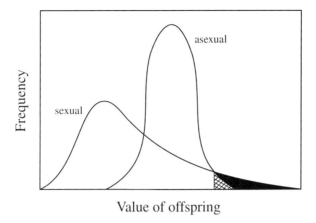

Value of offspring

Figure 13   The effect of variation in offspring value on the evolution of sex. Each curve is the distribution of offspring value in a family from a parent's point of view. Asexual families have a relatively higher mean because a parent transmits twice as many genes per offspring compared with sexual progeny. Sexual distributions have greater phenotypic variability because genetic information is mixed randomly between parents. Here, only the offspring above a cutoff survive—the hatched area among asexuals and the hatched plus shaded areas among sexuals. In this case the sexual strategy transmits more genes to future generations than the asexual strategy in spite of the twofold advantage of asexuality. In general, "intense competition favors variable response to the environment rather than high average response. Were this not so, I expect that the world would be much duller than is actually the case" (Haldane, 1932, pp. 177-178).

selection in which only the best individuals survive. In that case selection favors genotypes that produce highly variable traits rather than a high average value. If only the best individuals are picked, then a wide distribution with a relatively low mean has an advantage over a narrow distribution with a high mean (Fig. 13).

In the mutation clearance model each individual carries many deleterious alleles (Kondrashov, 1988). Asexual individuals have a narrow range of offspring quality because each offspring has approximately the same number of mutations as its parent. The stochastic processes of segregation and recombination in sexual genotypes produce a wide range in the number of mutants per offspring. The average value of offspring for a sexual genotype is reduced by the cost of sex, but sex can beat asexual genotypes because of the higher variance in quality. This requires sufficiently frequent mutations to produce a wide distribution of offspring quality, and strong selection that picks only the best offspring.

In a variable environment model the favored genotypes change from generation to generation. Sex increases the probability that a parent will have some offspring close to the favored genotype because sex increases the diversity of genotypes produced. Sex's ability to produce diversity and increase its chance to match a changing environment can outweigh the reproductive efficiency of asexuality. Sex is better at exploration of a changing environment, asexuality is better at exploitation of a fixed environment. (The way in which environments change can have an important effect. See Charlesworth, 1993.)

What could cause the environment to change sufficiently to favor sexual exploration at a twofold cost in efficiency? Several authors favor coevolving parasites (Levin, 1975; Jaenike, 1978; Hamilton, 1980). Parasite traits that avoid host immunity require counteradaptations by the host. Host-parasite interactions are a form of biotic challenge with coevolutionary feedback. In this view sex is a method to mix cooperative symbionts (genes) in search of good combinations against antagonistic symbionts (parasites): "Sex also creates true species in an otherwise straggling mess of clones: if the idea about parasites is right, species may be seen in essence as guilds of genotypes committed to free fair exchange of biochemical technology for parasite exclusion" (Hamilton, 1982, p. 271).

Opinions differ about the processes that favor sex. My purpose here was not to solve this puzzle, on which there is little agreement, but to show that certain types of argument recur in the evolution of adaptive systems. For the origin of sex, information decay and symbiosis dominate the arguments. The conflict between genomic parasites at the genic level of selection and survival of lineages at the cellular level also plays an important role. For the maintenance of sex, information decay and exploration versus exploitation divide the main theories in one dimension. Genic versus species level selection divide the theories in a second dimension. The genic level theory for exploration versus exploitation implicates antagonistic coevolution as a major challenge to the evolution of genetic systems.

VIII. Adaptive Systems as an Engineering Tool

Engineering faces many of the same challenges found in biological systems: information storage, complexity, and unpredictability. In recent years scientists have exploited adaptive systems to solve engineering problems. I present a few examples to show that engineering shares many challenges and solutions with genetics, learning, and development.

A. The Design of Biochemical Catalysts by Chemical Engineers

Three methods have been used to design catalysts (Benner, 1993). The first designs molecules based on the catalytic properties of functional groups and the predicted folding pattern of the components. The second method uses the adaptive properties of the vertebrate immune system to create catalytic antibodies (Lerner et al., 1991). The immune system can create an antibody that binds the rate-determining transition state of a reaction. Stabilizing the transition state lowers the activation energy and speeds the reaction.

A new adaptive method of variation and selection has recently been used to design RNA catalysts (ribozymes). Bartel and Szostak's (1993) goal was to create a ribozyme that catalyzed RNA replication. The discovery of self-catalyzing RNA would support the view that the origin and early evolution of life was based on a purely RNA system. In this theory RNA would be both the genotype encoding information and the phenotype controlling replication (Cech, 1993).

Although Bartel and Szostak's goal was basic understanding of the origin of life, their problem was one of engineering. They wished to find an efficient ribozyme that could enhance replication. They focused on ligation, the joining of separate RNA sequences into a single long sequence. They began their search with 10^{15} randomly generated RNA sequences. They then screened this large pool for those RNAs that extended themselves. The specific reaction was the attachment of a test sequence onto the end of the putative catalyst RNA. Those RNAs that attached to the test sequence under experimental conditions were selected for replication. The replication was conducted by a standard biochemical method, the polymerase chain reaction (PCR). The PCR process was intentionally modified to introduce mutations during replication, providing additional variability for the next round of selection. The methods of variation and selection were applied for 10 generations.

The final sequences enhanced the reaction rate by seven orders of magnitude relative to the random sequences at the start of the evolutionary process. This performance is not impressive if one uses natural systems as a benchmark. Natural ribozymes typically enhance reaction rates by three or four orders of magnitude more than Bartel and Szostak's selected ribozymes. Protein catalysts do even better, outperforming ribozymes by three to six orders of magnitude. On the other hand, the laboratory-evolved ribozymes beat the typical performance of both human-designed and antibody-selected enzymes, which typically cause reaction-rate enhancements of two to six orders of magnitude. The success of the ribozyme selection scheme is remarkable because both the designed and antibody-selected enzymes are composed of the 20 different amino acids, which provide a much wider range of biochemical properties for enzymes than the four nucleosides that compose RNA sequences (Benner, 1993).

Molecular design by natural selection is in its infancy. Although these methods will not displace all other approaches, adaptive design may play an important role in the future of biochemistry.

Adaptive systems are a tool used for discovery in molecular design. Once the exploration has finished and an efficient catalyst has been found, other more efficient techniques can be used to exploit that discovery. In this regard molecular design shares two general properties with other adaptive systems.

First, adaptive design is a process of discovery that follows cycles of broad exploration and efficient exploitation. The previous sections showed that exploration versus exploitation is a common theme in adaptive systems.

Second, I argued earlier that learning and vertebrate immunity are adaptive subsystems spawned by the genetic system to handle environmental challenge. Both learning and immunity have their processes of variation and their selective systems (goals) set by genetical evolution. From this point of view adaptive molecular design is a subsystem of variation and selection spawned by learning humans to handle an environmental challenge. This may seem an unnecessarily complicated way to describe a simple method. But in the sweep of evolutionary history, adaptive systems have occasionally discovered the use of subsystems of variation and selection to solve their problems. These subsystems greatly altered subsequent evolution.

It remains to be seen whether adaptive subsystems created by humans will be important. Adaptive design of catalysts and the following examples illustrate recent applications.

B. Genetic Algorithms and Protein Folding

A protein sequence is a string of amino acids. Biochemical tools can be used to read a protein sequence or to change particular amino acids within the string.

However, the ability to read and manipulate sequences has practical limitations because the structural and catalytic properties are determined by three-dimensional conformation. It has not been possible to predict conformation from the sequence.

A protein folds according to the energies of covalent bonds among the individual amino acids. The problem is to predict how particular bonds cause a linear sequence to fold into a three-dimensional shape. At each step in the folding process there are a large number of possible conformations. In principle, a computer program could search each of the possible conformations at each step and predict the folding pathway. However, this search process suffers from a common problem in computer optimization known as combinatorial explosion. The number of possible pathways is too large to search by any computer available now or in the future.

There are many computer techniques to search for solutions to large problems. The genetic algorithm is a popular method based on an analogy with natural selection (Holland, 1975; Forrest, 1993). Each potential solution is coded in a linear sequence of information (a chromosome). In the protein example, the goal is to find a minimum-energy conformation. For this case each "chromosome" represents a particular conformation.

The genetic algorithm is natural selection applied to a population of chromosomes. The initial population can be created randomly. Then in each generation selection, variation, and transmission occur. Selection chooses chromosomes for reproduction according to their fitness—in this case lower energy states have higher fitness. Selected chromosomes mutate with a probability set by the program. Chromosomes may also pair to mate and recombine. Recombination follows the biological process of swapping pieces of the chromosome. Enough progeny are produced for the next generation to form a new population. The cycle is repeated. The quality of the best solutions in the population usually improves for many generations and then levels off. The best solution (chromosome) during a run is the optimum discovered by the search process.

An interesting series of papers on genetic algorithms and protein folding illustrates the power and potential problems with adaptive search techniques (Judson et al., 1992, 1993; McGarrah and Judson, 1993; Tufféry et al., 1993). Unger and Moult (1993) compared the ability of a genetic algorithm and a "Monte Carlo" search method to find a minimum-energy conformation for a protein that folds in two dimensions. The Monte Carlo method has four components. (1) Start with a random conformation. (2) Make a single random change in the conformation. (3) Accept the change if the new conformation has lower energy. Otherwise accept the change with a probability that decreases as the energy of the new conformation rises. (4) Continue to test changes until some stopping criteria is met. The Monte Carlo method was the best search technique available in 1991 (Unger and Moult, 1993).

Unger and Moult's (1993) genetic algorithm encoded different protein conformations in the chromosomes of the evolving population. Conformations with lower energy states have higher fitness. Each mutation follows steps (2) and (3) of the Monte Carlo method. Recombination breaks two protein sequences at the same amino acid positions and swaps the fragments. The conformation of the fragments is maintained while swapping.

The genetic algorithm found lower energy conformations in shorter periods of time than the Monte Carlo method. Unger and Moult (1993) argue that the genetic algorithm succeeds because it naturally follows folding pathways (Judson, 1992). Real proteins are believed to fold in steps. Local regions of the chain fold first; a higher order structure forms by combination of these local conformations.

The genetic algorithm succeeds for problems in which subsets of the instructions can work well together in creating high fitness (Holland, 1975). In the case of protein conformation, the calculation of fitness (energy state) implicitly evaluates the quality in the population of every locally folded fragment with two amino acids, with three amino acids, with four amino acids, and so on. Fragments with low energy conformations increase the relative fitnesses of the conformations in which they reside; thus low energy fragments will increase in frequency in each generation.

The selection process causes the parents chosen for reproduction to have a higher than average fitness, and thus a set of relatively low energy fragments. Recombination of conformations is done by building offspring from fragments of the two parents. Thus recombination creates new conformations from good fragments.

The great power of the method is that the fitness calculation simultaneously evaluates the quality of fragments of all sizes. Initially, small fragments will contribute the most to fitness differences among conformations. As the better of the small fragments spread by selection and recombination, and most conformations contain them, fitness differences depend primarily on good combinations of small fragments and differences among the slightly larger size classes of fragments. Thus selection emphasizes conformations of increasingly larger fragments.

A simple genetic algorithm often performs reasonably well for a wide range of problems. However, for any specific case, specially tailored algorithms can often outperform a basic genetic algorithm. The tradeoffs are the familiar ones of general exploration versus exploitation of specific information (Newell, 1969; Davis, 1991).

For protein folding, McGarrah and Judson (1993) showed the superiority of a hybrid method that combines the genetic algorithm with a local search. The genetic algorithm is often good at broadly searching the space of possible conformations. However, because genetics includes the stochastic processes of mutation and recombination, the algorithm is inefficient at fine-tuning a conformation that is close to a local minimum. McGarrah and Judson used an alternating cycle of the genetic algorithm and an efficient local search (gradient descent).

The genetic algorithm provides a good spread of candidate conformations. The gradient descent uses these candidates as starting points and efficiently obtains the best local conformation. The fitness is assigned to each chromosome based on its conformation after gradient descent. Thus local optimization can be thought of as a period of learning during the phenotypic phase of the cycle. One can optionally use the phenotype after the learning period as the genotype for reproduction, providing a component of Lamarckian inheritance to the search process (Judson et al., 1992).

The work on protein folding follows a common pattern of growth in optimization studies. When a new problem is encountered, the first efforts use a general purpose method such as a basic genetic algorithm. Experience with the problem often shows that better performance can be obtained by enhancing the simple algorithm, that is, encoding problem-specific knowledge into the genetic program. The incorporation of problem-specific knowledge is similar to the Baldwin effect discussed earlier (see section on *Learning*).

At present, such problem-specific knowledge is usually inserted into the genetic algorithm by human intervention. One goal for optimization research is to mimic the Baldwin effect more closely, causing techniques learned by the evolving population to be incorporated into the search algorithm without intervention. This would allow general-purpose methods to evolve into problem-specific techniques by dynamically balancing further exploration versus exploitation of discovered knowledge.

C. Genetic Algorithms and Neural Nets

A brain is a network of neurons. Each individual processor (neuron) in a brain is relatively slow by engineering standards, with a response time measured in milliseconds. Yet an animal's neural network is capable of many tasks far beyond the success of the most powerful computers available today. Examples include vision and complex pattern recognition.

Neural nets achieve their power by massively parallel information channels. A net can have millions of simultaneously active, parallel connections, whereas most computers use only one or a few serial channels at any instant. Networks also have redundancy and fault tolerance. Cutting a few individual connections usually has very little effect. In serial architectures, loss of a few bits of information often causes total failure.

The admirable properties of computational networks were first studied in the 1940s (McCulloch and Pitts, 1943). This research has grown into a large enterprise focused on neural networks, sometimes called parallel distributed processing (McClelland et al., 1986; Rumelhart et al., 1986). Part of the research emphasizes models of real nervous systems. This is an extension of neurobiology. Another group has exploited the power of networks for engineering applications (Hertz et al., 1991). Examples include recognition of handwritten words, digital signal processing, and control systems in robots.

Constructing a neural net for an engineering application has three phases. First, the basic architecture of the net is chosen. This includes the number of neurons, the initial strength of interconnections, the detectors that pass information from the environment into the net, and the output system that signals the net's action in response to the environment. In the second phase the net is put through a training process. Inputs are provided, and the difference between the net's output and the desired goal are used to adjust the connection strengths among the neurons. Finally, the net is put to use when it can match inputs to desired outputs with sufficient accuracy, for example, if handwritten letters can be recognized within tolerable error limits.

The training method is usually deterministic. Thus a given architecture converges to a particular input-output response pattern. The quality of performance is therefore determined by the initial architecture. Although there are some guidelines about how architecture will affect performance (Hertz et al.,

1991; Kung, 1993), there is often a huge number of plausible structures. Testing the performance of each architecture is not possible. The difficulty is combinatorial explosion, just as in the protein-folding problem.

Genetic algorithms have had some success in the problem of network design (Harp and Samad, 1991; Harvey, 1991; Harvey et al., 1993). A chromosome represents a single architecture. In each generation a chromosome is translated into a net, the net is trained, and then performance is measured. The performance is fitness. The genetic algorithm then follows its usual cycle.

One problem is the developmental translation of linear information in the chromosome into a three-dimensional network (Harp and Samad, 1991; Harvey, 1991; Kitano, 1994). At present, each investigator specifies an *ad hoc* method for developmental translation. These range from direct coding of three-dimensional structure to a variety of clever generative rules that allow compression of structural information. Hemmi et al. (1994) have recently taken the next step, in which the generative rules are themselves encoded in the linear genome, allowing the developmental "language" to evolve along with the particular structural information. This active area of research may provide some interesting insights into generative rules, development and language in natural systems (Batali, 1994; Dellaert and Beer, 1994).

Natural networks may be wired by a program of developmental selection, although this remains an open question (see section on *Development*). If true, then chromosomes contain two types of information. First, there is the program of developmental selection. This information codes the processes of variation and selection that control the development of wiring patterns. Second, there is a set of initial conditions that provide the material for developmental selection. These initial conditions shape the final outcome via developmental selection.

Analogies with natural systems suggest some experiments with genetic algorithms. A chromosome that encoded a developmental selection program and initial conditions has two interesting features. First, each chromosome spawns an adaptive subsystem of developmental selection to create its phenotype. The nature of this subsystem will evolve in the usual cycle of the genetic algorithm. Second, a relatively small chromosome is needed to encode the developmental program and initial conditions when compared with chromosomes that encode the entire architecture.

Whether an experiment of this type is practical for engineering applications remains to be seen. These experiments would, however, provide insight into the power of developmental adaptive subsystems to store complex patterns in small chromosomes.

D. The Evolution of Robots

Robots require environmental detectors, motor controls, and computational machinery to link sensory input with motor output. Robots that perform simple, repetitive tasks are used in many applications. But current robots are not good at handling unpredictable conditions. Thus, several research groups have reasoned as follows: Animals handle unpredictability well. Animals evolved. Perhaps robots should be designed by evolutionary processes.

The Sussex research group has made an interesting start in this direction

(Harvey et al., 1993; Cliff et al., 1993). They believe that evolution can be a very effective design method, but that evolutionary complexity must be built with small steps. Robots cannot sweep the garage before they can avoid crashing into walls. The Sussex group has chosen effective maneuvering in space as a simple but crucial first step in robot evolution.

How to build a robot that avoids bumping into walls? Harvey et al. (1993) argue that the first design phases can be done entirely by computer simulation without the need to build costly prototypes. The problem for the robot is to avoid the walls while moving in a circular room with black walls and white floor and ceiling. The robot has visual sensors, an internal neural network, and two motorized wheels that can be controlled independently. The physics—location in the room, visual input, and motion in response to settings for the wheel motors—are tracked by computer simulation. A linear chromosome is used to encode the structure of the sensory system and the architecture of the neural network. At present both the network's structure and the connection weights are set by the genotype. Each robot could learn by adjusting connection weights as discussed in the previous example.

Evolutionary change follows the cycle of the genetic algorithm. An initial population of chromosomes is formed, each specifying the design of a robot. Each design is tested in the simulated room, the performance is scored and used as fitness. Chromosomes are chosen according to their fitness. Pairing, recombination, and mutation occur to form offspring for the next generation of the cycle.

Performance improves over the generations. Following the plan of incremental evolution, the next step is maneuvering in a cluttered room (Cliff et al., 1993). The technique of simulating the physical environment does not work very well for this problem because it is computationally very intensive. So Cliff et al. (1993) created a cluttered environment and a robot. The robot has visual detectors that it can move to scan its surroundings. The robot can also move itself. Fitness is determined by success at navigating through this environment. A genetically encoded neural network controls sensory scanning and does the computations that connect the sensory input with motor output. The network and the processing occur in software on a remote system, allowing the rapid evaluation of many different genetic programs (chromosomes). No results have been published with this system.

The coupling of sensory scanning and movement is particularly interesting in this system. Edelman (1987, 1992) has stressed the importance of this coupling in his theory of neural darwinism and has presented some simulations of his own with simple robots (see earlier section on *Learning*). Edelman's goal is to understand the functioning of real nervous systems; the Sussex group is trying to design efficient robots. It will be interesting to follow the parallel development of these two research programs.

E. Hierarchical Control and Learning in Robots

The examples of robot maneuvering illustrate one method of design by incremental evolution. Other research groups have taken a different evolutionary approach (Meyer et al., 1993; Cliff et al., 1994; Brooks and Maes, 1994). For example, Colombetti and Dorigo (1993) have emphasized the ability of an individual robot to learn. Their approach may be thought of as phenotypic evolution given a particular design (genotype), whereas the Sussex group focused on genotypic evolution without any phenotypic evolution.

Colombetti and Dorigo studied a hierarchy of independent behavioral components coordinated by a global integrator. For example, approaching, chasing, escaping, and eating are possible responses to a particular stimulus. The actual behavior depends on a resolution among the tendencies of each component, leading to suppression of one component by another or to an orderly sequence of behaviors. Issues of hierarchy and coordination are central problems of animal behavior (ethology) and were widely discussed in the 1950s and 1960s (e.g., Tinbergen, 1951; Dawkins, 1976b).

Each component behavior in Colombetti and Dorigo's robot learns by an extended genetic algorithm known as a classifier system (Holland et al., 1986). Classifiers are evolving populations of chromosomes in the genetic algorithm cycle, but each chromosome may use a portion of its coding for a series of condition-action rules that can control behavior. The condition part of the rule can be triggered by external sensors or the actions of other chromosomes; the actions can stimulate other chromosomes or activate output controls such as motors. Thus a population of classifiers forms an activation network.

Here is a simple behavioral hierarchy (Colombetti and Dorigo, 1993):

> **if** there is a predator
> > **then** Escape
> > **else if** hungry
> > > **then** Feed
> > > **else** Chase the moving object

Each behavior, escape, feed, and chase, has its own classifier system that evolves (learns) over time. The robot has sensory detectors that pass a message to each behavioral component. Each component generates a message in response. The response from each component is passed directly to the action controls or to the behavioral integrator, which is itself a classifier system. The integrator may then send a message to the action controls.

The robot learns by reinforcement or punishment, as in psychological conditioning experiments. Reinforcement notifies the classifier systems of success. Each classifier system assigns credit (high fitness) to the chromosomes that participated in the correct decision. The wrong behavioral choice leads to punishment and low fitness to participating chromosomes. These fitnesses are then used in a cycle of the genetic algorithm, with mating, recombination, and mutation to form a new population of chromosomes in each classifier system.

The hierarchical decomposition of this robot is set by the experimenters. It would be interesting to study how hierarchical decomposition evolves. This would require a mixture of the approaches by the Sussex group and Colombetti and Dorigo. The Sussex approach focuses on an evolving population of robots, where the genotype for each robot specifies a particular design. To study behavioral decomposition, the genotype must be able to encode a variety of components that divide environmental challenges in different ways. The phenotypic interactions for each genotype would follow Colombetti and Dorigo's approach: each behavioral component specified by the genotype spawns its own adaptive (classifier) subsystem in order to learn during the phenotypic stage of the life cycle.

F. Robot Symbiosis

Colombetti and Dorigo's (1993) classifier robot uses a distributed model of behavioral control. Each component is simple, mostly autonomous, and computes in parallel with other components. This is an internal symbiosis of cooperating components. Behavioral decomposition is a central tenet of many current research programs in robotics (Meyer et al., 1993; decomposition of complex problems arises in many fields, see Alexander, 1964; Simon, 1981; Minsky, 1985; Dennett, 1991).

Another design method emphasizes teamwork among a group of individual robots. Teams are useful for simple tasks that can be done in parallel, such as clearing a field of rocks. Teamwork can also boost efficiency for tasks that require division of labor and specialization, such as automated manufacturing, search and rescue, or surveillance (Parker, 1993).

Both internal symbiosis and teamwork must resolve the tension between the autonomy of components and control of the symbiotic group (Numaoka and Takeuchi, 1993). This is a difficult problem. A global control mechanism could assign tasks to components based on progress to the ultimate goal. But this global mechanism must be complex, difficult to design because it requires great foresight, and prone to failure. When a global controller fails, then the whole system fails.

On the other hand, each component may blindly pursue its own simple subgoal without regard for the success of the group. Efficient group behavior may emerge from pursuit of the individual subgoals. This is the strategy used by several research programs.

Parker (1993) proposed a model for division of labor and specialization among groups of "selfish" robots. For example, in janitorial service the team must empty the garbage, dust the furniture, and clean the floor. Each robot is controlled by a distributed hierarchy of behavioral controls as in the Colombetti and Dorigo study (see also Brooks, 1986). Several low-level controls deal with tasks such as collision avoidance. These are active at all times. Higher-level controls are grouped according to the garbage, furniture, and floor tasks. Only one of these task-specific groups is active at any one time. Each group is controlled by a motivational unit that receives sensory input, inhibitory feedback from other behaviors, and a variety of other connections.

There are also control units devoted to internal "behaviors" such as the competing factors of impatience and laziness. These set the goals that control the behavior of each robot. For example, two robots may be motivated to empty the same garbage can. One gets there first and begins; laziness in the other robot causes it to give way. However, if the task of emptying the garbage is not completed, the second robot grows increasingly impatient. After a while, it will step in and try to finish the task.

Parker uses the market economy approach to achieve group coordination and efficiency. Each robot desires that all tasks be accomplished; each is motivated to do a task with high supply and low demand.

It will be interesting to follow this "selfish" approach to teamwork. In biological examples of symbiosis, creating higher levels of organization from autonomous components has worked very well in a few cases, but there are also many inefficiencies caused by internal conflict (e.g., Hölldobler and Wilson, 1990; Hurst, this volume).

IX. Conclusions

The study of adaptive systems is composed of the individual puzzles in biology and engineering that made up my survey. This field is at a special time, when many of the puzzles have been defined, work has started, and the problems are just coming into focus. Much of the excitement is in the details of these puzzles and the ideas that are growing up simultaneously in traditionally separated academic disciplines.

What can be said beyond the listing of individual cases? I have argued throughout the chapter that a small set of challenges and responses have shaped adaptive systems (Tables 1 and 2). Classifications of this type can be problematic. On the positive side, they highlight simple, common features that can be obscured by details. On the negative side, classifications can be a semantic convenience that hides real differences. The balance often turns on matters of personal taste. My classification did bring some order to a diverse range of problems. I look forward to better classifications that will develop with a general theory of adaptive systems.

I turn now to a few speculations. First, I suggest that unpredictable challenge from coevolving systems has played a particularly important role in the history of adaptive systems. This is an old idea. What I find particularly interesting is that robotics provides new opportunities to test this idea.

The early evolution of robots will require much exploration. Adaptive systems influence two levels of design. At the hard-wired or genotypic level, evolutionary computation, such as a genetic algorithm, is used to search for effective architecture. This algorithm, which tests designs from a population of alternatives, shares many properties with genetical evolution. The good designs will proliferate and be modified, the bad designs will disappear.

Most of the early designs will be inefficient. But, for simple tasks such as cleaning office buildings and scraping barnacles from the bottoms of ships, the rate of architectural (genetic) evolution will slow as successful designs are discovered. Which brings up the interesting question: What types of challenge will favor continual evolution of architectures? Antagonistic coevolution seems the most likely answer; to use more common terms—war, combat, law enforcement, games of pursuit. Opponents will evolve to exploit design weaknesses, which require countermeasures to close the gaps. While shoring up defense, the search goes on for weakness in the opponents. And so on. Perhaps it is no surprise that the Office of Naval Research (USA) funded much of the early research on genetic algorithms.

My second comment is about cooperative symbioses that form in response to another kind of war. The battle is between humans and their parasites. In earlier sections I mentioned that host-parasite coevolution influences genetic polymorphism and that parasites are the challenge that shaped adaptive immunity. I also discussed the hypothesis that sex and the exploratory function of genetic mixing is shaped by parasitic challenge. There are two additional adaptive systems that humans use against parasites: learning and culture. As Mims (1987, p. 322) noted:

Vaccines have been of immense importance in the past and hold great promise for the future. *The evolution of a microorganism can be decisively terminated by the proper application of knowledge.* Smallpox, the most widespread and fatal disease in England in the eighteenth century and a major cause of blindness, has been totally eradicated from the earth. [italics added]

Genetics, adaptive immunity, learning, and culture have all been used in the battle against parasites. In addition, science (learning plus culture) has itself spawned new adaptive subsystems in the form of evolutionary computation.

This battle against parasites, waged by medical research, is an enormous cooperative symbiosis. Like all symbioses composed of autonomous agents, medical research is rife with internal conflict, for example, competition among research groups. The symbiosis is held together by a common external threat—parasites.

My final comment is about a different kind of symbiosis, in which the individual agents are themselves subsystems of a single evolutionary unit. For example, teams of robots may be the most effective way to solve complex problems. Although each robot makes its own behavioral decisions, the whole system is typically designed with a single purpose controlled externally by humans. I discussed some of the difficulties in my survey. First, how should complex tasks be divided into simpler subgoals, each subgoal achievable by single agents (robots)? Second, how can pursuit of subgoals be combined to solve a larger problem?

This work in robotics matches an approach that has recently been developed to study the human mind and the evolution of consciousness (e.g., Minsky, 1985; Dennett, 1991). According to this view, the mind has many nearly autonomous subsystems that handle particular tasks. A major feature of consciousness and focused attention is simply the temporary dominance of a particular subsystem. In some theories, the subsystems compete for control according to the importance of the challenges that they face. This is similar to Sachs et al.'s (1993) developmental selection in which the individual shoots of a single plant compete for root resources, or Parker's (1993) robot example, in which autonomy and controlled competition appear to be the only way to achieve workable complexity.

It will be interesting to follow the development of robotics and cognition. These fields have very different histories, but the recent "cognitive revolution" may break down barriers (Gardner, 1985). On the other hand, many people believe that natural selection, robotics, and "artificial" systems will teach us nothing very profound about the mind (references in Gardner, 1985; Dennett, 1991, 1995). Time will tell.

Acknowledgments

I thank R. W. Anderson, M. Antezana, R. M. Bush, H. Carstens, L. D. Hurst, G. V. Lauder, R. E. Lenski, and M. R. Rose for helpful comments. My research is supported by NSF grant DEB–9057331 and NIH grant GM42403.

References

Ackley, D., and Littman, M. (1992). Interactions between learning and evolution. In "Artificial Life II" (C. Langton, C. Taylor, J. D. Farmer, and S. Rasmussen, eds.), pp. 487-509. Addison-Wesley, NY.

Alcock, J. (1993). "Animal Behavior: An Evolutionary Approach," 5th edition. Sinauer, Sunderland, MA.

Alexander, C. (1964). "Notes on the Synthesis of Form." Harvard University Press, Cambridge, MA.

Anderson, R. M., and May, R. M. (1982). Coevolution of hosts and parasites. *Parasitology* 85, 411-426.

Anderson, R. W. (1995). Learning and evolution: a quantitative genetics approach. *J. Theor. Biol.* 175, 89-101.

Baldwin, J. M. (1896). A new factor in evolution. *Am. Nat.* 30, 441-451.

Barkow, J. H., Cosmides, L., and Tooby, J. (eds.) (1992). "The Adapted Mind: Evolutionary Psychology and the Generation of Culture." Oxford University Press, Oxford.

Bartel, D. P., and Szostak, J. W. (1993). Isolation of new ribozymes from a large pool of random sequences. *Science* 261, 1411-1418.

Batali, J. (1994). Innate biases and critical periods: combining evolution and learning in the acquisition of syntax. *In* "Artificial Life IV" (R. A. Brooks and P. Maes, eds.), pp. 160-171. MIT Press, Cambridge, MA.

Bell, G. (1993). The sexual nature of the eukaryote genome. *J. Hered.* 84, 351-359.

Benner, S. A. (1993). Catalysis: design versus selection. *Science* 261, 1402-1403.

Bernstein, H., Hopf, F. A., and Michod, R. E. (1988). Is meiotic recombination an adaptation for repairing DNA, producing genetic variation, or both? *In* "The Evolution of Sex" (R. E. Michod and B. R. Levin, eds.), pp. 139-160. Sinauer Associates, Sunderland, MA.

Boyd, R., and Richerson, P. J. (1985). "Culture and the Evolutionary Process." University of Chicago Press, Chicago.

Brandon, R. N. (1984). The levels of selection. *In* "Genes, Organisms, Populations: Controversies over the Units of Selection" (R. N. Brandon and R. M. Burian, eds.), pp. 133-141. MIT Press, Cambridge, MA.

Bremermann, H. J. (1983). Parasites at the origin of life. *J. Math. Biol.* 16, 165-180.

Bresch, C., Niesert, U., and Harnasch, D. (1980). Hypercycles, parasites and packages. *J. Theor. Biol.* 85, 399-405.

Brooks, R. A. (1986). A robust layered control system for a mobile robot. IEEE *J. Robot. Automation* RA-2, 14-23.

Brooks, R. A., and Maes, P. (eds.). (1994). "Artificial Life IV." MIT Press, Cambridge, MA.

Burnet, F. M. (1959). "The Clonal Selection Theory of Immunity." Vanderbilt University Press, Nashville, TN.

Cavalier-Smith, T. (1993). Concepts of Symbiogenesis: A Historical and Critical Study of the Research of Russian Botanists by L. N. Khakhina (Book Review). *Nature* 366, 641-642.

Cavalli-Sforza, L. L., and Feldman, M. W. (1981). "Cultural Transmission and Evolution: A Quantitative Approach." Princeton University Press, Princeton, NJ.

Cech, T. R. (1993). Fishing for fresh catalysts. *Nature* 365, 204-205.

Charlesworth, B. (1993). The evolution of sex and recombination in a varying environment. *J. Hered.* 84, 345-350.

Clark, C. W., and Harvell, C. D. (1992). Inducible defenses and the allocation of resources – a minimal model. *Am. Nat.* 139, 521-539.

Cliff, D., Husbands, P., and Harvey, I. (1993). Evolving visually guided robots. *In* "From Animals to Animats 2" (J.-A. Meyer, H. L. Roitblat, and S. W. Wilson, eds.), pp. 374-383. MIT Press, Cambridge, MA.

Cliff, D., Husbands, P., Meyer, J.-A., and Wilson, S. W. (eds.). (1994). "From Animals to Animats 3." MIT Press, Cambridge, MA.

Colombetti, M., and Dorigo, M. (1993). Learning to control an autonomous robot by distributed genetic algorithms. *In* "From Animals to Animats 2" (J.-A. Meyer, H. L. Roitblat, and S. W. Wilson, eds.), pp. 305-312. MIT Press, Cambridge, MA.

Davis, B. D., Dulbecco, R., Eisen, H. N., and Ginsberg, H. S. (1990). "Microbiology," 4th ed. Lippincott, Philadelphia.

Davis, L. (1991). A genetic algorithms tutorial. *In* "Handbook of Genetic Algorithms" (L. Davis, ed.), pp. 1-101. Van Nostrand Reinhold, New York.

Dawkins, R. (1976a). "The Selfish Gene." Oxford University Press, New York.

Dawkins, R. (1976b). Hierarchical organisation: a candidate principle for ethology. *In* "Growing Points in Ethology" (P. P. G. Bateson and R. A. Hinde, eds.), pp. 7-54. Cambridge University Press, Cambridge.

Dawkins, R. (1982). "The Extended Phenotype." Freeman, San Francisco.

Deamer, D. W., and Fleischaker, G. R. (eds.) (1994). "Origins of Life: The Central Concepts." Jones and Bartlett, Boston.

Dellaert, F., and Beer, R. D. (1994). Toward an evolvable model of development for autonomous agent synthesis. *In* "Artificial Life IV" (R. A. Brooks and P. Maes, eds.), pp. 246-257. MIT Press, Cambridge, MA.

Dennett, D. C. (1991). "Consciousness Explained." Little, Brown and Company, Boston.

Dennett, D. C. (1995). "Darwin's Dangerous Idea." Simon and Schuster, New York.

Drickamer, L. C. (1992). "Animal Behavior: Mechanisms, Ecology, and Evolution." Wm. C. Brown, Dubuque, IA.

Edelman, G. M. (1987). "Neural Darwinism: The Theory of Neuronal Group Selection." Basic Books, New York.

Edelman, G. M. (1988). "Topobiology: An Introduction to Molecular Embryology." Basic Books, New York.

Edelman, G. M. (1989). "The Remembered Present: A Biological Theory of Consciousness." Basic Books, New York.

Edelman, G. M. (1992). "Bright Air, Brilliant Fire: On the Matter of Mind." Basic Books, New York.

Edwardson, J. R. (1970). Cytoplasmic male sterility. *Bot. Rev.* 36, 341-420.

Eigen, M. (1971). Self-organization of matter and the evolution of biological macromolecules. *Naturwissenschaften* 58, 465-523.

Eigen, M. (1992). "Steps Towards Life: A Perspective on Evolution." Oxford University Press, Oxford.

Eigen, M., and Schuster, P. (1977). The hypercycle. A principle of natural self-organization. Part A: emergence of the hypercycle. *Naturwissenschaften* 64, 541-565.

Eigen, M., and Schuster, P. (1978a). The hypercycle. A principle of natural self-organization. Part B: the abstract hypercycle. *Naturwissenschaften* 65, 7-41.

Eigen, M., and Schuster, P. (1978b). The hypercycle. A principle of natural self-organization. Part C: the realistic hypercycle. *Naturwissenschaften* 65, 341-569.

Ewald, P. W. (1994). "The Evolution of Infectious Disease." Oxford University Press.

Feldman, M. W., and Zhivotovsky, L. A. (1992). Gene-culture coevolution: toward a general theory of cultural transmission. *Proc. Natl. Acad. Sci. USA* 89, 11935-11938.

Fisher, R. A. (1958). "The Genetical Theory of Natural Selection," 2nd ed. Dover, New York.

Fontanari, J. F., and Meir, R. (1990). The effect of learning on the evolution of asexual populations. *Complex Syst.* 4, 401-414.

Forrest, S. (1993). Genetic algorithms: principles of natural selection applied to computation. *Science* 261, 872-878.

Frank, S. A. (1989). The evolutionary dynamics of cytoplasmic male sterility. *Am. Nat.* 133, 345-376.

Frank, S. A. (1993). A model of inducible defense. *Evolution* 47, 325-327.

Frank, S. A. (1996). Models of parasite virulence. *Quart. Rev. Biol.* 71, 37-78.

French, R. M., and Messinger, A. (1994). Genes, phenes and the Baldwin effect: learning and evolution in a simulated population. *In* "Artificial Life IV" (R. A. Brooks and P. Maes, eds.), pp. 277-282. MIT Press, Cambridge, MA.

Gardner, H. (1985). "The Mind's New Science." Basic Books, New York.

Gell-Mann, M. (1994). Complex adaptive systems. *In* "Complexity: Metaphors, Models, and Reality" (G. Cowan, D. Pines, and D. Meltzer, eds.), pp. 17-45. Addison-Wesley, New York.

Golub, E. S., and Green, D. R. (1991). "Immunology: A Synthesis," 2nd ed. Sinauer Associates, Sunderland, MA.

Goodwin, B. C., Kauffman, S., and Murray, J. D. (1993). Is morphogenesis an intrinsically robust process? *J. Theor. Biol.* 163, 135-155.

Gouyon, P. H., and Couvet, D. (1985). Selfish cytoplasm and adaptation: variations in the reproductive system of thyme. *In* "Structure and Functioning of Plant Populations" (J. Haeck, and J. W. Woldendorp, eds.), Vol. 2, pp. 299-319. North-Holland Publishing Company, New York.

Haldane, J. B. S. (1929). The origin of life. Reprinted in "The Origin of Life" (J. D. Bernal, ed., 1967), pp. 242-249. Weidenfeld and Nicolson, London.

Haldane, J. B. S. (1932). "The Causes of Evolution." Cornell University Press, Ithaca, NY.

Hall, B. K. (1992). "Evolutionary Developmental Biology." Chapman and Hall, New York.

Hamilton, W. D. (1980). Sex versus non-sex versus parasite. *Oikos* 35, 282-290.

Hamilton, W. D. (1982). Pathogens as causes of genetic diversity in their host populations. *In* "Population Biology of Infectious Diseases" (R. M. Anderson and R. M. May, eds.), pp. 269-296. Dahlem Konferenzen, Berlin.

Hanson, M. R. (1991). Plant mitochondrial mutations and male sterility. *Annu. Rev. Genet.* 25, 461-486.

Harp, S. A., and Samad, T. (1991). Genetic synthesis of neural network architecture. *In* "Handbook of Genetic Algorithms" (L. Davis, ed.), pp. 202-221. Van Nostrand Reinhold, New York.

Harvell, C. D. (1990a). The evolution of inducible defense. *Parasitology* 100, S53-S61.

Harvell, C. D. (1990b). The ecology and evolution of inducible defenses. *Quart. Rev. Biol.* 65, 323-340.

Harvey, I. (1991). The artificial evolution of behavior. *In* "From Animals to Animats" (J.-A. Meyer and S. W. Wilson, eds.), pp. 400-408. MIT Press, Cambridge, MA.

Harvey, I., Husbands, P., and Cliff, D. (1993). Issues in evolutionary robotics. *In* "From Animals to Animats 2" (J.-A. Meyer, H. L. Roitblat, and S. W. Wilson, eds.), pp. 364-373. MIT Press, Cambridge, MA.

Hemmi, H., Mizoguchi, J., and Shimohara, K. (1994). Development and evolution of hardware behaviors. *In* "Artificial Life IV" (R. A. Brooks and P. Maes, eds.), pp. 371-376. MIT Press, Cambridge, MA.

Hertz, J., Krogh, A., and Palmer, R. G. (1991). "Introduction to the Theory of Neural Computation." Addison-Wesley, New York.

Hickey, D. A., and Rose, M. R. (1988). The role of gene transfer in the evolution of eukaryotic sex. *In* "The Evolution of Sex" (R. E. Michod and B. R. Levin, eds.), pp. 161-175. Sinauer Associates, Sunderland, MA.

Hinton, G. E., and Nowlan, S. J. (1987). How learning can guide evolution. *Complex Syst.* 1, 495-502.

Holland, J. H. (1975). "Adaptation in Natural and Artificial Systems." University of Michigan Press, Ann Arbor, MI.

Holland, J. H., Holyoak, K. J., Nisbett, R. E., and Thagard, P. R. (1986). "Induction: Processes of Inference, Learning, and Discovery." MIT Press, Cambridge, MA.

Hölldobler, B., and Wilson, E. O. (1990). "The Ants." Harvard University Press, Cambridge, MA.

Hurst, L. D. (1990). Parasite diversity and the evolution of diploidy, multicellularity and anisogamy. *J. Theor. Biol.* 144, 429-443.

Hurst, L. D. (1991). Sex, slime and selfish genes. *Nature* 354, 23-24.

Hurst, L. D., and Hamilton, W. D. (1992). Cytoplasmic fusion and the nature of sexes. *Proc. Roy. Soc. London.* B 247, 189-194.

Jaenike, J. (1978). An hypothesis to account for the maintenance of sex within populations. *Evol. Theor.* 3, 191-194.

Janeway, C. A., Jr. (1993). How the immune system recognizes invaders. *Sci. Am.* 269 (3), 73-79.

Jerne, N. K. (1955). The natural selection theory of antibody formation. *Proc. Natl. Acad. Sci. USA* 41, 849-857.

Judson, R. S. (1992). Teaching polymers to fold. *J. Phys. Chem.* 96, 10102-10104.

Judson, R. S., Colvin, M. E., Meza, J. C., Huffer, A., and Gutierrez, D. (1992). Do intelligent configuration search techniques outperform random search for large molecules? *Int. J. Quant. Chem.* 44, 277-290.

Judson, R. S., Jaeger, E. P., Treasurywala, A. M., and Peterson, M. L. (1993). Conformational searching methods for small molecules. II. Genetic algorithm approach. *J. Comp. Chem.* 14, 1407-1414.

Kagan, M. L., Novoplansky, N., and Sachs, T. (1992). Variable cell lineages form the functional pea epidermis. *Ann. Bot.* 69, 303-312.

Khakhina, L. N. (1992). "Concepts of Symbiogenesis: A Historical and Critical Study of the Research of Russian Botanists" (L. Margulis and M. McMenamin, eds., translated by S. Merkel and R. Coalson). Yale University Press, New Haven, CT.

Kitano, H. (1994). Neurogenetic learning: an integrated method of designing and training neural networks using genetic algorithms. *Physica* D 75, 225-238.

Klein, J. (1986). "The Natural History of the Major Histocompatibility Complex." Wiley, New York.

Kondrashov, A. S. (1988). Deleterious mutations and the evolution of sexual reproduction. *Nature* 336, 435-440.

Kondrashov, A. S. (1993). Classification of hypotheses on the advantage of amphimixis. J. Hered. 84, 372-387.

Kung, S. Y. (1993). "Digital Neural Networks." Prentice Hall, Englewood Cliffs, NJ.

Leigh, E. (1977). How does selection reconcile individual advantage with the good of the group? *Proc.*

Natl. Acad. Sci. USA 74, 4542-4546.

Lerner, R. A., Benkovic, S. J., and Schultz, P. G. (1991). At the crossroads of chemistry and immunology: catalytic antibodies. *Science* 252, 659-667.

Levin, D. A. (1975). Pest pressure and recombination systems in plants. *Am. Nat.* 109, 417-451.

Lewis, D. (1941). Male sterility in natural populations of hermaphrodite plants: the equilibrium between females and hermaphrodites to be expected with different types of inheritance. *New Phytol.* 40, 56-63.

Lindenmayer, A. (1971). Developmental systems without cellular interactions, their languages and grammars. *J. Theor. Biol.* 30, 455-484.

Lynch, M., Bürger, R., Butcher, D., and Gabriel, W. (1993). The mutational meltdown in asexual populations. *J. Hered.* 84, 339-344.

Margulis, L. (1981). "Symbiosis in Cell Evolution." Freeman, San Francisco.

Maynard Smith, J. (1978). "The Evolution of Sex." Cambridge University Press, Cambridge.

Maynard Smith, J. (1979). Hypercycles and the origin of life. *Nature* 280, 445-446.

Maynard Smith, J. (1983). Models of evolution. *Proc. Roy. Soc. London* B 219, 315-325.

Maynard Smith, J. (1987). When learning guides evolution. *Nature* 329, 761-762.

Maynard Smith, J. (1988). Evolutionary progress and levels of selection. *In* "Evolutionary Progress" (M. H. Nitecki, ed.), pp. 219-230. University of Chicago Press, Chicago.

Maynard Smith, J. (1992). Byte-sized evolution. *Nature* 355, 772-773.

Maynard Smith, J., and Szathmáry, E. (1993). The origin of chromosomes I. Selection for linkage. *J. Theor. Biol.* 164, 437-446.

Maynard Smith, J., and Szathmáry, E. (1995). "The Major Transitions in Evolution." Freeman, San Francisco.

McClelland, J. L., Rumelhart, D. E., and the PDP Research Group (eds.) (1986). "Parallel Distributed Processing: Explorations in the Microstructure of Cognition, Volume 2: Psychological and Biological Models." MIT Press, Cambridge, MA.

McCulloch, W. S., and Pitts, W. (1943). A logical calculus of ideas immanent in nervous activity. *Bull. Math. Biophys.* 5, 115-133.

McFarland, D., and Bösser, T. (1993). "Intelligent Behavior in Animals and Robots." MIT Press, Cambridge, MA.

McGarrah, D. B., and Judson, R. S. (1993). Analysis of the genetic algorithm method of molecular conformation determination. *J. Comp. Chem.* 14, 1385-1395.

McMichael, A. (1993). Natural selection at work on the surface of virus-infected cells. *Science* 260, 1771-1772.

Meinhardt, H. (1995). "The Algorithmic Beauty of Sea Shells." Springer-Verlag, New York.

Merzenich, M. M., Recanzone, G., Jenkins, W. M., Allard, T. T., and Nudo, R. J. (1988). Cortical representational plasticity. *In* "Neurobiology of Neocortex" (P. Rakie and W. Singer, ed.), pp. 41-67. Wiley, New York.

Meyer, J.-A., Roitblat, H. L., and Wilson, S. W. (eds.) (1993). "From Animals to Animats 2." MIT Press, Cambridge, MA.

Michod, R. E. (1993). Genetic error, sex, and diploidy. *J. Hered.* 84, 360-371.

Michod, R. E., and Levin, B. R. (eds.) (1988). "The Evolution of Sex." Sinauer, Sunderland, MA.

Mims, C. A. (1987). "The Pathogenesis of Infectious Disease," 3rd ed. Academic Press, London.

Mims, C. A., Playfair, J. H. L., Roitt, I. M., Wakelin, D., and Williams, R. (1993). "Medical Microbiology." Mosby, St. Louis, MO.

Minsky, M. (1985). "The Society of Mind." Simon and Schuster, New York.

Mitchison, A. (1993). Will we survive? *Sci. Am.* 269 (3), 136-144.

Muller, H. J. (1964). The relation of recombination to mutational advance. *Mut. Res.* 1, 2-9.

Murray, J. D. (1989). "Mathematical Biology." Springer-Verlag, New York.

Newell, A. (1969). Heuristic programming: ill-structured problems. *In* "Progress in Operations Research" (J. Arnofsky, ed.), pp. 363-414. Wiley, New York.

Nossal, G. J. V. (1993). Life, death and the immune system. *Sci. Am.* 269 (3), 53-62.

Nowak, M. A., Tarczy-Hornoch, K., and Austyn, J. M. (1992). The optimal number of major histocompatibility complex molecules in an individual. *Proc. Natl. Acad. Sci. USA* 89, 10896-10899.

Numaoka, C., and Takeuchi, A. (1993). Collective choice of strategic type. *In* "From Animals to Animats 2" (J.-A. Meyer, H. L. Roitblat, and S. W. Wilson, eds.), pp. 469-477. MIT Press, Cambridge, MA.

Oparin, A. I. (1924). "The Origin of Life [Proiskhozhdenie zhizny]." English translation *In* "The

Origin of Life" (J. D. Bernal, ed.), pp. 199-234. Weidenfeld and Nicolson, London.

Parker, L. E. (1993). Adaptive action selection for cooperative agent teams. *In* "From Animals to Animats 2" (J.-A. Meyer, H. L. Roitblat, and S. W. Wilson, eds.), pp. 442-450. MIT Press, Cambridge, MA.

Plotkin, H. C. (1987). Evolutionary epistemology as science. *Biol. Phil.* 2, 295-313.

Plotkin, H. C., and Odling-Smee, F. J. (1981). A multiple-level model of evolution and its implications for sociobiology. *Behav. Brain Sci.* 4, 225-268.

Ray, T. S. (1992). An approach to the synthesis of life. *In* "Artificial Life 2" (C. G. Langton, C. Taylor, D. J. Farmer, and S. Rasmussen, eds.), pp. 371-408. Addison-Wesley, Redwood City, CA.

Real, L. (1992). Information processing and the evolutionary ecology of cognitive architecture. *Am. Nat.* 140:S108-S145.

Real, L. A. (1993). Toward a cognitive ecology. *Trends Ecol. Evol.* 8, 413-417.

Richards, R. J. (1987). "Darwin and the Emergence of Evolutionary Theories of Mind and Behavior." University of Chicago Press, Chicago.

Rumelhart, D. E., McClelland, J. L., and the PDP Research Group (eds.) (1986). "Parallel Distributed Processing: Explorations in the Microstructure of Cognition, Volume 1: Foundations." MIT Press, Cambridge, MA.

Sachs, T. (1988). Epigenetic selection: an alternative mechanism of pattern formation. *J. Theor. Biol.* 134, 547-559.

Sachs, T. (1991). "Pattern Formation in Plant Tissues." Cambridge University Press, Cambridge.

Sachs, T., Novoplansky, A., and Cohen, D. (1993). Plants as competing populations of redundant organs. *Plant Cell Environ.* 16, 765-770.

Simon, H. A. (1981). "The Sciences of the Artificial," 2nd ed. MIT Press, Cambridge, MA.

Szathmáry, E. (1989a). The integration of the earliest genetic information. *Trends Ecol. Evol.* 4, 200-204.

Szathmáry, E. (1989b). The emergence, maintenance, and transitions of the earliest evolutionary units. *Oxf. Surv. Evol. Biol.* 6, 169-205.

Szathmáry, E., and Demeter, L. (1987). Group selection of early replicators and the origin of life. *J. Theor. Biol.* 128, 463-486.

Thompson, D. W. (1961). "On Growth and Form," abridged edition. Cambridge University Press, Cambridge.

Thomson, G. (1991). HLA populations genetics. Baillieres *Clin. Endocrinol. Metab.* 5, 247-260.

Tinbergen, N. (1951). "The Study of Instinct." Clarendon Press, Oxford.

Todd, P. M., and Wilson, S. W. (1993). Environment structure and adaptive behavior from the ground up. *In* "From Animals to Animats 2" (J.-A. Meyer, H. L. Roitblat, and S. W. Wilson, eds.), pp. 11-20. MIT Press, Cambridge, MA.

Tufféry, P., Etchebest, C., Hazout, S., and Lavery, R. (1993). A critical comparison of search algorithms applied to the optimization of protein side-chain conformations. *J. Comp. Chem.* 14, 790-798.

Unger, R., and Moult, J. (1993). Genetic algorithms for protein folding simulations. *J. Mol. Biol.* 231, 75-81.

Wagner, G. P., and Misof, B. Y. (1993). How can a character be developmentally constrained despite variation in developmental pathways? *J. Evol. Biol.* 6, 449-455.

Williams, G. C. (1975). "Sex and Evolution." Princeton University Press, Princeton, NJ.

Index